Introduction to Engine Valvetrains

Other SAE titles of interest:

Internal Combustion Engine Handbook—
Basics, Components, Systems, and Perspectives
By Richard van Basshuysen and Fred Schäfer
(Order No. R-345)

Advanced Three-Way Catalysts
By Joseph E. Kubsh
(Order No. PT-123)

For more information or to order a book, contact SAE International at
400 Commonwealth Drive, Warrendale, PA 15096-0001;
phone (724) 776-4970; fax (724) 776-0790;
e-mail CustomerService@sae.org;
website http://store.sae.org.

Introduction to Engine Valvetrains

Yushu Wang

SAE International™
Warrendale, Pa.

For permission and licensing requests, contact:

SAE Permissions
400 Commonwealth Drive
Warrendale, PA 15096-0001 USA
E-mail: permissions@sae.org
Tel: 724-772-4028
Fax: 724-772-4891

Library of Congress Cataloging-in-Publication Data

Wang, Yushu.
    Introduction to engine valvetrains / Yushu Wang.
    p. cm.
    Includes bibliographical references and index.
    ISBN-13: 978-0-7680-1079-4
    ISBN-10: 0-7680-1079-9
    1. Automobiles--Motors--Valves. I. Title.
 	
TL214.V3W36 2007
629.25--dc22   2006046471

SAE International
400 Commonwealth Drive
Warrendale, PA 15096-0001 USA
E-mail: CustomerService@sae.org
Tel: 877-606-7323 (inside USA and Canada)
          724-776-4970 (outside USA)
Fax: 724-776-1615

Copyright © 2007   SAE International

ISBN-10 0-7680-1079-9
ISBN-13 978-0-7680-1079-4

SAE Order No. R-339

Printed in the United States of America.

Society of Automotive Engineers is committed to preserving ancient forests and natural resources. We elected to print this title on 30% postconsumer recycled paper, processed chlorine-free. As a result, we have saved:

5 Trees (40' tall and 6-8" diameter)
1 Million BTUs of Total Energy
341 Pounds of Greenhouse Gases
2,076 Gallons of Wastewater
126 Pounds of Solid Waste

Society of Automotive Engineers made this paper choice because our printer, Thomson-Shore, Inc., is a member of Green Press Initiative, a nonprofit program dedicated to supporting authors, publishers, and suppliers in their efforts to reduce their use of fiber obtained from endangered forests.

For more information, visit www.greenpressinitiative.org

Environmental impact estimates were made using the Environmental Defense Paper Calculator. For more information visit: www.edf.org/papercalculator

Acknowledgments

I gratefully acknowledge my former colleagues in valvetrain engineering in Engine Air Management Operations at Eaton Corporation. My discussions with them, whether formal or casual, were invaluable in forming my knowledge base on this subject. These colleagues include Craig Bennett, Hope Bolton, Victoria Bouwens, Bryce Buuck, Rob Clark, James Dean, Mike Froehlick, Mike Guzak, Ed Hurlbert, George Hillebrand (retired), Larry Jenkins (retired), Jay Larson (retired), Doug Nielson, Sandy Schaefer, Brad Trine, and Steve Young (retired).

In addition, I gratefully acknowledge Gary Barber of Oakland University, Tim Lancefield of Mechadyne International, Michael Levin of Ford Motor Company, Heron Rodrigues of Engineering Sintering Components, Mark Theobald of General Motors Corporation, and Haoran Hu and Dong Zhu of Eaton Corporation Innovation Center.

Finally, I thank Kris Hattman and Jeff Worsinger of SAE International for their patience and assistance in helping me to prepare the manuscript.

Preface

In more than 100 years of internal combustion engine history, tremendous knowledge about valvetrains has been accumulated and published. However, much of the technical literature and many of the books dealing with the design, construction, and maintenance of various components of valvetrains are scattered, and the design engineer or student seldom has time to search through several sources of information for a solution to a problem or question. Therefore, this book aims to present a unified, precise, clear, and systematic description and explanation of the fundamentals of all essential components of valvetrains, as well as the valvetrain as a system. The objective is to introduce and explain fundamental valvetrain engineering concepts so that the reader can appreciate the design and material considerations and can understand the difficulties the engine designer faces in designing a valvetrain system to satisfy the functional requirements and the manufacturer's challenges in producing components that satisfy the designer's requirements. This book also provides up-to-date, broad-based, in-depth information devoted to the design, material and metallurgy, testing, tribology, and failure analysis of valvetrains. The completeness of the information given here should make the book useful as a reference source for design engineers and students alike.

The material within this book has come from many sources. The published sources have been acknowledged. Although great pains have been taken to avoid errors, it is impossible to eliminate them entirely in a work of this magnitude. I hope that readers who discover errors will kindly notify me or the publisher, so that those errors can be corrected at the first opportunity.

Contents

Chapter 1 Introduction and Overview of Engines 1
    1.1   Engine Fundamentals 1
          1.1.1   Definition of an Engine 1
          1.1.2   Types of Internal Combustion Engines 2
          1.1.3   Typical Internal Combustion Engine Structure 3
    1.2   Principles of the Four-Stroke Combustion Cycle 4
          1.2.1   Spark Ignition Engines 5
          1.2.2   Compression Ignition Engines 9
    1.3   Two-Stroke Engines 10
    1.4   Engine Displacement and Compression Ratio 11
    1.5   Engine Performance 13
          1.5.1   Torque 13
          1.5.2   Horsepower 16
    1.6   Internal Combustion Engine Efficiency 18
          1.6.1   Mechanical Efficiency 18
          1.6.2   Thermal Efficiency 18
          1.6.3   Engine Volumetric Efficiency 22
    1.7   References 23

Chapter 2 Valvetrain Systems 25
    2.1   Overview of Valvetrain Systems 25
    2.2   Valve Actuation 26
          2.2.1   Valve Lift and Duration 27
          2.2.2   Valve Timing and Overlap 27
          2.2.3   Effect of Valve Timing on Performance and Emissions 31
          2.2.4   Valvetrain System Timing 33
    2.3   Variable Valve Actuation—Cam Driven 34
          2.3.1   Overview of Variable Valve Timing 34
          2.3.2   Variable Valve Timing by Camshaft Phasing 36
          2.3.3   Variable Valve Lift by Cam Profile Switching 40
          2.3.4   Variable Valve Lift by Lost Motion 43
          2.3.5   Variable Valve Actuation by Cam Phasing and Variable Lift ... 46
          2.3.6   Variable Valve Actuation by Varying the Rocker Ratio 48
    2.4   Variable Valve Actuation—Camless System 52
          2.4.1   Electromagnetic Valve Actuation 53
          2.4.2   Electrohydraulic Valve Actuation 55
          2.4.3   Comparison of Various Variable Valve Actuation Systems 57

2.5   Engine Brake and Valvetrains..58
2.6   Types of Valvetrain..61
        2.6.1   Type I Valvetrain—Direct Acting, Overhead Cam....................64
        2.6.2   Type II Valvetrain—End-Pivot Rocker Arm, Overhead Cam...............64
        2.6.3   Type III Valvetrain—Center-Pivot Rocker Arm, Overhead Cam.............65
        2.6.4   Type IV Valvetrain—Center-Pivot Rocker Arm, Overhead Cam.............65
        2.6.5   Type V Valvetrain—Center-Pivot Rocker Arm, Overhead Valve.............66
        2.6.6   Ranking of Various Valvetrains...66
2.7   Valvetrain System Design...66
        2.7.1   Valvetrain Selection..67
        2.7.2   Valvetrain System Mechanics—Kinematics Analysis..........................69
        2.7.3   Valvetrain System Dynamics Analysis..96
2.8   References...118

Chapter 3 Valvetrain Components...121
3.1   Overview..121
3.2   Valves...121
        3.2.1   Valve Nomenclature and Construction.......................................121
        3.2.2   Valve Operating Characteristics..122
        3.2.3   Valve Design..124
        3.2.4   Valve Materials..148
3.3   Cams...174
        3.3.1   Cam Nomenclature and Design Considerations........................175
        3.3.2   Cam Profile Characteristics...176
        3.3.3   Cam Profile Design..179
        3.3.4   Considerations in Cam Profile Determination.............................190
        3.3.5   Lightweight Cam Design...200
        3.3.6   Cam Lobe Surface Finishes and Tolerances...............................201
        3.3.7   Cam Materials..203
3.4   Valvetrain Lash Compensators..206
        3.4.1   Lash and Mechanical Lash Adjusters...206
        3.4.2   Hydraulic Lash Compensators..208
        3.4.3   Hydraulic Lash Compensation Mechanism................................211
        3.4.4   Hydraulic Lifter Design Considerations.....................................213
        3.4.5   Dimensions, Tolerances, and Surface Finishes..........................222
        3.4.6   Materials and Stresses..228
3.5   Seat Inserts...230
        3.5.1   Nomenclature...230
        3.5.2   Insert Design Considerations..231
        3.5.3   Insert Materials...237
        3.5.4   Insert Material Properties..249
        3.5.5   Wear Resistance...252
        3.5.6   Machinability...254
        3.5.7   Corrosion Resistance..254
3.6   Valve Guides...255
        3.6.1   Valve Guide Types and Nomenclature..256

Contents

	3.6.2	Valve Guide Design	256
	3.6.3	Guide Materials	260
	3.6.4	Guide Material Properties	262
3.7		Rocker Arms	266
	3.7.1	Rocker Arm Configurations	266
	3.7.2	Design Guidelines	269
	3.7.3	Rocker Arm Materials	273
	3.7.4	Variable Actuation Rocker Arms	274
3.8		Valve Springs	276
	3.8.1	Introduction	276
	3.8.2	Design Considerations	281
	3.8.3	Spring Performance	285
	3.8.4	Valve Spring Materials	295
3.9		Valve Stem Seals	299
	3.9.1	Introduction	299
	3.9.2	Design Considerations	299
	3.9.3	Seal Materials	310
3.10		Keys	310
	3.10.1	Single-Bead Keys	311
	3.10.2	Multiple-Bead Keys	312
	3.10.3	Dimensions and Tolerances	313
	3.10.4	Key Materials	313
3.11		Retainers	313
	3.11.1	One-Piece Retainers	314
	3.11.2	Two-Piece Retainers	316
	3.11.3	Dimensions and Tolerances	317
	3.11.4	Retainer Materials	317
3.12		Other Components	317
	3.12.1	Valve Rotators	318
	3.12.2	Pushrods	320
	3.12.3	Valve Bridges	322
	3.12.4	Crosshead Rocker Arms	323
3.13		References	324
Chapter 4		Valvetrain Testing	329
4.1		Introduction	329
4.2		Materials and Testing	329
	4.2.1	Material Behavior	330
	4.2.2	Mechanical Tensile Testing	348
	4.2.3	Hardness Testing	354
	4.2.4	Fracture Toughness Testing	363
	4.2.5	Fatigue Testing	371
	4.2.6	Friction Testing	383
	4.2.7	Wear Testing	386
	4.2.8	Corrosion Testing	391

4.3   Valvetrain Component Bench Testing 395
    4.3.1   Component Testing Overview 395
    4.3.2   Cam and Follower Wear Testing 396
    4.3.3   Valve Seat and Insert Wear Testing 396
    4.3.4   Valve Stem and Guide Wear Testing 398
    4.3.5   Flow Testing 399
    4.3.6   Hydraulic Lifter Leak-Down Testing 399
    4.3.7   Thermal Shock Testing 402
    4.3.8   Stem Seal Oil Metering Testing 402
    4.3.9   Strain Gage Testing 404
4.4   Nondestructive Testing 404
    4.4.1   Overview of Nondestructive Testing 404
    4.4.2   X-Ray Inspection 404
    4.4.3   Ultrasonic Inspection 408
    4.4.4   Liquid Penetrant Inspection 411
    4.4.5   Magnetic Particle Inspection 411
    4.4.6   Eddy Current Inspection 414
4.5   Engine Testing 418
    4.5.1   Valvetrain System Testing Overview 418
    4.5.2   Valvetrain Dynamic Testing 418
    4.5.3   Lash Measurement 427
    4.5.4   Valve Temperature Measurement 429
    4.5.5   Valvetrain Friction Measurement 430
4.6   References 431

Chapter 5 Valvetrain Tribology 437
5.1   Introduction to Tribology 437
    5.1.1   Valvetrain Tribology Overview 437
    5.1.2   Surface Topography and Contact Mechanics 438
    5.1.3   Friction 443
    5.1.4   Wear 448
    5.1.5   Lubrication and Lubricants 456
5.2   Engine Lubrication and Lubricants 467
    5.2.1   Engine Lubrication 467
    5.2.2   Engine Lubricants 472
    5.2.3   Synthetic Engine Base Oil and Additives 473
    5.2.4   Lubricant-Related Engine Malfunctions 477
5.3   Valvetrain Friction Loss or Energy Consumption 481
    5.3.1   Overview 481
    5.3.2   Speed Effects 481
    5.3.3   Effects of Valvetrain Type 483
    5.3.4   Temperature Effects 485
    5.3.5   Spring Load Effects 485
    5.3.6   Rolling Element Effects 486
    5.3.7   Lubricant Type Effects 487
    5.3.8   Oil Flow Rate Effects 490

5.4	Valvetrain Wear	491
	5.4.1 Introduction	491
	5.4.2 Cam and Follower Interface	491
	5.4.3 Valve Seat and Seat Insert Interface	494
	5.4.4 Valve Stem and Guide Interface	494
	5.4.5 Valve Tip Wear	495
5.5	Valvetrain Lubrication	495
	5.5.1 Film Thickness	495
	5.5.2 Stress Effects	498
	5.5.3 Viscosity Effects	500
	5.5.4 Anti-Wear Additive Effects	501
	5.5.5 Temperature Effects	503
	5.5.6 Engine Oil Degradation	504
	5.5.7 Soot and Carbon Effects	505
	5.5.8 Lubricant and Material Reactions	505
5.6	Surface Engineering of Valvetrain Components	507
	5.6.1 Overview	507
	5.6.2 Selective Surface Hardening	507
	5.6.3 Diffusion Surface Hardening	510
	5.6.4 Thin Film Coatings	517
	5.6.5 Thick Film Coatings	521
	5.6.6 Other Surface Treatments	526
5.7	References	531
Chapter 6	Valvetrain Failure Analysis	537
6.1	General Failure Analysis Practice	537
	6.1.1 Background Information	538
	6.1.2 Preliminary Examination	538
	6.1.3 Macroscopic Examination and Analysis	539
	6.1.4 Metrology Measurement	539
	6.1.5 Microscopic Examination and Analysis	539
	6.1.6 Scanning Electron Microscopy and Energy Dispersive X-Ray Spectrometry Analysis	540
	6.1.7 Mechanical Testing and Simulation Testing	541
	6.1.8 Determination of Failure Mechanisms	541
	6.1.9 Root Cause of Failure	543
	6.1.10 Writing the Failure Analysis Report	544
6.2	Valve Failures	546
	6.2.1 Overview	546
	6.2.2 Valve Face Failures	547
	6.2.3 Valve Seat Failures	550
	6.2.4 Valve Fillet Failures	553
	6.2.5 Stem-Fillet Blend Area Failures	553
	6.2.6 Stem Failures	553
	6.2.7 Keeper Groove Failures	554
	6.2.8 Tip Failures	554

6.3   Cam Failures 554
6.4   Lifter Failures.. 555
6.5   Insert Failures.. 557
6.6   Guide Failures 558
6.7   Spring Failures 558
6.8   Stem Seal Failures 559
6.9   Rocker Arm Failures 560
6.10   Retainer Failures 560
6.11   Key Failures.. 561
6.12   References 561

Index... 563

About the Author.. 586

Chapter 1

Introduction and Overview of Engines

1.1 Engine Fundamentals

1.1.1 Definition of an Engine

The first law of thermodynamics states that energy can be neither created nor destroyed. Only the form in which energy exists can be changed (e.g., heat can be transformed into mechanical energy). An engine is a device that converts energy into useful work. The goal of the engine is to utilize that energy repeatedly, efficiently, and cost effectively. There are two basic types of engines: (1) external combustion engines, and (2) internal combustion engines (Figure 1.1). Internal combustion (IC) engines generate power by converting the chemical energy bound in

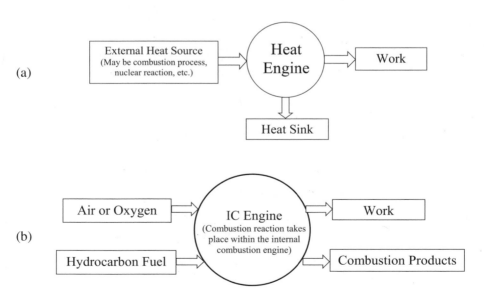

Figure 1.1 *Principles of internal combustion and external combustion engines: (a) the heat engine process, and (b) the internal combustion engine process.*

the fuel into heat, and the heat thus produced into mechanical work. In an external combustion engine, the products of the combustion of air and fuel transfer heat to a second fluid, which then becomes the motive or working fluid for producing power. In an internal combustion engine, the products of combustion are directly the working fluid.

Most engines currently used in transportation are of the internal combustion type. The reciprocating internal combustion engine by far has been or will be in the foreseeable future the most common form of engine or prime mover because of its high work output with high efficiency. Its feature of simplicity and a resulting high thermal efficiency make it one of the most lightweight power-generating units known; therefore, transportation is its field of greatest application. Today, the manufacture of combustion engines for automobiles, boats, airplanes, tractors, tanks, trains, and small power units is one of the largest industries in the world.

1.1.2 Types of Internal Combustion Engines

There are four general types of internal combustion engines:

1. Four-stroke cycle engines

2. Two-stroke cycle engines

3. Rotary engines

4. Continuous combustion gas turbine engines

Four-stroke engines are based on the reciprocating piston principle, as are two-stroke cycle engines; thus, they are the most common. Four-stroke engines, including both compression ignition (CI) and spark ignition (SI) types are used in most automotive applications. Depending on the cylinder arrangement, four-stroke engines can be subcategorized as in-line, V (vee), opposed-piston, and radial types. Two-stroke spark ignition engines commonly are used in lightweight applications (e.g., outboard motors, motor scooters, motorcycles, snowmobiles, and chainsaws), whereas two-stroke compression ignition engines are used in on-highway trucks, city buses, low-speed marine applications, and some railroad diesel applications.

Typically, four-stroke internal combustion engines in commercial vehicles are identified in four ways:

1. By displacement (e.g., 3.8L)

2 By the type of fuel used (e.g., gasoline versus diesel versus compressed natural gas [CNG])

3. By block configuration (e.g., in-line versus V versus opposed versus rotary)

4 By valvetrain configuration (e.g., overhead valve [OHV] versus overhead cam [OHC] versus number of valves per cylinder)

Discretionary information such as the cooling system (e.g., air versus liquid versus adiabatic), fuel delivery (e.g., direct injection versus port injection versus carburetor), and variable valve timing mechanisms (VTEC versus VVT-I versus Valvetron®*) sometimes are added to the description of an engine.

1.1.3 Typical Internal Combustion Engine Structure

The modern internal combustion engine (referred to as "engine" from this point onward) is a complex machine consisting of many types of mechanisms, systems, and structures. Even with the same type of configuration, the specific structure can vary significantly. A typical engine consists of the block assembly, the cranking and valvetrain mechanisms, the fuel delivery system, the ignition system, the cooling system, the lubrication system, and the starting system (Figure 1.2). These systems will be described in the following paragraphs.

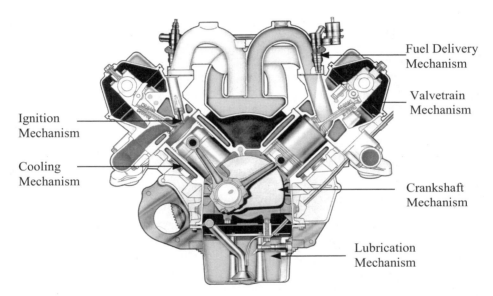

Figure 1.2 *This engine assembly illustrates the structure (Ford, 4.6L, V8).*

The block assembly consists of the cylinder head, block, and oil pan. The block assembly functions as the assembly basis for each mechanism and system, while many of its own parts are part of the valvetrain, fuel delivery, cooling, and lubrication systems. The cylinder walls in the block and head compose part of the combustion chamber and are subject to high temperature and pressure. In structural analysis, the block assembly usually is listed in the crankshaft mechanism.

The crankshaft mechanism includes the piston, connecting rod, and camshaft with the flywheel attached. This is the mechanism by which the engine generates power by transferring the piston linear reciprocating motion to the crankshaft rotating motion.

---

* VTEC, VVT-I, and Valvetron are registerd trademarks of Honda, Toyota, and BMW, respectively.

The valvetrain mechanism includes the intake valve, exhaust valve, camshaft, valve stem seal, and valve spring. Additional valvetrain components include the lash adjuster (lifter, tappet), pushrod, rocker arm, valve seat insert, and valve guide, depending on the type of valvetrain. Its function is to control the intake of the air or fuel/air mixture into the cylinder and the removal of exhaust gas.

The fuel delivery system includes the gas tank, gas pump, fuel filter, carburetor or fuel injector, air filter, intake and exhaust manifolds, muffler, and so forth. Its function is to mix the fuel and air into optimum charge in the cylinder for combustion and to direct the exhaust gas away from the engine.

The role of the ignition system is to ensure that the compressed charge is ignited in a timely way. The system includes a motor and a battery that supplies low-voltage current.

The cooling system mainly includes the water pump, radiator, fan, water hose, block, water relief valve, cylinder, and head water jackets. Its function is to release the heat from the engine parts to the atmosphere and to ensure the normal operation of the engine.

The lubrication system includes the oil pump, oil filter, pressure regulator, lubrication channel, and oil cooler. Its function is to provide lubricating oil to the parts with a tribological contact to reduce friction between the surfaces, to reduce surface wear, and to partially cool and clean the friction surfaces.

The starting system includes a starter and its auxiliaries. The system starts the engine from inert.

1.2 Principles of the Four-Stroke Combustion Cycle

Among all internal combustion engines that have been invented to date, few have enjoyed as much commercial success as the four-stroke reciprocating diesel and spark ignition engines. Figure 1.3 shows a typical engine schematic and nomenclature of associated parts. The location where the piston reaches its uppermost position is called top dead center (TDC), while the lowest piston position is called bottom dead center (BDC). The distance between TDC and BDC is called the piston stroke (S). The distance between the center of the crankshaft and the center of the low connecting rod end is called the crankshaft radius (R). Piston travel of one stroke equals the crankshaft rotation of 180° from TDC to BDC, or vice versa. The distance the piston travels is referred to as the stroke of the engine. The piston stroke usually equals twice the crankshaft radius, or S = 2R. The diameter of the cylinder is referred to as its bore (D), and the combination of the bore and stroke determines the displacement of the cylinder.

The cylinder is sealed opposite from the moving piston by the cylinder head. In most engines, the intake and exhaust valves are located in the cylinder head, as shown in Figure 1.3. The following sections illustrate the principles of the spark ignition engine and the compression ignition engine.

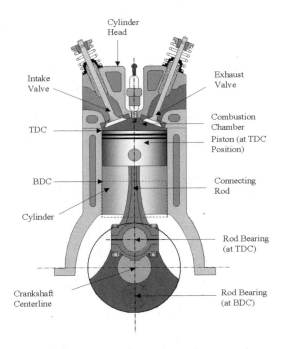

Figure 1.3 *Engine schematic and nomenclature.*

1.2.1 Spark Ignition Engines

The four-stroke spark ignition engine includes four piston strokes:

1. The intake stroke

2. The compression stroke

3. The power stroke

4. The exhaust stroke

To analyze the relationship of gas pressure P and cylinder clearance volume V in relation to piston position, engine cycle working diagrams often are used. These diagrams describe the pressure change inside the cylinder at different piston positions. The area enclosed by the curves in the working diagram represents the work done by gas in a single cylinder of the engine during the whole working cycle. Figure 1.4 shows a schematic of a four-stroke spark ignition engine.

Figure 1.5(a) shows the corresponding pressure–volume diagram that is illustrated in Figure 1.4. For comparison, an ideal pressure–volume diagram also is shown side by side (Figure 1.5(b)).

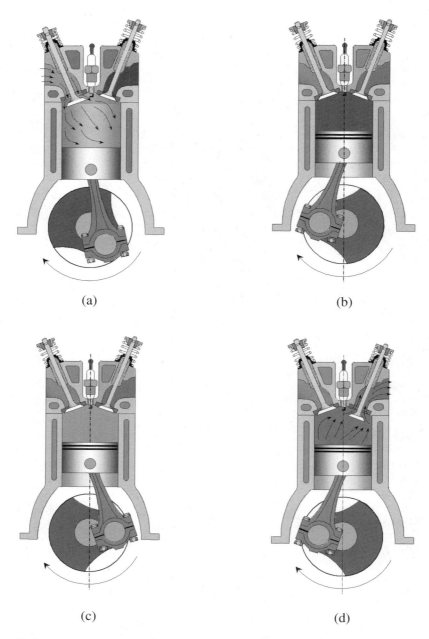

(a)   (b)

(c)   (d)

Figure 1.4 *Schematic of a four-stroke engine: (a) the intake stroke, (b) the compression stroke, (c) the power stroke, and (d) the exhaust stroke.*

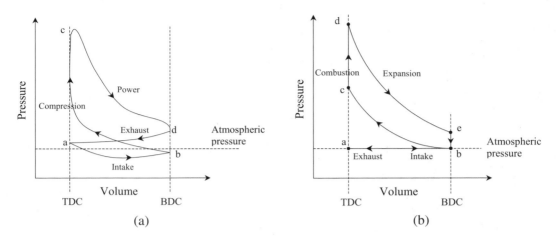

Figure 1.5 *Pressure–volume diagram for a four-stroke engine: (a) typical pressure–volume diagram of a four-stroke engine, and (b) idealized pressure–volume diagram for a four-stroke engine (Stone).*

1.2.1.1 Intake Stroke

In Figure 1.4(a), the intake valve is open, and the piston travels down the cylinder, drawing in a premixed charge of air and fuel, or only air in the case of a direct injection engine. The exhaust valve is closed for most of the stroke. As the piston moves from TDC to BDC, the cylinder volume above the piston increases, and the pressure inside the cylinder decreases to less than atmospheric pressure, thus creating a vacuum suction force. Therefore, the flammable mixture gas passes through the intake manifold and valve into the cylinder. At the end of the intake stroke, the pressure inside the cylinder is approximately 0.075~0.09 MPa due to resistance from the intake manifold.

The flammable mixture gas entering the cylinder can reach up to 370~400K because of contact with the cylinder wall and the piston top, and mixing with the residual high-temperature exhaust gas.

In the pressure–volume state diagram (Figure 1.5(a)), the intake stroke is represented by curve a-b. The curve locates below the atmospheric pressure line, and the difference between the curve and the atmospheric pressure represents the degree of vacuum inside the cylinder (exaggerated in the figure to illustrate the point).

1.2.1.2 Compression Stroke

For the mixture gas that is drawn into the cylinder to burn rapidly enough to produce higher pressure and to make the engine generate higher output, the mixture gas must be compressed before combustion (Figure 1.4(b)). In the compression process, both the intake and exhaust valves are closed. The stroke in which the crankshaft pushes the piston from BDC toward TDC is called the compression stroke.

In the pressure–volume state diagram shown in Figure 1.5(a), the compression stroke is represented by curve b-c. At the end of the compression stroke, the piston reaches TDC. At this moment, the mixture is compressed into a small space above the piston (i.e., the combustion chamber). The pressure of the mixture gas typically is elevated to 0.6~1.2 MPa, and the temperature reaches 600~700K.

1.2.1.3 Power Stroke

In the power stroke process (Figure 1.4(c)), the intake and exhaust valves are still closed. When the piston reaches TDC, a spark plug on the cylinder head ignites a spark, which starts the gas to burn. The whole chamber of gas does not burn instantaneously. The flame spreads from the spark plug, moving across the combustion chamber. It takes approximately 1/350 of a second to complete this flame travel in an average automobile engine cylinder. The burning gas releases a significant amount of heat. Therefore, the burning gas pressure and the temperature increase rapidly, and the peak pressure can reach 3~5 MPa with a corresponding temperature of 2200~2800K. The piston is pushed downward from TDC toward BDC by high-pressure high-temperature gas, and this rotates the crankshaft via the connecting rod, thus transferring the thermal energy into mechanical energy. Some portion of this energy is used to maintain engine function, and some is used to overcome the internal friction. The remainder is used for output work.

In the pressure–volume state diagram shown in Figure 1.5(a), curve c-d indicates the downward movement of the piston, while the volume inside the cylinder increases and the gas pressure and temperature decrease. At the end of the power stroke, the pressure decreases to 0.3~0.5 MPa, and the temperature decreases to 1300~1600K.

1.2.1.4 Exhaust Stroke

The gas/air mixture becomes exhaust gas after combustion and must be expelled from the cylinder for the next intake stroke to begin again (Figure 1.4(d)).

Before the power stroke expansion reaches the end, the exhaust valve opens, and exhaust gas is expelled under the pressure inside the cylinder. When the piston moves toward TDC after reaching BDC, the exhaust gas is forced farther out of the cylinder to the atmosphere through the exhaust manifold system. The exhaust stroke ends when the piston reaches near TDC. This stroke is represented as curve d-a in the pressure–volume diagram shown in Figure 1.5(a). In the exhaust stroke, the pressure inside the cylinder is slightly higher than that of the atmosphere (0.101 MPa), or 0.105~0.115 MPa. The exhaust gas temperature is 900~1200K at the end of the exhaust stroke.

Because the combustion chamber occupies a certain amount of volume, the exhaust gas cannot be expelled completely at the end of the exhaust stroke. Residual exhaust gas remains in the chamber.

In summary, the four-stroke gasoline engine requires the intake, compression, power, and exhaust strokes to complete one full cycle. During this cycle, the piston reciprocates four strokes between TDC and BDC, and the crankshaft rotates two cycles. Automotive four-stroke spark ignition engines usually have cylinder volumes in the range 50–500 cm^3, with the total swept volume rarely being greater than 5000 cm^3. Engine outputs typically are 45 kW/L, a value that can be increased sevenfold by tuning and turbocharging.

1.2.2 Compression Ignition Engines

Similar to a gasoline engine, a four-stroke diesel engine (compression ignition) also experiences the four strokes of intake, compression, power, and exhaust for each working cycle. However, because diesel fuel has a higher viscosity and a lower autoignition temperature than those of gasoline, diesel is not susceptible to vaporizing. Therefore, the formation and ignition of the combustion mixture differ from those of a gasoline engine.

A diesel engine draws in pure air instead of a fuel/air mixture as in a gasoline engine (except for the direct injection gasoline engine) in the intake stroke. Near the end of the compression stroke, diesel is injected into the cylinder through the injector under more than 10 MPa pressure raised by a fuel pump. The injected fuel mixes with high-temperature compressed air in a short time, thus forming the combustion charge. Because the compression ratio in a diesel engine usually is high (16~22), the air pressure inside the cylinder can reach 3.5~4.5 MPa at the end of the compression stroke. Meanwhile, the temperature can reach as high as 750~1000K, greatly exceeding the autoburning temperature of diesel fuel. Therefore, the diesel fuel mixes with air soon after being injected into the cylinder, autoignites, and starts the combustion process. The gas pressure inside the cylinder increases rapidly and reaches 6–9 MPa, while the temperature reaches 2000–2500K. Under high gas pressure, the piston pushes the crankshaft rotation and generates output power. Under the exhaust stroke, the piston expels the exhaust gas to atmosphere through the exhaust manifold.

Comparing the gasoline engine to the diesel engine, the gasoline engine has a higher speed (maximum-speed 6000–7000-rpm racing and motorcycle engines have even higher maximum revolutions per minute), less weight, lower noise, easier starting, and lower costs in manufacturing and maintaining. However, fuel consumption rates are higher, and the fuel efficiency is lower. The gasoline engine is used primarily for motorcycles, passenger cars, pickup trucks, and sports utility vehicles (SUVs). However, the diesel engine has an average lower consumption rate (usually 30% lower) due to its higher compression ratio. Generally, trucks weighing more than 7 tons use a diesel-fueled engine. The diesel engine has a relatively lower speed (maximum speed of 2500–3000 rpm, but the maximum speed for some automotive vehicles that run on diesel fuel can reach as high as 5000 rpm).

The spark ignition engine also is referred to as the petrol, gasoline, or gas engine (based on its typical fuels), and the Otto engine, after its inventor. The compression ignition engine also is referred to as the diesel or oil engine; the fuel likewise is named after its inventor. Although spark ignition and compression ignition engines have many features in common, they differ in the manner in which the fuel and air are mixed and ignited. The fundamental difference between spark ignition and compression ignition engines lies in the type of combustion that occurs. In the spark ignition engine, the fuel is ignited by a spark. In the compression ignition engine, the rise in temperature and pressure during compression is sufficient to cause spontaneous ignition of the fuel. The spark ignition engine usually has a premixed fuel/air mixture using a spark to ignite the mixture, whereas the compression ignition engine uses fuel injection to ignite the fuel under high pressure. With premixed combustion, the fuel/air mixture must always be close to stoichiometric for reliable ignition and combustion. To control the power output, a spark ignition engine is throttled, thus reducing the mass of fuel and air in the combustion chamber and thereby reducing the cycle efficiency. In contrast, for the compression ignition engine with fuel injection, the mixture is close to stoichiometric only at the flame front. The output of the compression ignition engine can be varied by controlling the amount of fuel injected. This accounts for the superior part-load fuel economy of the compression ignition engine.

In the four-stroke engine (either diesel or gasoline), only one piston stroke of four delivers power to the crankshaft. Obviously, the crankshaft rotates faster in the power stroke than in the other three strokes, which results in unstable engine operation. To keep the crankshaft turning more steadily between power strokes, the flywheel must be attached and must be made with considerable momentum. By doing so, the engine weight and dimensions are increased. Obviously, a single-cylinder engine vibrates during operation. The multicylinder engine can counterbalance this drawback of the single-cylinder engine. Four-, six-, and eight-cylinder engines are the most commonly used engines.

In every cylinder of a multicylinder four-stroke engine, the operating process is the same and follows the same sequence as described here previously, although the power stroke does not occur at the same time. In a four-cylinder engine, for example, there is one power stroke in every half turn of the crankshaft; for an eight-cylinder engine, there is one power stroke in every quarter turn of the crankshaft. The greater the number of cylinders, the smoother the engine will operate. However, the complexity, dimensions, and weight generally increase as the number of cylinders increases.

1.3 Two-Stroke Engines

An alternative to the four-stroke cycle engine is the two-stroke cycle engine shown in Figure 1.6. As implied, a complete operating cycle is achieved with every two strokes of the piston (one revolution of the crankshaft). The compression and power strokes are similar to those of the four-stroke engine; however, the gas exchange occurs as the piston approaches BDC, in what is termed the scavenging process. During scavenging, the intake and exhaust passages simultaneously are open, and the engine relies on an intake supply pressure maintained higher than the exhaust pressure to force the combusted products out and to fill the cylinder with fresh air or air/fuel mixture.

In the two-stroke diesel engine, the incoming air is pressurized with a crankshaft-driven compressor and enters through ports near the bottom of the cylinder. Light-duty two-stroke engines often are crankcase supercharged. In such engines, each time the piston moves upward in the cylinder, a fresh charge (mixed with lubricating oil) is drawn into the crankcase. As the piston moves downward, the crankcase is sealed, and the mixture is compressed. Then the mixture is transferred from the crankcase through the intake ports as the piston approaches BDC.

In the two-stroke spark ignition configuration, the engine suffers the disadvantage of sending a fresh air/fuel mixture out with the exhaust during each scavenging period. Much recent

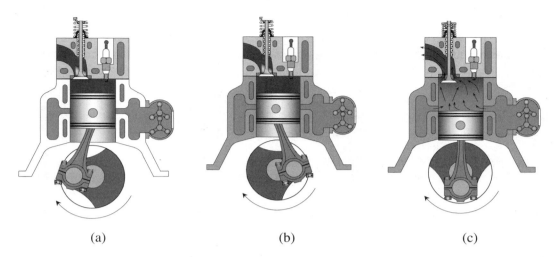

(a)   (b)   (c)

Figure 1.6 *The two-stroke operating cycle: (a) compression, (b) power, and (c) scavenging.*

attention is being given to engines that overcome this problem by injecting the fuel directly into the cylinder after the ports have been sealed.

1.4 Engine Displacement and Compression Ratio

The engine size or displacement or the engine working volume refers to the total space or volume in the cylinders when the pistons move from the top (TDC) of the stroke to the bottom (BDC) of the stroke (Figure 1.7).

The volume swept by a piston from TDC to BDC is called the cylinder working volume or the cylinder clearance volume, represented by the symbol V_S. The engine working volume, engine clearance volume, or engine displacement is represented by the symbol V_E, which is the total working volume of each cylinder,

$$V_E = V_S \times i = \frac{\pi \times D}{4} \times S \times i \tag{1.1}$$

where

    D = cylinder diameter
    S = piston displacement
    i = number of cylinders

The units are cubic inches or liters (1 L = 61 in.^{3}).

The compression ratio is a ratio of the mixture volume inside the cylinder when the piston is at BDC to the cylinder volume when the piston reaches TDC (Figure 1.7),

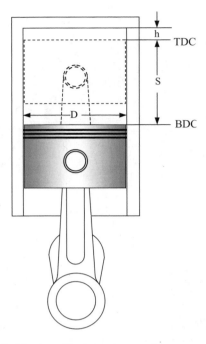

Figure 1.7 Engine displacement and compression ratio.

$$r_V = \frac{V_{BDC}}{V_{TDC}} = \frac{S+h}{h} \qquad (1.2)$$

where

  S = stroke
  h = distance from the piston top at TDC to the cylinder top

To a certain degree, the higher the compression ratio, the more power the engine will produce and the more efficient the engine. In practicality, however, compression ratios range from 5:1 to slightly more than 11:1 for gasoline engines. Diesel engine compression ratios range from 17.5:1 to 22.5:1. The reason for not allowing increased compression beyond a certain point is that when a fuel mixture is compressed, heat energy is developed. The temperature at which the air/fuel mixture will ignite limits the amount of fuel that can be compressed satisfactorily. In addition, effective emissions control has required that compression ratios not be too high.

When the stroke length and the bore diameter are equal, the stroke-to-bore ratio of the engine is said to be square. If the stroke is smaller than the cylinder diameter, the stroke-to-bore ratio is said to be oversquare; if the stroke is larger than the cylinder diameter, the engine is said to be undersquare. The stroke-to-bore ratio for various engines can range from 0.6:1 to 1.4:1.

Oversquare engines are more suitable for gasoline passenger car engines, whereas undersquare engines are better utilized for large diesel engines. Oversquare engines result in lower piston speed, thus prolonging the life of the cylinders, pistons, and rings. An undersquare or long-stroke small-diameter cylinder more closely approaches the minimum surface area of a sphere than a short-stroke large-diameter cylinder of the same cylinder displacement. Thus, a long stroke-to-bore ratio produces a smaller surface-to-volume ratio than a short stroke-to-bore ratio engine. Consequently, long-stroke engines are preferred to minimize heat losses and the formation of hydrocarbon, as opposed to short-stroke large-diameter cylinders. However, peak cylinder pressure tends to decrease as the stroke-to-bore ratio becomes more undersquare.

1.5 Engine Performance

Engine performance usually is assessed by the torque and horsepower the engine can produce. Specifically, the rated maximum horsepower and maximum torque, the shape of the torque-speed curve, and the specific output (horsepower per liter [hp/L]) typically are used to determine engine performance.

1.5.1 Torque

Typically, torque increases as the engine speed increases. Torque starts to decrease as the engine speed reaches a certain point, due to the decreased volumetric efficiency at the higher engine speed. Horsepower is a product of torque and engine speed but continues to increase as the engine speed increases. This is shown in Figure 1.8, which is an example of the engine torque and horsepower rate for a Nissan 3.0L V6 engine versus engine speed.

Engine torque is the ability of the crankshaft to impart a twisting or turning force. It should not be confused with work, although it shares the same units as work in pound-force foot (lbf-ft, English) and Newton meter (N-m, SI units). Figure 1.9 illustrates the definition of torque as

$$\tau = F \times r \qquad (1.3)$$

where

τ = engine torque, N-m
F = force applied to crank, N
r = effective crank-arm radius, m

Mechanical work is defined as the product of a force and the displacements of the force (Figure 1.10),

$$W = F \times \left(X_2 - X_1\right) \qquad (1.4)$$

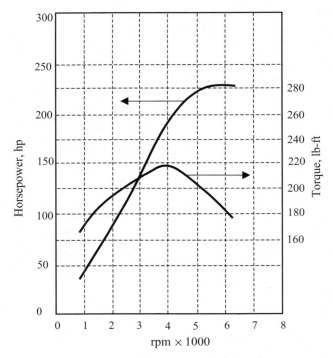

Figure 1.8 *Horsepower and torque graph*
(Nissan 3.0L V6, 1 HP = 746 W, 1 N-m = 1.356 lb-ft).

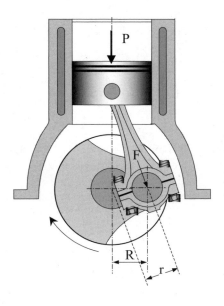

Figure 1.9 *Definition of torque:* $\tau = F \times r$.

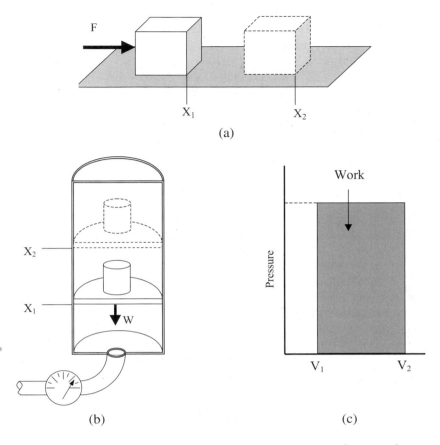

Figure 1.10 *Definition of work, where (a)* $W = F \times (X_2 - X_1)$, *(b)* $W = F \times (X_2 - X_1) = P \times V$, *and (c) pressure–volume work diagram (Stockel et al.).*

If in a pressure vessel such as in a piston-cylinder situation in an engine, the work is defined as

$$W = F \times (X_2 - X_1)$$

where

$$F \quad = \quad \text{pressure} \times \text{area}$$

$$(X_2 - X_1) = \frac{(V_2 - V_1)}{\text{Area}}$$

then

$$W = P \times (V_2 - V_1) \tag{1.5}$$

1.5.2 Horsepower

Power is a measure of the time rate at which work can be done (work/time) and is the product of torque and speed. Units are horsepower (English) and watts (SI). In an engine, brake power is the power measured at the output shaft, and indicated power is the power computed from the work done by the pistons. The work done by a piston during the compression stroke is given by the area enclosed by the curve, from TDC to BDC, and between the cylinder pressure line and P_{atm}. The work done by the piston during the power stroke is given by the area enclosed by the curve, from TDC to BDC, and between the cylinder pressure line and P_{atm}. The net work done on the piston as a result of the compression and expansion strokes is given by the area enclosed by the compression and expansion cylinder pressure curves, from TDC to BDC. Pumping work is from the piston to the charge and is subtracted from the net compression/expansion work,

$$HP = T \times N = T(Nm) \times N(rpm)\left(\frac{1 \text{ min}}{60 \text{ sec}}\right) \times \left(\frac{2\pi}{\text{rev}}\right) \times 10^{-3} = \frac{T \times N}{9551} \qquad (1.6)$$

where

HP = horsepower, kW
T = torque, N-m
N = rpm

Indicated horsepower is a measure of the power developed by the burning fuel in the cylinders. Indicated mean effective pressure (IMEP) is the constant pressure (P_{IMEP}) acting over the same displacement volume that would produce the same work output, as shown in Figure 1.11,

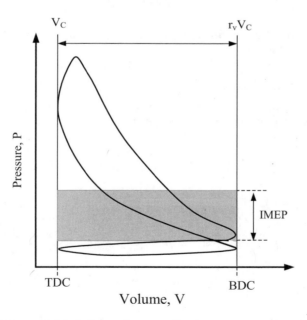

Figure 1.11 *Indicated mean effective pressure (Stone).*

$$P_{IMEP} = \frac{W_I}{V_S} \qquad (1.7)$$

where

W_I = indicated work output, N-m
V_S = swept volume per cylinder, m^3

Once the IMEP is determined, the following formula is used to determine the indicated horsepower:

$$HP_{IND} = \frac{P_{IMEP} \times S \times A \times N \times i}{60} \qquad (1.8)$$

where

HP_{IND} = indicated horsepower, kW (1 hp = 0.746 kW)
P_{IMEP} = indicated mean effective pressure, kPa
S = stroke, m
A = cylinder area, m^2
N = power strokes per minute, $\frac{rpm}{2}$ for a four-stroke engine
i = number of cylinders

Brake horsepower is a measure of the actual usable horsepower delivered at the engine crankshaft,

$$HP_B = \frac{P_{BMEP} \times S \times A \times N \times i}{60} \qquad (1.9)$$

where P_{BMEP} is brake mean effective pressure (BMEP).

Brake mean effective pressure is the work output of the engine normalized by displacement. It is the constant pressure acting over the displacement volume that would produce the same amount of work.

The relationship of indicated, brake, friction, pumping, and gross horsepower/pressure are shown in the following:

Net IMEP (or IMEP) = (Gross Indicated MEP) – (Pumping MEP)

Brake MEP (BMEP) = (Net IMEP) – (Friction MEP)

Frictional horsepower is a measure of the horsepower lost to engine friction. Net brake horsepower ratings are taken with all normal accessories on the engine. Gross brake horsepower figures are calculated with accessories such as the fan, air cleaner, and alternator removed.

Engine brake horsepower increases with engine speed. Engine torque increases with speed up to the point where the engine is drawing in the maximum amount of fuel mixture, after all factors are considered. Torque is greatest at this point, and any additional increase in speed will cause torque to diminish.

1.6 Internal Combustion Engine Efficiency

The second law of thermodynamics states that heat cannot be completely converted to another form of energy (e.g., mechanical work). Engine efficiency provides another criterion in judging the performance of an engine, and there are many ways to define engine efficiency. The most often used are mechanical efficiency, thermal efficiency, and volumetric efficiency.

1.6.1 Mechanical Efficiency

Indicated horsepower is a measure of the power developed by the burning fuel within the cylinders; it is not the power delivered by the crankshaft. Crankshaft horsepower is called brake horsepower and is measured with a dynamometer. The ratio of brake to indicated power is the mechanical efficiency of the engine,

$$\eta_m = \frac{HP_B}{HP_I} \tag{1.10}$$

where

HP_B = brake horsepower
HP_I = indicated horsepower

1.6.2 Thermal Efficiency

Engine thermal efficiency (i.e., heat efficiency) is based on how much of the energy (i.e., ability to do work) of the burning fuel is converted into useful horsepower and is expressed as

$$\eta_T = \frac{HP_B \times 33,000}{778 \times C_F \times \rho_F} \tag{1.11}$$

where

η_T = brake thermal efficiency
HP_B = brake horsepower, hp
C_F = fuel heat value, BTU/lb
ρ_F = weight of fuel burned per minute

The heat generated by the burning fuel drives the piston downward during the power stroke. Much of this heat is lost to the cooling system, some is lost to the lubrication system, and a great deal is lost to the exhaust system. The thermal efficiency of an average engine is approximately 25%. The following illustrates the derivation of the theoretical thermal efficiency limit for spark and compression ignition engines. The processes of the Otto cycle as shown in Figure 1.12 are as follows (Stone):

1-2   isentropic compression of air through a volume ratio V_1/V_2 and the compression ratio r_V

2-3   addition of heat Q_{23} at constant volume

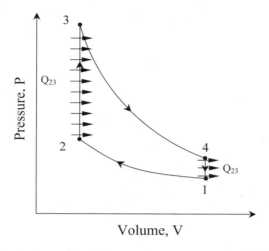

Figure 1.12 *Ideal air standard Otto cycle pressure–volume diagram (Stone).*

3-4   isentropic expansion of air to the original volume

4-1   rejection of heat Q_{41} at constant volume to complete the cycle

The efficiency of the Otto cycle is

$$\eta_{Otto} = \frac{W}{Q_{23}} = \frac{Q_{23} - Q_{41}}{Q_{23}} = 1 - \frac{Q_{41}}{Q_{23}} \tag{1.12}$$

By considering air as a perfect gas, we have constant specific heat capacities, and for mass m of air, the heat transfers are

$$Q_{23} = mc_v \left(T_3 - T_2\right)$$

$$Q_{41} = mc_v \left(T_4 - T_1\right)$$

Thus,

$$\eta_{Otto} = 1 - \frac{T_4 - T_1}{T_3 - T_2} \tag{1.13}$$

For the two isentropic processes 1-2 and 3-4, $TV^{\gamma-1}$ is a constant. Thus,

$$\frac{T_2}{T_1} = \frac{T_3}{T_4} = r_V^{\gamma-1}$$

where γ is the ratio of gas specific heat capacities, c_P/c_V. Thus,

$$\eta_{Otto} = 1 - \frac{1}{r_V^{\gamma-1}} \qquad (1.14)$$

The value of η_{Otto} depends on the compression ratio, r_V, and not on the temperatures in the cycle. To make a comparison with a real engine, only the compression ratio must be specified.

The diesel cycle or compression ignition cycle has heat addition at constant pressure, instead of heat addition at constant volume as in the Otto cycle. With the combination of high compression ratio to cause self-ignition of the fuel and constant-volume combustion, the peak pressures can be high. Figure 1.13 shows the four nonflow processes constituting the cycle:

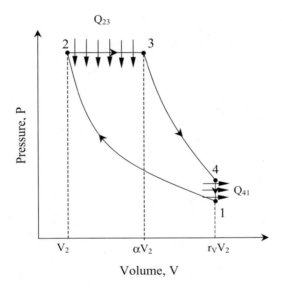

Figure 1.13 *Ideal air standard diesel cycle (Stone).*

1-2   isentropic compression of air through a volume ratio V_1/V_2, the compression ratio r_V

2-3   addition of heat Q_{23} at constant pressure while the volume expands through a ratio V_3/V_2, the load or cutoff ratio α

3-4   isentropic expansion of air to the original volume

4-1   rejection of heat Q_{41} at constant volume to complete the cycle

The efficiency of the diesel cycle is

$$\eta_{Diesel} = \frac{W}{Q_{23}} = \frac{Q_{23} - Q_{41}}{Q_{23}} = 1 - \frac{Q_{41}}{Q_{23}} \qquad (1.15)$$

By considering air as a perfect gas, we have constant specific heat capacities, and for mass m of air, the heat transfers are

$$Q_{23} = mc_p(T_3 - T_2)$$

$$Q_{41} = mc_p(T_4 - T_1)$$

Note that the process 2-3 is at constant pressure. Thus,

$$\eta_{\text{Diesel}} = 1 - \frac{1}{\gamma}\frac{T_4 - T_1}{T_3 - T_2} \qquad (1.16)$$

For the isentropic process 1-2, $TV^{\gamma-1}$ is a constant,

$$T_2 = T_1 r_V^{\gamma-1}$$

For the constant pressure process 2-3,

$$\frac{T_3}{T_2} = \frac{V_3}{V_2} = \alpha$$

Thus,

$$T_3 = \alpha r_V^{\gamma-1} T_1$$

For the isentropic process 3-4, $TV^{\gamma-1}$ is a constant,

$$\frac{T_4}{T_3} = \left(\frac{V_3}{V_4}\right)^{\gamma-1} = \left(\frac{\alpha}{r_V}\right)^{\gamma-1}$$

Therefore,

$$\eta_{\text{Diesel}} = 1 - \frac{1}{\gamma}\frac{\alpha^\gamma - 1}{\alpha r_V^{\gamma-1} - r_V^{\gamma-1}} = 1 - \frac{1}{r_V^{\gamma-1}}\left[\frac{\alpha^\gamma - 1}{\gamma(\alpha - 1)}\right] \qquad (1.17)$$

The efficiency of the diesel engine is not solely dependent on compression ratio r_V, but also is dependent on the load ratio α. The load ratio lies in the range $1 < \alpha < r_V$ and thus is always greater than unit. Consequently, the diesel cycle efficiency is less than the Otto cycle efficiency for the same compression ratio. This is shown in Figure 1.14, where efficiencies have been calculated for a variety of compression ratios and load ratios. There are two limiting cases. The first is as $\alpha \to 1$, then $\eta_{\text{Diesel}} \to \eta_{\text{Otto}}$. The second limiting case is when $\alpha \to r_V$ and point $3 \to 4$ in the cycle, and the expansion is wholly at constant pressure; this corresponds to the maximum work output in the cycle. Figure 1.14 also shows that as the load increases, with a fixed compression

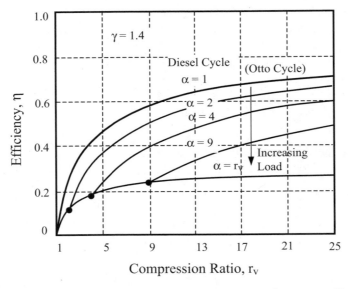

Figure 1.14 Diesel cycle efficiency for different load ratios, α *(Stone).*

ratio the efficiency decreases. The compression ratio of a compression ignition engine usually is greater than for a spark ignition engine; thus, the diesel engine usually is more efficient. The arbitrary overall efficiency of an internal combustion engine is the product of thermal efficiency and mechanical efficiency, $\eta_0 = \eta_m \eta_T$.

1.6.3 Engine Volumetric Efficiency

Volumetric efficiency, sometimes referred to as breathing ability, is the measure of the ability of an engine to draw the air or air/fuel mixture into the cylinders. It is determined by the ratio between what is actually drawn in and what could be drawn in if all cylinders were filled completely,

$$\eta_V = \frac{V_T}{V_E} \tag{1.18}$$

where

η_V = volumetric efficiency
V_T = total volume of the charger
V_E = engine displacement

As engine speed increases beyond a certain point, the piston speed becomes so fast that the intake stroke is of such short duration that less and less fuel mixture is drawn into the cylinders. Because torque is greatest when the cylinders receive the largest amount of fuel mixture, the torque will drop off as the volumetric efficiency decreases.

Many factors influence volumetric efficiency, including engine speed, temperature, throttle position, intake system design, valve size and position, exhaust system configuration, and atmospheric pressure. The negative effect of temperature on volumetric efficiency is illustrated in the following perfect gas equation. A higher temperature in the cylinder results in higher pressure, which diminishes the effectiveness of the intake stroke,

$$PV = mRT \qquad (1.19)$$

where

P	=	pressure of gas
V	=	volume of gas
m	=	mass of gas
T	=	absolute temperature of gas
R	=	gas constant

Volumetric efficiency can be improved by adding a supercharger; a straighter, smoother, and ram-length intake manifold; larger intake valves; a more efficient exhaust system; and other modifications to the induction and exhaust systems. To increase the specific power output of an engine (power output per unit of displacement), some form of precompression often is considered. This is rapidly becoming standard practice in diesel engines and often is seen in high-performance spark ignition engines.

The engine that draws fresh charge into the cylinder at atmospheric pressure and exhausts directly to the atmosphere is termed naturally aspirated. A crankshaft-driven compressor may be added to elevate the pressure of the air (or mixture) prior to drawing it into the cylinder. This allows more mixture to be burned in a given cylinder volume. The crankshaft-driven device generally is referred to as a supercharger.

Recognizing that the exhaust gases leaving the engine still contain a significant quantity of energy that we were unable to access as work, an alternative is to use a portion of this energy to drive the compressor. This configuration is the turbocharged engine.

Whenever air is compressed, its temperature increases. The density of the air can be increased further (and still more air is forced into the cylinder) if it is cooled after compression. The charge air cooler, variously termed intercooler (cooling between stages of compression) and aftercooler (cooling after compression), may be used with either the turbocharger or the supercharger.

1.7 References

Stone, R., *Introduction to Internal Combustion Engines, Second Edition*, Society of Automotive Engineers, Warrendale, PA, 1995.

Chapter 2

Valvetrain Systems

2.1 Overview of Valvetrain Systems

The function of the valvetrain mechanism is to use the intake and exhaust valves to control in a timely way the entry of charge and the exit of the exhaust gas in each individual cylinder following each cycle of engine operation. The valves must respond quickly to the valvetrain actions, and they must seal against the combustion pressures and temperatures. The poppet type of valve has a low lubrication requirement, low friction, simplified sealing, easy adjustment, and low cost. Therefore, it has been the most commonly used valve in four-stroke engines. Throughout this book, without mentioning the poppet, the terminology of valve or valvetrain implies the poppet valve and its train system.

The valvetrain system consists of a camshaft, one or more transfer elements between the camshaft and valves, valve springs, retainers, keepers, valve stem seals, and the actual valves (Figure 2.1). The transfer elements usually include the rocker arm, pushrod, and lash compensator (adjuster, lifter, tappet, or bucket). In addition, the valve guide and valve seat insert are included as valvetrain components because of their significant impact on its performance.

The valve in the valvetrain system directly controls the gas exchange process. The intake and exhaust valves allow the air or air/fuel mixture to enter the combustion chamber during the intake stroke, to seal the chamber during the compression and combustion strokes, and to allow exhaust gas to escape during the exhaust stroke.

A valve spring is used to close the valve and store valve-opening energy during each cycle. The stored energy is used during each cycle to keep the camshaft and transfer elements in contact at all times. The spring retainer is used to fix the movable end of the valve spring to the valve to maintain a working spring height. The retainer is secured to the valve keeper grooves by the keepers or keys.

Valve stem seals are used to seal the valve guide in the cylinder head to permit a metered amount of oil to flow down the guide as required for valve stem lubrication, while providing acceptable oil economy and emissions.

Transfer elements such as rocker or follower arms are used to transfer the reciprocating motion imparted by the camshaft lobe to the intake and exhaust valves. Generally, a hydraulic unit

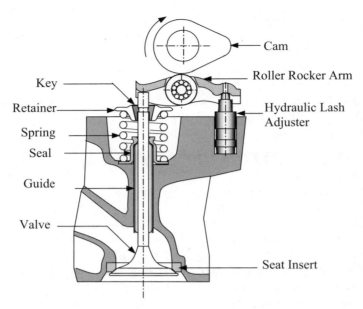

Figure 2.1 Valvetrain schematic and nomenclature, end-pivot rocker arm or Type II valvetrain.

such as a tappet (valve lifter) or a lash adjuster is placed in the valvetrain to provide a suitable means of compensation of lash in the length variation of valvetrain components during engine operation.

A typical camshaft must include lobes, bearing journals, and a thrust face to prevent fore and aft motion of the camshaft. The camshaft can include a gear to drive the distributor and an eccentric to drive a fuel pump. The camshaft lobes supply the motion to open and close the valves through a transfer element (tappets, pushrods, or rocker and follower arms).

The valvetrain system is a significant part of the powertrain function and contributes to a power loss detrimental to fuel economy. It often causes noise, vibration, and harshness (NVH), emissions, and durability issues. The challenges for valvetrain system designers are that the system must operate robustly with a variety of fuels, changing lubricants, environmental conditions, and customer operating conditions. The mechanism must perform its task in small sizes, without service, quietly, and at low cost.

2.2 Valve Actuation

Valve actuation includes valve timing, valve lift, and duration. Valve lift and duration depend solely on the cam shape. However, valve timing is controlled by both the cam shape and the relative position of each cam to each other. Valve actuation determines the flow dynamics of the air or air/fuel mixture entering and exiting internal combustion engines and is one of the controlling factors of engine performance, fuel economy, and emissions.

2.2.1 Valve Lift and Duration

For the maximum volumetric efficiency, the valve should be lifted fully to its maximum as soon as the valve starts opening, and it should be closed in the same fashion. The ideal valve lift profile plotted versus cam angle should be rectangular in shape, as shown in Figure 2.2. Realistically, however, under cam driving, the cam profile determines the valve lift curve shape. Velocity and acceleration of valve opening and closing predominantly influence the cam profile and the valve lift profile. Valve lift equals the cam lift in a direct-acting valvetrain system; it equals the cam lift times the rocker ratio in the case where a rocker follower is used.

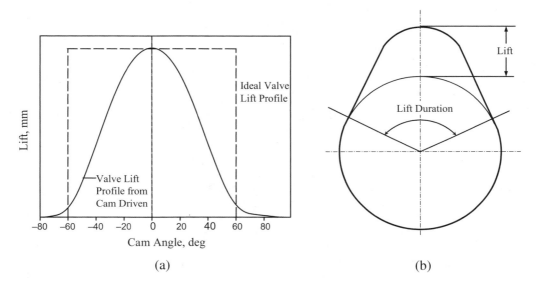

Figure 2.2 *Valve lift and duration: (a) ideal and cam-driven valve lift profile, and (b) typical cam profile.*

2.2.2 Valve Timing and Overlap

Valve timing is the valve opening and closing position relative to the piston position. Valve overlap occurs when the exhaust valve opens (EO) into the intake stroke, and the intake valve opens (IO) before the piston reaches top dead center (TDC) on the exhaust stroke. Valve overlap is designed to make use of the outgoing exhaust gas momentum leaving behind a depression that induces a fresh charge to enter the combustion chamber when the piston is near TDC. The valve timing and valve overlap angle affect volumetric efficiency and thus engine performance at both high and low speeds. There are three typical timing diagrams: (1) the theoretical timing diagram; (2) the typical valve timing diagram of a conventional engine, including compression ignition and spark ignition engines; and (3) the aggressive valve timing diagram, which is typical of a high-performance spark ignition engine or supercharged engine (Heisler).

Figure 2.3 shows a theoretical valve timing diagram. The intake valve opens and the exhaust valve closes (EC) at TDC to start the induction stroke (crank angle of 360–540°). The intake

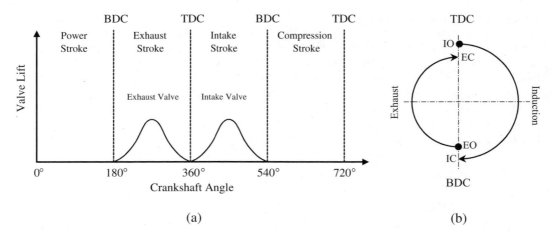

Figure 2.3 *Theoretical valve actuation diagram: (a) valve timing, lift, and duration, and (b) circular valve timing diagram.*

valve closes (IC) at bottom dead center (BDC), and the piston upswings to start the compression stroke (540–720°). Both the intake and the exhaust valves are closed during the combustion stroke (0–180°). The exhaust valve then is opened at BDC to start the exhaust stroke (180–360°). The cycle repeats when the piston reaches TDC again after a 720° crank angle turn.

Figure 2.4 shows a regular valve timing diagram. Instead of the theoretical opening and closing of the intake and exhaust valves at dead center and staying open during 180° of crank travel, the valves always open before and close after the respective dead centers in the valve timing design (before top dead center [BTDC], after top dead center [ATDC], before bottom dead

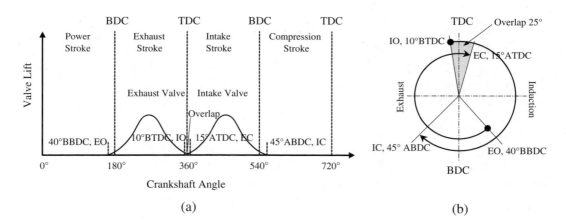

Figure 2.4 *Regular valve timing diagram for naturally aspirated engines: (a) valve lift and timing diagram, and (b) circular valve timing diagram.*

center [BBDC], and after bottom dead center [ABDC]). The objective of this early opening and late closing is to increase the average valve lift and to reduce the gas velocities, and thus their flow resistance in the intake and exhaust ports.

The opening of the intake valve occurs only slightly before TDC (typically 15° BTDC). Thus, by the start of the induction stroke, there is a large effective flow area. It is preferable to have the intake valve fully opened at the points of high piston speeds. Engine performance is fairly insensitive to intake valve opening in the range 10–25°. The intake valve closure is always retarded considerably after BDC (approximately 45° ABDC) to make use of the high velocity of the air induced in the intake pipe and thus to obtain a greater charge into the cylinder. The intake valve remains open during part of the compression stroke to take advantage of the dynamics of the charge and to allow more time for cylinder filling, because at BDC, the cylinder pressure usually still is below the intake manifold pressure. Typically, a late intake valve closure reduces the volumetric efficiency at low speeds. In contrast, at high speeds, an early intake valve closure leads to a greater reduction in volumetric efficiency, and this limits the maximum power output.

The conditions are reversed for the exhaust valve. The exhaust valve must open considerably earlier before BDC (40° BBDC) to allow most of the gases to escape in the blowdown process and to reduce the pressure in the cylinder to as near to atmospheric pressure as possible before the exhaust stroke starts. Exhaust valve closure invariably occurs after TDC (ATDC), and the higher the boost pressure in turbocharged engines, or the higher the speed for which the engine performance is optimized, then the later the exhaust valve closure. The exhaust valve usually is closed in the range of 10–20° ATDC. The aim is to avoid any compression of the cylinder contents toward the end of the exhaust stroke. The exhaust valve closure time does not seem to affect the level of residuals trapped in the cylinder or the reverse flow into the intake manifold. However, for engines with in-manifold mixture preparation, a late exhaust valve closure can lead to fuel entering the exhaust manifold directly. The correct timing is a function of engine speed. With increased engine speed, the intake valve should be closed later and the exhaust valve opened earlier. In addition to the consideration of speed, the best valve timing can be determined only by actual tests, because it depends greatly on the design of the intake and exhaust passages. When determining the closing timing for the intake valve, the designer must consider that at a low engine speed, especially at starting, a late closing reduces the compression ratio so much that it can be difficult to start the engine. Traditionally, the diameter of the exhaust valve is somewhat smaller than that of the intake valve, owing to the fact that the output flow starts at a higher pressure, which accelerates the flow. Also with regard to the exhaust side, the aim with naturally aspirated engines is to achieve the lowest possible pressure drop.

The overlap area is preferred over overlap degrees because it gives a more relevant "volume of flow" past both the intake and exhaust valves in the overlap region. As shown in Figure 2.5, overlap degrees as a measure of overlap can be misleading because the same overlap degrees can have many different overlap areas and vice versa, depending on the lift profiles.

Figure 2.6 shows a supercharged or high-performance engine valve timing diagram. With super-charged engines, the conditions are rather different from those of naturally aspirated engines because the air is fed into the cylinder under a pressure above atmospheric. The main difference between supercharged and naturally aspirated engines is the larger overlap of the exhaust and intake valve timing. Compared with the valve timing of the naturally aspirated engine, the intake

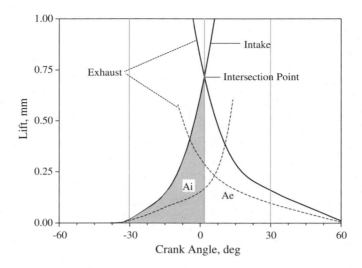

Figure 2.5 *Overlap area; Ai is the overlap area from the intake, and Ae is the overlap area from the exhaust.*

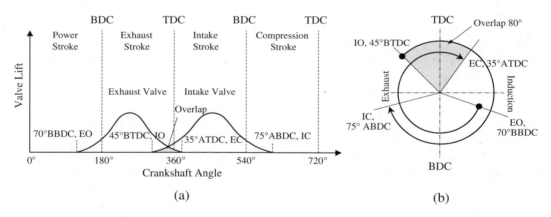

Figure 2.6 *Aggressive valve timing diagram for supercharged or high-performance engines: (a) valve lift and timing diagram, and (b) circular valve timing diagram.*

valve of the supercharged engine opens ~45° earlier (BTDC) and closes ~75° later (ABDC), and its exhaust valve opens ~70° earlier (BBDC) and closes ~35° later (ATDC). The beneficial influence of a larger overlap is a thorough scavenging of the entire combustion space. Consequently, residual gases thus are effectively removed, and the excess air has a cooling effect on the parts concerned, resulting in higher volumetric efficiencies and improved performance. However, the disadvantage is that valve recesses in the piston top and space must be made to provide for the additional valve movement. This may have a negative influence on the combustion efficiency.

The early opening of the exhaust valve leads to a reduction in the effective expansion ratio and expansion work, but this is compensated for by reduced exhaust stroke pumping work. In the case of turbocharged engines, some of the expansion work that is lost by earlier opening of the exhaust valve is recovered by the turbine.

2.2.3 Effect of Valve Timing on Performance and Emissions

In spark ignition engines with a large valve overlap, the part throttle and idling operation suffers because the reduced induction manifold pressure causes backflow of the exhaust. Furthermore, full-load economy is poor because some unburned mixture will pass straight through the engine combustion chamber when both valves are open at TDC. These problems are avoided in a turbocharged engine with in-cylinder or direct fuel injection. In general, the benefits of delaying the exhaust closing while opening the intake valve earlier, so that both the intake and exhaust opening periods overlap each other, are better cylinder clearing and filling in the mid- to upper speed range of the engine. The extended valve overlap improves cylinder volumetric efficiency, owing to the high exit velocity of the exhaust gases. This greatly assists in drawing in fresh air or mixture from the induction manifold even before the intake stroke starts.

In compression ignition engines, the valve overlap at TDC often is limited by the piston-to-cylinder-head clearance. Also, the intake valve must close soon after BDC; otherwise, the reduction in compression ratio may make cold starting too difficult. The exhaust valve opens approximately 40° BBDC to ensure that all the combustion products have sufficient time to escape. This entails a slight penalty in the power stroke, but 40° BBDC represents only approximately 12% of the engine power stroke. Also note that 5° after starting to open, the valve may be 1% of fully open. After 10°, it may be 5% of fully open, and it may not be fully open until 120° after starting to open. Highly turbocharged compression ignition engines have lower compression ratios to limit the peak pressures and temperatures. The lower compression ratios increase the clearances at TDC, permitting greater valve overlaps. In the case of large engines with quiescent combustion systems, cutouts in the piston to provide valve clearance have a less serious effect on the combustion chamber performance. Thus, large turbocharged engines designed for specific operating conditions can have a valve overlap of 150° or so. At a full-load operating condition, the boost pressure from the compressor will be greater than the backpressure from the turbine. Consequently, the large valve overlap allows a positive flow of air through the engine, and this ensures excellent scavenging and cooling of the components with high thermal loadings (i.e., exhaust valves, turbine, and combustion chamber). However, such a turbocharged engine running at reduced speeds or at part load is likely to have a boost pressure that is lower than the turbine backpressure. In this case, a reduced valve overlap probably is desirable to reduce the backflow of the exhaust into the induction system. Too much reverse flow can lead to combustion deposits fouling the intake port and thereby throttling the airflow. In summary, minimum overlap during idling is a guarantee for smooth operation. The advanced intake closing at low and medium speeds results in high volumetric efficiencies. The big valve overlap improves the internal exhaust gas recirculation rate, thereby lowering the nitrogen oxide (NO_x) emissions. The delayed intake closing at high engine speeds corresponds to the basic setting again and yields maximum airflow rates. Figure 2.7 and Table 2.1 summarize the valve timing and overlap for typical gasoline, diesel, and high-performance engines and their effect on application, performance, and emissions.

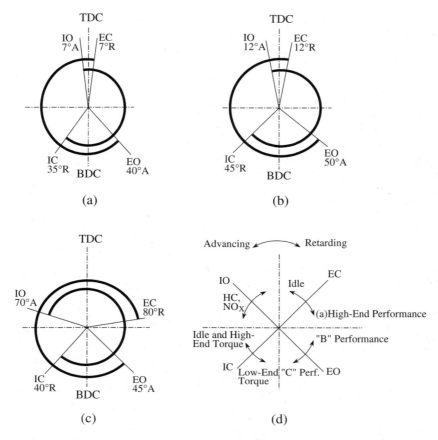

Figure 2.7 *Effect of timing on performance and emissions: (a) low speed, (b) high speed, (c) turbocharged, and (d) effect of timing on engine performance and emissions. "B" represents with backpressure, and "C" represents without backpressure. "7°A" means 7° advanced, and "7°R" means 7° retarded.*

TABLE 2.1
VALVE INJECTION AND
IGNITION TIMING COMPARISONS

	Intake Valve		Exhaust Valve		Injection		Ignition
	Open BTDC (deg)	Close ABDC (deg)	Open BBDC (deg)	Close ATDC (deg)	Begin (deg)	End (deg)	BTDC (deg)
Diesel Engine	0–30	30–50	30–55	5–40	10–25 BTDC	0–20 ATDC	
Performance Engine	50–80	20–50	45–70	50–70	10–25 BTDC	0–20 ATDC	
Gasoline Engine	0–20	30–55	30 - 55	5–20	15–35 BTDC	5–15 ATDC	TDC–45

2.2.4 Valvetrain System Timing

After the individual valve actuation profile is determined, the valve timing events taking place in relation to crank position, cam position, stroke, piston position, and valve overlap in each cylinder must be laid out. Figure 2.8 illustrates such a relationship within one engine cylinder. Although the events shown are for cylinder #1, because of the cam and crank gear timing marks, the relationship between the valve event and piston position would be the same for all cylinders.

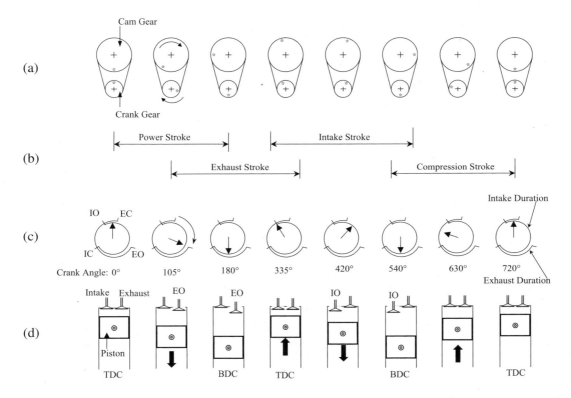

Figure. 2.8 *Valve event related to (a) crank gear and cam gear, (b) combustion event, (c) crank angle, and (d) piston position.*

Figure 2.9 shows the relationship of valve events in a typical six-cylinder engine. The firing order is 1-4-2-5-3-6. This figure shows the start, end, and duration of each valve opening relative to crank angles in each cylinder. The relationships for valve events of other cylinders also are timed from the #1 cylinder firing. The crankshaft gear turns twice for each camshaft revolution: 1 cam degree equals 2 crank degrees, and 360 camshaft degrees equals 720 crankshaft degrees. Camshaft events such as intake open (IO), intake closed (IC), exhaust open (EO), and exhaust closed (EC) are expressed in terms of crankshaft degrees because these events control gas flow and are delineated primarily by piston position. All camshaft events are timed from the #1 cylinder firing (0 degrees).

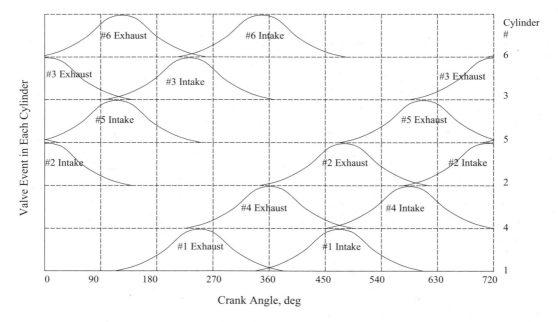

Figure 2.9 *Exhaust and intake valve open duration and power stroke for a six-cylinder engine.*

Figure 2.10 shows the relationship of the intake (or exhaust) cam lobe phasing in each cylinder. The firing order is 1-3-4-2 for an in-line four-cylinder engine, 1-4-3-6-2-5 for a V6 overhead valve (OHV) engine, and 1-3-7-2-6-5-4-8 for a V8 OHV valve engine. The power stroke or the exhaust or intake stroke occurs when the cam rotates every 90° for the in-line four-cylinder engine, every 60° for a V6 engine, and every 45° for a V8 engine. For a V-type engine with overhead cam (OHC), more than one camshaft is required, and cam phasing may differ from the ones shown in Figure 2.10(b) but is still timed with cylinder #1, depending on the cylinder firing order. The determining factor of choosing the firing order for each cylinder for an engine is to maintain the balance of the engine during firing.

2.3 Variable Valve Actuation—Cam Driven

2.3.1 Overview of Variable Valve Timing

Variable valve actuation (VVA) takes account of variable valve timing (VVT), lift, and duration. In conventional engines with fixed valve timing, the timing of the intake valve is a compromise between the two contradictory requirements for low and high speeds. Engines that produce high torque at low speeds have lower overlap between the closing of the exhaust valve and the opening of the intake valve. Small overlap allows for little communication between the exhaust gases and the incoming fresh charge, limiting the amount of uncontrolled mixing. This leads to stable operation. However, at high speeds, the inertia of the gases requires a greater period of overlap to allow for gas exchange. Many approaches have been proposed and tried in attempts to optimize

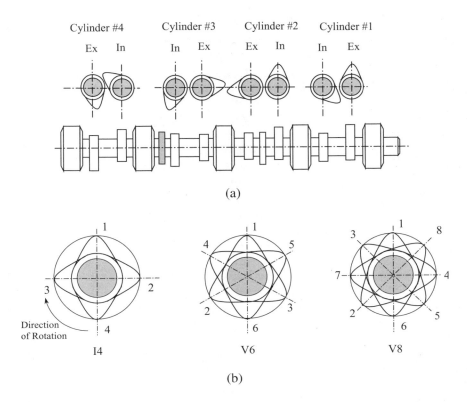

Figure 2.10 *Relative intake (or exhaust) cam lobe phasing for each cylinder: (a) four-cylinder engine cam lobe phasing, and (b) cam lobe phasing projection of a four-, six-, and eight-cylinder engine.*

the flow processes. The benefits gained by increasing the valve overlap are realized only in the upper speed and load range, and not in starting and low-speed running conditions. With a large valve overlap, some fresh unburned charge will be expelled into the exhaust manifold, which, accordingly, increases the hydrocarbon exhaust emissions. Conversely, residual exhaust gas may be drawn into the induction manifold under part throttle and idling conditions; therefore, it contaminates and dilutes the fresh mixture entering the cylinder. For simplicity, economy, and durability, most conventional four-stroke engines use the tried-and-true fixed camshaft systems that have constant phase (when the valves are opened). However, most engines must work over a wide speed and load range, making it difficult to achieve optimum efficiency over more than a narrow part of that range. An engine that produces high torque for its capacity at low engine speeds usually gives poor torque at high engine speeds, and vice versa. Therefore, VVT is critical to the performance of an engine at both high and low engine speeds.

The effective expansion ratio is determined by the timing of the exhaust valve opening. Usually, the exhaust valve begins to open well in advance of BDC to provide enough time for the cylinder blowdown at high engine speeds. This makes the effective expansion ratio less than it could be if the timing were close to BDC. At low speeds, however, there is much more time for the blowdown, and the excessively early exhaust valve opening is wasteful. With variable exhaust valve timing, this deficiency can be eliminated by retarding the exhaust valve opening at low speeds and, in general, optimizing the timing as a function of engine speed. Increased expansion stroke work increases torque and, especially at low engine speeds, improves engine fuel efficiency.

Exhaust gas recirculation (EGR) directly affects the residual gas fraction in the cylinder. A higher quantity of residual gas lowers the peak combustion temperature and reduces the quantity of NO_X produced during combustion. The ability to vary the timing of the exhaust valve closing eliminates the need for external recirculation. Advancing the exhaust valve closing to BTDC permits retention of the last portion of the gases leaving the cylinder. On the other hand, when the exhaust valve closing is substantially retarded, a certain quantity of exhaust gas is sucked back into the cylinder from the exhaust port by the downward-moving piston on its early part of the intake stroke. The later the valve closure, the more exhaust gas is sucked into the cylinder. Thus, varying the timing of closure controls the quantity of residual gas in the cylinder.

Camshaft-driven engines usually produce a torque curve with a distinctive peak value in the mid-speed range. The fixed one-value valve actuation compromises the performance for both low and high speeds. Optimization of valve actuation yields a flatter torque curve due to improvement in volumetric efficiency. Engine torque at high speeds can be increased when the intake valve closing is delayed to take full advantage of ram-charging. At low speeds, the closing delay should be eliminated to retain the maximum effective compression ratio. An electronic control system senses the changes in the engine speed and continuously adjusts the valve closure to achieve the best compromise between ram-charging and compression ratio at each speed. With proper tuning of the intake closing timing for best ram-charging effect, the engine peak torque can be improved considerably.

Figure 2.11 shows the ideal valve timing and lift for various engine speeds to produce a flatter torque curve while achieving less fuel consumption and lower emissions. Some of the VVT mechanisms currently in production are described in the following section.

2.3.2 Variable Valve Timing by Camshaft Phasing

Most typically, VVT is implemented by shifting the phase angle of the whole camshaft forward or backward by means of a hydraulic actuator attached to the end of the camshaft. The timing is calculated by the engine management system with engine speed and acceleration. The system can have continuously variable valve timing across a wide range of up to 60°, or only two stages at preset angles. The most frequently used cam phasing mechanism is the camshaft sprocket drive coupled to drive through a helical spline such as that used by Alfa Romeo with the mechanism shown in Figure 2.12(a) controlling the phasing of the intake camshaft.

The main components of the sprocket drive unit in the camshaft for VVT are a timing piston with internal straight splines and external helical splines, an internal helical splined sleeve sprocket, an external straight splined hubshaft mounted on the camshaft, a control plunger, and an external control solenoid. When the engine is running and the solenoid stem is withdrawn, engine lubricant under pressure enters the camshaft from the bearing pedestal and flows axially

Valvetrain Systems

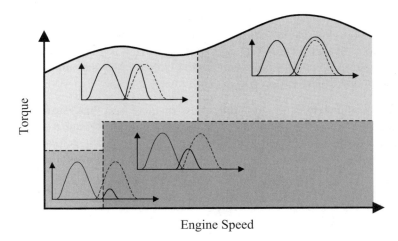

Figure 2.11 *Optimal valve events in engine operation map (Brüstle and Schwarzenthal).*

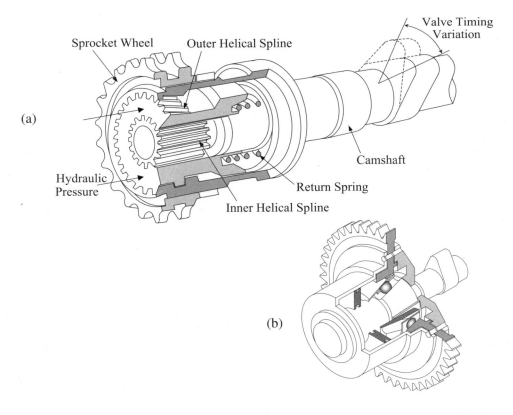

Figure 2.12 *Variable valve timing cam phasing devices: (a) Alfa Romeo cam phasing mechanism (Heisler), and (b) Mechadyne cam phaser with rolling contact in inclined grooves to reduce friction.*

toward the hub end of the camshaft. Then it is diverted radially to the surface of the hubshaft and the bore of the timing piston. Oil then passes between the clearance space formed between the meshing piston and the hubshaft straight splines. Then it is directed radially inward, passes through the center of the control plunger, and finally escapes to the sump via the relief hole in the side of the plunger. The valve overlap is minimal.

When engine speed increases, the hydraulic pressure increases, thus pushing the helical spline along the camshaft axis and consequently rotating the camshaft through the mating helical gear. At some predetermined speed and load condition, the solenoid is actuated by the electronic fuel injection and ignition management system. This extends the solenoid stem, pushing the control-plunger into the end plate until it covers the relief hole in its side. Immediately, the oil flow from the plunger is blocked, which therefore pressurizes the stepped side of the timing piston. As a result, the timing piston is forced to move toward the flanged end of the hubshaft. As it does, the external helical piston splines progressively screw the matching internal helical sleeve splines, causing the sleeve and sprocket to rotate in a forward direction relative to the hubshaft that is attached to the camshaft. Accordingly, the intake valve lead will increase, and the intake valve lag will decrease by the amount of relative angular twist imposed on the sleeve and sprocket as the piston slides along the external sleeve splines from the left to the right end. The valve overlap is maximum.

The latest cam phasing device employs the principle of rolling contact in inclined grooves to achieve angular movement of the camshaft with respect to the input drive sprocket (Figure 2.12(b)). The rolling contact reduces friction and improves response time—an important factor in reacting to transient conditions that now are becoming a greater consideration in emissions control work. The low operating force also means the unit is less demanding of the hydraulic oil supply because it is able to function with lower pressure or lower flow rates than existing cam phasers, which can pose significant pressure and flow demands unless constraints on speed of operation are accepted. The rolling element phaser can be used on the intake camshaft alone or on both the intake and exhaust.

An advantage of this phasing control is that at light load and low speed (significant conditions in urban driving cycles), it is beneficial to delay both the intake valve closure and the intake valve opening. The delayed intake valve closure reduces the throttling losses, and the reduced valve overlap at light load also improves combustion. Phasers are the simplest system to integrate on existing engines and may be used for intake for performance enhancement or for exhaust for fuel consumption and emissions control. The characteristics of this variable intake valve opening point mechanism are that the total induction period remains constant, but the start and finish points are variable (Figure 2.13). One drawback of this type of variable intake valve timing mechanism is that when the intake valve opening lead and valve overlap are increased, it is at the expense of the intake valve closing lag, which is correspondingly reduced by the same amount. Thus, at high speed and wide-open throttle (WOT) conditions, when a delayed intake valve closure is desirable to take advantage of the inertia of the fresh charge, the reverse happens—the intake valve closes earlier, immediately after BDC. Thereby it cuts short the final cylinder filling when the piston inward movement is still small relative to the crank angle displacement. At the same time, the larger intake valve closing lag reduces the effective compression ratio at idling speed, which is beneficial for good low-speed stability.

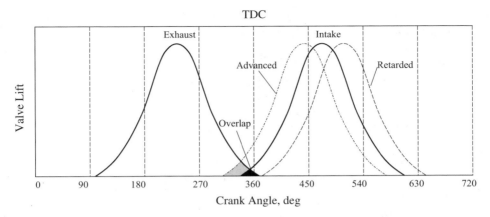

Figure 2.13 *Valve lift characteristics for intake valve cam phasing.*

Variable valve timing for intake and exhaust valves takes advantage of both performance improvements and better emissions control. Figure 2.14 shows the BMW Double VANOS variable camshaft control system VVT sprocket drive timing diagram.

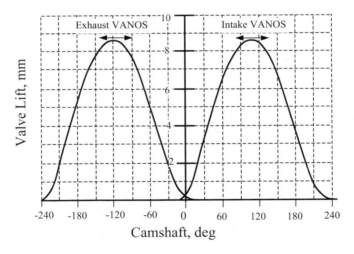

Figure 2.14 *BMW Double VANOS cam phasing system timing diagram (Flierl and Klüting).*

The advantages of a larger overlap and advanced valve opening grow with increased engine speeds and make higher power outputs possible. Figure 2.15 presents three experimentally obtained torque curves at full load. The first curve is the result of revolutions-per-minute dependent optimized intake valve timing, the second with a constantly advanced intake valve

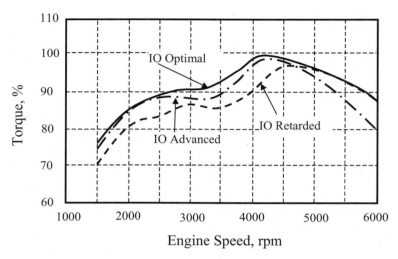

Figure 2.15 *Torque comparison with constant valve timing and with fully variable adjustment (Grohn and Wolf).*

opening, and the third with a constantly retarded intake valve opening. It can be seen clearly that by switching from the constantly advanced to the constantly retarded intake valve timing at 4600 rpm, it is possible to achieve practically the torque curve of the revolutions-per-minute optimized intake valve timing.

Nissan claimed a 7–10% improvement in WOT torque in the low engine speed range by advancing the intake cam phasing by 20 crank degrees for the intake cam phasing engine (Maekawa *et al.*). However, if the goal is to use the engine at the most fuel-economical speed (i.e., intermediate speed), then the overlap must be reduced to adapt to lower engine speeds.

2.3.3 Variable Valve Lift by Cam Profile Switching

In camshaft phasing, if the phase of a valve event is altered (e.g., advancing the intake valve opening to an earlier crankshaft angle), then the closure of that intake valve also is advanced. In many cases, this causes a reduction in the amount of charge that can enter the engine. To overcome this situation, the duration of the valve event may be altered. Consequently, as the engine speed and valve overlap are increased, the period that the intake valve remains open is extended to delay the closing.

The peak lift of valves is designed to accommodate gas flow at maximum engine speeds. At low engine speeds, the velocity of incoming gases through the valve curtain will produce less turbulence and may lead to lower torque than would be achieved with a smaller valve opening. By varying the valve lift with the engine speed, the torque may be enhanced over the entire operating range of the engine. Additionally, reduced valve lift at lower speeds reduces the frictional losses of the valvetrain and reduces valvetrain wear. Cam profile switching is one way to achieve the variable valve lift. Honda's two-stage VTEC, now developed to three-stage i-VTEC, is one example based on the concept of cam lobe profile switching. The i-VTEC system chooses among three different cam profiles activating one full-load and two part-load cams, depending on the respective load and speed range (Figure 2.16). A synchronizing piston connects or disconnects the three intake rocker arms. Hydraulic pressure against a timing piston moves the synchronizing pistons in one direction, while stopper pistons and return springs move the synchronizing pistons back when the hydraulic pressure is reduced.

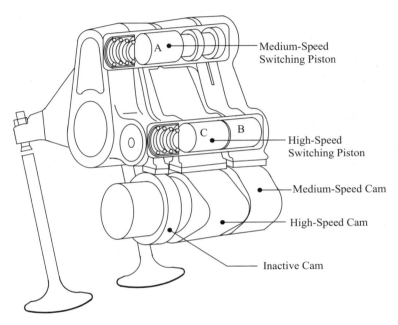

Figure 2.16 *Honda three-stage i-VTEC cam profile switching schematic (Matsuki et al.).*

At low speed, the primary and secondary rocker arms are not connected to the middle rocker arm but are driven separately by left (secondary) and right (primary) cam lobes at different timing and lift (Figure 2.17). The lift of the secondary cam lobe is small, so that one intake valve opens slightly (one-valve control). Although the middle rocker arm is following the center cam lobe with the lost motion assembly, it has no effect on the opening and closing of the valves in the low-revolutions-per-minute range. Therefore, at low engine speed, the three pieces of the rocker arms move independently, and the primary intake valve (right) operates at normal lift, while the secondary intake valve (left) opens only slightly to prevent fuel accumulation in the intake port.

The left rocker arm, which actuates the secondary intake valve, is driven by the left low-lift cam (Stage 1 valve lift curve in Figure 2.18). The right rocker arm, which actuates the primary intake valve, is driven by the right medium-lift cam (Stage 2 valve lift curve in Figure 2.18).

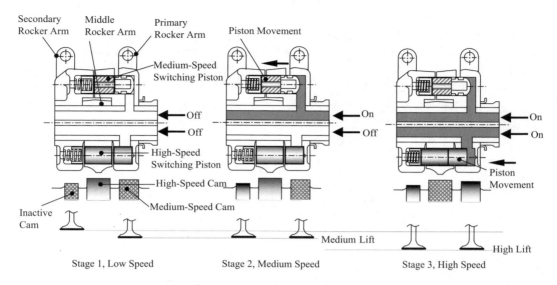

Figure 2.17 *The Honda three-stage i-VTEC variable valve actuation mechanism (Matsuki et al.).*

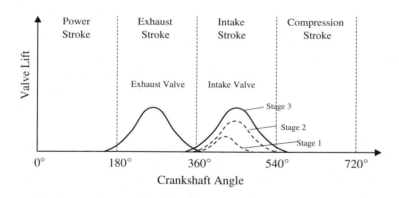

Figure 2.18 *Honda i-VTEC timing diagram.*

The timing of both cams is relatively slow compared with that of the middle cam, which actuates no valve in the first stage.

Because the two intake valves have unequal lifts, the air flows into the cylinder in two unequal streams, which provide better tumble and swirl effect and, when mixed, can provide a variety of in-cylinder flow patterns, depending on the relative strength of each stream. The end results will be better fuel efficiency, better combustion stability, improved idle, and lower emissions.

At medium engine speed or Stage 2, hydraulic pressure pushes the synchronizing piston A and connects the rocker arms in the left and right together, leaving the middle rocker arm and cam to run on their own. Because the right cam is larger than the left cam, those connected rocker arms are driven by the right cam. As a result, both intake valves obtain slow timing but medium lift (Stage 2 valve lift curve in Figure 2.18).

At high engine speed or Stage 3, hydraulic pressure pushes the synchronizing pistons B and C as well as A and connects all three rocker arms together. Because the middle cam is the largest, both intake valves actually are driven by that fast cam. Therefore, fast timing and high lift are obtained in both valves. At high speed, the timing piston moves in the direction shown by the arrow in the figure. As a result, the primary, secondary, and middle rocker arms are linked by three synchronizing pistons, and the three rocker arms move as a single unit. In this state, all rocker arms are driven by the center cam lobe, opening and closing the valves at the valve timing and valve lift set for high operation (Stage 3 valve lift curve in Figure 2.18).

The i-VTEC engine has a flat torque curve all the way to its revolution limit and a 40% increase in power output over the VTEC engine (Figure 2.19).

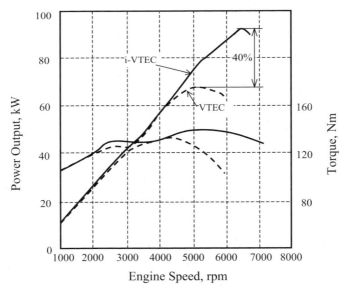

Figure 2.19 *Engine output power comparison of the Honda i-VTEC and VTEC (Matsuki et al.).*

2.3.4 Variable Valve Lift by Lost Motion

Another mechanism to discretely change the valve lift is by lost motion of the cam follower/ lifter. Mercedes' cylinder cutout system shuts off four of eight cylinders by deactivating their valves (*Automotive Industry*, March 1999). The system retains the double-rockershaft design of the base engine but replaces its one-piece rockers with a pair of unique arms. One arm follows the cam lobe for valve opening and closing, while the other controls the deactivation of the valves. In full V8 mode, hydraulic pressure forces a tiny piston to lock both arms together. In deactivated four-cylinder mode, electromagnetic shift valves force the locking pistons against the return springs, shutting the valves (both the intake and exhaust sides) and thus the cylinder. Valve reactivation occurs when the hydraulic pressure releases the pistons, which again locks up both arms. To keep the walls of the deactivated cylinders warm during cutout, the system closes the exhaust valves immediately after the power stroke. They open first during reactivation. The system apparently deprives the base V8 of approximately 7 hp (299 hp versus 306 hp). However, the fuel efficiency (on the new European driving cycle) is claimed to increase 7%. At a steady 56 mph, it is improved by 15%.

Honda's Hyper-VTEC system is a bucket tappet mechanism that is fed by an adjacent oil gallery in the cylinder head (Figure 2.20). An electronically controlled spool valve within the bucket follower allows oil pressure to engage a slide pin, which engages the bucket to move the pair of inoperative valves.

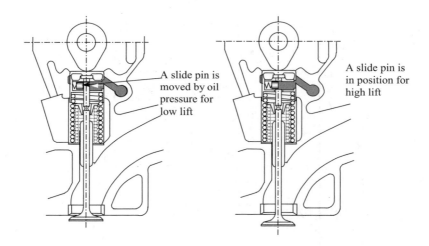

Figure 2.20 *Honda's Hyper-VTEC (Automotive Industry, March 1999).*

Figure 2.21 shows the valve lift-timing diagram for cam phasing plus the variable lift system. Depending on the requirement of low lift, either a switch-over or a switch-off system can be realized. Or both options can be combined, which would allow two-line engines to run on a one-cylinder bank at low speeds and low loads.

The variable lift also can be realized by a latching mechanism for a hydraulic lifter. The switch-over process is triggered hydraulically by the engine oil pressure. Figure 2.22 illustrates the General Motors Displacement-on-Demand mechanism. The mechanism uses the existing oil pump to provide hydraulic pressure to activate the system. The actuators are special hydraulic

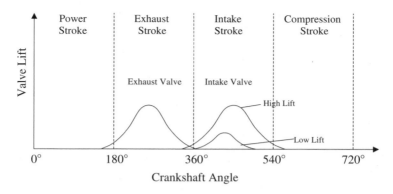

Figure 2.21 *Valve lift-timing diagram for cam phasing plus a dual lift system.*

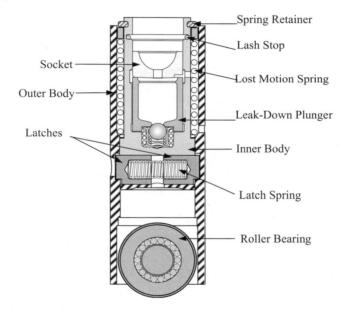

Figure 2.22 *The General Motors Displacement-on-Demand roller lifter* (Automotive Engineering International, *January 2002*).

lifters, each with a spring-loaded locking pin that deactivates the cylinders. In normal operation, when the camshaft rotates against the hydraulic lifter, it opens and closes either an intake or exhaust valve. The special lifter is designed so that one section can collapse, or telescope, into the other section. The two sections can be coupled or uncoupled to each other by means of the locking pin. For cylinder deactivation, hydraulic pressure dislodges the locking pin and collapses the lifter, closing the valve. Shutting off the hydraulic pressure causes the locking pin to return to its latched position and to restore the normal function of the lifter. In V4 mode, every other cylinder in the firing order is deactivated. In a V8, the process affects the outer two cylinders on one bank and the inner two cylinders on the opposite bank.

2.3.5 Variable Valve Actuation by Cam Phasing and Variable Lift

In addition to the systems that use what is called the "phase-shift" function, other variants exist in which the valve timing is controlled by varying the valve lift and duration. In an engine with a conventional cam-driven valvetrain, the lifts of the intake and exhaust valves are selected to ensure good volumetric efficiency and minimum exhaust stroke pumping during high-speed operation. At lower engine speeds, lower valve lifts are adequate and sometimes are desirable. Because the energy consumed by the valvetrain decreases with reduction in the valve stroke, varying the valve lift as a function of engine speed can improve fuel efficiency at low speeds. Reduced intake valve lift at low engine speeds has another potential benefit: it increases the intake air velocity. This leads to a faster burn rate and improved idle stability. The ability to tune the valve overlap as well as valve lift offers an opportunity to lower idle speed and thus achieve a significant reduction in fuel consumption. A VVT and lift mechanism is used so the engine achieves both low fuel consumption and high output. With such a system, a great lift difference is provided for the intake valve according to the engine speed range. In combination with the fuel injection air control system, this has achieved high combustion efficiency and low fuel consumption in the low-speed range of the engine, while maintaining high output in the high-speed range of the engine.

One such variable lift mechanism in addition to the cam profile switching approach is the Porsche controllable-bucket-tappet approach. Porsche uses a bucket tappet of a two-in-one design: a larger bucket contains a smaller one that acts directly on the valve and meshes with the center cam (Figure 2.23). The Porsche design provides adjustable valve timing plus two camshaft profiles and two sets of tappets to vary the valve lift and duration. In normally aspirated engines, VarioCam alters the timing by using camshaft chain tension to switch the intake camshaft between two positions. Each intake camshaft includes two sets of lobes to provide two camshaft profiles: (1) a 3-mm low-lift, short-duration profile, and (2) a 10-mm high-lift, long-duration profile. The high-lift "lobe" is actually a pair of lobes—one on each side of the low-lift lobe. Likewise, there are two sets of hydraulic tappets, but in this case, the arrangement is a tappet within a tappet. The two parts of the tappet can move independently or simultaneously when locked together via an oil-pressure-driven pin. With the pin engaged, the tappet follows the high-lift lobes. With the pin disengaged, the doughnut section simply slides along the inner section. Thus, the low-lift lobe determines valve actuation. For valve-lift variation, the center cam shows a lower lift height corresponding to no more than 40% of the full-load lift. During triggering, the buckets simply lock with each other, that is, the two outer cams, which are in permanent engagement with the larger bucket, and transfer the lift movement to the valve via the inner bucket. The switch-over process is triggered hydraulically by the engine oil pressure (0.8 bar). In addition, VarioCam Plus continuously varies camshaft timing via a helically splined camshaft pulley and camshaft.

Figure 2.24 shows the typical friction loss behaviors of a conventional tappet valvetrain and the VarioCam Plus valvetrain. In high valve lift mode, VarioCam Plus shows a higher friction loss up to 1500 rpm than the conventional tappet system. This is due to the crowned cam/tappet contact

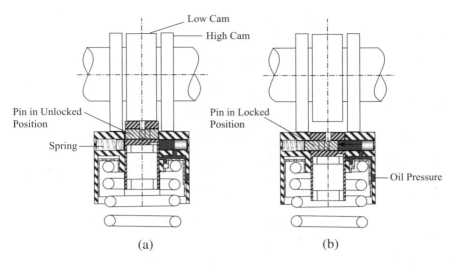

Figure 2.23 *Porsche VarioCam Plus system. (a) Inner tappet is in the unlocked position, inner and outer tappet can move separately. The amount of lift is controlled by the low cam. (b) Inner tappet is locked with the outer tappet by a pin due to high oil pressure. The amount of lift is controlled by the high cam. (Brüstle and Schwarzenthal).*

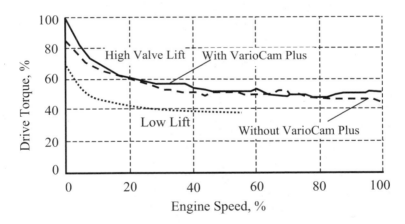

Figure 2.24 *Friction loss comparison of variable valve lift (Brüstle and Schwarzenthal).*

surface, which has a more distinct mixed friction area and a higher normal force contribution of the bucket guide. This disadvantage is more than compensated for by the lower spring forces in short-valve-lift mode, which clearly reduce the friction losses of the VarioCam Plus system.

Another example of variable valve lift is the Toyota VVT-i system, which uses a single rocker arm follower to actuate the intake valves (or both intake and exhaust valves) (Figure 2.25). It also has two cam lobes acting on that rocker arm follower. The lobes have different profiles:

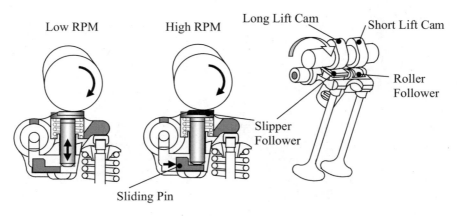

Figure 2.25 *Toyota VVT-i system—cam phasing plus a dual lift system (Moriya et al.).*

one with a long valve opening duration profile for high speed, and another with a short valve opening duration profile for low speed. At low speed, the slow cam actuates the rocker arm follower via a roller bearing (to reduce friction). The high-speed cam does not have any effect on the rocker follower because there is sufficient spacing beneath this hydraulic tappet. When the speed has increased to the threshold point, the sliding pin is pushed by hydraulic pressure to fill the spacing. The high-speed cam becomes effective. Note that the fast cam provides a longer valve opening duration, whereas the sliding pin adds valve lift. Combining cam lobe changing, cam phasing, and variable lift could satisfy the requirement of both top-end power and flexibility throughout the entire speed range.

2.3.6 Variable Valve Actuation by Varying the Rocker Ratio

Many mechanical devices have been proposed to achieve the fully and continuously variable valve timing, lift, and duration through varying the rocker ratio. The BMW Valvetronic is one example of such a system. The Valvetronic uses a so-called intermediate lever positioned vertically above each rocker arm and pivoting about its center, while the intake camshaft acts on the side of the lever, causing it to swing back and forth. A foot at the base of the lever swings across the rocker arm, similar to a pendulum, and opens the valve. By moving the position of the intermediate lever pivot point closer to or away from the camshaft, it changes the rocker ratio, which in turn changes the degree of valve opening or valve lift. Adjustment is made by an eccentric cam operated by a worm gear and electric motor (Figure 2.26).

The Valvetronic variable rocker ratio system, combined with a cam phasing device for both intake and exhaust (Double VANOS), provides the continuously variable valve lift of 0–9.7 mm and continuously variable valve timing up to 60°. Figure 2.27 illustrates the Valvetronic valve lift and timing diagram.

Pumping loss is a major factor affecting the efficiency of the spark ignition engine. The problem occurs at part load with the throttle partially closed. On each induction cycle, the engine is forced

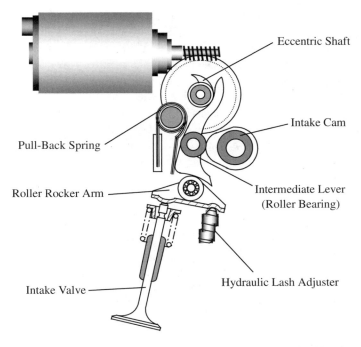

Figure 2.26 *BMW Valvetronic system layout and nomenclature (Flierl and Klüting).*

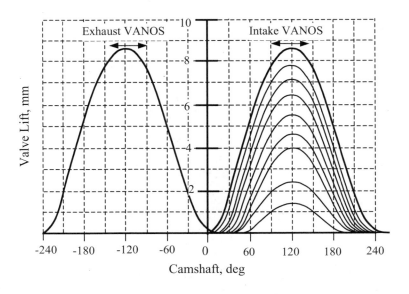

Figure 2.27 *Full variable timing diagram.*

to suck in air through a restricted airway that consumes power and increases fuel consumption. When an engine is running throttled at part load, it is forced to do work in creating a depression in the intake manifold during each induction cycle. The energy expended is never recovered because the intake valve eventually closes, and the intake manifold pressure is equalized by the external atmospheric pressure. Therefore, controlling the engine load below its full power using intake air throttling creates the pumping loss that reduces useful work. The ability to control the intake valve lift can change this situation. In that case, the engine throttle would remain open, regardless of load. To reduce the engine load below the maximum, the variable timing, lift, and duration are used to reduce the volume of air in the cylinder. The engine airflow thus is reduced without resorting to throttling. This largely eliminates the pumping loss and improves fuel efficiency (Figure 2.28).

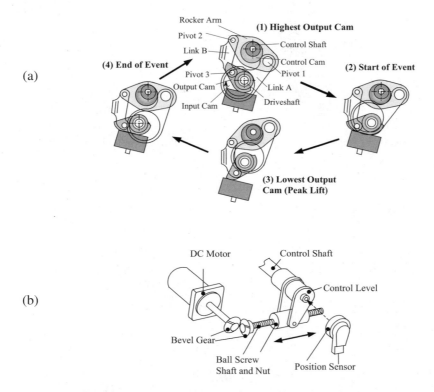

Figure 2.28 *Nissan continuously variable valve event and lift control mechanism: (a) lift control, and (b) timing control (Nakamura et al.).*

Engine efficiency is influenced greatly by the cylinder pressure before combustion. Thus, if the compression ratio is raised, the thermal efficiency will increase, and vice versa. With a gasoline engine, the cylinder pressure will be at its maximum at full throttle when there is no restriction on the cylinder compression pressure just before combustion. However, as the throttle valve is steadily closed, there is a corresponding increase in the cylinder and manifold depression during the induction stroke. This means that during the early part of the compression stroke, the cylinder pressure still will be below the atmospheric pressure, and, effectively, compression will not commence until the cylinder pressure has changed from a negative depression to a positive pressure that is greater than the atmospheric pressure. Thus, the effective compression ratio of the engine will decrease as the butterfly throttle valve is moved from the fully open to the almost closed position, which correspondingly causes a reduction in the thermal efficiency of the engine. At the same time, the increasing intake depression causes the engine to use more of the power to pull in the air/fuel mixture.

A diesel engine controls its power output by the quantity of metered fuel per injection; therefore, it does not need to throttle the air intake. Consequently, each cylinder receives the same amount of air per induction stroke, so that the compression pressures remain roughly the same throughout the speed range of the engine. In contrast to the gasoline engine, there is no power loss due to throttle restrictions as the load and speed are reduced.

However, because of the added components, such as the intermediate lever, eccentric cam, additional spring, and the gear between the cam and cam follower, the valvetrain dynamic and vibration characteristics at high engine speeds are changed significantly. The efficiency of Valvetronic engines drops rapidly above 6000 rpm because stronger valve springs are required. The stronger springs create higher friction losses. Typically, the Valvetronic system realizes a savings of 10% in cruise condition and 20% in idle condition (Flierl and Klüting).

Another example of a continuously variable valve actuation mechanism is the Nissan VEL system, as shown in Figure 2.28. The driveshaft is the equivalent of a conventional camshaft. This shaft is synchronized with the crankshaft and rotates at half its speed. An eccentric input cam (drive cam) is fixed to the driveshaft by a pin. Therefore, the input cam rotates together with the driveshaft. On the outer periphery of the eccentric input cam, the axis cylinder of the large end of link A is supported. In addition, a pin fitted into the end of the rocker arm supports the other axis cylinder of the small end of link A (pivot 1). Consequently, as the input cam rotates, it moves link A up and down, oscillating the rocker arm.

Link C oscillates around the fulcrum cylinder located in its center. One end of link B, which is supported by a pin fitted into the other end of the rocker arm (pivot 2) moves up and down. At this time, due to the existence of the rocker arm rocker ratio, the up-and-down movement of link B is enlarged. As a result, the output cam oscillates largely through a pin (pivot 3), and the valve is lifted. Figure 2.28 shows how the states of the mechanism change during one rotation of the driveshaft. Step 1 in Figure 2.28 shows the condition of the output cam at the highest oscillating position. At this time, no valve lift occurs. Next, as the driveshaft rotates clockwise, a ramp lift passes as shown in step 2, after which the event starts and the lift increases rapidly, approaching the peak lift shown in step 3. Furthermore, when the driveshaft rotates, the lift drops. In step 4, the event is finished, a ramp interval passes, and the lift becomes zero, returning to the original phase condition.

Variable valve lift is controlled by adjusting the phase angle of the control shaft. The control shaft phase is changed continuously over an approximately 80° range. As a result, the amount of valve lift and event can be changed continuously. The distance between the center of the rocker arm oscillation and the driveshaft axis (output cam oscillation center) changes. As a result, it was utilized for the lift and event change method. Using this method, by merely making a slight change in the position of the rocker arm oscillation center, the state of the output cam can be changed greatly by the large up-and-down movement of the position of pivot 2, due to the high rocker ratio. As a result, a wide range of variation can be realized.

Figure 2.29 shows the results of average driveshaft torque indicating friction loss. The results show a slight increase in total friction loss for the VEL system, compared to that of the conventional valvetrain at the same valve lift of 8.3 mm, due to an increased number of elements in the VEL system.

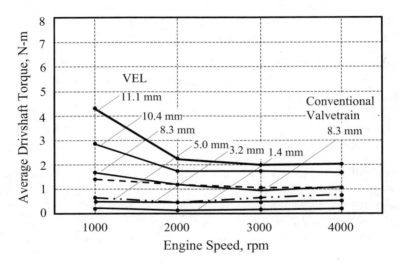

Figure 2.29 *Comparison of measured driveshaft torque for the VEL system at various lifts and conventional valvetrain (Nakamura et al.).*

Figure 2.30 shows another example of variable valve actuation systems using variable rocker ratio mechanisms. The systems maintain the ramp functions (although rescaled) at all lift variations. However, added valvetrain weight and flexibility can limit engine speed.

2.4 Variable Valve Actuation—Camless System

Typical piston-type internal combustion engines use mechanically driven camshafts for intake and exhaust valve operation. Such conventional mechanical valvetrains generally have fixed values for valve lift, timing, and duration. These fixed valve events represent a compromise among the conflicting requirements for various operating conditions. Camshaft-based VVA mechanisms previously discussed offer measurable improvements at some engine operating points, but they are only a fraction of what could be achieved if the valve timing, lift, event duration, and other

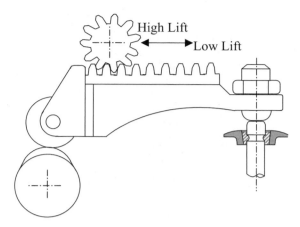

Figure 2.30 *Another variable valve actuation mechanism proposed by varying the rocker ratios (Riley).*

parameters of the valve motion were optimized selectively for each operating condition. The greatest degree of variability in the valvetrain is achieved when each individual valve has its own timing system, enabling it to be opened and closed as and when required. If the crankshaft could be freed from the camshaft, the engine performance could be improved radically. With electronically controlled valve actuation, the crankshaft would spin freely without the drag caused by the camshaft. The camless systems (i.e., electromagnetic or electrohydraulic) combined with mechanical systems allow fully variable operation of the intake and exhaust valves without the camshaft.

In an engine with a conventional cam-driven valvetrain, the valve opening and closing is a gradual process that lasts the entire valve event. The valve velocity is related rigidly to the crankshaft speed and cam profile, so that the valves move faster only when the crankshaft turns faster. In contrast to this, the valve movements in a camless engine are rapid occurrences governed only by forces (hydraulic or electromagnetic) that are dependent on neither the crankshaft speed nor the cam profile. The opening and closing occupy only small portions of the entire valve event. Consequently, the valve lift profile, as plotted versus crank angle, has a nearly rectangular shape at low speeds (Figure 2.31). Even at high speeds, the profile can appear trapezoidal if properly programmed. Such form of the intake valve event, in which the valve remains at maximum opening during a substantial part of the event, favors better volumetric efficiency at high speeds. This contributes to higher torque.

2.4.1 Electromagnetic Valve Actuation

Electromagnetic actuators can open and close the valves under direct control from the engine management computer. It potentially offers the ultimate flexibility of timing, duration, and lift for each valve. The actuators are designed on the basis of electromechanical theory. A movable armature is guided between an upper and a lower magnet. When no magnetic force

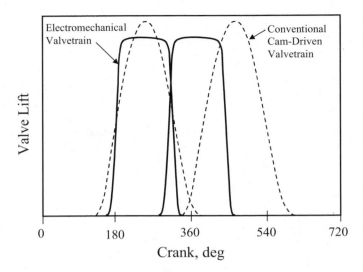

Figure 2.31 _Camless valvetrain timing diagram._

exists, the armature is held by an upper and a lower spring in the middle position between the two magnets. This condition corresponding to the valve half-open position occurs when the engine is shut off. During engine operation, a current in the coil of the upper magnet is used to hold the armature against the upper magnet so that the valve is in the closed position. To open the valve, the current is interrupted, and the armature is moved by the spring forces to the lower magnet. Providing a current to the coil of the lower magnet compensates for the losses during the movement, and the valve is held in the open position. To close the valve, the current is interrupted in the lower magnet and is reapplied to the coil of the upper magnet. The valve seating velocity and the velocity of the armature upon contact with the magnet have a significant effect on wear and acoustics. These velocities are determined by the shape of the current curve during armature movement.

The intake and exhaust valves are actuated by a single control unit per valve, consisting of the actuator and a sensor for registering the operating condition at any given time. The actuator is controlled by a special electronic valve timing system, which determines the required current flow for the control units. The energy requirement is covered by a high-efficiency generator in the form of a crankshaft-mounted starter-generator. Figure 2.32 illustrates the design principle of an actuator and a typical operating cycle when opening a valve. An armature regulated by a spring is moved between two solenoids that hold it in the fully open or closed positions. The start of the valve opening process is initiated by a signal from the engine management system for the corresponding valve to be opened. The electronic valve control system implements this signal by controlling the electrical supply to the two solenoids: the upper coil holding the voltage is switched off, the compressed spring accelerates the armature, and the valve is opened. When the armature approaches the maximum open position, the lower coil is activated, and the armature is held in the open position. When the specified valve opening time has been

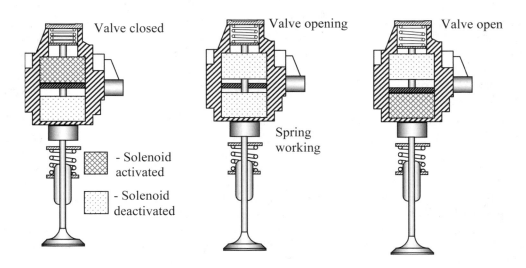

Figure 2.32 *Electromechanical camless valvetrain (Flierl and Klüting).*

reached, the control unit transmits a corresponding closure command, and the valve is closed in a similar manner.

One of the major advantages of individual valvetrain actuation lies in its high degree of flexibility. The flexible process method in conjunction with the steep valve-opening ramps means that cylinder filling at low and medium engine speeds can be improved by controlled energizing and systematic utilization of charge-cycle dynamics. The disadvantages are that valve motion is not precisely controlled, either in opening or in closing, resulting in high impacts on the valve. The impacts cause high noise and excessive wear, and they increase valvetrain stresses. Energy losses are higher than for standard valvetrains because of high actuator power requirements, low actuator efficiency, and losses in other parts of the electrical system. The cost and complexity are prohibitively high.

2.4.2 Electrohydraulic Valve Actuation

Another camless valve actuation mechanism is electrohydraulic, using electrohydraulically controlled actuator pistons to open and close the engine valves. The electrohydraulic camless valvetrain exploits the elastic properties of a compressed hydraulic fluid, which, acting as a liquid spring, accelerates and decelerates each engine valve during its opening and closing motions. This is the principle of the hydraulic pendulum. Similar to a mechanical pendulum, the hydraulic pendulum involves conversion of potential energy into kinetic energy and then back to potential energy with minimal energy loss. During acceleration, potential energy of the fluid is converted into kinetic energy of the valve. During deceleration, the energy of the valve motion is returned to the fluid. This takes place during both valve opening and valve closing. Recuperation of kinetic energy is the key to the low energy consumption of this system.

Figure 2.33 illustrates the electrohydraulic valvetrain actuation concept. The system incorporates high- and low-pressure reservoirs. A small double-acting piston is fixed to the top of the engine valve that rides in a sleeve. The volume above the piston can be connected to either a high- or low-pressure source. The pressure area above the piston is significantly larger than the pressure area below the piston. The engine valve opening is controlled by a high-pressure solenoid valve that is open during the engine valve acceleration and closed during deceleration. Opening and closing of a low-pressure solenoid valve controls the valve closing. The system also includes high- and low-pressure check valves. During the valve opening, the high-pressure solenoid valve is open, and the net pressure force pushing on the double-acting piston accelerates the engine valve downward. When the solenoid valve closes, pressure above the piston drops, and the piston decelerates, pushing the fluid from the lower volume back into the high-pressure reservoir. Low-pressure fluid flowing through the low-pressure check valve fills the volume above the piston during deceleration. When the downward motion of the valve stops, the check valve closes, and the engine valve remains locked in the open position. The process of the valve closing is similar in principle to that of the valve opening. The low-pressure solenoid valve opens, the pressure above the piston drops to the level in the low-pressure reservoir, and the net pressure force acting on the piston accelerates the engine valve upward. Then the solenoid valve closes, the pressure above the piston rises, and the piston decelerates, pushing the fluid from the volume above it through the high-pressure check valve and back into the high-pressure reservoir.

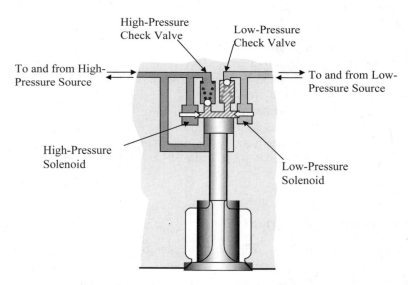

Figure 2.33 *Electrohydraulic valvetrain, pendulum mechanism (Schechter and Levin).*

The system uses neither cams nor springs, which reduces the engine height and weight. Hydraulic force both opens and closes the valves. During valve acceleration, the potential energy of compressed fluid is converted into kinetic energy of the valve. During deceleration, the energy of the valve motion is returned to the fluid. Recovery of kinetic energy is the key to the low energy consumption. The system offers continuously variable and independent control of virtually all parameters of valve motion. This permits optimization of valve events for each operating condition without any compromise.

The advantages of electrohydraulic valve actuation, as in electromagnetic valve actuation, are the complete freedom for all aspects of valve actuation. The valvetrain usually is mechanically simple, having fewer mechanical elements in the valvetrain than most standard systems have. The disadvantages are that to achieve useful engine speeds, this technique requires the expense and complexity of very fast switching of the hydraulic flows and a high-pressure oil supply. Valve motion is not precisely controlled, either in opening or in closing, resulting in high impacts of the valve. The impacts cause high noise, excessive wear, and increased valvetrain stresses. Fluid compressibility also is a limiting factor, and high energy losses are expected. In addition, the cost is prohibitively high.

2.4.3 Comparison of Various Variable Valve Actuation Systems

There is a significant benefit in fuel economy, emissions reduction, and performance improvement in using VVA. However, the tradeoff in using a VVA system is that it increases the cost and complexity. Table 2.2 compares the benefits versus the tradeoffs among various VVA systems.

TABLE 2.2
COMPARISON OF VARIABLE VALVE
ACTUATION MECHANISMS

Function	Fuel Economy	Emission	Power	Torque	Simplicity	Cost
Cam Phasing: Two-Step	+	++	++	++	-	- -
Continuous	+	++	++	++	- -	- -
Variable Valve L	+	0	++	++	- - -	- -
Cam Profile Switching	+	+	+++	+++	- -	- -
Variable Valve Lift with Cam Phasing	++	+++	+++	+++	- - -	-
Continuously Variable Lift with Cam Phasing	+++	+++	+++	+++	- - - -	- - - -
Lost Motion or Deactivation	++	0	++	++	- - -	- -
Electrohydraulic Valve Actuation	++	++++	++	++	- - - -	- - - -
Electromagnetic Valve Actuation	++	++++	++	++	- - - -	- - - -

Notes: + Plus or beneficial
         - Minus or detrimental

Typically, a 5–10% savings in fuel consumption is realized by using variable valve timing. The timing of valve events also impacts exhaust emissions. For example, the control of valve overlap can be used to control the level of exhaust residuals, thereby regulating the nitrogen oxide (NO_X)

emissions. The level of exhaust residuals trapped in the cylinder has a significant effect on the cycle-by-cycle variations in combustion and NO_X emissions. As with exhaust gas recirculation (EGR), high levels of exhaust residuals lead to lower emissions of NO_X and greater cycle-by-cycle variations in combustion. In addition to improved engine performance in torque and power output, engine idle stability also is improved by using a VVA system. Variable valve actuation systems have seen a steady increase in applications in production in recent years. Table 2.3 shows some examples of VVA devices currently used in production engines.

2.5 Engine Brake and Valvetrains

The role of the engine brake is to convert a power-producing engine into a power-absorbing retarding mechanism, and the valvetrain is a significant part of its mechanism. The engine braking mechanism for a four-stroke diesel engine is as follows. On the normal intake stroke, the intake valve opens, and air is forced into the cylinder by boost pressure from the turbocharger. Then, in the compression stroke, air is compressed to approximately 3.45 MPa by the engine piston. The energy required to compress this air is produced by the driving wheels of the vehicle. Near TDC, the engine brake opens the exhaust valves, venting the high-pressure air and dissipating the stored energy through the exhaust system. On the downward stroke, essentially no energy is returned to the piston (and to the driving wheels). There is a loss of energy, and this loss is how the retarding work is accomplished. Figure 2.34 illustrates the principle of the engine exhaust braking mechanism.

Without an engine brake, all the valves remain closed at the top of the compression stroke and are ready to start the power stroke. In a no-fuel condition, the air pressure exerts a force on the piston and returns most of the power to the crankshaft. With an engine brake active, the exhaust valves are opened near TDC of the compression stroke, and the energy stored in the air exits through the exhaust system. Theoretically, none of the power is returned to the crankshaft, and the engine is able to provide retarding power for the vehicle. The timing of the valve-opening event is important because the piston at TDC has done the maximum amount of work. If the valve is opened early, not much power will be absorbed. Similarly, if the valve is opened after TDC, some power has been returned to the crank by the compressed air. As the valve-opening event moves away from TDC, the brake becomes less effective. In practice, it is necessary to use a cam or rocker motion that occurs close to TDC. This motion will be picked up by a master piston and transmitted through a hydraulic circuit to a slave piston to open the exhaust valve.

One common characteristic of all engine brakes is that retarding horsepower increases as engine revolutions per minute increase. In general, the same things that affect retarding performance are engine displacement, compression ratio, turbo boost, and the timing of the valve-opening event. In its simplest description, the engine brake converts a diesel engine into an air compressor. All other factors being equal, the larger the engine displacement, the more powerful an air compressor it can make, and the higher the retarding performance. Compression ratios of diesel engines typically are approximately 15:1. At the top of the piston stroke, the air will occupy 1/15 of the original volume and will be at approximately 3.45 MPa. It takes power to compress air to high pressures; the engine crankshaft supplies this power. Higher compression ratios producing higher cylinder pressures would result in higher retarding performance.

TABLE 2.3
EXAMPLES OF VARIABLE VALVE ACTUATION
IN CURRENT APPLICATIONS

Company	System	Application	Timing	Lift	Duration
BMW	VANOS	2.5L, 3.0L, 3.2L I6	I, Cont.		
	Double VANOS	4.4L, 5.0L V8	B, Cont.		
	Valvetronic	4.4L, 6.0L V8	I, Cont.	I, Cont.	I, Cont.
DCX	VVT	2.3L I4			
	Deactivation	5.0L V8		2-Stage	
Fiat	Super Fire		I, 2-Stage		
Ford	Zetec SE	2.0L I4	I, 2-Stage		
GM	VVT-i VVTL-i	1.8L I4	I, Cont.		
	VVT	4.2L I6			
Honda	VTEC	1.0L, 1.3L, 1.7L, 2.0L, 2.3L, 2.4L I4, 3.0L, 3.2L, 3.5L V6	I, Cont.	I, 2-Stage	I, 2-Stage
	i-VTEC	2.0L, 2.4L I4, 3.0L V6	I, Cont.	I, 3-Stage	I, 3-Stage
Isuzu	VVT	4.2L I6			
Jaguar	VVT	2.5L, 3.0L V6, 4.0L, 4.2L V8			
Mazda	VVT	3.0L V6			
Mitsubishi	MIVEC			2-Stage	2-Stage
Nissan	Neo VVL	1.8L I4, 3.5 V6, 4.5 V8		B, 3-Stage	B, 3-Stage
Porsche	VacioCam	2.7L, 3.2L, 3.6L Bxr-6, 4.5L V8	I, 3-Stage		
	VarioCam Plus	3.6L Bxr-6, 4.5L V8	I, 2-Stage	I, 2-Stage	I, 2-Stage
Suzuki	VVT		E, Cont.		
Toyota	VVT-i	1.5L, 1.8L, 2.0L, 2.4L I4, 3.0L I6, 3.0L V6, 4.3L V8	I, Cont.		
	VVTL-i	1.8L I4	I, Cont.	B, 2-Stage	B, 2-Stage
Volvo	VVT	2.3L I4, 2.5L I5	I		

Notes:

    I = Intake

    E = Exhaust

    B = Both intake and exhaust

    Cont. = Continuous

(a)

(b)

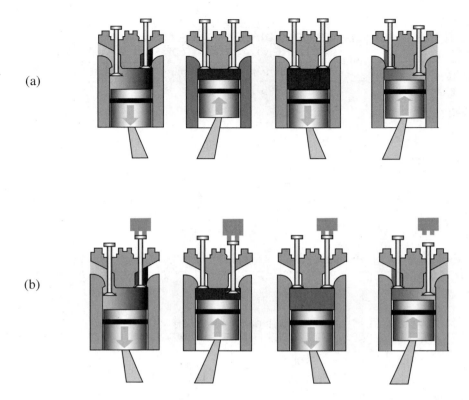

Figure 2.34 *Comparison of an engine operation with and without the engine brake: (a) normal four-stroke engine operation, and (b) engine brake operation principle.*

Figure 2.35 shows another example of an engine brake system from Volvo. To open the exhaust valves and to increase and reduce the pressure in the cylinder, the exhaust cam has two additional cams: one charging cam, and a decompression cam. The lifting height of the additional cams is not higher than the valve lash or clearance. This means that the additional cams never can open the exhaust valves during normal engine operation. For the compression brake to function correctly, the valve clearance must be eliminated. The lubrication system of the engine is used for this purpose. It activates a power piston in the exhaust rocker arm by means of a control valve, and this piston eliminates the valve clearance. The exhaust rocker arm has a built-in hydraulic piston—a so-called power piston. The lower part of the piston has a ball socket and rests against the shim on the valve crosshead. The pressure side of the piston is constantly in contact with the engine oil system. During normal operation, the control valve limits the oil pressure to ensure that it does not affect the power piston. When the Volvo engine brake (VEB) is activated, the oil pressure is increased by means of a valve arrangement built into the exhaust rocker arm. The pressure builds up on top of the power piston, which is pushed downward and eliminates the valve lash. The camshaft now is in direct contact with the exhaust valves, and the extra cams can start the compression brake. A pressure-limiting valve is built into the power piston and opens if the oil pressure becomes abnormally high.

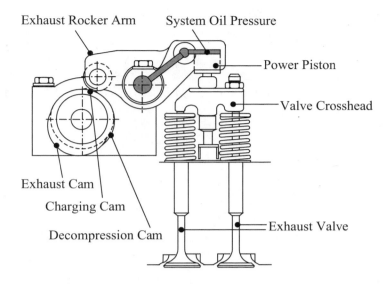

Exhaust Rocker Arm   System Oil Pressure

Power Piston

Valve Crosshead

Exhaust Cam

Charging Cam

Decompression Cam

Exhaust Valve

Figure 2.35 *Schematic of the Volvo engine brake (VEB) (Carlström).*

The technical feature that gives the VEB an extra increase in pressure is an extra cam on the exhaust cam of the camshaft. This opens the exhaust valves briefly before and at the start of the compression stroke. During the short period of time in which the exhaust valves are open, the cylinder is filled with the higher pressure created by the exhaust brake in the exhaust manifold. Thus, a higher compression than normal is achieved during the compression stroke. Immediately before the compression stroke is completed, just before the piston reaches its TDC, a decompression cam on the exhaust cam opens the exhaust valves again and releases the high compression pressure into the exhaust manifold (Figure 2.36). This takes the "sting" out of the energy that otherwise would give unwanted power to the crankshaft.

The unique feature of the VEB compression engine brake is that the compression pressure in the cylinder is increased during the compression stroke and is amplified by the exhaust valves opening and admitting the higher pressure that is in the exhaust manifold because the exhaust brake is engaged. This means that the retarding power to the crankshaft will be higher. The compression stroke is completed by the compression pressure being released into the exhaust manifold by the exhaust valves opening again.

2.6 Types of Valvetrain

There is no industry standard to classify the types of valvetrain for cam-driven engines. However, many classifications exist, based on cam position, valve position, and rocker arm types. According to a valvetrain architect, valvetrains can be classified into three general categories: (1) direct-acting type, (2) center-pivot rocker arm type, and (3) end-pivot rocker type. The center-pivot rocker arm type can have variations of the OHC and pushrod OHV type. Increasing numbers

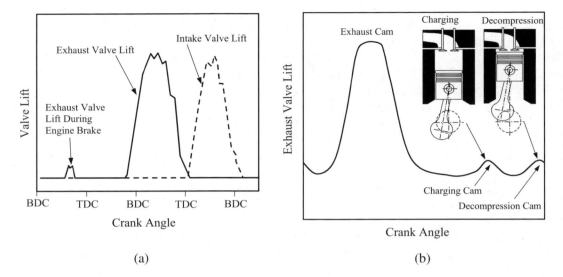

Figure 2.36 *Volvo engine brake (VEB) valve timing and lift curve: (a) typical valve lift curve for an engine brake, and (b) VEB exhaust valve lift curve (Carlström).*

of literature report the use of a five-type classification of valvetrain since 1982 (Buuck), and these types are shown schematically in Figure 2.37. The cam is located in the cylinder head for Types I, II, III, and IV valvetrains or OHCs, while the cam is located in the block for the Type V valvetrain (i.e., OHV).

The five types encompass both cam-in-block and cam-in-cylinder head arrangements. There can be variations of these five basic types. For Type I, referred to as a direct-acting valvetrain, it can be either mechanical or hydraulic tappet as the cam follower. Type II refers to the finger follower type or end-pivot type valvetrain, can either be flat or roller rocker arm, and can use a lash adjuster or shaft as the end pivot. In Types III and IV, referred to as center-pivot OHV types of valvetrain, a flat face or roller hydraulic lifter can be used as the cam follower, and a hydraulic capsule or elephant foot at the valve tip end. Type V, referred to as the center-pivot OHV type valvetrain, can have radius or roller followers in contact with the cam lobe as well as at the valve tip. Types III, IV, and V can have roller bearings, as well as sliding bushings at the center of rotation of the rocker arms.

All of these valvetrain configurations have been applied to automotive engine designs. Depending on the engine application emphasis, different designs have preferable valvetrain types because some design attributes are inherent with a particular valvetrain arrangement. Table 2.4 depicts the major design attributes of each system.

This evaluation assumes that Types II through V use rolling element followers and pivots and that Type I is a slipper follower. The Type I valvetrains, which fit reasonable package envelopes, do not support the application of roller followers due to the radius of curvature constraints on

Valvetrain Systems

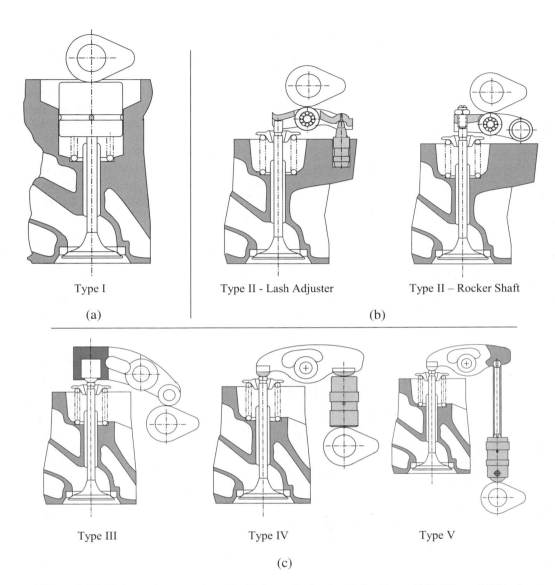

Type I

(a)

Type II - Lash Adjuster   Type II – Rocker Shaft

(b)

Type III   Type IV   Type V

(c)

Figure 2.37 *Valvetrain types (cam is in the cylinder head for Types I, II, III, and IV valve-trains or OHCs, while the cam is located in the block for the Type V valvetrain, or OHV): (a) direct acting, (b) end-pivot rocker arm, and (c) center-pivot rocker arm.*

the cam lobe profile when the valvetrain has a rocker ratio of one. The Type I valvetrain is typical of flat-head and OHC engines. Types II through IV are used on OHC engines, and Type V represents the cam in block, pushrod-actuated OHV configuration.

TABLE 2.4
COMPARISON OF DESIGN ATTRIBUTES
FOR VARIOUS TYPES OF VALVETRAIN

	Type I	Type II	Type III	Type IV	Type V
Natural Frequency (Hz)	2000–3000	1200–1600	900–1400	900–1400	400–700
Effective Mass @ Valve (g)	140–160	80–120	120–160	130–170	240–290
Maximum Speed (rpm)	6500++	6500+	6000+	6000+	4000–6000
Friction	E	A	B	C–D	C–D
Overall Engine Packaging	D–E	D–E	B	C	A

Notes:
    A = Best
    E = Worst

2.6.1 Type I Valvetrain—Direct Acting, Overhead Cam

The Type I valvetrain typically is a direct-acting OHC valvetrain in which the cam is positioned directly over the valve tip. The cam activates the valve through an inverted lifter or bucket. This type of valvetrain is found in hydraulic and mechanical versions. However, this type is particularly difficult to adjust mechanically because shims are required to vary lash between the bucket and cam or the bucket and valve tip. Another application of the Type I valvetrain is in "L-head" engines, commonly used in small industrial applications. These engines usually have mechanically adjusted lash, with the adjustment obtained by filing the correct gap between the bucket cam contact face and the valve tip. Examples of direct-acting Type I valvetrains are the Porsche 3.6L Flat 6 and the Toyota 2.4L I4 engines. Other engines with direct-acting Type I valvetrains include the Ford/Mazda I4/I5 1.8L, 2.0L, and 2.3L (mechanical); Lexus V8 4.0L (mechanical tappet); BMW 3L I6; and Audi 1.8L I5 and 2.7L V6.

The Type I valvetrain has the highest stiffness among all valvetrain types, and because it has fewer components, it is suitable for high-speed applications. It has the lowest friction (compared to other types of valvetrains without roller bearings). However, it requires double camshafts in cross flow or multiple valve heads. The height of the engine is large, which is a drawback in engine packaging. It cannot incorporate the roller bearings; thus, it has relatively high friction compared to that of other valvetrains with rollers.

2.6.2 Type II Valvetrain—End-Pivot Rocker Arm, Overhead Cam

The Type II valvetrain, also referred to as a finger follower, typically is an OHC valvetrain with an end-pivot rocker arm. The cam is positioned above the rocker arm in the middle. This type of valvetrain typically uses a hydraulic unit called a lash adjuster in the pivot position. In some cases, a rocker shaft is used in the pivot position. A roller interface can be adapted easily between the rocker and the cam. Examples of Type II valvetrain engines are the Ford V8 4.6L engine using a finger follower and the Honda VTEC I4 2.0L engine using a rocker shaft. Other engines with Type II valvetrains include the Ford 5.4L/5.7L V8 and 3.0L V6, DaimlerChrysler 2.2L, GM Vortec 4200 4.2L I6, and Honda 2.0L I4.

The Type II valvetrain has low valvetrain friction in the roller configuration. High rocker ratios are possible. It also has easy adaptability of the hydraulic lash adjustment without including additional weight in the moving mass system, and it has a low hood height. Disadvantages of the Type II valvetrain are lower stiffness, more components, and a higher side load on the valve than with Type I. It requires relatively wide cylinder heads, especially in multiple-valve configurations, and it is sensitive to cam concavity. It has the highest friction without a roller rocker arm.

2.6.3 Type III Valvetrain—Center-Pivot Rocker Arm, Overhead Cam

The Type III valvetrain is an OHC (or in-head cam) valvetrain with a center-pivot rocker, which is activated directly by a camshaft located in the cylinder head. A Type III valvetrain can be either mechanical or hydraulic. When hydraulic capsules are used, they typically are located in the rocker arm, directly above the valve. This valvetrain design frequently is used in industrial engines with mechanical adjustment such as the Detroit Diesel Series 149 and Series 60 and the Caterpillar 3406E. Many passenger car engines also have used this valvetrain. Hyundai and several Honda engines, including the VTEC, use this type of valvetrain with hydraulic compensation. The hydraulic unit used in this valvetrain is referred to as a capsule, similar to Type I, and usually is placed over the valve tip. Other engines with Type III valvetrains include the Mercedes-Benz Modular V6 and V8 engines, 2.0–5.0L.

The Type III valvetrain has low friction in the roller follower mode and low engine height. It has a simple design and relatively high stiffness (similar to Type II). However, it increases the mass at the valve in the hydraulic configuration and is difficult to place the cam in the optimal location in the cylinder head. (It can interfere with ports and water jackets.) Furthermore, it is difficult to incorporate the hydraulic adjustment, and its friction can be high if a shaft is used as the pivot without a rolling element.

2.6.4 Type IV Valvetrain—Center-Pivot Rocker Arm, Overhead Cam

The Type IV valvetrain is an OHC valvetrain with a center-pivot rocker and cam follower directly acting on the rocker arm without a pushrod. The cam follower can be a flat face or roller hydraulic lift. The Ford Escort 1.9L CVH engine utilizes this valve gear with a hydraulic cam follower. Few applications have been found for this type of valvetrain. The hydraulic device used in this application is referred to as a hydraulic lifter. The lifter also may incorporate a roller.

The Type IV valvetrain allows adaptation of traditional hydraulic technology to a Type III style valvetrain. It has improved stiffness over a Type V valvetrain design and is versatile in the placement of the valves. However, it has largest number of components and the highest valvetrain weight of the OHC valvetrain types. Likewise, it increases the cylinder head size and decreases the stiffness over any of the other OHC designs. In addition, it has higher friction due to the shaft pivot.

2.6.5 Type V Valvetrain—Center-Pivot Rocker Arm, Overhead Valve

The Type V valvetrain has an OHV, cam in block, using either a mechanical or hydraulic lifter. Many engines in the United States continue to use this type of valvetrain. Examples include the small-block and large-block Chevrolet and Ford V8 engines, the Chrysler 5.2L V8 and 8.0L V10, the General Motors "Iron Duke" in-line four cylinder, and many others. This design lends itself to V-configuration engines such as V8s and V6s but is common among in-line configurations, too. It is simple, allows the cam to be located near the crankshaft, and is the only valvetrain design that allows only one cam in a V-configuration engine. The cam follower in this design oscillates up and down in a bore machined in the engine block. The hydraulic device this application is called a hydraulic lifter or hydraulic tappet. The follower also may incorporate a roller or be of the mushroom or nailhead variety. Examples of this type of valvetrain application are the General Motors 5.7L and Ford 5.0L V8 engines. Other engines with Type V valvetrains include the General Motors 5.3L V8, DaimlerChrysler 3.3L V6, and Ford 3.0L V6.

The Type V has the simplest cam drive next to the crankshaft, and it is easy to adapt to hydraulic lash compensation and roller followers. It has a low hood height and relatively low cost. However, it has a large number of components and relatively high friction. Furthermore, it has the lowest valvetrain stiffness, which limits high-speed applications. It also is not easy to adapt to multiple-valve applications and has higher friction due to the shaft pivot.

2.6.6 Ranking of Various Valvetrains

The preceding discussion of valvetrain characteristics leads to a direct comparison of each type of valvetrain. Table 2.5 lists the major performance characteristics of each system, along with camless valvetrains such as the electromagnetic and electrohydraulic systems, where "1" represents the worst ranking and "5" represents the best. Discretion should be applied because the robustness of the design and the development of new materials or coatings can change or reverse the rankings shown in the table.

2.7 Valvetrain System Design

One of the most difficult and critical design tasks within the internal combustion engine is the design of a valvetrain. Poor valvetrain dynamic characteristics and abnormal valvetrain vibrations significantly affect the performance, durability, and noise of an engine. The ability to design a valvetrain system and prove it on bench tests prior to incorporation into an engine test program has enormous effects on the total cost and development time of an engine program. Criteria of valvetrain system design include minimum contact pressure between the cam and follower, maximum rigidity of the camshaft bearing mounting, minimum valvetrain inertia, maximum valvetrain stiffness, minimum acceleration and deceleration during valve opening and closing,

TABLE 2.5
COMPARISON OF PERFORMANCE AND ATTRIBUTES
FOR VARIOUS VALVETRAIN SYSTEMS

| | Valvetrain Type | | | | | | | | | |
| | Direct Acting | End-Pivot Rocker Arm | | Center-Pivot Rocker Arm | | | | | Camless | |
Parameter	I	II (Flat)	II (Roller)	III (Flat)	III (Roller)	IV	V (Flat)	V (Roller)	EM	EH
VT Friction/Wear	4	1	5	3	5	2	3	5	3	3
VT Effective Weight	4	5	5	1	1	3	2	2	5	5
VT Stiffness	5	4	4	3	3	2	1	1	5	5
Valve Lift	2	4	4	5	5	3	1	1	5	5
Valve Acceleration	5	4	4	3	3	2	1	1	5	5
Valve Event	4	3	3	5	5	5	4	4	5	5
Valve Side Load	5	1	3	3	4	3	3	4	5	5
System Cost	3	3	3	2	2	4	5	5	1	1
Maximum Speed	5	4	4	3	3	2	1	1	1	1
Cam/Follower Wear	4	2	5	3	5	3	3	5	NA	NA
Packaging	2	3	3	4	4	4	5	5	1	1
VVA Capability	3	2	2	1	1	4	4	4	5	5

Notes:

(a) Valve event: Type I is limited only by the diameter of the follower. Types II and III are limited by cam lobe concavity. Types IV and V are limited by the diameter of the follower and valvetrain stiffness.

(b) Friction: Type I has the lowest friction potential without any roller element. The roller follower provides low friction potential.

(c) Event control: Types II and III have timing variation sensitivity due to tolerance stackup. All types have improved control when using hydraulic lash adjusters.

(d) Roller followers have a beneficial effect on the valve event and friction.

(e) There is a distinct design tradeoff between the speed capability and the valve event for all types. Type I would have the maximum design freedom due to its stiffness.

NA = Not applicable.

minimum valve seating velocity, minimum spring load without causing the valve to jump at high engine speeds, and so forth.

2.7.1 Valvetrain Selection

The valvetrain system plays an integral part in developing an efficient, durable, high-performance engine with the fewest harmful emissions. The key considerations in the selection of a valvetrain system are engine application, emissions, fuel economy, packaging, dynamic characteristics, NVH, durability, reliability, serviceability, and cost. The engine performance depends on the flow capacity or the amount of air or air/fuel mixture during the intake cycle. The more air or air/fuel mixture that flows in, the higher the horsepower possible from the engine. The design of a system that provides a high flow capacity with acceptable valvetrain dynamics is paramount to accomplishing most engine objectives. Many factors affect the flow capacity, but the valve structure and timing are considered critical in affecting the flow and subsequently the engine performance. The noise of the valvetrain is a major part of engine NVH and is due in part to deviation from ideal valvetrain dynamics. The type, weight, stiffness, natural frequency of the valvetrain and spring, and cam lobe profiles are the controlling factors of valvetrain noise. Cam lift, velocity, and acceleration curves combined with the natural frequency of the valvetrain shall demonstrate the adequacy level of noise and dynamics. The durability of the valvetrain depends on the valvetrain geometry, layout, and dynamics and is controlled by the weak link of the system. Any components and the interfaces between the components in the system that fail to perform their functions may result in valvetrain malfunction and may lead to the catastrophic failure of the engine.

To illustrate the valvetrain system design approach with focuses on performance, noise, and durability, Figure 2.38 shows a flow-chart of a new valvetrain system design. When the speci-fications of a new engine are issued for the specific power output or performance, a minimum volumetric flow rate is required for that performance. The flow rate is dictated primarily by the valve lift and valve geometry, including the size, shape, and number of valves. Valve size and location evolve from preliminary layouts and analytical engine performance analysis. Then, the valvetrain system type is determined, based on package constraints, weight, cost, performance, NHV, maximum speed, and durability requirements. The next step would be to evaluate the valvetrain weight and to estimate the valvetrain frequency for camshaft and valve spring design. After valvetrain geometry and dimensions are determined, valvetrain weight and frequency can be estimated, and camshaft and spring design can be determined. Therefore, valvetrain motion evaluation or dynamics analysis can be performed. Based on the evaluation and the feedback from the motion evaluation, valvetrain weight, dimensions, camshaft design, spring design, and valvetrain frequency can be readjusted if they are unsatisfactory. Then the valvetrain motion is

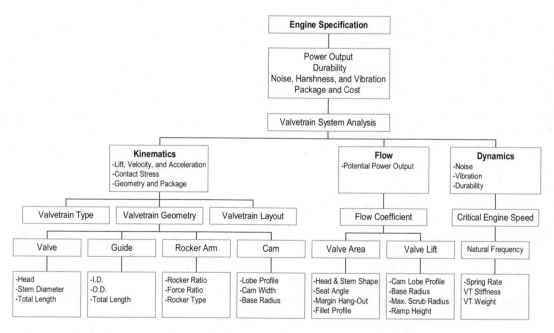

Figure 2.38 *Engine valvetrain system design flow-chart.*

reevaluated, and there is feedback of the data into the engine design for optimization. The loop continues until the design is optimized.

A high-quality valvetrain design method minimizes the time and cost of establishing a satisfactory design and has a high probability that the evolved design will prove out on actual engine design verification tests. Ideally, the method could be accomplished only by computer simulation procedures and not require experimental input. However, current computer simulation programs require extensive empirical input and are not proven in their ability to provide estimates of noise generation, not to mention the durability of the valvetrain.

2.7.2 Valvetrain System Mechanics—Kinematics Analysis

Mechanics may be defined as the science that describes and predicts the conditions of rest or motion of bodies under the action of forces. The valvetrain system analysis is focused primarily on the dynamics of mechanics rather than the statics of it. Dynamics includes kinematics and kinetics. Kinematics, which is the study of the geometry of motion, is used to relate displacement, velocity, acceleration, and time, without reference to the cause of the motion. Kinetics, which is the study of the relations existing among the forces acting on a body, the mass of the body, and the motion of the body, is used to predict the motion caused by given forces or to determine the forces required to produce a given motion. Valvetrain kinematics analysis includes detailed analysis of the cam profile and the valvetrain system geometry, layout, contact pattern, and stress. The cam profile dictates the valve lift, velocity, and acceleration for optimized valve timing and engine performance. The valvetrain system geometry and layout provide for motion stability, and the contact stresses at the valvetrain interfaces determine durability. Valvetrain kinematics analysis results provide engine designers first-order values for loads and motions of the mechanisms, based on geometric and applied load inputs. The results from the analysis could help a design engineer optimize the valvetrain geometry, port, and combustion chamber designs.

2.7.2.1 Lift, Velocity, and Acceleration—Direct-Acting Valvetrains

In valvetrain system kinematics analysis, an understanding of the relationship between the valve and cam is essential, both in terms of determining valve motion and in calculating the loads transferred throughout the valvetrain. A cam is a mechanical member for transmitting a desired motion to a follower by direct contact. The follower in an internal combustion engine valvetrain can be a lifter (mechanical or hydraulic) or rocker arm, flat face, or roller contact in configuration. The rotational movement of the cam is transformed to reciprocating movement of the follower, thus achieving the opening and closing of the valve. The motion of a cam-operated engine valve can be described as valve displacement, velocity, and acceleration.

The displacement (y) of the follower is given by

$$y = f(\theta), \text{ mm} \tag{2.1a}$$

where θ is the cam angle rotation in radians.

Or, because the cam rotates at a constant angular velocity, the displacement can be expressed as

$$y = g(t) \quad \text{and} \quad \theta = \omega t \tag{2.1b}$$

where t is the time for the cam to rotate through angle θ, and ω is the cam angular velocity.

The velocity (v) is established as the instantaneous time rate of change of displacement or slope of the displacement curve at angle θ or time t,

$$v = \frac{dy}{dt}, \text{ mm/s} \tag{2.2a}$$

$$v = \frac{dy}{dt} = \frac{d\theta}{dt}\frac{dy}{d\theta} = \omega\frac{dy}{\theta}, \text{ mm/s} \tag{2.2b}$$

The acceleration is the instantaneous time rate of change of velocity or slope of the velocity curve at angle θ or time t,

$$a = \frac{d^2y}{dt^2} = \frac{dv}{dt}, \text{ mm/s}^2 \tag{2.3a}$$

$$a = \frac{d^2y}{dt^2} = \frac{d^2\theta}{dt^2}\frac{d^2y}{d\theta^2} = \omega^2\frac{d^2y}{d\theta^2}, \text{ mm/s}^2 \tag{2.3b}$$

The shape and values of the acceleration curves are of critical concern for moderate to high-speed engines. From it, analysis can be made regarding the shock, noise, wear, vibration, and general performance of a cam follower system. It will be shown later that for the best action, the acceleration curve should be smooth and have the smallest maximum values possible.

Another term, pulse (commonly called "jerk"), is used to define the instantaneous time rate of change of acceleration or the slope of the acceleration curve at angle θ or time t,

$$p = \frac{d^3y}{dt^3} = \frac{da}{dt}, \text{ mm/s}^3 \tag{2.4}$$

For high-speed engines, it will be shown later that the maximum values of the pulse should not be too large. Vibrations then will be kept to a minimum.

The distance the valve travels off its seat is called valve lift (L_V). The valve is lifted from its seat and accelerated to maximum velocity, at which point the cam lobe follower contact is approximately halfway through the flank of the cam. From this point, the valve is decelerated from maximum to zero velocity at the top of the lobe (maximum lift point), with the acceleration reaching its maximum negative value. The decelerating force is provided by the valve spring, which opposes the movement of the valve in its opening motion. At this point, the valve is decelerated by the cam and finally is brought to rest. Figure 2.39 shows typical valve lift, velocity, and acceleration curves. In a direct-acting system, the valve lift is the same as the cam lift. The valve lift required for a given engine involves many factors and variables. The ideal condition for a valve would be to make the lift as high and as fast as possible. Variables and limitations will occur while attempting to meet the ideal condition, namely, the valve weight, the maximum diameter of the valve head, the spring rate, and the distance between the intake and exhaust valves.

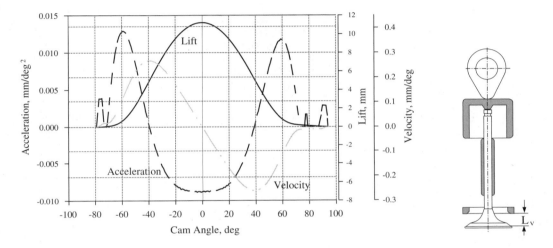

Figure 2.39 *Example of valve lift, velocity, and acceleration curve.*

The velocity of the opening ramp should remain constant so that minute deflections in the valvetrain system (clearances, or slack, and manufacturing deviations) at the point of juncture at the base circle and the ramp can be taken up before the load (actual lift) is applied to the system. The velocity at the end of the closing side must be low to prevent a "hammering" or impact noise as the valve approaches its seat. The velocity at the point of maximum lift must be zero; if it is not, the tappet will tend to lift off the cam contour, resulting in a hammering noise. The maximum velocity on the opening and closing sides is important for flat followers due to its relationship with tappet diameter, as shown in Figure 2.40. The tappet velocity (V) at any part of the lift cycle is the vertical velocity of the contact point (P_C) at distance q or eccentricity (mm) from the tappet centerline. Translation of the angular velocity (ω) of the cam lobe to the linear velocity of the tappet is

$$V = \omega q = \frac{2\pi q}{360} = 0.0174q, \ \ mm/deg \tag{2.5}$$

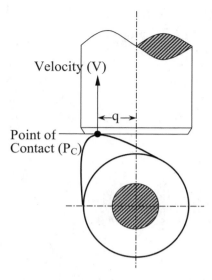

Figure 2.40 *Valve maximum velocity for a direct-acting type of valvetrain.*

This says that at a chosen cam angle for which the translating velocity of a flat follower is V, the radius or distance to the center of the contact patterns is

$$q = \frac{V}{0.0174}$$

The maximum lift velocity allowable can be found by

$$V_{max} = 0.0174 q_{max} = 0.0174 \times \left(\frac{d}{2} - \tau \right) \qquad (2.6)$$

where d is the tappet diameter and τ is the margin of safety.

This suggests that the tappet radius must be larger than q_{max}, or the edge of the tappet would dig into the lobe. The maximum acceleration for a particular cam contour is dependent on the type of valvetrain. A system with few components, such as the direct-acting OHC, can tolerate a higher maximum acceleration value compared to that of a pushrod system (having many components), which would require a lower permissible maximum acceleration. Because the valve spring force (f) required to hold the valve closed and to hold the tappet on the cam is in direct proportion to the mass (m) of the system and its acceleration (a), $(f = m \times a)$. Therefore, the maximum acceleration in the pushrod system must be lower than that in the direct-acting system to offset the greater number of components, the greater mass, and the higher flexibility. Increasing the maximum cam acceleration results in increased area under the lift curve, which improves engine performance characteristics and reduces cam-tappet maximum contact stresses due to the larger nose radius obtained from increasing the area under the lift curve.

2.7.2.2 Lift, Velocity, and Acceleration—Rocker Followers

For valvetrain systems with rocker arm followers, the valve lift equals the cam lift multiplied by the instantaneous rocker arm ratio. The valve velocity, acceleration, and jerk curves can be derived the same way from the valve lift curve as described in the preceding discussion about the direct-acting type.

The rocker ratio (RR) is the ratio of the valve lift (L_V) divided by the cam lift (L_V), that is,

$$RR = L_V/L_C = R_1/R_2$$

where R_1 is the distance from the rocker pivot center to the valve centerline at the valve tip, and R_2 is the distance from the rocker pivot center to the cam/follower contact center. Figure 2.41 illustrates examples of the instantaneous rocker ratios for center-pivot and end-pivot rocker arms.

In a direct-acting valvetrain, no rocker arm is involved, and the valve lift equals the cam lift. Therefore, the rocker and force ratios are treated as constants (i.e., 1). However, the rocker ratios are not constants for other types of valvetrains during valve lifting, as shown in Figure 2.41. They vary, depending on the cam lobe contact position and the amount of valve lift. Figure 2.42 illustrates how the rocker ratio varies with the cam angle and valve lift position during valve motion for an end-pivot rocker arm valvetrain. It applies to center-pivot rocker arm types of valvetrains as well. A rocker ratio typically is greater than 1; therefore, to achieve the same amount of valve lift as the direct-acting valvetrain system, the maximum cam lift is proportionally smaller than the cam lift used in the direct-acting valvetrain.

Because the rocker follower can be a flat surface or roller surface in contact with the cam, the limiting velocity also must consider the roller follower. Equations 2.1 and 2.2 also are applicable in calculating valve lift velocity for a roller follower. However, roller followers have different requirements because of geometry, which results in different camshaft contact patterns over the cam event. The eccentricity "q" variance over the cam event for a roller is quite small compared to that of a flat follower (Figure 2.43). For the roller follower, the maximum valve lifting velocity is reduced due to the fact that q is reduced,

$$V_{max} = 0.0174q_{max}$$

Table 2.6 lists typical rocker ratios for the five basic types of valvetrain. The values listed are based on experience with existing valvetrains and are not boundaries or limiting values. Successful valvetrains can be designed with values other than those listed in the table.

2.7.2.3 Valvetrain System Geometry, Layout, and Contact Patterns

In general, the geometry of a valvetrain system must be designed correctly to prevent dynamic, durability, and NVH concerns. The geometry must be designed for stable loading throughout the valve event. To prevent durability issues, geometrical stackups must not cause unacceptable levels of stress or loads. The weight of the valvetrain components and the resulting system kinematics velocities and loads must not cause dynamic and NVH concerns.

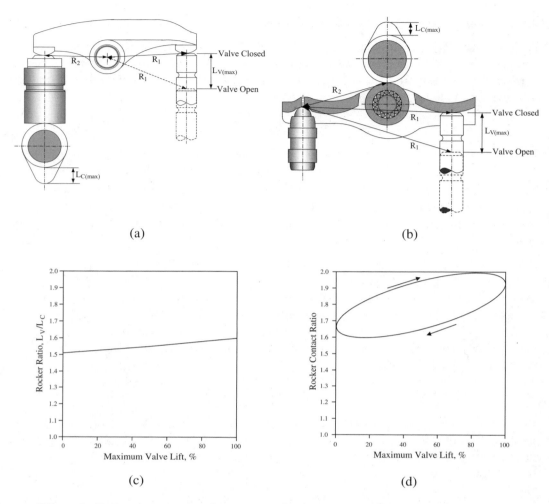

Figure 2.41 *Instantaneous rocker ratios for end-pivot and center-pivot rocker arms: (a) center-pivot rocker arm, (b) end-pivot rocker arm, (c) change of rocker ratio versus percentage of valve lift for center-pivot rocker arm, and (d) change of rocker ratio versus percentage of valve lift for end-pivot rocker arm.*

2.7.2.3.1 Direct-Acting Valvetrain System with Bucket Tappet

Figure 2.44 shows the hydraulic bucket tappet system and mechanical bucket for the direct-acting valvetrain system. The diameter of the bucket lifter is calculated from the maximum designed cam velocity (Eq. 2.5). Additional compensation is made for a margin (to prevent edge loading) and a beveled corner break (for ease of cam/follower interface, typical values for margins, and the relationships necessary to define the diameters). Maximum allowable velocity is determined by the tappet diameter (Eq. 2.6). If the cam profile is designed with a velocity that is too high

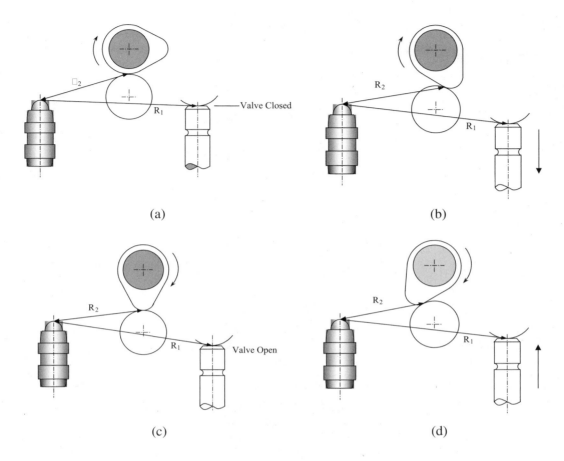

(a)

(b)

(c)

(d)

Figure 2.42 *Rocker ratio variation as a function of valve lift for an end-pivot rocker arm: (a) on base circle, (b) opening event, (c) at maximum lift, and (d) closing event.*

for the given tappet diameter, the contact point at maximum velocity will fall off the edge of the tappet, and the corner of the tappet will dig into the cam face.

Rotation of a follower is beneficial to extend the life of its wear face. This probably is the result of improved oil film formation, less sliding motion, and spreading the cyclic loading over a much greater area. Rotation of the follower can be induced through the proper design of the cam/follower interface. Two techniques are in common use: (1) a radius on the follower in conjunction with a tapered cam lobe, and (2) for flat followers, an offset of the cam lobe centerline with respect to the centerline of the follower. Both techniques move the contact point away from the follower centerline, thus providing a torque that induces the follower to rotate.

For the systems shown in Figure 2.44(b) (flat shim on top of the tappet), the shim is flat with no lobe taper. However, in the longitudinal view, the centerline of the lobe and the centerline of

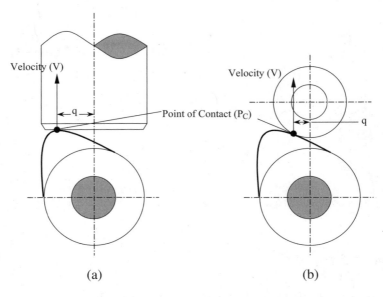

Figure 2.43 *Cam and follower contact and maximum velocity: (a) flat follower, and (b) roller follower.*

TABLE 2.6
TYPICAL ROCKER RATIOS FOR DIFFERENT
VALVETRAIN TYPES

Valvetrain Type	Rocker Ratio
Type I, OHC Direct Acting	1:1
Type II, OHC w/End-Pivot Rocker	1:1.6–1.9
Type III, OHC w/Center-Pivot Rocker	1: 1.2–1.5
Type IV, OHC w/Rocker and Lift	1: 1.5–1.7
Type V, Pushrod, Gas	1: 1.2–1.5
Type V, Pushrod, Diesel	1: 1.5–1.7

the tappet are offset (OS) (Figure 2.45). For aluminum tappet bodies with a steel top shim, the centerline of the cam lobe and the valve centerline are in line with the centerline of the tappet offset approximately 0.5–0.7 mm relative to them.

For flat tappet systems, the maximum velocity is directly related to the usable tappet contact surface. Maximum velocity has a large effect on how much lift is achievable from a given event length profile. The reduction in lift and lift area due to limited velocity, limited deceleration, and the 1:1 rocker ratio is offset partially by the higher accelerations and jerks allowable with this stiffer and more lightweight system.

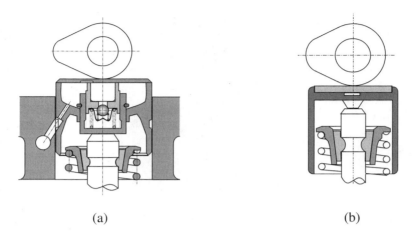

Figure 2.44 *Direct-acting valvetrain system: (a) hydraulic bucket tappet system, and (b) mechanical bucket tappet system.*

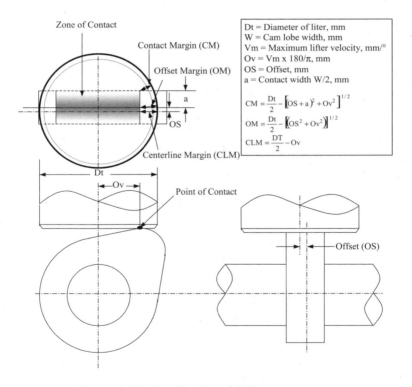

Dt = Diameter of liter, mm
W = Cam lobe width, mm
Vm = Maximum lifter velocity, mm/°
Ov = Vm x 180/π, mm
OS = Offset, mm
a = Contact width W/2, mm

$$CM = \frac{Dt}{2} - \left[(OS + a)^2 + Ov^2 \right]^{1/2}$$

$$OM = \frac{Dt}{2} - \left[(OS^2 + Ov^2) \right]^{1/2}$$

$$CLM = \frac{DT}{2} - Ov$$

Figure 2.45 *Cam/flat-faced follower contact.*

Even those lifters designed as "flat" have some small crown to ensure that production tolerances never result in a concave lifter face. Those designs that use a radius to induce rotation have a considerably larger crown. The crown usually is measured as the height of the lifter at its centerline relative to its height at a gage diameter. A typical specification for a flat follower would be flat to 0.0125-mm crown at 20-mm gage diameter, whereas a typical spherical-faced lifter would have 0.048–0.071-mm crown at 20-mm gage diameter. Figure 2.46 shows the relationships among offset, lifter spherical face radius, and lobe taper.

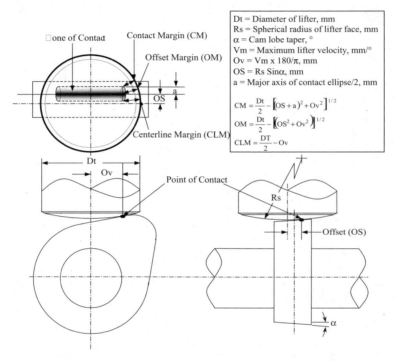

Figure 2.46 *Contact between the cam and spherical surface follower.*

Followers that have a spherical face radius are always used with camshafts that have a taper on their lobes. This combination results in the offset of the point of contact used to induce rotation of the follower. Offset values typically are 1.0–3.5 mm.

Another important criterion with the bucket tappet is controlling the size of the nose radius of the cam profile to control the maximum contact stress between the cam and tappet. This is done by limiting the amount of deceleration during cam profile design. Again, this tends to limit the lift and lift area that can be achieved.

Cam profile designs for mechanical tappets have longer constant velocity ramps on the opening and closing sides (versus hydraulic systems) to gently take up lash in the system for durability and NVH concerns. For mechanical systems, it is important to establish cold and hot lash requirements to properly design the ramps. With mechanical systems, the engine does not see the theoretical cam lift due to the lash in the system. If lash decreases over time due to valve seat recession or wear, the valve lift will increase, which might help performance but hurt emissions and idle quality due to the growth of the overlap area.

For either hydraulic or mechanical direct-acting systems with a crowned tappet surface and tapered lobe, the secondary lobe of the same cylinder should be offset and the lobe tapered left of the centerline to balance the thrusting motion imparted to the camshaft by the primary lobe. The tappet should be offset to the low end of the lobe taper to promote tappet rotation.

2.7.2.3.2 End-Pivot Rocker Arm Valvetrain System

Figure 2.47 shows a layout for an end-pivot rocker arm valvetrain. The centerline of the valve, finger follower, and lash adjuster or ball stud should fall into one plane, and coplanar geometry prevents unusual component side loading and the resultant wear. As an initial starting point, locate the center of the finger follower roller or pad horizontally halfway between the center of the socket and the center of the valve pad radius. This point will need to be adjusted later to optimize the effect of lift on pressure angle, slip-slide, and force balances.

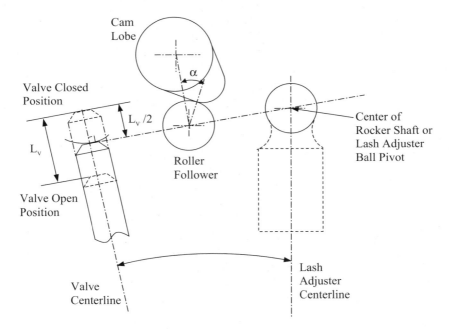

Figure 2.47 End-pivot rocker arm valvetrain layout.

The combustion chamber should be defined enough to dictate the valve head sizes and the valve angles extended through the cylinder head. The valve centerline points to the cylinder head face, and the valve gage line to the head face should be defined. Typically, a layout or concept sketch is required to package the valves, spark plugs, runners, and so forth as the initial start-ing point within the combustion chamber. The valve tip location is determined by allowing for adequate valve guide support as a function of valve head diameter and overhang of the valve from the bottom of the guide. The valve keeper groove location is selected to allow for sufficient retainer-to-seal clearance at maximum anticipated valve lift and adequate spring installed height for an optimum spring design. The valve tip then is determined, while allowing for acceptable clearance between the retainer and the pad end of the finger follower.

The motion that the follower arm valve pad makes as it slides fore and aft across the valve tip is referred to as slip-slide and is illustrated in Figure 2.48. For typical end-pivot rocker arm valvetrain geometries, the contact point on the valve tip should be at point F (inboard from the centerline, that is, close to the pivot point) when the valve is closed. The contact point travels outboard and passes the centerline when the valve opens. At approximately half the valve lift, a line AH is perpendicular to the axis of the valve. In this lift position, contact point H is located

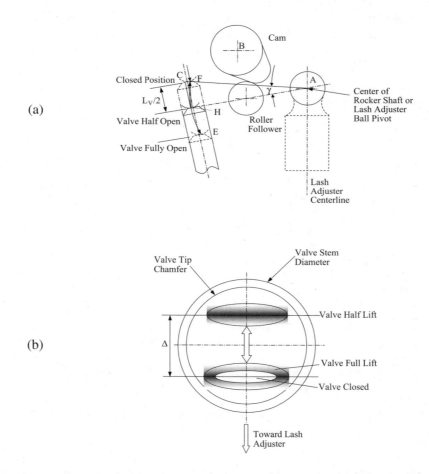

Figure 2.48 *Typical slip-slide pattern (Type II, finger follower): (a) valvetrain layout for an end-pivot rocker arm valvetrain, and (b) contact patterns on the valve tip.*

most outboard (farthest away from the pivot point A). Then the contact point travels inboard and passes the centerline again to the original position when the valve fully opens, point E. Based on the geometry, the slip-slide distance (Δ) on the valve tip can be calculated as

$$\frac{L_V/2}{AH} = \frac{\Delta}{L_V/2}$$

Therefore,

$$\Delta = \frac{\left(L_V/2\right)^2}{AH}$$

For a rocker arm with AH = 36 mm and L_V = 10 mm, the slip-slide distance Δ is approximately 0.7 mm.

The scrub or sliding velocity is the differential in the rate of change of the contact point travel on the flat valve tip and curved follower arm pad. High scrub velocities without adequate lubrication can result in excessive valve tip wear.

Figure 2.48b shows schematically the slip-slide patterns on the valve tip for the end-pivot rocker arm type of valvetrain. This pattern represents the total point contact travel between the follower arm pad and the valve tip and historically has been called slip-slide. The total pattern is the sum of the contact pattern from the centerline of the valve closed position and the centerline of the valve to the valve open position. The actual contact pattern between the arm and the valve tip is seen to be elliptical in shape because of the intersection contact point between two parts with curved surfaces. It also shows that as the maximum valve lift is approached, the width of the contact pattern is increased because of the increase in the valve spring load.

The design intent relative to the valve tip/follower arm pad contact is to ensure that the contact pattern lies entirely on the valve tip over the valve event. It should limit large differences in the scrubbing velocity between the valve tip and the rocker arm tip pad to prevent wear (i.e., allow more of a rolling action between parts).

The cam lobe base circle radius and the roller follower radius should be selected based on the following reasons. First, depending on the valve lift and the event length required, these sizes typically will allow for cam lobe profile designs without a negative radius of curvature. Second, the combination of the two radii will allow for acceptable contact stresses. Third, this combination allows for a fairly compact valvetrain design in overall height for packageability.

It is recommended that the rocker contact ratio be within the range of 1.6:1 to 2:1 for the end-pivot rocker arm valvetrain to maintain adequate stiffness and rigidity of the arm, and to minimize the amplification of any errors in cam lobe manufacture. A greater rocker ratio leads to larger slip-slide, which results in higher scrub velocity at both ends of the inboard and outboard, and thus greater side loads to the valve and lash adjuster. Valve tips in Type II valvetrains are more prone to tip wear; thus, a Type II system should be checked at the maximum valve lift to look at the slip-slide pattern on the valve tip and to verify that none of the contact pattern has run off the valve tip surface.

The pressure angle is the angle formed by the line through the centerline of the cam to the centerline of the roller and from the center of the ball pivot (at nominal operating position) to the centerline of the roller. Figure 2.49 shows the layout of the end-pivot rocker arm valvetrain, illustrating the effect of the pressure angle on the side load of the lash adjuster.

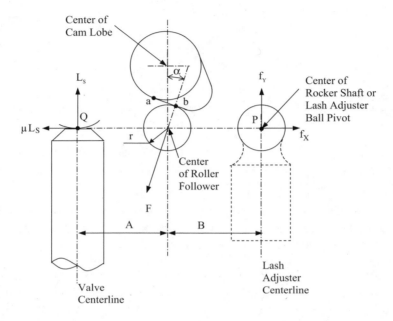

Figure 2.49 _Effect of the pressure angle on the side load of the lash adjuster._

For static equilibrium, the sum of the forces along the horizontal and vertical axes is

$$\sum F_X = 0 = f_X - F \cdot \sin\alpha - \mu L_S \tag{2.7}$$

$$\sum F_Y = 0 = f_Y - F \cdot \cos\alpha + L_S \tag{2.8}$$

From statics, the sum of the moments at points P and Q is

$$\sum M_P = 0 = L_S(A+B) - F \cdot B \cdot \cos\alpha \tag{2.9}$$

$$\sum M_Q = 0 = f_Y(A+B) - F \cdot A \cdot \cos\alpha \tag{2.10}$$

where

μ = coefficient of friction between the rocker pad and valve tip
F = force exerted from the cam through the centerline rocker arm roller
L_S = spring load during lifting

A = distance between the valve and roller follower centerline
B = distance between the roller follower and lash adjuster centerline
f_X = side load on the lash adjuster
f_Y = vertical force on the lash adjuster
α = pressure angle

Solving Eqs. 2.7 through 2.10 inclusive for the side load on the lash adjuster gives

$$f_X = \mu \cdot L_S + L_S \left(1 + \frac{A}{B}\right) \cdot tg\alpha \qquad (2.11)$$

Equation 2.11 suggests that the side load on the lash adjuster is a function of the valve spring load, the coefficient of friction (μ) at the valve tip, the rocker ratio $\left(\dfrac{A + B}{B}\right)$, and the pressure angle range (α). In the end-pivot rocker arm valvetrain (Type II), the hydraulic lifter or the lash adjuster acts as a pivot point and subjects it to high side load when the cam opens and closes the valve. The side load experienced by the lash adjuster depends on the engine speeds. It can vary from 500 N at 1000 rpm to 900 N at 7000 rpm engine speed. To minimize the side load on the lash adjuster and the valve stem/guide, the pressure angle range (α) and the rocker ratio $\left(\dfrac{A + B}{B}\right)$ should be kept as small as possible. However, to achieve the same valve lift requires a larger cam lift for the smaller rocker ratio, which unavoidably increases the side load on the lash adjuster. The spring load is the maximum when the valve is fully open; therefore, it is recommended that the pressure angle be as close to 0° as possible when the valve is fully open. Then the pressure angle at the base circle would be depend on the ramp length and roller follower radius, that is, $\theta r = ab$ or $\theta = ab/r$, where a is the contact point in the base circle, b is the contact point near maximum lift, r is the roller follower radius, and θ is the pressure angle at the base circle.

2.7.2.3.3 Center-Pivot Rocker Arm Valvetrain System

Most of the design principles outlined for the end-pivot rocker arm valvetrain system, such as coplanar layout, contact pattern, and pressure angle, are applicable to the center-pivot rocker arm valvetrain system as well. However, some unique features are found in the latter system. For purposes of explanation, a Type V pushrod valvetrain is assumed (Figure 2.50). The axis of the valve, the centerline of the fulcrum fastener, and the axis of the pushrod must converge at a common apex at one-half valve lift. This is extremely important with a stud mounting fulcrum, but not as important with the cylindrical fulcrum because this design has a large seat area and can remove some side and inboard/outboard loading. Contact travel of the valve tip and rocker arm pad as a function of valve lift percentage is plotted in Figure 2.50(b), based on the center-pivot rocker arm configuration shown in Figure 2.50(a). Contact travel on the valve tip is $\Delta x_0 = R_1 (\cos\theta - \cos\theta_0)$; contact travel on the rocker pad is $\Delta\rho_0 = 0.443 R_2 (\theta_0 - \theta)$ (Voorhies; Turkish).

At approximately one-half valve lift, the included angle (λ) between a line perpendicular to the valve centerline and a line perpendicular to the pushrod centerline at one-half valve lift should

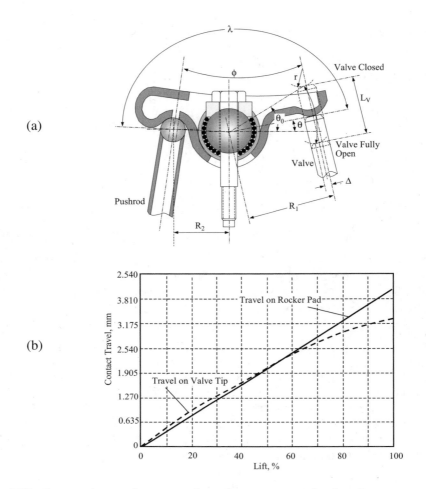

Figure 2.50 *Center-pivot rocker arm valvetrain geometry and valve tip contact travel:*
(a) center-pivot rocker arm valvetrain layout, and (b) graph of contact travel
on the rocker arm and valve tip (Voorhies; Turkish).

be 140–170°. Also, a line perpendicular to the axis of the valve must pass through the contact point between the rocker arm pad radius and the valve tip, and through the rocker arm pivot point. A line through the pushrod ball center and through the rocker arm pivot should form a perpendicular to the pushrod axis. The technical basis for these guidelines is to allow the distribution of the rocker arm tip contact pattern over the valve tip such that the pattern does not scrub in one spot and result in a wear concern. Small differences in scrubbing velocity between the valve tip and the rocker arm tip pad allow more of a rolling action between parts. They also minimize valve side loading and rocker arm pad and valve tip runoff. The rocker ratio should range from 1.5:1 to 1.9:1, which is slightly lower than that of the end-pivot rocker arm. Higher rocker arm ratios cause loss of rigidity and greater magnification of the effect of deviations in cam manufacture.

The slip-slide contact pattern of the center-pivot rocker arm to the valve tip should be offset from the valve centerline inboard when the valve is closed. At one-half valve lift, a line through the pivot center and valve tip contact point should be perpendicular to the valve centerline. The contact point at one-half valve lift should be at the outboard position (farthest away from the pivot point). When the valve fully opens, the contact point travels back to the inboard position. To minimize the potential for valve tip wear, there must be no runoff of the contact pattern between the rocker arm pad and the valve tip. Offsetting the slip-slide pattern inboard allows the contact pattern to be contained within the valve tip area at maximum lift. The total slip-slide pattern travel width is to be held to as small as possible to minimize the scrubbing velocity and, consequentially, tip wear.

The primary concern for the Type V valvetrain is that the pressure angle and subsequent forces in the follower may jam the roller follower. The maximum pressure angle establishes the cam size, torque, load, acceleration, wear life, and other pertinent factors. The side thrust or jamming effect on all flat-faced followers in the center-pivot rocker arm valvetrain is insignificant compared to that on the roller follower. Figure 2.51(a) shows the translating flat-faced follower with normal forces N_1 and N_2. It is obvious that their magnitudes change, depending on the distance q. It follows that the pressure angle is zero at all times, allowing the cam to be much smaller. Jamming of the follower is caused by the net effect of the opposing moments of forces F and μF. For best performance, the ratio B/A should be kept large, the coefficient of friction μ and the eccentricity q should be kept small, and the rod rigidity should be as high as possible.

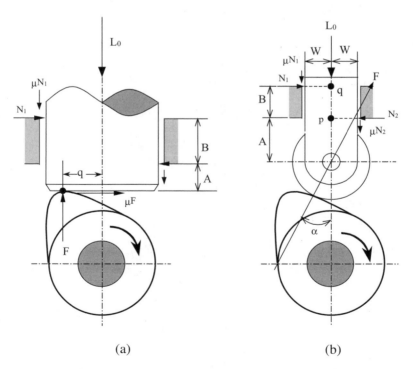

(a)   (b)

Figure 2.51 *Maximum pressure angle versus side thrust for (a) a flat-faced follower and (b) a roller follower.*

Figure 2.51(b) shows the direction of cam rotation, the normal forces on the follower, and the frictional forces opposing the motion of the roller follower. For static equilibrium, the sum of the forces along the vertical axis is

$$\sum F_y = -L_0 + F\cos\alpha - \mu N_1 - \mu N_2 \qquad (2.12)$$

Let points p and q be the intersection of N_1 and N_2 on the line of follower motion. From statics, the sum of the moments is

$$\sum M_p = 0 = -FA\sin\alpha + N_1 B - \mu N_1 W + \mu N_2 W \qquad (2.13)$$

$$\sum M_q = 0 = -F(A+B)\sin\alpha + N_2 B - \mu N_1 W + \mu N_2 W \qquad (2.14)$$

where

μ	=	coefficient of friction between the roller lift body and its guide
N1 and N2	=	forces normal to the roller lift body, kg
F	=	force normal to the cam profile, kg
L_0	=	total external load on the follower, including weight, spring, force, inertia, friction, etc., kg
A	=	follower overhang, mm
B	=	follower bearing length, mm
α	=	pressure angle, deg
W	=	radius of the follower body, mm

Solving Eqs. 2.12 through 2.14 for the force normal to the cam by assuming $\mu N_1 W$ and $\mu N_2 W$ equal zero because they are negligible gives

$$F = \frac{L_0}{\cos\alpha - \mu\left(\dfrac{2A + B}{B}\right)\sin\alpha} \qquad (2.15)$$

The normal force F is a maximum (i.e., equals infinity), which means that the follower will jam in its guide when the denominator of Eq. 2.15 equals zero. Therefore,

$$\cos\alpha_m - \mu\left(\frac{2A + B}{B}\right)\sin\alpha_m = 0$$

The maximum pressure angle for locking the follower in its guide is

$$\alpha_m = \tan^{-1}\frac{B}{\mu(2A+B)} \qquad (2.16)$$

Equation 2.16 suggests that in practice, the coefficient of friction, the follower overhang A, and the backlash must be kept as small as possible, with the bearing length B as large as possible.

In addition, the follower stem should be made rigid. Fulfilling these requirements will give the largest pressure angle and the smoothest follower action. Generally, the safe limiting pressure angle in practice is 30°. The maximum pressure angle also depends on the cam and roller follower size.

2.7.2.4 Valvetrain Contact Mechanics

Cam follower surfaces, including the tappet face, rocker arm pads, and valve tips, are subject to tribological contact and may fail in scoring, pitting, or flaking, or any combination of these. Such failures may be attributed to several factors: (a) the stress in the tappet face, (b) the materials used in contact, (c) the lubrication, (d) the rubbing velocity, and (e) the surface finish and treatment of contacting surfaces. Contact pressure is a force applied to a surface and is measured as force per unit area in the direction of loading. Contact stress also is the force per unit area of surface, but the area of surface can be chosen arbitrarily.

2.7.2.4.1 Elements of Elasticity Theory

The displacement of a point in a strained body from its position in the unstrained state is represented by the vector

$$\mathbf{u} = \left[u_x, u_y, u_z \right] \tag{2.17}$$

The components u_x, u_y, and u_z represent projections of $\mathbf{u}$ on the x, y, and z axes, respectively.

In linear elasticity, the nine components of strain are given in terms of the first derivatives of the displacement components; thus,

$$e_{xx} = \frac{\partial u_x}{\partial x}$$

$$e_{yy} = \frac{\partial u_y}{\partial y} \tag{2.18}$$

$$e_{zz} = \frac{\partial u_z}{\partial z}$$

and

$$e_{yz} = e_{zy} = \frac{1}{2} \left(\frac{\partial u_y}{\partial u_z} + \frac{\partial u_z}{\partial u_y} \right)$$

$$e_{zx} = e_{xz} = \frac{1}{2} \left(\frac{\partial u_z}{\partial u_x} + \frac{\partial u_x}{\partial u_z} \right) \tag{2.19}$$

$$e_{xy} = e_{yx} = \frac{1}{2} \left(\frac{\partial u_x}{\partial u_y} + \frac{\partial u_y}{\partial u_x} \right)$$

The three strains defined in Eq. 2.18 are the normal strains and represent the fractional change in the length of elements parallel to the x, y, and z axes, respectively. The six components defined in Eq. 2.19 are the shear strains. In elasticity theory, an element of volume experiences forces via stresses applied to its surface by the surrounding material. A complete description of the stresses requires not only specification of the magnitude and direction of the force but also of the orientation of the surface, for as the orientation changes, so, in general, does the force. Consequently, nine components must be defined to specify the state of stress. These are shown with reference to an elemental cube aligned with the x, y, and z axes in Figure 2.52. The component σ_{ij}, where i and j can be x, y, or z, is defined as the force per unit area exerted in the $+i$ direction on a face with outward normal in the $+j$ direction by the material outside upon the material inside. For a face with outward normal in the $-j$ direction (i.e., the bottom and back faces shown in Figure 2.52(b)), σ_{ij} is the force per unit area exerted in the $-i$ direction. For example, σ_{yz} acts in the positive y direction on the top face and the negative y direction on the bottom face.

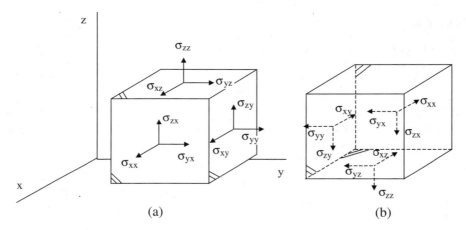

(a)   (b)

Figure 2.52 *Components of stress acting on the surfaces of an elemental cube: (a) top and front faces, and (b) bottom and back faces.*

The six components with $i \neq j$ are the shear stresses. By considering moments of forces taken about the x, y, and z axes placed through the center of the cube, it can be shown that the rotational equilibrium of the element (i.e., net couple = 0) requires $\sigma_{yz} = \sigma_{zy}$, $\sigma_{zx} = \sigma_{xz}$, and $\sigma_{xy} = \sigma_{yx}$. Thus, the order in which the subscripts "i" and "j" are written is immaterial. The three remaining components σ_{xx}, σ_{yy}, and σ_{zz} are the normal components. From the preceding definition, a positive normal stress results in tension, and a negative one results in compression. The effective pressure acting on a volume element therefore is

$$p = -\frac{\sigma_{xx} + \sigma_{yy} + \sigma_{zz}}{3} \qquad (2.20)$$

The relationship between stress and strain in linear elasticity is known as Hooke's Law, in which each stress component is linearly proportional to each strain. For isotropic solids, two proportionality constants are required:

$$\sigma_{xx} = 2Ge_{xx} + \lambda\left(e_{xx} + e_{yy} + e_{zz}\right)$$

$$\sigma_{yy} = 2Ge_{yy} + \lambda\left(e_{xx} + e_{yy} + e_{zz}\right)$$

$$\sigma_{zz} = 2Ge_{zz} + \lambda\left(e_{xx} + e_{yy} + e_{zz}\right) \qquad (2.21)$$

$$\sigma_{xy} = 2Ge_{xy}$$

$$\sigma_{yz} = 2Ge_{yz}$$

$$\sigma_{zx} = 2Ge_{zx}$$

Although λ and G are constants, G is known more commonly as the shear modulus. Other elastic constants are used frequently, the most useful being Young's modulus, E, and Poisson's ratio, ν. Under uniaxial normal loading in the longitudinal direction, E is the ratio of longitudinal stress to longitudinal strain, and ν is the ratio of lateral strain to longitudinal strain. These constants are interrelated:

$$E = 2G(1 + \nu)$$

$$\nu = \frac{\lambda}{2(\lambda + G)} \qquad (2.22)$$

Typical values of E and ν for metallic and ceramic solids lie in the ranges 40–600 GNm^{-2} and 0.2–0.45, respectively.

2.7.2.4.2 Normal Contact of Elastic Solids—Hertz Theory

Equations available for calculating the contact pressure, a prerequisite of calculating the contact stress, are Hertzian equations. The following assumptions were made in deriving the equations:

- Two contact materials are absolutely homogeneous.
- Deformation during contact is within the elastic limit, with no residual deformation.
- The contact surfaces are smooth.
- No friction force or tangential load exists.
- The contact area is much smaller than the bodies in contact.
- No lubricant film exists between the contact surfaces.

In the reality of the cam and follower contact, neither the material of the cam nor the follower is homogeneous, plastic deformation does occur at the asperity level in the contact interface, the surfaces are not smooth, friction or tangential force does exist, and usually there is lubrication.

Nevertheless, Hertzian equations are still the closest to reality and are widely used equations in calculating the contact stress at the cam/follower interface.

The stress in the tappet face when it is on the cam nose is the criterion of the success or failure of the tappet/cam design. It depends on the cam nose radius, the cam width, the valve open spring load, the elastic modulus and Poisson's ratio for the materials in contact, and the geometry of the tappet face such as flat and crown or spherical.

Table 2.7 (Johnson) summarizes Hertz elastic contact pressure formulas.

Equivalent modulus:

$$E^* \equiv \left(\frac{1 - v^2}{E_1} + \frac{1 - v^2}{E_2} \right)^{-1} \tag{2.23}$$

Equivalent radius:

$$R \equiv \left(\frac{1}{R_1} + \frac{1}{R_2} \right)^{-1} \tag{2.24}$$

(a) Line Contacts (load P per unit length)

Semi-contact width:

$$a = \left(\frac{4PR}{\pi E^*} \right)^{1/2} \tag{2.25}$$

Maximum contact pressure:

$$p_0 = \frac{2P}{\pi a} = \left(\frac{PE^*}{\pi R} \right)^{1/2} \tag{2.26}$$

Maximum shear stress at $x = 0$, $z = 0.78a$,

$$\tau_1 = 0.30p_0 \tag{2.27}$$

(b) Circular Point Contacts (load P)

Radius of contact circle:

$$a = \left(\frac{3PR}{4E^*} \right)^{1/3} \tag{2.28}$$

TABLE 2.7
FORMULAS FOR STRESS DUE TO PRESSURE ON OR BETWEEN ELASTIC BODIES

Contact Configurations	Formulas	Detail Contact Classification	Applications
Sphere	$a = 0.721 \sqrt[3]{P K_D C_E}$ $Max\ \sigma_c = 1.5\,\dfrac{P}{\pi a^2} = 0.918 \sqrt[3]{\dfrac{P}{K_D^2 C_E^2}}$ If $E_1 = E_2 = E$ and $v_1 = v_2 = 0.3$, then $a = 0.881 \sqrt[3]{\dfrac{P K_D}{E}}$ $Max\ \sigma_c = 0.616 \sqrt[3]{\dfrac{P E^2}{K_D^2}}$ $Max\ \sigma_t \approx 0.133(max\ \sigma_c)$ *at the edge of contact area* $Max\ \tau \approx \dfrac{max\ \sigma_c}{3}$, $a/2$ *below the contact surface*	Sphere on a flat plate, $K_D = D_2$ Sphere on a sphere $K_D = \dfrac{D_1 D_2}{D_1 + D_2}$	Valve tip/ lift/arm for Type 1, 3, 4 & 5 Valve tip/ lift/arm for Type 1, 3, 4 & 5 if both surfaces are spherical.
Cylinder	$b = 1.60 \sqrt{p K_D C_E}$ $Max\ \sigma_c = 0.798 \sqrt{\dfrac{p}{K_D C_E}}$ If $E_1 = E_2$ and $v_1 = v_2 = 0.3$ $b = 2.15 \sqrt{\dfrac{p K_D}{E}}$ $Max\ \sigma_c = 0.591 \sqrt{\dfrac{p E}{K_D}}$	Cylinder on a flat plate $K_D = D_2$ Max $\tau \approx (max\ \sigma_c)/3$ at 0.4b below the surface Cylinder on a cylinder $K_D = \dfrac{D_1 D_2}{D_1 + D_2}$	Cam & flat face follower for Type 1, 2, 3, 4 & 5 Cam & roller follower, Type 2, 3 & 5

Notation: P = total load; p = load per unit length; a = radius of circular contact area for spherical contact; b = width of rectangular contact area for cylinder contact; $C_E = \dfrac{1-\upsilon_1^2}{E_1} + \dfrac{1-\upsilon_2^2}{E_2}$; E = modulus of elasticity. ν = Poisson's ratio;

Maximum contact pressure:

$$p_0 = \frac{3P}{2\pi a^2} = \left(\frac{6PE^{*2}}{\pi^3 R^2}\right)^{1/3} \qquad (2.29)$$

Maximum deformation depth:

$$\delta = \frac{a^2}{R} = \left(\frac{9P^2}{16RE^{*2}}\right)^{1/3} \qquad (2.30)$$

Maximum shear stress at $r = 0$, $z = 0.48a$:

$$\tau_1 = 0.31p_0 \qquad (2.31)$$

Maximum tensile stress at $r = a$, $z = 0$:

$$\sigma_r = \frac{1}{3}(1-2v)p_0 \qquad (2.32)$$

(c) Elliptical Point Contacts (load P)

A = major semi-axis; b = minor semi-axis; $c = (ab)^{1/2}$; R' and R'' are major and minor relative radii of curvature, respectively; equivalent radius of curvature $R_e = (R' R'')^{1/2}$.

$$\frac{a}{b} \approx \left(\frac{R'}{R''}\right)^{2/3}$$

$$c = (ab)^{1/2} = \left(\frac{3PR_e}{4E^*}\right)^{1/3} F_1\left(\frac{R'}{R''}\right) \qquad (2.33)$$

Maximum contact pressure:

$$p_0 = \frac{3P}{2\pi ab} = \left(\frac{6PE^{*2}}{\pi^3 R_e^2}\right)^{1/3} \left[F_1\left(\frac{R'}{R''}\right)\right]^{-2/3} \qquad (2.34)$$

Maximum deformation depth:

$$\delta = \left(\frac{9P^2}{16RE^{*2}}\right)^{1/3} F_2\left(\frac{R'}{R''}\right) \qquad (2.35)$$

The functions $F_1\left(\frac{R'}{R''}\right)$ and $F_2\left(\frac{R'}{R''}\right)$ are plotted in Figure 2.53. To a first approximation, it may be taken to be unity.

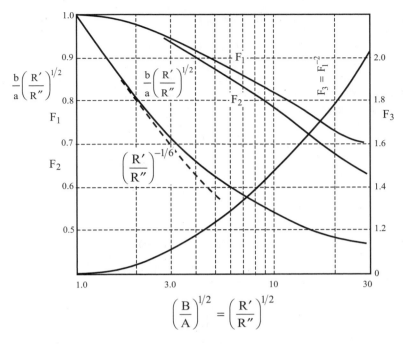

Figure 2.53 *Contact of bodies with general profiles. The shape of the ellipse* $\frac{b}{a}$ *and the functions* F_1, F_2, *and* F_3 *in terms of the ratio* $\frac{R'}{R''}$ *of relative curvatures (Johnson).*

The following material data can be used for stress calculations:

ν = 0.28 for all cam and tappet materials

E = 20.7×10^4 MPa for chilled iron, nodular iron, and steel; 18×10^4 MPa for hardenable cast iron

As a rule of thumb, the calculated Hertzian contact pressure at the cam/follower interface should not exceed the following limits at any speed within the operating range.

Non-roller followers:

Types II and III (slider interface)	690 MPa
Rotating follower—diesel	827 MPa
Rotating follower—gasoline	1240 MPa

Roller followers:

Steel roller/hardenable iron cam	1240 MPa
Steel roller/nodular iron cam	1480 MPa
Steel roller/steel or PM cam	1724 MPa

2.7.2.4.3 Tangential Loading and Sliding Contact

A normal force P pressing the bodies together gives rise to an area of contact, which, in the absence of friction forces, would have dimensions given by the Hertz theory. Thus, in a friction-less contact, the contact stresses would be unaffected by the sliding motion. However, a sliding motion (or any tendency to slide) of real surfaces introduces a tangential force of friction Q, acting on each surface, in a direction that opposes the motion (Figure 2.54).

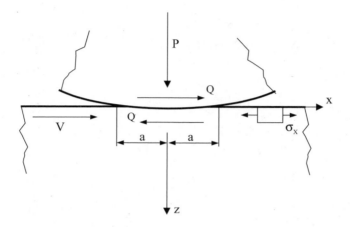

Figure 2.54 *Sliding contact of a cylinder on a plane.*

When a cylinder sliding perpendicular to its axis is on a plane, and both have the same elastic properties, the width of the contact strip 2a and the normal pressure distribution are given by the Hertz theory (Johnson):

$$p(x) = \frac{2P}{\pi a^2}\left(a^2 - x^2\right)^{1/2} \tag{2.36}$$

where P is the normal force per unit axial length pressing the cylinder into contact with the plane. Then, the tangential traction is

$$q(x) = \mp\frac{2\mu P}{\pi a^2}\left(a^2 - x^2\right)^{1/2} \tag{2.37}$$

where the negative sign is associated with a positive velocity V, as shown in Figure 2.55.

The stress parallel to the surface $\left(\sigma_x\right)_q$ can be shown as

$$\left(\sigma_x\right)_q = \frac{q_0}{a}\left[n\left(2 - \frac{z^2 - m^2}{m^2 + n^2}\right) - 2x\right] \tag{2.38}$$

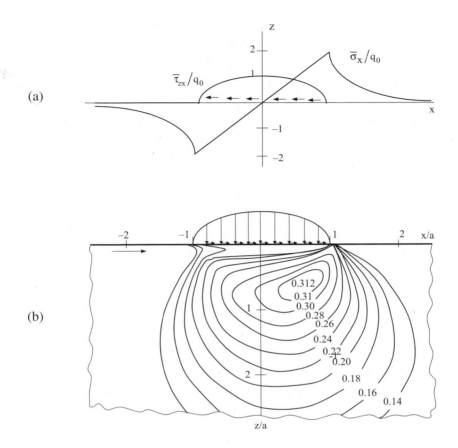

Figure 2.55 *Maximum shear and tensile stress under contact: (a) surface stresses*

$$q = q_0\left(1 - \frac{x^2}{a^2}\right)^{1/2} \quad \textit{due to frictional traction, and (b) contours of the principal shear}$$

stress τ_1 beneath a sliding contact (Johnson).

At the surface $z = 0$, the expression for the direct stress reduces to

$$(\bar{\sigma}_x)_q = -\frac{2q_0 x}{a}, \qquad |x| \le a \tag{2.39a}$$

$$(\bar{\sigma}_x)_q = -2q_0\left[\frac{x}{a} \mp \left(\frac{x^2}{a^2} - 1\right)^{1/2}\right], \qquad |x| > a \tag{2.39b}$$

The surface stresses in the moving plane are shown in Figure 2.55 using Eq. 2.39.

Tappet face failures in the form of pitting or flaking are of a fatigue nature. The stresses normal to the surface on the contact area are compressive and are the ones usually calculated. However, just beyond the confines of this area, the stresses are tensile in the plane of the tappet face. Below the surface and contact area, the stresses are compressive and vary in magnitude for three coordinate directions. Because of the existing differences in the compressive stresses, planes of shear can be located beneath the surface of the material. It is evident that each point on and just beneath the tappet face goes through stress cycles each time the valve is operated. Such stress will inherently increase fatigue failure.

In a cam and roller follower contact design, the contact of the cylindrical cam on the cylindrical roller follower usually results in higher contact pressure, compared to the contact of the cylindrical cam on the flat lifter, according to Hertzian equations. However, both tensile stresses at the contact trailing location and the shear stresses beneath the contact surface are lower due to lower friction at the rolling contact surfaces. Therefore, contact fatigue failures such as pitting or spalling of the roller followers are fewer compared to those for flat followers.

2.7.3 Valvetrain System Dynamics Analysis

Kinetics is the study of the relation existing among the forces acting on a body, the mass of the body, and the motion of the body. It is used to predict the motion caused by the given forces or to determine the forces required to produce a given motion. Valvetrain system dynamics or kinetics analysis studies the impact of a particular cam event profile on the mass-elastic system.

The input for the valvetrain dynamics analysis consists of geometric descriptions, mass parameters for the components, valve spring discrete mass and geometric parameters, hydraulic lash adjuster physical properties, valve or cam lift profiles, and so forth. The result of this analysis was a series of system-specific valvetrain dynamic models that could predict the dynamic response of the valvetrain when calibrated with the results of an initial valvetrain dynamics test.

2.7.3.1 Overview of Vibration

All bodies possessing mass and elasticity, including valvetrains, are capable of vibration. A mechanical vibration generally results when a system is displaced from a position of stable equilibrium. The system tends to return to this position under the action of restoring forces (either elastic forces, as in the case of a mass attached to a spring, or gravitational forces, as in the case of a pendulum). However, the system generally reaches its original position with a certain acquired velocity that carries it beyond that position. Because the process can be repeated indefinitely, the system keeps moving back and forth across its position of equilibrium. The time interval required for the system to complete a full cycle of motion is called the period of the vibration. The number of cycles per unit time defines the frequency, and the maximum displacement of the system from its position of equilibrium is called the amplitude of the vibration. Vibrations in valvetrains are undesirable because of the increased stresses and energy losses that accompany them. Therefore, they should be eliminated or reduced as much as possible by appropriate design.

There are two classes of vibration: (1) free vibration without external forces, and (2) forced vibration with external forces exciting the system. When the motion is maintained by only the restoring forces, the vibration is said to be a free vibration. When a periodic force is applied to the system, the resulting motion is described as a forced vibration. When the effects of friction may be neglected, the vibrations are said to be undamped. However, all vibrations are damped to some degree. Damping dissipates energy from the system. If a free vibration is only slightly damped, its amplitude slowly decreases until, after a certain time, the motion comes to a stop. But damping may be large enough to prevent any true vibration; the system then slowly regains its original position. A damped forced vibration is maintained as long as the periodic force that produces the vibration is applied. The amplitude of the vibration, however, is affected by the magnitude of the damping forces.

2.7.3.1.1 Vibration Without Damping

Consider a body of mass m attached to a spring of constant k (Figure 2.56(a)). When the mass is in static equilibrium, the forces acting on it are its weight W and the force T exerted by the spring, of magnitude

$$t = k\delta_{st}$$

where δ_{st} denotes the elongation of the spring, and k denotes the spring rate. Therefore,

$$W = k\delta_{st} \qquad (2.40)$$

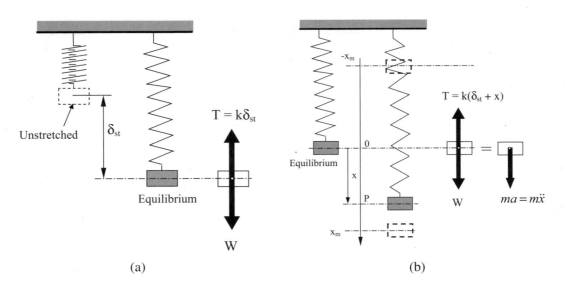

(a)

(b)

Figure 2.56 *Single vibration and harmonic motion: (a) in equilibrium position, and (b) in harmonic motion/vibration.*

Suppose now that the body is displaced through a distance x_m from its equilibrium position and is released with no initial velocity. If x_m has been chosen smaller than δ_{st}, the body will move back and forth through its equilibrium position; a vibration of amplitude x_m has been generated. The vibration also may be produced by imparting a certain initial velocity to the body when it is in its equilibrium position $x = 0$ or, more generally, by starting the body from any given position $x = x_0$ with a given initial velocity v_0.

When the body is in a position P at some arbitrary time t (Figure 2.56(b)), the forces acting on the body are its weight W and the force T exerted by the spring, which, in this position, has a magnitude of $t = k(\delta_{st} + x)$. Therefore, the magnitude of the resultant F of the two forces (positive downward) is

$$F = W - k(\delta_{st} + x) = -kx \qquad (2.41)$$

where x is the displacement OP measured from the equilibrium position O (positive downward). Thus, the result of the forces exerted on the body is proportional to the displacement OP measured from the equilibrium position, and F is always directed toward the equilibrium position O. Substituting for F into the fundamental equation $F = ma$ and a is the second derivative of x with respect to t,

$$m\ddot{x} + kx = 0 \qquad (2.42)$$

The same sign convention should be used for the acceleration a and for the displacement x, namely, positive downward. Equation 2.42 is a linear differential equation of the second order, setting $\omega^2 = \dfrac{k}{m}$; thus,

$$\ddot{x} + \omega^2 x = 0 \qquad (2.43)$$

The motion defined by Eq. 2.43 is called simple harmonic motion. It is characterized by the fact that the acceleration is proportional to the displacement and of opposite direction. The general solution of Eq. 2.43 for the displacement, velocity, and acceleration of P is

$$x = x_m \sin(\omega t + \phi) \qquad (2.44)$$

Differentiating, the velocity and acceleration at time t are obtained:

$$v = \dot{x} = x_m \omega \cos(\omega t + \phi) \qquad (2.45)$$

$$a = \ddot{x} = -x_m \omega^2 (\sin(\omega t + \phi)) \qquad (2.46)$$

The displacement-time curve is represented by a sine curve (Figure 2.57), and the maximum value x_m of the displacement is called the amplitude of the vibration. The angular velocity ω is known as the circular frequency of the vibration and is measured in radians per second (rad/s), while the angle ϕ that defines the initial position is called the phase angle.

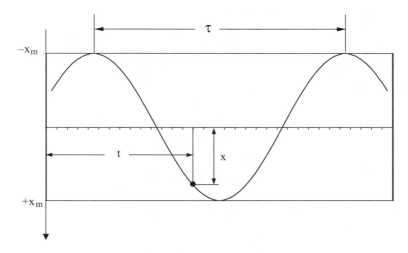

Figure 2.57 *Harmonic motion schematic: displacement versus time curve.*

A full cycle is described after the angle ωt has increased by 2π rad, and the corresponding value of t, denoted by τ, is called the period of the vibration and is measured in seconds,

$$\tau = \frac{2\pi}{\omega} \tag{2.47}$$

The number of cycles described per unit of time is denoted by f and is known as the frequency of the vibration,

$$f = \frac{1}{\tau} = \frac{\omega}{2\pi} = \frac{1}{2\pi}\sqrt{\frac{k}{m}} \tag{2.48}$$

The unit of frequency is the number of cycles per second (i.e., 1/s or s^{-1}). It is called a hertz (Hz) in the SI system of units. In applications involving angular velocities expressed in revolutions per minute (rpm),

$$1 \text{ rpm} = \frac{1}{60}s^{-1} = \frac{1}{60}\text{Hz}$$

The period and the frequency are independent of the initial conditions and of the amplitude of the vibration but are dependent on the mass m. The velocity-time and acceleration-time curves may be represented by sine curves of the same period as the displacement-time curve, but with different phase angles. The maximum values of the magnitudes of the velocity and acceleration are

$$v_m = x_m\omega; \qquad a_m = x_m\omega^2 \tag{2.49}$$

2.7.3.1.2 Damped Free Vibrations

The vibrating systems considered previously were assumed free of damping. Actually, all vibrations are damped to some degree by friction. Assuming that the friction force is directly proportional to the speed of the moving body, the equation of motion is

$$m\ddot{x} + c\dot{x} + kx = 0 \tag{2.50}$$

The general solution to Eq. 2.50 is in the form of

$$x = x_m e^{-\frac{c}{2m}t} \sin(qt + \phi) \tag{2.51}$$

where c is the damping coefficient, and where q is defined by the relation

$$q^2 = \frac{k}{m} = \left(\frac{c}{2m}\right)^2 \tag{2.52}$$

The motion defined by Eq. 2.51 is vibratory with diminishing amplitude. Although this motion does not actually repeat itself, the time interval $\tau = \dfrac{2\pi}{q}$, corresponding to two successive points where the curve touches one of the limiting curves shown in Figure 2.58, is commonly referred to as the period of damped vibration. The period of damped vibration, τ, is larger than that of the corresponding undamped system.

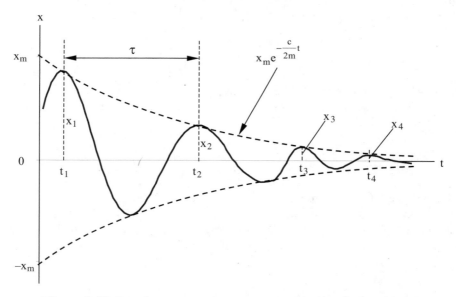

Figure 2.58 *Displacement versus time curve for damped vibration.*

2.7.3.2 Valvetrain Vibration and Natural Frequency

To assess the vibration of a valvetrain system, it is necessary to calculate the natural frequency of the system, and this requires knowing the mass and stiffness of the valvetrain. The frequency of the vibration shown in Eq. 2.48 as

$$f = \frac{1}{\tau} = \frac{\omega}{2\pi} = \frac{1}{2\pi}\sqrt{\frac{k}{m}}$$

is called the natural frequency of the system and depends on the constant k of the spring and the mass m of the body. It shows that the lighter the valvetrain weight and the stiffer the valvetrain, the higher the valvetrain natural frequency.

2.7.3.2.1 Valvetrain Weight Calculations

Single- and multiple-mass models can be used to calculate the valvetrain weight. Figure 2.59 shows the schematic comparison of single-mass versus multiple-mass in a center-pivot rocker arm valvetrain (pushrod Type V). The single-mass model is simple, and the data are easy to acquire. The single-mass model is not accurate in the details, has a good macro perspective, and is a good tool for initial valvetrain design evaluation. The multiple-mass model is more complex, and accurate data are more difficult to define. Detail and accuracy are increased, there is improved potential for microanalysis, and the multiple-mass model is a good tool for optimizing valvetrain designs.

For a simplified single-mass valvetrain model, there are two ways to evaluate the weight of a valvetrain system. In addition to the absolute weight of the various components, an effective weight can be calculated at either the valve or cam side of the rocker arm. The effective weight takes into account the effect of the rocker arm ratio. For a Type V pushrod engine, the effective mass is

$$M_{effective} = M_{valve} + M_{retainer\ and\ keys} + \frac{M_{spring}}{2} + \frac{M_{rocker\ arm}}{3} + \frac{M_{pushrod}}{RR^2} + \frac{M_{tappet}}{RR^2} \quad (2.53)$$

Table 2.8 shows examples of the calculation method used to establish weights of the various types of valvetrains. The information shown in Table 2.8 is to be used only to calculate system weights for each system type and does not reflect actual weight comparisons. Each valvetrain type would have its own valve, lifter weight, and so forth, and these specific weights are not reflected in Table 2.8. Note that the total effective weight at the cam can be obtained directly by multiplying the effective weight at the valve by the rocker arm ratio squared.

2.7.3.2.2 Valvetrain Stiffness

Any study of valvetrain dynamics must begin with an estimate of valvetrain weights and stiffness. From the weight and stiffness of a chosen valvetrain, the natural frequency of the valvetrain can be estimated. The required valve spring loads are estimated from the theoretical negative

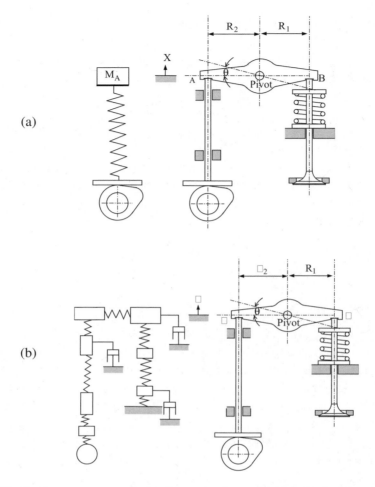

Figure 2.59 *Single-mass versus multiple-mass schematic: (a) single-mass assumption, and (b) multiple-mass assumption.*

camshaft acceleration and system weights. The allowable camshaft profile parameters of acceleration limits, pulse width, and so forth, as well as the required ramp heights, all depend on the valvetrain stiffness.

The effective valvetrain stiffness can be obtained by measuring the valvetrain deflection or force under load. This static deflection test can be applied to any valvetrain and will give an approximation of deflection or force encountered under actual engine operating conditions. For a valvetrain with a rocker arm, a load is applied to the rocker arm at the valve tip, and corresponding deflections are obtained on indicators. Analysis of the resulting deflection will give

TABLE 2.8
EXAMPLES OF VALVETRAIN ACTUAL AND EFFECTIVE WEIGHT CALCULATION COMPARISON

Component	Type I Direct Acting		Type II End-Pivot Rocker		Type III Center-Pivot Rocker		Type IV Center-Pivot Rocker		Type V Pushrod	
	Act. wt.	Effect. wt. at valve	Act. wt.	Effect. wt. at valve	Act. wt.	Effect. wt. at valve	Act. wt.	Effect. wt. at valve	Act. wt.	Effect. wt. at valve
Valve	100	100	100	100	100	100	100	100	100	100
Retainer	30	30	30	30	30	30	30	30	30	30
Keys	5	5	5	5	5	5	5	5	5	5
Spring[a]	80	1/2 (80)	80	1/2 (80)	80	1/2 (80)	80	1/2 (80)	80	1/2 (80)
R/A[b,c]	0	0	120	1/3(120)	120	1/3(120)	120	1/3(120)	120	1/3(120)
Pushrod	0	0	0	0	0	0	0	0	60	$60/(1.6)^2$
Tappet	110	110	0	0	0	0	110	$110/(1.6)^2$	110	$110(1.6)^2$
Total	325 gm	285 gm	335 gm	215 gm	335 gm	215 gm	445 gm	258 gm	505 gm	281 gm

Notes:
(a) Use 50% of the valve spring weight.
(b) R/A ratio = 1.6.
(c) Use 1/3 of the rocker arm weight.

each component stiffness from which the overall valvetrain stiffness can be obtained. A typical example of the Type V valvetrain stiffness calculation is shown as

$$\frac{1}{K} = \frac{1}{S_{CT}} + \frac{1}{S_{pr}} + \frac{1}{S_{ra}} + \frac{1}{S_{CH}}$$ (2.54)

where

S_{CT} = cam/tappet stiffness
S_{pr} = pushrod stiffness
S_{ra} = rocker arm stiffness
S_{CH} = cylinder head pedestal/fulcrum stiffness
K = overall valvetrain stiffness

Table 2.9 lists typical valvetrain stiffness and natural frequencies for various types of valvetrains.

TABLE 2.9
TYPICAL VALVETRAIN STIFFNESS AND NATURAL
FREQUENCY RANGES FROM EXPERIENCE

Valvetrain Type	Effective Weight (grams)	Stiffness (N/mm)	Natural Frequency (Hertz)
I	200–300	17,000–23,000	1200–1700
II	200–230	8,000–12,000	940–1230
III	200–230	6,000–10,000	810–1130
IV	240–270	4,000–6,000	610–800
V	260–300	2,500–5,000	460–710

2.7.3.2.3 *Valvetrain Natural Frequency*

Any consideration of valvetrain dynamics must begin with an estimate or measurement of the valvetrain natural frequency. If the effective valvetrain stiffness or spring rate is known and an estimate of the effective valvetrain weight can be made, the natural frequency can be obtained from the equation

$$f = \frac{1}{\tau} = \frac{1}{2\pi}\sqrt{\frac{k}{m}}$$

A more accurate and preferred approach for obtaining the valvetrain natural frequency is through measurements of the valve stem, pushrod force, or rocker arm strain or force taken during actual valvetrain operation and can be calculated from

$$f = nN\frac{X_{360}}{X_C}\frac{1}{60}, \quad Hz \qquad (2.55)$$

where

f = valvetrain system natural frequency
n = number of impulses within X_C; measurements are made from peak to peak
N = cam revolutions per minute = 1/2 engine revolutions per minute
X_C and X_{360} = strain or force measurement, and can be either angular or linear dimension

Figure 2.60 shows a typical representation of valve stem force and rocker arm strain. In Figure 2.60(a), it ran at engine speed 2400 rpm, n = 10 within X_C (90°) cam angle; therefore, the natural frequency of the valvetrain can be calculated as

$$f = 10 \times 1200 \times \frac{360°}{90°} \times \frac{1}{60} = 800 \text{ Hz}$$

In Figure 2.60(b), from the measured rocker arm strain data, the natural frequency of the valvetrain can be calculated as

$$f = 4 \times 350 \times \frac{193 \text{ mm}}{13 \text{ mm}} \times \frac{1}{60} = 346 \text{ Hz}$$

For accuracy, only the cycles occurring during the negative cam acceleration interval should be used. At higher speeds, the cycles during the positive acceleration pulse are forced and tend to give a lower frequency.

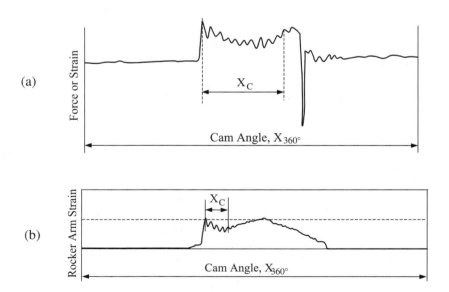

Figure 2.60 *Determining valvetrain frequency from measured data: (a) from valve stem force data, 2400 rpm, and (b) from rocker arm strain data, 700 rpm (Akiba and Kakiuchi).*

2.7.3.2.4 *Valve Spring Natural Frequency and Surge*

When one end of a compression spring is held against a flat surface and the other end is disturbed, a compression wave is created that travels back and forth from one end to the other, in the same way as a swimming-pool wave. This effect is called spring surge. If the disturbance occurs at a frequency close to the natural frequency of the spring, the spring may go into severe resonance, resulting in spring failure. In addition to valvetrain system dynamic analysis, an engine valve spring surge dynamics analysis is critical in enabling the optimization of the spring design.

For helical springs having one end against a flat plate and the other end driven with a harmonic motion, the basic equation of natural frequency is

$$f = \frac{1}{\tau} = \frac{1}{2\pi}\sqrt{\frac{k}{m}}$$

where k is the spring rate, and m is the spring mass. The spring rate can be calculated from the equation

$$k = \frac{Gd^4}{8D^3N}$$

where

G = shear modulus of elasticity
d = wire diameter
D = mean spring diameter
N = active number of coils

It can be derived as

$$f = \frac{2.15 \times 10^7 d}{60ND^2}, \quad Hz \tag{2.56}$$

The solutions to the spring surge can be the use of spring dampers, multiple springs each having its own natural frequency and controlling each other by damping, or nonlinear springs.

2.7.3.3 Valve Seating Velocity

The valvetrain comprises a dynamic multi-spring-mass system that is driven by cam profile input. This system has several resonant frequencies, the most important of which (in terms of closing dynamics) is the "valvetrain natural frequency," as it traditionally is called. Excitation of this mode causes deviation of the valve motion with respect to the cam, such that the valve seats at high velocity at various engine speeds. Valve closing dynamic performance is an important area of concern because of its many effects on valve stem fatigue life, valve timing, valve bounce, keeper groove wear and fatigue, seat recession, limiting speed, hydraulic lifter pump-up, noise, engine performance, and so forth.

If the valve follows the cam design profile at low revolutions per minute, the closing ramp of the cam velocity profile slows the valve before closure to ensure smooth quiet seating at engine idle speed. This ramp has a design rate measured in terms of millimeters per degree (mm/deg) at the valve. The valve seating velocity in terms of millimeters per second (mm/sec) is directly proportional to the cam rotational speed, up to approximately 2000–4000 crank revolutions per minute where valvetrain dynamics becomes influential. This is explained graphically in Figure 2.61 and mathematically in Eq. 2.57 as

$$V_{CV} = RAMP \times 3 \times RPM \tag{2.57}$$

where

V_{CV} = closing velocity at the valve, mm/sec
RAMP = closing ramp rate at the valve, mm/deg
RPM = crankshaft revolutions per minute, rpm

In automotive practice, this ramp is designed to close the valve at a rate of approximately 0.11 mm/deg or a velocity of approximately 25 mm/sec at engine idle speed. The closing ramp height (in millimeters) is defined as the design lift at the point where the acceleration profile becomes zero. This height is designed to account for the valvetrain deflections from valve-closed spring load, valve deceleration inertia load, hot lash of the mechanical lifter, hydraulic lifter leak-down, fluid compression (aeration), and valve cock. These factors can significantly affect the valve seating velocity at engine speeds greater than 4000 rpm (Figure 2.61(b)).

Theoretical and experimental studies show that valve closing is an impact phenomenon and, as such, involves a momentum transfer as the valve head strikes the valve seat. These studies indicate that the valve stem stresses are linearly proportional to the closing velocity (Figure 2.62).

Excessive valve closing velocities can be produced by several conditions, including excessive engine speed, cam design, excessive lash, and in-phase vibration of the valvetrain at the time of closure. Valve closing velocity consists of the theoretical closing velocity from the cam profile plus the vibration-induced velocity. The cam and valvetrain vibrate at a system natural frequency that generally is 500–1200 Hz. This vibration is excited by the cam profile harmonics, and the vibration results in a sinusoidal-like velocity component that is added to the design (non-vibratory) valve velocity. Depending on the event length and the phasing of the exciting cam harmonics, this vibration component can add to or subtract from the design valve closing velocity. Thus, the resulting closing velocity changes with engine speed in a semi-periodic fashion, as depicted in Figure 2.63. Due to the complex relationships of the valvetrain natural frequency, cam harmonic phase angles, and event lengths, valve closing velocities currently are determined experimentally.

2.7.3.4 Critical Engine Speed

In general, valve closing loads increase with engine speed, causing increasing deflection of the valvetrain, which results in progressively earlier seating of the valve. The critical engine speed

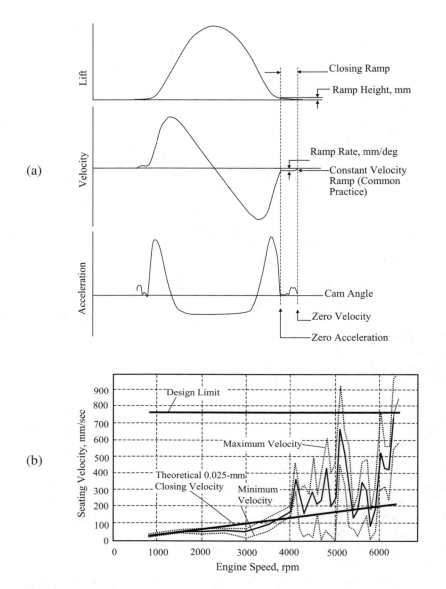

Figure 2.61 *Valve closing velocity: (a) valve lift, velocity, and acceleration curves, and (b) example of exhaust valve closing velocity versus engine speed.*

is defined as the speed at which one cycle of valvetrain vibration takes place during the opening acceleration pulse and depends on the positive acceleration period (Ap) and the natural frequency of the valvetrain system (f).

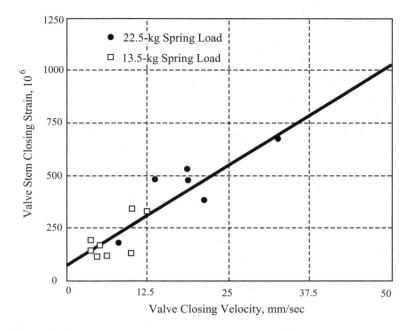

Figure 2.62 *Valve closing strains with velocity (Worthen and Rauen).*

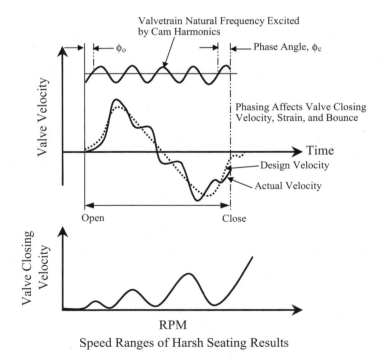

Figure 2.63 *Valvetrain natural frequency effects on valve closing velocity (Worthen and Rauen).*

2.7.3.4.1 Quasi-Static Deflection at Closure

The valvetrain deflections that contribute to early closure at high velocity can be classified as dynamic deflections or quasi-static deflections. Dynamic deflections are the result of transient response or resonance of valvetrain natural frequency and/or valve spring surge. Quasi-static deflections are from valvetrain inertia/stiffness, valve spring load, hydraulic lifter deflections (leak-down, aeration, pump-up), mechanical lifter deflections, hot lash, and so forth. When the valve "runs out of ramp," it begins to seat at progressively higher velocities as it "climbs up" the velocity profile (Figure 2.64).

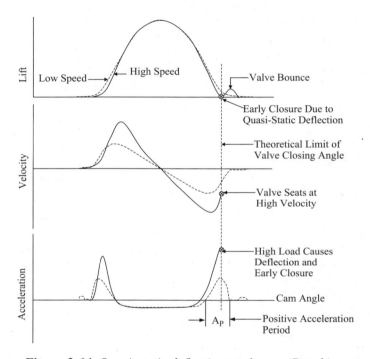

Figure 2.64 *Quasi-static deflection at closure (Buuck).*

The most influential of the quasi-static deflections in terms of high-speed operation is the valvetrain inertia effect. The inertia force at closure is controlled by the positive acceleration period, Ap. Its amplitude increases as the square of the speed, unlike the other quasi-static effects that remain relatively constant or even decrease at high speeds. Thus, Ap is considered to be the primary cause of high closing velocities, even when truly dynamic effects are ignored.

The quasi-static deflection, δ, at the valve during closure has components of inertia, spring load, lifter aeration, and hot lash, that is,

$$\delta = \frac{mA_{PK}RR(3\ RPM)^2 + PL}{k} + RR(LA + HL) \tag{2.58}$$

where

RPM	=	speed when the valve closes at the peak of Ap or crank revolutions per minute
A_{PK}	=	amplitude of the positive acceleration period
k	=	valvetrain stiffness
m	=	valvetrain mass
LA	=	lifter aeration
HL	=	hot lash
RR	=	rocker ratio
PL	=	spring pre-load

The derivative of Eq. 2.58 plus the valve design seating velocity gives the valve seating velocity with a quasi-static deflection. However, because the relationship between the deflection and cam angle is not defined clearly, the valve seating velocity may not be easily derived.

A limit is believed to exist at the peak of the positive acceleration period. Beyond this point (at earlier cam angles), the acceleration is less, which leads to less inertia force and therefore less valvetrain deflection. This creates later closure, rather than earlier closure. Hence, it is conjectured that the valve can never seat before the peak of the positive acceleration period, for this special case of pure quasi-static deflection. This implies that beyond the speed where the peak acceleration point is reached, valve closing velocity will increase at a linear rate proportional to RPM and design velocity curve magnitude at this cam angle.

For the valve to close at the peak of Ap, the valvetrain deflection, δ, must equal the design lift at Ap, designated as Lap. Replacing δ with Lap in Eq. 2.58, the speed when the valve closes at the peak of Ap is

$$RPM = \frac{1}{3}\sqrt{\frac{k(Lap-LA-HL) - PL}{m \times A_{PK}}} \tag{2.59}$$

2.7.3.4.2 Shock Response

The critical engine speed or shock response is the dynamic response of the valvetrain. At high engine speeds, the Ap or angle of the cam profiles behaves similarly to a shock impulse to the valvetrain. The dynamic overshoot and subsequent undershoot of this system (i.e., at mis-motion of the valve) can be calculated with classical vibration theory. This type of analysis is known as "transient response" (Figure 2.65).

Two types of valve mis-motion are generated. One is the primary response, and the other is the residual response. The primary response is the "overshoot" of the valve with respect to the peak acceleration input of the cam. This causes the largest force during the lift event at high speeds. Thus, it contributes to the largest acceleration-induced valvetrain deflection, which promotes early valve closure at high seating velocity as well as high cam stress and wear potential. The

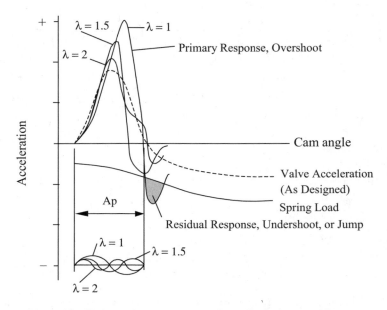

Figure 2.65 *Transient response of a valvetrain (Rothbart).*

residual response is the "undershoot" or peak deceleration of the valve at the end of the Ap. This can cause the inertia force of the valve to exceed the spring force (unloading), that is, the valve and spring move in the same direction, and the valve accelerates faster than the spring, leading to valve toss and hydraulic lifter pump-up.

As shown in Figure 2.66, actual valve acceleration response curves are compared to the as-designed valve acceleration curve. The acceleration response curve is related to the number of full cycles of natural valvetrain vibration occurring during the initial positive acceleration pulse (Ap). The positive opening acceleration pulse (Ap) has been shown to generate severe valvetrain vibration when the period of the open acceleration coincides with a half-cycle vibration of the natural frequency of the valvetrain,

$$\frac{Ap}{360\lambda} = \frac{RPM/2}{f} \quad \text{or} \quad RPM = \frac{Ap \times f}{180 \times \lambda} \qquad (2.60)$$

where

Ap	=	positive acceleration pulse, deg
RPM	=	critical engine speed or crank revolutions per minute
f	=	valvetrain natural frequency, cycles per minute
λ	=	cycles during the positive acceleration period

Equation 2.60 shows that, in general, to raise engine speeds, the valvetrain natural frequency (VTNF) or the width of Ap must be increased. This method of analysis is based on only the

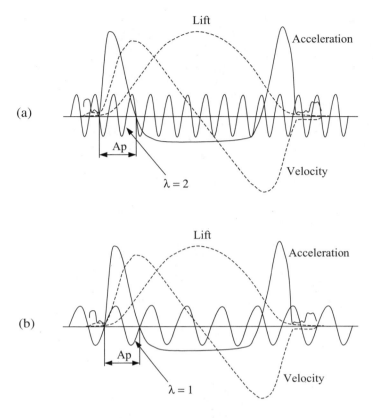

Figure 2.66 *Cam harmonic effect on valvetrain dynamics: (a) low speed and good dynamic response, and (b) high speed and severe dynamic response.*

first positive acceleration period of the cam; the remainder of the profile is ignored. Expected dynamic response from various λ values is shown in Figure 2.66 and Table 2.10.

TABLE 2.10
EXPECTED DYNAMIC RESPONSE FROM
VARIOUS λ VALUES

λ	Expected Dynamic Response
>1.33	Good
1.22–1.33	Good/Acceptable
1.10–1.22	Acceptable
1.05–1.10	Acceptable/Severe
1.00–1.05	Severe
<1.00	Very Severe

Figure 2.67 shows that the actual acceleration response curve can fall below the available compression spring load and actually result in a non-follow (jump) condition or hydraulic lifter pump-up. To prevent the occurrence of severe valvetrain vibration within the engine speed range, a minimum Ap is required and is dependent on the valvetrain natural frequency.

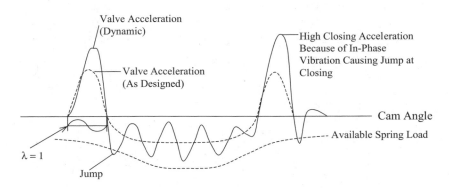

Figure 2.67 *Typical valvetrain natural frequency versus acceleration pulse.*

The design of a valvetrain system with the valve response as shown in Figure 2.68(a) has a significant improvement in performance, durability, and noise characteristics over a design similar to that shown in Figure 2.68(b). Figure 2.68(b) shows the acceleration curve of a camshaft generating excessive valve vibration. The cam has a high opening rate of rise, high peak acceleration (AM_0), and a reduced acceleration pulse width (Ap). The resulting valve acceleration force shows a wide variation in the response curve, which overcomes available spring load and causes a "jump" condition (i.e., momentary separation of the valvetrain components).

Figure 2.68(a) shows a valve acceleration curve that has improved high-speed characteristics. Note that the valve acceleration forces do not exceed the available spring load; consequently, the valvetrain system will hold together at this level of engine revolutions per minute. The valve acceleration curve is seen to have a reduced rate of rise, a lower maximum acceleration (AM_0), and a wider positive acceleration pulse (Ap). It is equally obvious that there must be control over the acceleration factors to generate a camshaft design capable of operating satisfactorily in any given valvetrain system. This valvetrain vibration also can produce high closing acceleration when it is in phase with the acceleration pulse (Figure 2.68(b)).

Care must be taken in selecting the values of Ap and the valve spring rate. Failure to adequately describe these parameters will result in a critical engine speed occurring below the desired engine speed when the negative valve acceleration exceeds the valve spring acceleration. The opening pulse must be broad enough to keep this resonant point above the operating range of the engine. Engine float (hydraulic tappet pump-up) speed will increase or decrease in proportion to the opening acceleration pulse-width. When hydraulic lifters are used, any momentary valvetrain separation can result in pump-up (a gain in lifter length for each cycle) because the function of a lifter is to compensate for lash.

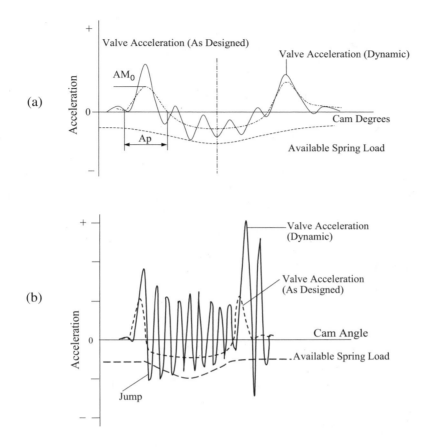

Figure 2.68 *Valve acceleration response comparison: (a) with good high-speed characteristics, and (b) excessive valvetrain vibration characteristics.*

2.7.3.4.3 Harmonic Excitation of Valvetrains

The preceding analysis focuses on the width of the positive acceleration period, Ap. Actually, the equally important criterion is the amplitude of the Ap. The purpose of the harmonic analysis is to analyze one major cycle of a waveform for its harmonic content. The amplitude of the waveform can be calculated numerically for each individual harmonic and then compared to spring load characteristics. If the amplitude exceeds the spring limit at a certain harmonic timing with engine speed, the critical engine speed must be reevaluated. Harmonic analysis considers the natural frequency f of the valvetrain system a "base-excited" spring-mass-damper system. The transfer function of this type of system in terms of mis-motion between the mass (valve) and the base (cam) is plotted against frequency in Figure 2.69. This dynamic system is excited by cam profile harmonics. When the harmonic amplitude amplifies and the phase shifts, the motion of the valve thus is distorted, causing harsh closing dynamics at various speeds.

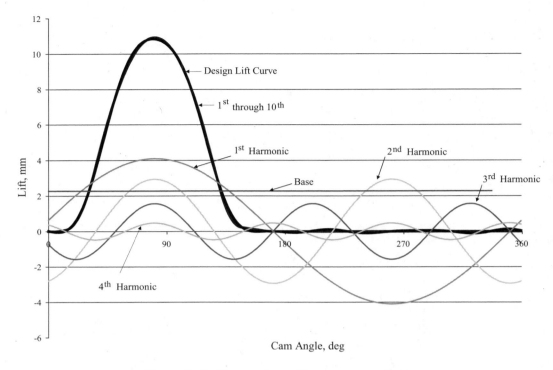

Figure 2.69 *Harmonic and harmonic amplitude.*

Any periodic motion can be represented by the Fourier equation as

$$x(t) = \frac{a_0}{2} + a_1 \cos\omega_1 t + b_1 \sin\omega_1 t + a_2 \cos 2\omega_2 t + b_2 \sin 2\omega_2 t + \cdots$$

$$+ a_n \cos N\omega_N t + b_N \sin N\omega_N t$$

Practically, this means that a valve lift profile can be approximated by

$$x(\theta) = A_0 + \sum_{n=1}^{\infty} A_n \cos(n\theta + \phi_n) \tag{2.61}$$

Coefficients A_n and ϕ_n can be determined through numerical methods.

Control of the motion of a valve by means of forces applied by a cam through a lifter, follower arm, and so forth is a complicated vibratory problem. The valve moves in the same general direction as the cam directs, but it will vibrate throughout the valve open period. As long as the vibratory accelerating forces on the valve do not overcome the available valve spring load, the system will hold together. Any given cam harmonic will coincide with the valvetrain system natural frequencies at specific engine speeds. The critical issue is whether a high-amplitude harmonic will coincide with a system frequency within the operating range of the engine. The engine speed is

$$\text{RPM} = \frac{f}{H} \times 60 \times 2 \tag{2.62}$$

where

RPM = engine speed
H = cam harmonic number
f   = natural frequency of the valve spring or valvetrain

Figure 2.70 shows graphically the relationship of engine speed, valvetrain natural frequency, and cam harmonic number based on Eq. 2.62.

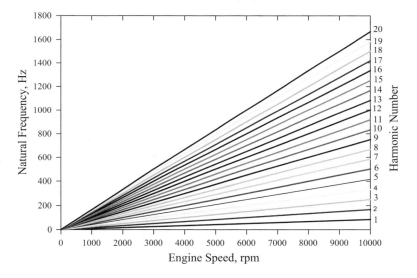

Figure 2.70 *Relationship of engine speed, valvetrain natural frequency, and cam harmonic number, based on Eq. 2.62.*

Therefore, the goal of harmonic analysis is to determine the harmonic number at high amplitude (>0.075 mm). Thus, it can determine the critical engine speed from the known valvetrain natural frequency. An example is provided graphically in Figure 2.71 and indicates that the critical engine speed exceeds 9000 rpm.

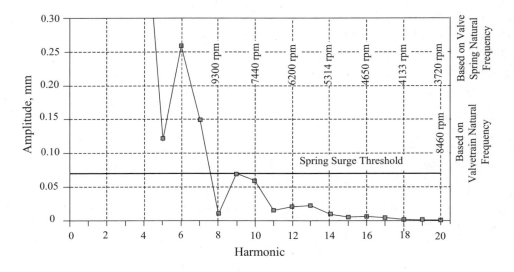

Figure 2.71 *Harmonic amplitude and spring surge.*

2.8 References

Akiba, K. and Kakiuchi, T., "A Dynamic Study of Engine Valving Mechanisms: Determination of the Impulse Force Acting on the Valve," SAE Paper No. 880389, Society of Automotive Engineers, Warrendale, PA, 1988.

Automotive Engineering International, January 2002.

Automotive Industry, March 1999.

Brüstle, C. and Schwarzenthal, D., "VarioCam Plus—A Highlight of the Porsche 911 Turbo Engine," SAE Paper No. 2001-01-0245, Society of Automotive Engineers, Warrendale, PA, 2001.

Buuck, B., "Elementary Design Considerations for Valve Gears," SAE Paper No. 821574, Society of Automotive Engineers, Warrendale, PA, 1982.

Carlström, P., "Volvo High Power Engine Brake," *Truck and Commercial Vehicle International '94,* ed. by P. Kennett, Sterling Publications Limited, London, U.K., 1994.

Flierl R. and Klüting, M., "The Third Generation of Valvetrains—New Fully Variable Valvetrains for Throttle-Free Load Control," SAE Paper No. 2000-01-1227, Society of Automotive Engineers, Warrendale, PA, 2001.

Grohn, M. and Wolf, K., "Variable Valve Timing in the New Mercedes-Benz Four-Valve Engines," SAE Paper No. 891990, Society of Automotive Engineers, Warrendale, PA, 1989.

Valvetrain Systems

Heisler, H., *Advanced Engine Technology*, Society of Automotive Engineers, Warrendale, PA, 1997.

Johnson, K., *Contact Mechanics*, Cambridge University Press, Cambridge, U.K., 1985.

Maekawa, K., Ohsawa, N., and Akasake, A., "Development of a Valve Timing Control System," SAE Paper No. 890680, Society of Automotive Engineers, Warrendale, PA, 1989.

Matsuki, M., Nakano, K., Amemiya, T., Tanabe, Y., Shimizu, D., and Ohmura, I., "Development of a Lean Burn Engine with a Variable Valve Timing Mechanism," SAE Paper No. 960583, Society of Automotive Engineers, Warrendale, PA, 1996.

Moriya, Y., Watanabe, A., Uda, H., Kawamura, H., and Yoshioka, M., "A Newly Developed Intelligent Variable Valve Timing System—Continuously Controlled Cam Phasing as Applied to a New 3-Liter Inline 6 Engine," SAE Paper No. 960579, Society of Automotive Engineers, Warrendale, PA, 1996.

Nakamura, M., Hara, S., Yamada, Y., Takeda, K., Okamoto, N., Hibi, T., Takemura, S., and Aoyama, S., "A Continuous Variable Valve Event and Lift Control Device (VEL) for Automotive Engines," SAE Paper No. 2001-01-0244, Society of Automotive Engineers, Warrendale, PA, 2001.

Riley, M., "Variable Valve Lift Mechanism for Internal Combustion Engine," U.S. Patent #5,572,962.

Rothbart, H., *Cams Design, Dynamics, and Accuracy*, John Wiley and Sons,, New York, 1956.

Schechter, M. and Levin, M., "Camless Engine," SAE Paper No. 960581, Society of Automotive Engineers, Warrendale, PA, 1996.

Turkish, M., *Valve Gear Design, A Handbook for Designers and Engineers to Aid in the Design of Cams, Tappets and Springs for the Valve Gear of Internal Combustion Engines*, Eaton, Detroit, MI, 1946.

Voorhies, C., "Valve-Gear Research as Applied to Diesel Engines," SAE Paper No. 420112, Society of Automotive Engineers, Warrendale, PA, 1942.

Worthen, R. and Rauen, D., "Measurement of Valve Temperatures and Strain in a Firing Engine," SAE Paper No. 860356, Society of Automotive Engineers, Warrendale, PA, 1986.

Chapter 3

 Valvetrain Components

3.1 Overview

For the poppet valve and its train system, the types of components used depend on the types of valvetrain system chosen: cam-driven or camless valvetrain. In the cam-driven valvetrain system, the valvetrain components may include the valve, cam, lifter, stem guide, seat insert, key, spring, spring retainer, rocker arm, and pushrod. Depending on whether the type of valvetrain is direct acting (Type I), end-pivot rocker arm (Type II), or center-pivot rocker arm (Types III, IV, and V), some of these components may not be needed. For example, a rocker arm or pushrod would not be necessary in a Type I valvetrain system. In the camless valvetrain, depending on the electrical hydraulic system or the electrical magnetic system, the cam is replaced with an electrical hydraulic or electrical magnetic drive system. In addition to the cam, the hydraulic lifter and rocker arm also are eliminated from the camless valvetrain system. Depending on the application and preference, all components come with different designs and materials. This chapter will cover the various designs and materials of these components, with a focus on cam-driven valvetrain components.

3.2 Valves

3.2.1 Valve Nomenclature and Construction

Engine valves control combustion chamber induction and exhaust flow in reciprocating combustion engines. Figure 3.1 illustrates the valve nomenclature and its typical construction. A typical intake valve is a one-piece construction with hardenable steel. However, the construction of an exhaust valve is more complex. Most exhaust valves use two-piece construction. Because the exhaust valve head is exposed constantly to hot exhaust gas, the valve head usually uses heat-resistant material, typically austenitic steels or nickel-based superalloys. Hardenable martensitic steels usually are used for the stem in two-piece construction because of their low cost and low-temperature applications. For additional wear resistance at the valve tip and seat, wear-resistant alloys sometimes are welded to the tip or seat. For high-speed engines that overheat the valves, hollow internally cooled valves are used. Oscillating coolant inside the valve can effectively transfer the heat from the valve head to the guide, but then the heat is carried away by the engine cooling system.

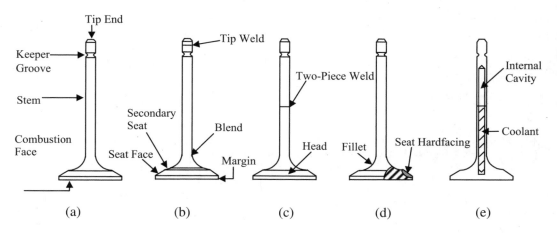

Figure 3.1 *Valve nomenclature and typical construction: (a) one piece, (b) wafer or tip welded, (c) two piece, (d) seat welded, and (e) internally cooled (Larson* et al.*).*

3.2.2 Valve Operating Characteristics

Engine valves are required to operate satisfactorily over long periods of time at elevated temperatures, under high stress, and in corrosive environments. This is especially true for exhaust valves because they operate at elevated temperatures that degrade alloy strength and promote oxidation and corrosion processes. Valve material and design must satisfy the following requirements: seal the port, meet the flow requirements, and meet the durability requirements of fatigue, corrosion, and wear.

3.2.2.1 Temperature Distribution

The valve operating temperature is a function of the specific output of the engine, the relative efficiency of the combustion process, the effectiveness of the engine cooling system, the shape of the valve head, the cylinder head valve layout, and the relationship of the stem-guide and seat insert. The temperature profile or distribution of the exhaust valve is particularly dependent on the temperature of the exhaust gases or the fuel/air ratio and compression ratio. The higher engine speed and higher exhaust gas temperatures that are typical of spark ignition (gasoline) engines result in significantly higher heat input into the port side of the exhaust valve, producing higher valve temperatures in this area than in relatively lower-speed compression ignition engines. Figure 3.2 shows the typical temperature profiles for the intake and exhaust valves for both spark and compression ignition engines.

3.2.2.2 Stress Distribution

Every combustion event imposes high pressures on the combustion chamber side of the valve head, generating cyclical tensile stresses on the port side of the valve head. The magnitude of these stresses is a function of the peak combustion pressure of the engine and is considerably

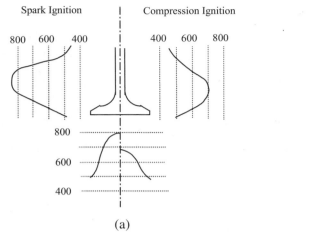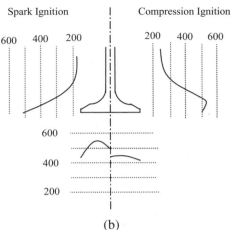

Figure 3.2 *Valve temperature distributions in spark ignition and compression ignition engines: (a) exhaust valve temperature (°C), and (b) intake valve temperature (°C) (Giles; Larson et al.).*

higher for compression ignition engines than for spark ignition engines. The valve seating event imposes cyclical tensile stresses at the junction of the stem and fillet on the port side of the head. The stress profiles of the valve head either can be measured by strain gaging the valve or can be calculated by the finite element method (FEM). The latter usually is preferred due to the availability of the sophisticated FEM program and good correlation with measurements.

Thermal stress is another factor that causes valve radial cracks to originate from the valve seat, especially when the valve seats are hardfaced. The high temperature gradients and different thermal expansion coefficients between the hardfacing material and the valve head material could initiate radial cracks, leading to guttering and valve head fracture. Figure 3.3 shows an example of the valve head stress distribution. Valve design can significantly affect the magnitude and shape of the stress distribution.

3.2.2.3   Corrosive Environment

The environment of the exhaust valve application is exhaust gas; however, deposition of compounds from the environment on the valve surfaces also may occur. Such deposits, referred to as ash, can cause alloys to react with the environment at substantially increased rates. This oxidation of alloys in the presence of ash deposits is called hot corrosion. When the ash bears sulfur, the process also is referred to as sulfidation corrosion. Exhaust valve materials selected on the basis of their oxidation resistance perform well for gasoline engine applications but may not be resistant to corrosion under a sulfidation condition. Therefore, it may not be appropriate for a diesel engine application when the fuel contains high sulfur. The exhaust valve head is exposed to products of combustion often containing highly corrosive constituents. The potential severity of corrosive attack is a function of the chemical composition of the valve alloy, its metallurgical

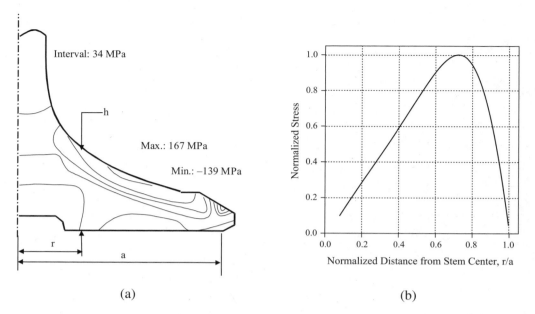

Figure 3.3 *An example of the valve head stress distribution under 15.5-MPa combustion pressure: (a) valve head stress contours, and (b) fillet surface bending stress (Larson et al.).*

characteristics such as grain size and microconstituent phases, and the valve operating environment. The principal constituents causing corrosion at high temperatures are sulfur, sodium, and vanadium. Molten salts frequently are formed from the combustion products of these contaminants, and it is believed that these compounds initiate the high rates of corrosive attack.

3.2.3 Valve Design

The physical dimensions of a valve must be derived from the basic dimensions of the engine. The intake and exhaust ports are designed to give maximum volumetric efficiency. The port diameter for large engines usually is based on a maximum gas velocity of approximately 60 m/s. For supercharged and high-speed naturally aspirated engines, this velocity may be exceeded considerably, and the port diameters then are made as large as physical dimensions allow. The port diameter leads to the valve head diameter, and the proportioning of the valve then proceeds in a logical manner. The basic valve dimensions are the head diameter, the overall valve length, the stem diameter, and the seat width.

3.2.3.1 Valve Geometry

3.2.3.1.1 Valve Sizes and Tolerances

As with any other engine components, valve sizes and tolerances are the result of many years of experience in satisfying the durability requirements of many different types of engines and the economic requirements to manufacture a valve at a competitive cost. Valve design trends are toward a larger valve head, especially a larger ratio of the intake and exhaust head for volumetric efficiency; a smaller stem diameter for lighter weight; and a narrow seat width to meet the requirements of performance. Table 3.1 and Figure 3.4 show a typical valve geometry relationship for poppet valve design.

TABLE 3.1
VALVE DIMENSION RELATIONSHIPS

	Intake	Exhaust
Head O.D.	1.21	1.00
Stem Diameter/O.D.	0.16	0.20
Head O.D./Seat Gage	1.06	1.06
Seat Width/Head O.D.	0.06	0.06
Head Thickness/Head O.D.	0.08	0.11
Stem/Keeper Diameter	1.22	1.22

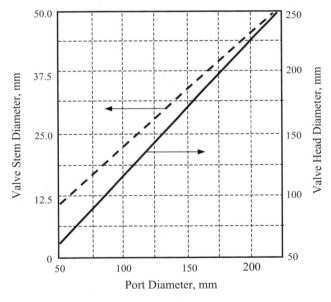

Figure 3.4 *Relationship of port, valve head, and stem diameter (Newton and Allen).*

In general, the tolerances shown in Table 3.2 and Figure 3.5 are applicable to most valves. However, modifications to these tolerances may be appropriate, depending on the application and method of manufacturing.

TABLE 3.2
TYPICAL VALVE TOLERANCES
(IN MILLIMETERS UNLESS SPECIFIED)

Location	Tolerance
Gage to Tip	± 0.125
Gage to Combustion Face	± 0.125
Head Thickness	± 0.25
Keeper Groove Width	± 0.075
Stem Diameter	± 0.0075
Keeper Groove Diameter	± 0.075
Head Diameter	± 0.125
Seat Angle	± 15°
Chamfer Angle	± 2°
Fillet Angle or Relief Angle	± 2°
Seat Face Runout	± 0.03
Keeper Groove Runout	± 0.1
Head O.D. Runout	± 0.25
Tip Surface Runout	± 0.015
Stem Straightness	± 0.0075
Stem Roundness	± 0.005
Seat Face Microfinish, Ra Maximum, μm	± 0.65
Stem Microfinish, Ra Maximum, μm	± 0.45
Tip Microfinish, Ra Maximum, μm	± 0.40

3.2.3.1.2 Valve Head Design

After the port diameter and the number of valves per cylinder are fixed, the valve head diameter becomes a function of the seat width. A good valve head design allows a wide seating surface with little overhang of the valve face beyond the seating surface. This is to limit exposure to exhaust gas heat and to obtain maximum cooling at the seat. A satisfactory design would be to allow approximately 0.75 mm above and below the seating surface. In some cases, it may be advisable to allow slightly more extension above the seat for regrinding in the field. A radius at the margin (outer corner of the valve head) will tend to minimize glow points that form at sharp corners and to reduce the amount of heat picked up from the combustion products. The thickness of the flat part of the valve head should be enough to preclude wide temperature fluctuations that may lead to thermal fatigue failure at the outside diameter (O.D.). A nominal thickness of 3 mm can be used as a guideline for valves with a 75-mm head diameter. This can increase on larger valves in proportion to the increases in the seat width. Concave valve heads are specified to reduce weight, but weight is of secondary importance in relation to strength. The design of the concave head should be considered carefully to avoid head cupping and breakage in service.

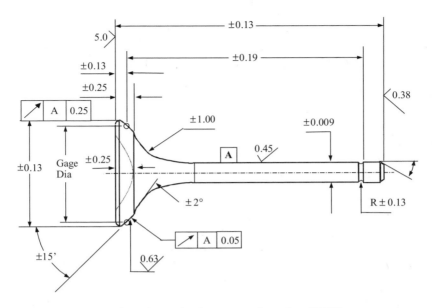

Figure 3.5 *Typical tolerances of a valve (TRW).*

The valve face, fillet, and neck shapes vary and are shown in Figure 3.6. In the interest of optimizing volumetric efficiency, valve head diameters usually are as large as possible. In general, the ratio of the exhaust head diameter to the intake head diameter is in the range of 0.7–0.9:1 for naturally aspirated applications. The ratio can be closer to 1:1 in supercharged or turbocharged applications. Coolant passages in the cylinder head also influence valve sizing. Coolant passages should provide 360° of coolant to the seat areas. Adequate coolant in the bridge area (between the seats) is particularly important to promote uniform heat transfer out of the seat area. Uniform cooling minimizes thermal distortion of the seat area and promotes good valve sealing. Poor sealing will shorten valve life because of localized overheating, and it will reduce engine efficiency. The use of valve seat inserts can further limit valve sizing. The counterbore area around the insert must be of adequate section thickness to retain the insert and still provide 360° of coolant.

The fillet shape of a valve significantly influences flow. As the fillet angle increases (from flat to tulip shaped), the valve weight increases. To compensate for the increase in weight, a cup can be added to the combustion chamber side of the valve (Figures 3.6(c) and 3.6(d)). In diesel applications, the top of the valve head is left flat for optimizing the compression ratio and for providing adequate valve head section to withstand the firing pressures.

The corner between the valve seat and the fillet angle can have a dramatic effect on flow. This is particularly important on intake valves. Secondary seat angles (or back angles) in this area can produce significant improvements in flow. These valve head shapes should be investigated through flow models and confirmed with engine testing.

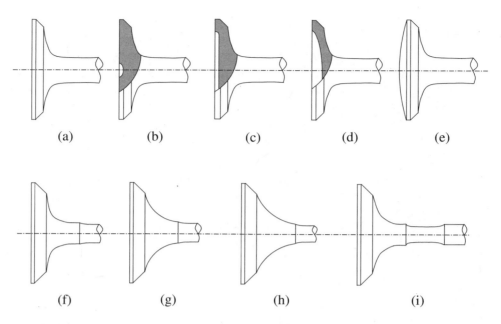

Figure 3.6 *Typical valve head shapes: (a) flat face, (b) with dimple, (c) and (d) with cup, (e) domed, (f) through (h) increased radius at fillet, and (i) with carbon scraper (Tunnecliffe and Jenkins).*

The blend radius between the stem and the fillet also can influence flow, but more importantly, it is sized to maintain a uniform distribution of stress levels throughout the fillet area. In general, these radii are larger on exhaust valves than on intake valves.

For improved durability considerations, a tapered section can be added to the exhaust gas impingement area (Figures 3.6(f) through 3.6(h)). This zone is very sensitive to corrosion and fatigue failures. The added section provided by this tapered zone can extend valve life without the added weight and cost penalty of increasing the stem size. This also may avoid the need for using a more expensive alloy.

Difficulty often is experienced as a result of deposits in the end of the exhaust valve guide and on the exhaust valve stem. These deposits usually are hard and tenacious and frequently lead to valve burning by interfering with the proper seating of the valve. A carbon scraper on the valve stem (Figure 3.6(i)) is an effective way to minimize the effects of these deposits in the guide and on the stem.

3.2.3.1.3 Valve Stem Design

The valve stem diameter is influenced primarily by the valve head diameter and the engine application. In general, the ratio between the diameter of the intake head and the diameter of the stem should be in the range of 5.5–6.0:1. For simplification of cylinder head machining, exhaust valves and intake valves usually utilize the same nominal stem size. However, durability considerations or other engine design factors can require two different stem sizes.

In general, the industry practice is to establish the nominal guide/stem diameters in increments of 5, 5.5, 6, 7, 7.5, 8, 8.5, or 9 mm, and so forth. Availability of valve locks in these sizes makes utilization of these standard stem sizes desirable. Commercial availability of guide reamer tooling influences the actual guide diameters. For example, an 8.00-mm guide usually is specified as 8.000–8.025 mm. The stem-to-guide clearance requirements then are controlled by the diameter specifications. Clearances are influenced by position (intake or exhaust), application, valve material, and guide and stem tolerances. The selection of the minimum clearance should be based on engine design experience with specific applications, in conjunction with valve and guide material combinations. Typical stem clearances are 0.025–0.050 mm for intake valves and 0.050–0.080 mm for exhaust valves. When changes in valve materials are made, the thermal expansion characteristics of the new material should be compared to the old material to determine if any clearance changes are necessary.

Another design feature, the stem taper, often is employed on exhaust valves where stem wear and scuffing are occurring at the port end of guide travel. Specifying a stem taper (i.e., smaller at the port end of guide travel by approximately 0.025 mm) will help compensate for the thermal gradients and the differences in coefficients of thermal expansion. Exhaust valve alloys typically have higher thermal expansion coefficients than that of cast iron. The exhaust stem area, which is exposed to exhaust gas at valve opening and returns to the inside of the guide at valve closing, can be as much as 260°C hotter than the guide at the corresponding location.

3.2.3.1.4 Valve Seat Angle

In designing valve seat angles, several factors must be considered. Figure 3.7 shows the valve seat angle and its effect on seating force, flow, and sealing or leakage. Most valve designs utilize a 45° seat angle. A 45° seat angle has demonstrated its ability to provide good sealing characteristics, as well as adequate flow and seat wear resistance. Reducing the seat angle, as measured from the valve face shown in Figures 3.7(a) and 3.7(b), generally is done to reduce seat wear. Seats with 30° angles often are used in applications where the combustion products have little or no lubricating properties. Turbocharged diesel engines often employ reduced seat angles (15–20°) on the intake valves for improved wear resistance.

A 45° seat angle often is used for exhaust valves to obtain higher sealing pressures between the valve seat and the insert. The seating force, F, is

$$F = p\sin\alpha = \frac{PA}{a}\sin\alpha$$

where

P = combustion pressure
A = valve face area
a = seating area

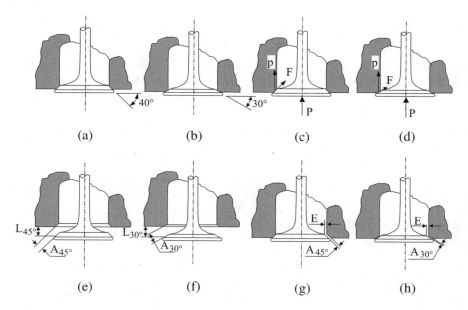

Figure 3.7 *Seat angle and its effect on seating force, flow, and sealing: (a) 45° seat angle, (b) 30° seat angle, (c) force from a 45° seat, (d) force from a 30° seat, (e) valve lift area at a 45° seat angle, (f) valve lift area at a 30° seat angle, (g) higher leakage potential at a 45° seat angle due to eccentricity, and (h) less leakage potential at a 30° seat angle due to eccentricity (Newton and Allen).*

Calculations indicate that the seating force is approximately 40% greater with a 45° seat angle than with a 30° seat angle (Figures 3.7(c) and 3.7(d)). This advantage often dictates the use of a 45° seat angle in applications where higher seating force is required due to combustion products resulting in heavy deposits on the valve seating surfaces.

In addition to affecting wear and sealing, seat angles can influence the flow and the valve operating temperatures. Also, as the seat angle decreases, the valve loses its ability to crush deposits at the valve seat/insert interface. The valve lift must be selected so that the valve will present the minimum amount of restriction to gas flow. Often, the valve lift area in the valve open position can be made to equal the throat area. However, just as often, the engine designer is limited by the proximity of the piston and is forced to compromise on valve lift. For the same lift, a 30° face angle gives greater lift area than a 45° seat angle. A 45° seat angle requires approximately 20% greater lift for a given valve open area (Figures 3.7(e) and 3.7(f)). This volumetric advantage of the 30° seat angle is particularly effective on intake valves during the early stages of valve lift. Some engines may require turbulence of the intake air charge for proper combustion, and in those instances, a 45° seat angle may be preferred. Extensive testing often is required to determine the most satisfactory valve and port design to promote thorough scavenging of the cylinder. Calculations indicate that for a given amount of seat eccentricity (E), the leakage area of a 45° seat angle is approximately 40% greater than with a 30° seat angle (Figures 3.7(e) and 3.7(h)). This type of eccentricity is the result of uneven expansion of the valve seating area due to nonsymmetrical design. The amount of inherent seat eccentricity present during engine operation depends on the success of uniform cooling around the valve seat. Poor casting practices and seat insert installations greatly aggravate seat eccentricity. When deposits do not build up sufficiently on the seat to break away and cause blowby, a 30° seat angle or smaller on exhaust valves may improve valve life by reducing leakage caused by mismated seating surfaces.

3.2.3.1.5 Relationship of the Valve Seat and Seat Insert

Figure 3.8 shows the relationship of the valve seat and the seat insert. The valve head O.D. should overhang the finished seat by approximately 0.5–1.0 mm. The valve face angle versus the seat angle should be tolerance stacked to provide line contact at the seat O.D. The interference angle (α) is in the range of 0.5–1.5°. Under no circumstances should the tolerance stack of angles produce a negative interference angle or a crevice opening toward the combustion chamber. For best results, finish machining of the valve seats should be done with three angles. The top cut and throat cut will provide uniform seat contact widths. Uniform seats promote uniform valve head cooling and more predictable wear characteristics. Contact widths are in the range of 1.00–1.50 mm on intake valves and 1.50–2.00 mm on exhaust valves. These values would apply to typical passenger car valve sizes. The valve head diameter and application will influence these seat-finishing guidelines.

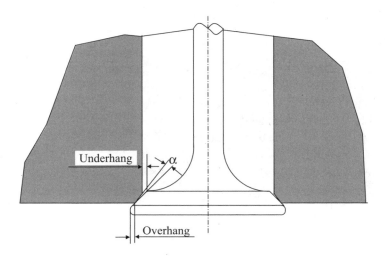

Figure 3.8 *Valve seat and insert relationship.*

3.2.3.1.6 Keeper Groove Design

For keeper groove designs, there are single-groove, double-groove, and multiple-groove designs. In the single-groove design, there are rectangular-groove and radius-groove designs (Figures 3.9(a) and 3.9(b)). The rectangular-groove design is used most commonly in the

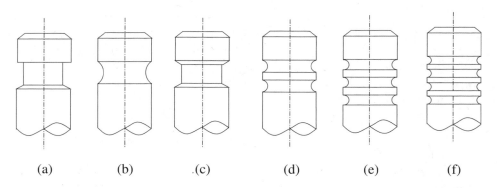

(a)   (b)   .(c)   (d)   (e)   (f)

Figure 3.9 *Keeper groove designs: (a) rectangular-groove design, (b) radius-groove design, (c) chamfered-groove design, (d) double-radius-groove design, (e) three-bead design, and (f) four-bead design (Tunnecliffe and Jenkins).*

U.S. automotive industry, with 8.5-mm stem diameters and larger. The single-radius groove design can be used on virtually any stem size but is used most commonly on 8.00-mm stems and smaller. The primary advantage of this latter design is the reduced notch sensitivity that the radius-groove design provides compared to a rectangular-groove design. This is of particular importance in highly stressed applications utilizing small stem diameters. The double-radius-groove design commonly is used in stem sizes of 9.00 mm and larger (Figure 3.9(d)). The radial shape of the grooves provides low-notch sensitivity, and the two grooves offer additional resistance to bead shear.

Three- and four-bead versions of this key design can be used as shown in Figures 3.9(e) and 3.9(f). These keys are designed to butt against each other when assembled, rather than clamp against the valve stem. This system supports the valve at the bead/groove interface, providing the valve with the freedom to rotate independently of the key and cap assembly. This "free valve" system allows the valve to rotate by virtue of normal engine vibration and/or scrubbing action imparted by the rocker arm or cam follower. These keys are case hardened, and the groove area of the valve stem is fully hardened to provide the required wear resistance at the groove/bead interface.

3.2.3.2 Valve Construction

3.2.3.2.1 One-Piece Valves

The most basic type of intake valve construction is a one-piece design (Figure 3.10(a)). This is used most commonly with martensitic alloys that have good hardenability—usually SAE 1547 or Sil 1. The high hardenability will provide Rockwell "C" (HRc) hardness levels of more than HRc 50 on the tip surface, which typically is required for resistance to spalling and wear.

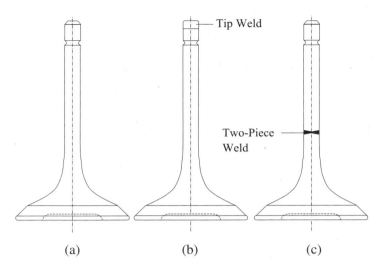

Figure 3.10 *Examples of valve construction: (a) one piece, (b) tip weld, and (c) two-piece stem weld.*

3.2.3.2.2 Tip Weld

Valves with resistance or projection-welded tips most commonly are used in exhaust positions or in applications that require the use of austenitic steels (Figure 3.10(b)). These materials provide the required high-temperature strength but are not hardenable to the required HRc 50-plus hardness. Projection or resistance welding of a hardenable material (e.g., SAE 3140 or 1457) on the tip provides a means of achieving the HRc 50-plus hardness. The tip weld material thickness usually is in the range of 2.0–2.5 mm.

3.2.3.2.3 Two-Piece or Stem-Welded Valves

Another means of providing HRc 50-plus tip hardness on a valve is stem welding. A hardenable steel alloy, such as SAE 4140 or 8645, is friction welded to the head section. In addition to providing the necessary tip hardness, this construction provides additional fatigue strength through the keeper groove area. Exhaust valve designs that employ multiple-bead butting locks for a "free valve" system require stem-welded construction. Hardness of HRc 50-plus through the groove area is recommended to provide the necessary wear resistance at the keeper groove/lock bead interface. The stem weld is located such that it remains inside the guide at maximum valve lift (Figure 3.10(c)).

3.2.3.2.4 Hollow and Internally Cooled Valves

One way to improve valvetrain dynamics is to reduce the valve weight. Hollow valves can considerably reduce valve weight, thus improving valvetrain dynamics. For valves of more than approximately 50 mm in diameter or even for valves of smaller size that are subjected to severe conditions (as in aircraft), internal cooling may be necessary. Figure 3.11 shows internally cooled exhaust valves. The hollow space in the valve is half filled with sodium or sodium potassium, which liquefies at valve temperature and "sloshes" back and forth in the valves as the valve moves. Thus, the heat conductivity from the exposed surface of the valve to the seat and stem is greatly increased. Hollow valves usually are made by drilling. The valve head and stem are drilled and then welded together, such as in Figure 3.11(a) or in a one-piece valve drilled from the face and ball/face welded (Figures 3.11(b) and 3.11(c)). The Ultralight™ valve (an Eaton trademark) is made by deep drawing and is hollow in the stem and head. Shells are drawn and caps are stamped, and they are welded together (Figure 3.11(d)). Internally cooled valves are expensive and are used only under circumstances where valve performance otherwise would be unsatisfactory. They are used most frequently in the case of engines of high specific output, such as aircraft, locomotive, racing, and military engines.

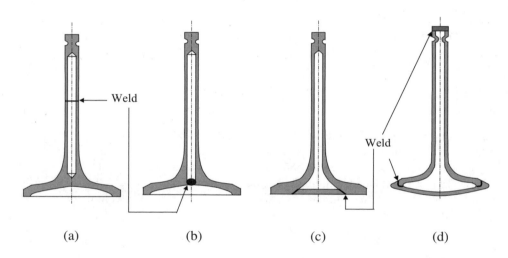

(a)	(b)	(c)	(d)

Figure 3.11 Examples of hollow valve construction: (a) the valve head and stem are drilled and welded together, (b) and (c) a one-piece valve is drilled from the face and ball/face welded, and (d) the valve is deep drawn, with a hollow stem and head, and the shells are drawn and caps stamped and then welded together.

3.2.3.2.5 Seat Hardfacing

Valves are hardfaced to ensure an adequate margin of safety against mechanical loads, thermal loads, and corrosive environments. In the absence of such protection, failures caused by corrosion, wear, guttering, and fatigue can occur. The need for hardfacing also may depend on designer choice, durability requirements, and design limitations in the cylinder head (Figure 3.12). The seat weld or hardfacing thickness usually is a minimum of 0.5 mm. To minimize the dilution effect from base material, a thicker layer of 1.0 mm minimum is needed. Sometimes, the double-pass process is preferred because it minimizes the dilution effect. For many years, Stellite™

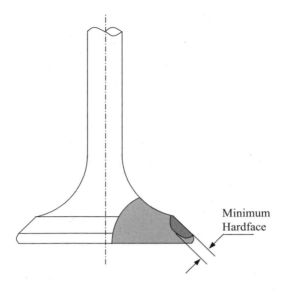

Figure 3.12 *Valve seat hardfacing.*

alloys have been the major facing alloys. These contain high cobalt additions. The cost and strategic nature of this element resulted in other alloy systems being developed. The iron-based hardfacing material has been used most successfully in a variety of diesel and lead-free gasoline engines. The oxyacetylene pudding process was used almost exclusively to deposit the various Stellite alloys. However, with the development of cobalt-free alloys, new hardfacing processes also have been introduced. Plasma transferred arc techniques using wire and powder now are being used where applicable to deposit the new iron-, nickel-, and cobalt-based facing alloys. These processes have been introduced to improve the quality, performance, and economics of the hardfacing techniques.

3.2.3.2.6 Stem Surface Treatment

As the valve operates between its open and closed positions, lateral movement is restricted by the use of a guide. The guide can be made as an integral part of the cylinder head, or it can be made as a separate component and pressed in the cylinder head. The critical feature of this component is its relationship to the cylinder head seat on which the valve closes. This concentricity of guide to insert must be held closely to avoid non-uniform valve seating and side loading of the valve stem and tip. To allow free axial movement of the valve stem, a clearance must be maintained between the guide and stem. The continual sliding motion of the stem along the length of the valve guide can result in wear of the stem and/or the guide. Similar problems can occur if the valve is sliding in a non-axial manner because of head distortion, high runout between the valve and seat insert, or if the load transfer to the valve tip is non-uniform, resulting in side loading from the actuating mechanism. Wear also can occur if there is a high degree of friction between the stem and the guide. Therefore, the resulting requirement is to provide a minimum coefficient of friction between the two surfaces. In the past, allowing oil to run down the valve stem has been effective. However, because of emissions requirements, this passage of oil is restricted by using various forms of stem seals. Thus, there is a need for stem coatings, combined with oil metered down from the stem seal, to reduce the coefficient of friction and to provide a hard wear-resistant surface.

Electrolytic chrome plating has been effective in providing the valve stem with the low coefficient of friction, and it is a hard wear-resistant surface. Various chrome thicknesses, from 0.75–20 μm, can be used, depending on engine requirements. Chrome plating frequently possesses high residual tensile stresses on the surface. This condition can result in a substantial reduction in fatigue strength if located in areas subject to mechanical bending stresses. As a general rule, flash chrome plating can be deposited in retainer groove areas without detriment. However, a thicker coating, such as the triple flash and full or heavy chrome plating, should be avoided in retainer grooves, tips, and the fillet stem blend area. In short, it should be confined to areas within the valve guides. Heavy chrome plating develops nodules that may increase guide wear caused by their cutting or abrasive action. In most instances, this may require a finishing operation to eliminate these nodules.

Nitriding is a surface treatment that produces a hard wear-resistant surface on the base material. The outer layer is a compound layer of iron and chrome nitrides that provides excellent tribological properties (i.e., low friction and high wear resistance). A diffusion layer of nitrogen below the compound layer provides further support to the compound layer and improves fatigue strength due to the nature of the compressive residual stress.

The thickness of the compound and the diffusion layer for the exhaust and intake valves vary, depending on the valve materials and engine applications. The salt bath nitriding process may be limited to iron-based alloys. Nickel-based materials, such as Inconel 751, do not develop a sufficient diffusion and compound layer to improve wear resistance.

3.2.3.3 Valvetrain Design for Flow Capacity

Engine torque or power is directly proportional to the weight of air burned in the unit time under the conditions of the constant fuel/air ratio or the compression ratio and ignition timing. The mass of airflow into the cylinder or the flow capacity is proportional to the air density, velocity, and area through which the gas flows, that is,

$$W = \rho A v \tag{3.1}$$

where

W = flow mass, kg/sec
A = area through which the gas flows, m^2
ρ = flow density, kg/m^3
v = flow velocity, m/s

In addition to these factors, the details of the valve timing, specifically the valve opening characteristics, affect the flow characteristics and engine performance. In the case of direct injection engines, it is necessary for the intake ports to generate an adequate swirl and tumble effect within the cylinder at the end of induction, to give efficient combustion when fuel injection later takes place, with a minimum reduction of volumetric efficiency.

3.2.3.3.1 Flow Area

The most obvious factor that affects the gas flow and output of an engine is the flow area (A). It is essential to pay attention to the intake valve and port design that routes air to the working cylinders, and to the exhaust system through which it must be discharged after combustion.

The valve capacity in the cylinder head depends on the number of valves and the valve arrangement. The first step in designing a valve is to establish the valve head diameter. The primary factors influencing valve head sizing are the bore size, number of valves per cylinder, valve geometry, spark plug or injector location, use of inserts, and cylinder head design.

To determine the valve head size, it is necessary to know how many valves are in a cylinder and the cylinder head shape. The valve capacity is the maximum feasible valve-diameter-to-bore ratio for flat cylinder heads with an exhaust-to-intake area ratio of 0.87. Figure 3.13 and Table 3.3 both show the typical relationship of the intake, exhaust valve head, and port size for a four-cylinder engine.

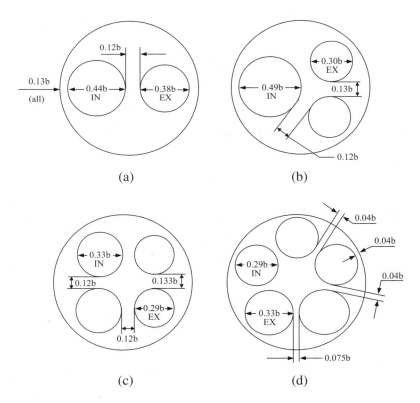

Figure 3.13 *Relationship of valve head and port size in a flat cylinder head, b = bore diameter: (a) two valves per cylinder, (b) three valves, (c) four valves, and (d) five valves (Taylor).*

TABLE 3.3
VALVE AND PORT SIZE RELATIONSHIP (TAYLOR)

Class	SI or Diesel	Intake				Exhaust			
		No.	D_i/b Min.–Max.	A_i/A_p Min.–Max.	No.	D_e/b Min.–Max.	A_e/A_p Min.–Max.	A_e/A_i Min.–Max.
Small	SI	1	0.58	0.34		0.53	0.28	0.82
Passenger Car	SI	1	0.39–0.49	0.15–0.24	1	0.36–0.41	0.13–0.17	0.83–0.85
Automotive	D	1	0.45–0.49	0.20–0.24	1	0.31–0.39	0.10–0.15	0.70–0.75
Diesel	D	2	0.34	0.23	2	0.34	0.23	1.0
Locomotive	D	2	0.32–0.35	0.20–0.24	2	0.31–0.35	0.19–0.24	0.83–1.0
Large Gas	SI	2	0.30–0.31	0.18–0.19	2	0.30–0.31	0.18–0.19	1.0
Midsize	D	2	0.30–0.32	0.18–0.20	2	0.30	0.18	1.0
Diesel	D	1	0.40–0.42	0.16–0.18	1	0.37–0.40	0.14–0.16	0.78–1.0
Auto	SI	1	0.53–0.63	0.28–0.40	1	0.55–0.63	0.30–0.40	0.72–1.0
Racing	SI	2	0.37–0.39	0.27–0.30	2	0.35–0.36	0.25–0.26	0.85–0.92

Notes:

D_i = Outside diameter, intake valve $A_i = \pi D_i^2/4$

D_e = Outside diameter, exhaust valve $A_p = \pi b^2/4$

b = Cylinder bore $A_e = \pi D_e^2/4$

Four valves and two valves per cylinder are the most common configurations. But three and five valves per cylinder also are often seen in applications. The five-valve engine (three valves for intake and two valves for exhaust) claims to give top-notch performance with low fuel and exhaust levels (Figure 3.14). The five-valve engine owes its effectiveness to the intensive ventilation obtained through five valves per cylinder and the electronic distribution of the ultra-point fuel injection system. This technology guarantees a high level of efficiency, even at low speeds. Where a four-valve head is used, a central or near-central fuel injector or spark plug also must be accommodated, and the maximum possible intake throat diameters are approximately 0.33b; for the exhaust, up to 0.29b can be used.

Three valves (two valves for the intake and one for the exhaust) per cylinder give better control for reducing cold-start emissions. Engine test results show that six- and eight-cylinder engines cannot generate sufficient heat during the warm-up phase. As a result, the catalytic converter achieves its full efficiency only after some delay. The smaller exhaust port area of the three-valve unit makes it possible to keep heat losses in the exhaust flow to a minimum. As a result, the difference in temperature compared to the four-valve engine is approximately 70°C; thus, the underfloor catalytic converter reaches its operating temperature of 300°C much earlier. The shorter response time of the catalytic converter following a cold start reduces pollutant emissions.

For two vertical valves that operate within the cylinder bore b without interference, the maximum intake port diameter (i.e., the throat diameter at the minimum cylinder head port diameter) is from 0.43–0.46b diameter, and the corresponding exhaust port size is 0.35–0.38b diameter.

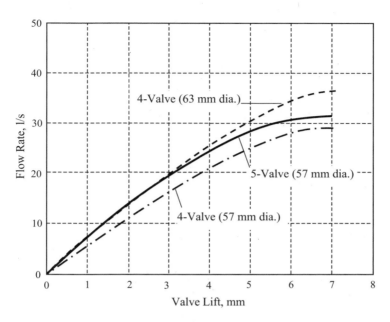

Figure 3.14 *Flow rate comparison: five valves versus four valves (Aoi et al.).*

Table 3.3 shows the relationship of the intake and exhaust valve sizes and the cylinder bore size for various applications.

Where maximum valve capacity is required, as in racing engines and aircraft engines, the cylinder head can be "domed." Figure 3.15 shows that the valve area/piston area can be increased drastically by "doming" the head of a two-valve cylinder. The disadvantages of doming include more complex valvetrain geometry and higher costs of manufacturing.

In general, the intake valve has a larger diameter than the exhaust valve because a pressure drop during induction has a more detrimental effect on performance than a pressure drop during the exhaust stroke. For a flat, twin-valve cylinder head, the maximum intake valve diameter typically is 44–48%, and the maximum exhaust valve diameter typically is 40–44% of the bore diameter. With pent-roof and hemispherical combustion chambers, the valve sizes can be larger. The intake and exhaust passages should converge slightly to avoid the risk of flow separation with its associated pressure drop. At the intake side, the division between the two valve ports should have a well-rounded nose; this will be insensitive to the angle of incidence of the flow. A knife-edge division wall would be very sensitive to flow breakaway on one side or the other. For the exhaust side, the division wall can taper out to a sharp edge.

The first step is to determine the largest valve sizes that can be accommodated while maintaining adequate cooling between them and the core thicknesses that permit sound castings. Practically all diesel engines use vertical valves due to the need to have very close piston-to-head clearances (approximately 10% of the stroke), in order to have a compact combustion chamber and

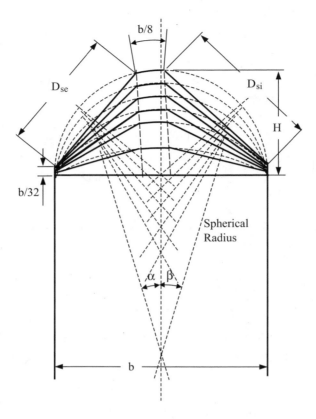

Figure 3.15 *Valve-area ratios for domed cylinder heads (Taylor).*

the necessary high compression ratio needed for good cold starting and light-load high-speed operation. It is usual to make exhaust valves slightly smaller in diameter than the intake ones because a slightly higher pressure drop is permissible without penalizing engine breathing and output.

With the use of 45° valve seats, the valve head diameter becomes approximately 1.09 times the throat (gage) diameter, d; the corresponding size is 1.11d when 30° valve seats are used. The valve should be opened as quickly as other aspects, usually dynamic problems, of the valvetrain design permit. Obviously, the lift in excess of 0.25d has little value, but the increase of area under the curve at lower lifts is valuable when valvetrain dynamics permit.

The controlling velocity in a compressible flow system is the velocity in the smallest cross section. Considering typical engine designs, it appears that the smallest cross section in the intake and exhaust flow system usually is the valve opening. Aside from heat-transfer effects, valve timing, and cam contour design, the key valvetrain design factor having important effects on volumetric efficiency is the valve flow area. For a given valve and cylinder head port design, the air or exhaust gas passing through the whole port/valve combination at a range of lifts from zero to the full designed lift is limited by the bottleneck of the conical frustum flow area at the valve seat. Figure 3.16 shows a valve lifted from its seat.

At low lifts, shown on the right of Figure 3.16, the flow area normal to the seat faces is that of the surface for part of a frustum of a cone. The diameters for the cone are d_1 and d_L. Then XY is the slant height,

$$XY = L\cos\alpha = S$$

where L is the valve lift and α is the valve seat angle. Now

$$d_L = d_1 + 2XZ = d_1 + 2XY\sin\alpha = d_1 + 2L\cos\alpha\sin\alpha$$

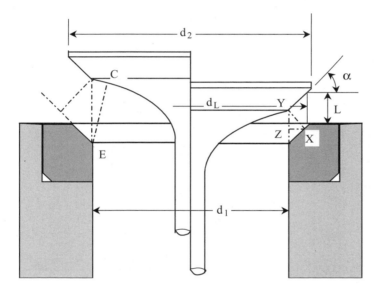

Figure 3.16 *Valve area with conical seats (Lilly).*

The surface area of a right circular cone is

$$A = \frac{\pi}{2}S(d_1 + d_L) \tag{3.2}$$

This gives the valve opening area per valve

$$A_V = \pi L\cos\alpha(d_1 + L\cos\alpha\sin\alpha) = \pi d_1^2\cos\alpha\left(1 + \frac{L}{d_1}\cos\alpha\sin\alpha\right)\frac{L}{d_1} \tag{3.3}$$

A non-dimensional lift $\dfrac{L}{d_1}$ is convenient when comparing flow characteristics for differing valves and ports. Figure 3.17 compares the valve seat angle effect on the flow area at low lifts, assuming the bore diameter $d_1 = 54$ mm. It shows that a 30° seat angle valve has greater flow area than a 45° seat angle valve.

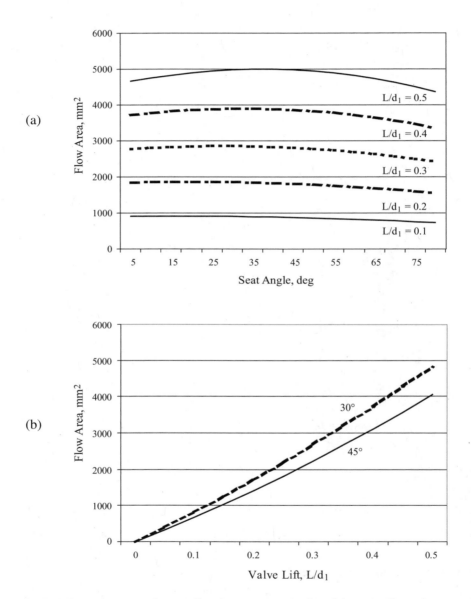

Figure 3.17 *Effect of seat angle on valve flow area under low lift: (a) effect of seat angle on the flow area at a different lift, and (b) effect of valve lift on the flow area for two seat angles.*

At higher lift, however, when the valve lift becomes such that the perpendicular from the inner corner of the valve head seat, Y, falls outside the O.D. of the stationary head seat, the opening area is no longer that of the frustum surface. As indicated in the left side of Figure 3.16, a line EC must be found for which the flow area at the lift concerned is the minimum. When lift increases to a certain degree, the minimum flow area will be the annulus between the valve stem diameter and the valve inner seat diameter. This value of valve lift is $\dfrac{L}{d_1} = 0.2946$ for a 45° valve and $\dfrac{L}{d_1} = 0.2491$ for a 30° valve. In practice, when flow coefficients are being determined, the frustum area often is used beyond the lift for which its truth is strictly accurate, the associated coefficients being calculated on this assumption.

Two other approaches to nominal area are in use. The first is to assume that the nominal flow area is the vertical collar area based on the valve throat diameter, that is, $\pi d_1 L$ for the lift L being considered. The second is a similar vertical collar area based on the valve head diameter d_2, that is, the area is $\pi d_2 L$.

3.2.3.3.2 Flow Velocity

It is common practice to assume incompressible flow for the intake valve because the pressure drop is small. The fundamental equation for the flow is

$$v = c\sqrt{2gh} \tag{3.4}$$

where

- v = flow velocity, m/sec
- c = coefficient of discharge
- g = gravitational acceleration, ≈ 9.8 m/sec^2
- h = head of fluid causing the flow, m

For many years, the mean intake gas velocity has been used as a rule-of-thumb assessment of maximum operating speed. Typical values at maximum power output are 76 m/s for automotive types and up to approximately 52 m/s for the larger stationary engine. If the stroke is L meters and the engine crankshaft speed is N rev/s, then the mean piston speed is $v_p = 2LN$. Therefore, with an intake throat diameter of d meters and a cylinder bore diameter of b, the mean intake valve gas velocity is

$$v_g = 2LN\frac{A_P}{A_V} = 2LN\left(\frac{b}{d}\right)^2 \tag{3.5}$$

Ideally, the area represented by the valve stem diameter should be subtracted from the valve throat area; however, because its value is small, it usually is neglected when making these empirical design comparisons. The valve stem diameter typically is 0.20–0.22 times the valve throat diameter.

To have the same nominal area as the valve throat, the valve should be lifted by a quarter of the valve throat diameter $\left(\dfrac{d}{4}\right)$. In practice, an attempt is made to make the maximum valve lift at least equal to 0.26d. This usually can be done on low-speed engines, but valvetrain dynamics may make this difficult for pushrod-operated systems in small high-speed automotive engines.

3.2.3.3.3 Flow Coefficient

Figure 3.18 shows the flow characteristics of a sharp-edged intake valve. At low lift, the jet fills the gap and adheres to both the valve and the insert seat. At intermediate lift, the flow will break away from one of the surfaces; at high lift, the jet breaks away from both surfaces to form a free jet. The transition points will depend on whether the valve is opening or closing.

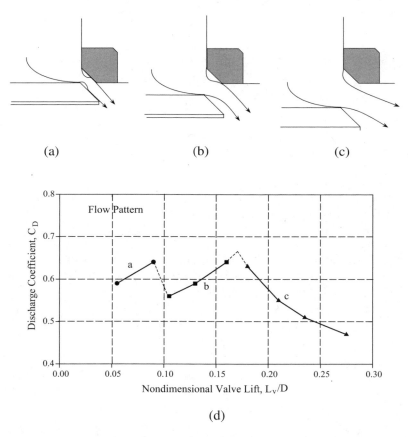

Figure 3.18 *Flow characteristics of a sharp-edged intake valve: (a) low lift, (b) intermediate lift, (c) high lift, free jet formed, and (d) discharge coefficient versus valve lift (Annand and Roe).*

Eliminating the sharp edges of an intake insert can appreciably increase the flow coefficient for intake valves (Figure 3.19). The effect of valve lift on the discharge coefficient is much smaller for the exhaust. The range of pressure ratios across the exhaust valve is much greater than that across the intake valve, but the effect on the discharge coefficient is small. The design of the exhaust valve appears less critical than the intake valve, and the seat angle should be 30–45°.

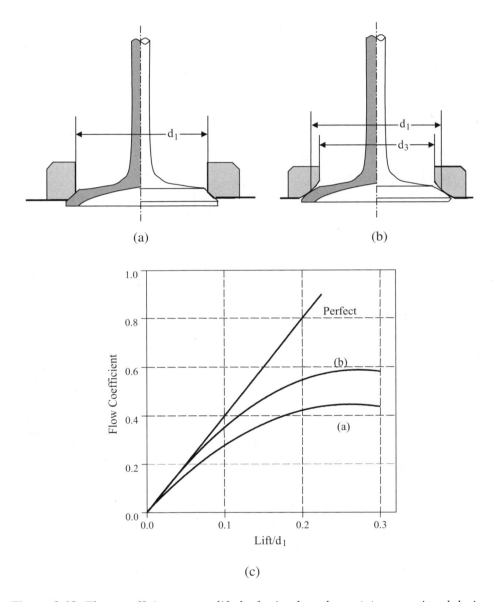

Figure 3.19 *Flow coefficient versus lift d_1 for intake valves: (a) conventional design; (b) modified valve and insert, $\dfrac{d_3}{d_1} = 0.8$; and (c) flow coefficient versus valve lift (Taylor).*

3.2.3.3.4 Pressure Differential

For intake valves, it is common practice to assume incompressible flow because the pressure drop is small, in which case the ideal velocity would be (Lilly)

$$v = \sqrt{2RT_0 \frac{(r-1)}{r}} \tag{3.6}$$

For compressible exhaust gas,

$$v = \sqrt{\frac{2\gamma RT_0}{\gamma - 1}\left[1 - \frac{\gamma - 1}{r\gamma}\right]} \tag{3.7}$$

where

v	=	frictionless velocity
R	=	gas constant, J/kgK
T_0	=	upstream temperature, K
r	=	$\dfrac{p_0}{p_c}$
p_0	=	upstream pressure, N/m^2
p_c	=	cylinder pressure or downstream, N/m^2
γ	=	ratio of specific heats = 1.4 for air

In the four-stroke engine, the pressure drop across the intake valve is created by the suction generated by the descending piston in the case of the naturally aspirated version. When the engine is supercharged, the boost pressure is added, usually by a turbocharger or a supercharger.

The flow velocity of the exhaust gas is a function of the ratio of the cylinder pressure and exhaust port pressure. The backpressure from the muffler and catalytic converter can increase the downstream pressure, thus reducing the $\dfrac{p_0}{p_c}$ ratio and leading to the reduction of exhaust gas flow.

Much effort has been put into modeling the complete engine working cycle, including the complete gas flow phenomena in intake and exhaust systems, using computers. Although not yet completely satisfactory, fair agreement now is possible between the predicted and measured pressures within the cylinder and for the pressure fluctuations in the intake and exhaust systems.

3.2.3.3.5 Flow Density

Volumetric efficiency is used to describe the air capacity. Volumetric efficiency (η_v) is a measure of the effectiveness of the induction process and is defined as the rate of mass (M) of air inhaled per cylinder per cycle over mass (M_0) of air to occupy the swept volume per cylinder at ambient pressure (p) and temperature (T). Volumetric efficiency depends on the density of the chargers at the end of the induction process, which depends on the temperature and pressure of the charge. The higher the pressure and the lower the temperature, the higher the power output. Volumetric efficiency for a naturally aspirated engine can be greater than 0.9 but usually is less than 1. It could exceed 1 for a tuned induction system such as a supercharged engine.

The flow mass is directly proportional to the density of the flow fluids. For engines with an exhaust gas recirculation (EGR) device, the recirculated gas must be cooled before entering the cylinder to increase the density and mass of the gas, thus maximizing combustion efficiency.

3.2.3.3.6 Valve Timing Factor

Valve timing can significantly affect flow and volumetric efficiency, as discussed in Chapter 2. However, in practice, many factors limit an engine in achieving its maximum volumetric efficiency. For small compression ignition engines, the piston-to-head clearance is small at firing top dead center (TDC). For this reason, the small high-speed diesel cannot use as much valve overlap in either period or area as is conventional for gasoline engines. As a result, the intake valve cannot be opened earlier than approximately 12° before TDC and the exhaust valve closed later than 12° after TDC without the risk of the valves hitting the flat-topped pistons. The machining of cavities in the piston crown to accommodate the valve heads with wider timings distorts the combustion chamber too much to be acceptable in small engines. In small high-speed automotive engines, the selection of intake closing timing is a compromise between making as much use of inertia ram in the intake system as possible at top speeds and maintaining a high effective compression ratio at low starting speeds to assist cold starting. In general, the intake valve closing will occur at approximately 40–45° after bottom dead center (after BDC) for this type of engine. In large slow-speed engines, the intake closing tends to occur earlier at approximately 35° after BDC.

Large direct injection engines use a larger-diameter shallower combustion chamber cavity in the piston crown, partly because lower compression ratios are adequate for cold starting and partly because they use a lower swirl ratio in conjunction with a larger number of fuel injection sprays to obtain satisfactory combustion. With large turbocharged engines, wide overlap periods of up to 150° crank are used to pass scavenge air through the cylinder, giving some cooling to the cylinder head and valves. Therefore, the mean operating temperature of the turbocharger is kept down to reduce thermal stresses and creep in the turbine rotor. The greater the boost used, the more important this becomes.

The exhaust valve opening ideally should occur at a timing such that atmospheric pressure is reached within the cylinder at an equal number of degrees after BDC as the exhaust opening occurred before BDC. If this is done, the work lost on the expansion stroke and the negative work done during the early part of the exhaust stroke are minimized. In practice, the exhaust opening for the small high-speed engine usually is 55–45° before BDC.

Similar timings for the exhaust opening usually are adopted for engines that are moderately turbocharged. The actual timings often are modified to obtain the best effect from the exhaust pulses used to improve the exhaust turbine work. As the boost pressure is increased, more work is required for the exhaust turbine of the turbocharger. The tendency is to open the exhaust valve earlier to increase the pressure drop across the exhaust turbine. Although some engine cylinder expansion work is sacrificed, the higher boost pressure obtainable as a consequence benefits the overall engine output.

Cams have been designed in which the exhaust valve almost closes at exhaust TDC and then partly reopens again before finally closing, in order to minimize the machining of the piston crown to avoid valve/piston interference. With this approach, the intake valve is opened early, partly closed until the piston starts to descend, and then finally lifted to its normal full lift.

Cam ramps are used to ensure that valve opening and closing takes place at a known, usually constant, velocity. They allow some margin for wear, for wind-up in the valvetrain due to its flexibility, and for changing clearances due to relative expansions or contractions during transient speed and load changes. Ramp heights should not be excessive to avoid possible valve burning at the exhaust opening and excessive gas leakage at the intake valve closure during cold starting.

3.2.4 Valve Materials

3.2.4.1 Valve Alloy Chemistry

Valves are manufactured from iron-, nickel-, and cobalt-based metallic alloys that often are welded together in various combinations. Usually, martensitic steels are used for intake valves, and austenitic alloys and superalloys are used for exhaust valves. Special iron-, nickel-, and cobalt-based alloys are welded to many of the valve head alloys to improve seat face wear and corrosion resistance. Table 3.4 lists the nominal chemistries of the major valve alloys. Table 3.5 lists applications for the listed materials.

3.2.4.2 Valve Alloy Microstructure

The structure of a material determines its properties. Microstructure is the arrangement in a material of grains (polycrystalline materials) and discrete regions having different atomic and defect structures; regions with different atomic structures typically are called phases (multiphase materials). Often, multiphase materials have phases with different chemical compositions as well as different atomic structures. The microstructure is classified by its morphology: the size, shape, amount, and distribution of the discrete regions. Table 3.6 summarizes important metallurgical phases or microconstituents, their crystalline structures, and their characteristics.

3.2.4.2.1 Intake Valve Alloys

Intake valve alloys range in composition from plain carbon steels to high-alloy stainless steels. In some applications, exhaust valve materials also are being used for intake valve applications. Because of the low operating temperatures of intake valves, especially in automotive applications, the low-carbon materials have adequate strength and oxidation resistance to satisfy most applications. With increasing temperatures and load, oxidation and/or creep may become a problem. Under these conditions, the high-alloy carbon steels or austenitic material or even nickel-based alloy become necessary. The following are typical intake valve alloys:

- SAE 1547 is a medium carbon or carbon-manganese steel that can be cold extruded and warm forged. It is suitable for passenger car intake valves.

TABLE 3.4
VALVE ALLOY CHEMISTRY
(MEAN WEIGHT PERCENTAGE [%])

SAE	Commercial	C	Mn	Si	Cr	Ni	Co	W	Fe	Others
Intake Valve—Martensitic Alloys										
NV 1	1541H	0.41	1.50	0.25	–	–	–	–	Bal.	
NV 2	1547	0.47	1.50	0.25	–	–	–	–	Bal.	
NV 4	3140	0.40	0.80	0.25	0.65	1.25	–	–	Bal.	
NV 5	8645	0.45	0.90	0.25	0.50	0.55	–	–	Bal.	Mo:0.20
NV 6	5150H	0.50	0.80	0.25	0.80	–	–	–	Bal.	
NV 7	4140H	0.40	0.90	0.25	1.0	–	–	–	Bal.	Mo:0.20
HNV 3	Sil 1	0.45	<0.8	3.0	9.0	–	–	–	Bal.	
HNV 6	Sil XB	0.85	<0.8	2.2	20.0	–	–	–	Bal.	
HNV 8	422 SS	0.22	<1.0	<0.75	11.8	0.75	1.0		Bal.	Mo:1.0, V:0.23, Cu<0.5
	Alloy C	0.85	<1.5	<1.0	17.5	<0.5	–	–	Bal.	Mo:2.25, V:0.45, Cu<0.22
	SUH 3	0.40	<0.6	2.2	10	<0.6	–	–	Bal.	Mo:1.0
	SUH 11M	0.51	<0.6	1.5	8.5	<0.6	–	–	Bal.	
Exhaust Valve—Fe-Based Austenitic Alloys										
EV-4	21-12N	0.2	1.25	1.0	21.0	11.5	–	–	Bal.	N:0.20
EV-8	21-4N	0.53	9.0	<0.25	21.0	4.0	–	–	Bal.	N:0.43
EV-12	21-2N	0.55	8.5	<0.25	21.0	2.1	–	–	Bal.	N:0.30
EV-16	23-8N	0.33	2.5	<0.75	23.0	8.0	–	<0.5	Bal.	N:0.30, Mo<0.5,
XEV-F	Mod. 21-4N	0.50	9.0	<0.45	21.0	4.5	–	2.2	Bal.	N:0.50, Nb+Ta:2.20
	SUH 35	0.53	9.0	<0.35	21.0	4.0	–	–	Bal.	N:0.43
	Ni30	0.06	0.5	0.5	14.0	30.0	–	–	Bal.	Ti:2.6, Al:1.9
Exhaust Valve—Nickel-Based Superalloys										
HEV-3	Inconel 751	0.07	<0.5	<0.5	15.5	Bal.	<1.0	–	7.0	Ti:2.3, Al:1.2, Nb+Ta:1.0, Mo<0.5
HEV-5	Nimonic 80A	0.07	<1.0	<1.0	19.5	Bal.	<2.0	–	<3.0	Ti:2.3, Al:1.4, Zr:0.07, Cu<0.2
HEV-6	Nimonic 90	0.10	–	–	19.5	Bal.	18.0	–	1.5	Ti:2.4, Al:1.4, Mo:1.0
	Wasploy	0.07	<1.0	<0.75	19.5	Bal.	13.5	–	<2.0	Ti:3, Al:1.4, Mo<4.5, Cu<0.5
	Pyromet 31V	0.05	<0.2	<0.2	22.5	Bal.	<1.0	–	17.0	Ti:2.3, Al:1.3, Nb:0.9, Mo:2.0
Valve Seat Hardfacing Materials										
VF 2	Stellite 6	1.2	<1.0	<1.5	29	<5.0	Bal.	4.5	<6.0	Mo<1.5
VF 3	Eatonite	2.4	<0.5	0.75	29	Bal.	10.0	15.0	<8.0	
VF 4	X-782	2.0	<1.0	<0.5	26	Bal.	<0.5	9.0	<4.0	
VF 5	Stellite F	1.8	<1.0	1.2	25	22	Bal.	12.0	<6.0	Mo<1.0
VF 6	Stellite 1	2.5	<1.0	1.2	30	<3.0	Bal.	12.5	<6.0	Mo<1.0
VF 7	Stellite 12	1.4	<1.0	1.2	30	<3.0	Bal.	8.25	<3.0	Mo<1.0
VF 8	Tribaloy 400	<0.08	–	2.6	9.0	–	Bal.	–		Fe+Ni<3.0, Mo: 30
VF 9	Eatonite 3	2.0	<1.0	1.0	29	Bal.	–	–	4.5	Mo:5.0
VF 10	Eatonite 5	2.0	<1.0	1.0	29	Bal.	–	–	4.5	Mo:8.0
VF 11	VMS 585	2.3	–	1.1	24	11	–	–	Bal.	Mo:5.5
	Tribaloy 800	<0.08	–	3.4	18	<1.5	Bal.	–	<1.5	Mo:29
	Eatonite 6	1.8	<1.0	1.3	28	17	–	–	Bal.	Mo:4.5

TABLE 3.5
VALVE MATERIAL APPLICATIONS

Materials	Light-Duty Gasoline		Heavy-Duty Gasoline		Light-Duty Diesel		Heavy-Duty Diesel		Natural Gas	
	Intake	Exhaust	Intake	Exhaust	Intake	Exhaust	Intake	Exhaust	Intake	Exhaust
1541H	√		■		■					
1547	√		√		√					
3140										
8645			■							
5150H	■									
4140H	■									
Sil 1			■		■		√		√	
Sil XB			■							
422 SS							■			
Alloy C							■			
SUH 3	■						■			
SUH 11M							■			
21-12N						■		■		
21-4N		■		√						
21-2N		√		√	■	√				√
23-8N					■	√	■	■		
Mod. 21-4N				■				■		
SUH 35		■						■		
Ni30		■		■						
Inconel 751				■				■		
Nimonic 80A				■		■				
Nimonic 90				■				■		
Wasploy				■				■		
Pyromet 31V				■				■		
Stellite 6				√				√	√	√
Eatonite								■		
X-782				■						
Stellite F				√		√				
Stellite 1				√						
Stellite 12							■	■		
Tribaloy 400								■		
Eatonite 3								■		
Eatonite 5								■		
VMS 585		■		√	■		■	■		
Tribaloy 800							■	■		
Eatonite 6					■	■	■	■	■	■

√ = Used frequently

■ = Used

TABLE 3.6
VALVE ALLOY MICROSTRUCTURE AND
ITS CHARACTERISTICS

Phase (Microconstituent)	Crystal Structure of Phase	Characteristics
Ferrite (α-Iron)	bcc	Relatively soft low-temperature phase; stable equilibrium phase.
δ-ferrite (δ-Iron)	bcc	Isomorphous with α-iron; high-temperature phase; stable equilibrium phase.
Austenite (γ-Iron)	fcc	Relatively soft medium-temperature phase; stable equilibrium phase.
Cementite (Fe_3C)	Complex orthorhombic	Hard metastable phase.
Graphite	Hexagonal	Stable equilibrium phase.
Pearlite	bcc and orthorhombic	Metastable microconstituent; lamellar mixture of ferrite and cementite.
Martensite	bct (supersaturated solution of carbon in ferrite)	Hard metastable phase; lathe morphology when <0.6 wt%C; plate morphology when >1.0 wt%C and mixture of those between them.
Bainite	bct (supersaturated solution of carbon in ferrite)	Hard metastable microconstituent; nonlamellar mixture of ferrite and cementite on an extremely fine scale; upper bainite formed at higher temperatures has a feathery appearance. The hardness of bainite increases with decreasing temperature of formation.

Notes:
    bcc = body-centered cubic
    bct = body-centered tetragonal
    fcc = face-centered cubic

- SAE 8645 is a medium carbon and low-alloy steel (i.e., chromium-molybdenum-manganese-carbon steel). This alloy is used extensively as the stem material for welded, two-piece exhaust valves.

- Sil 1 is a medium carbon, chromium silicon steel that has excellent elevated-temperature air oxidation resistance and good strength to 500–600°C. Sil 1 is the primary alloy used for medium- to heavy-duty intake valve applications. SUH 3 and SUH 11M have similar chemistries, microstructure characteristics, and mechanical properties.

- SAE 4140H is also a medium-carbon low-alloy steel, that is, a nickel-manganese-chromium-carbon steel used as a tip material for exhaust valve applications that require a higher-hardness wear surface. This alloy can be resistant welded and induction hardened to provide a higher-hardness wear surface on the tip of austenitic iron-based and nickel-based exhaust valves.

- Sil XB is a high-carbon, chromium silicon-nickel steel that has good elevated-temperature air oxidation resistance and strength. This alloy is used for medium- to heavy-duty intake valve applications.

- 422 SS is a martensitic stainless steel that has good elevated temperature air oxidation resistance and strength. This alloy is used for medium- to heavy-duty intake valve applications.

3.2.4.2.2 Exhaust Valve Alloys—Austenitic Steels

Austenite does not exist at room temperature in plain carbon and low-alloy steels, except as small amounts of retained austenite that did not transform during rapid cooling. However, in certain high-alloy steels, such as the precipitation-hardening austenitic stainless steels, austenite is the microstructure. In these steels, sufficient quantities of alloying elements that stabilize austenite at room temperature are present (e.g., manganese [Mn] and nickel [Ni]). The crystal structure of austenite is face-centered cubic (fcc) as compared to ferrite, which has a body-centered cubic (bcc) lattice. An fcc alloy has certain desirable characteristics; for example, it has superior low-temperature toughness, excellent weldability, and corrosion resistance and is nonmagnetic. Disadvantages are its expense (because of the alloying elements), its susceptibility to stress-corrosion cracking (certain austenitic steels), its relatively low yield strength, and the fact that it primarily can be strengthened only by cold working, interstitial solid-solution strengthening, or precipitation hardening.

The precipitation-hardened stainless steels are by far the most widely used alloys for exhaust valves. These alloys derive their corrosion resistance from the combined effects of chromium, nickel, and manganese additions. High strengths at elevated temperatures are obtained from the high level of interstitial carbon and nitrogen. These alloys are most often used after the solution has been treated and aged. However, in many applications, these alloys can be used in the aged-only condition. These alloys also are weldable, which permits valve design modification to meet specific demands.

Austenite, similar to ferrite, can be strengthened by interstitial elements such as carbon and nitrogen. However, carbon usually is excluded because of the deleterious effect associated with the precipitation of chromium carbides on austenite grain boundaries (a process called sensitization). These chromium carbides deplete the grain-boundary regions of chromium, and the denuded boundaries are extremely susceptible to corrosion. Such steels can be desensitized by heating to high temperatures to dissolve the carbides and placing the chromium back into solution in the austenite. On the other hand, nitrogen is soluble in austenite and is added for strengthening. To prevent nitrogen from forming deleterious nitrides, manganese is added to lower the activity of the nitrogen in the austenite, as well as to stabilize the austenite.

The following are typical exhaust valve alloys:

- 21-2N is the most widely used precipitation-hardening stainless steel for exhaust valves. Approximately 80% of the North American market in light- to medium-duty engine applications is satisfied by this material. The alloy represents a near-optimum balance in performance and economics. 21-2N maintains a cost advantage over other candidate alloys, such as 21-4N, 23-8N, and Inconel 751, while providing good resistance to lead oxide and sulfidation corrosion and good elevated-temperature strength.

- 21-4N or SUH 35 is used in light-, medium-, and heavy-duty engine applications. Similar to 21-2N, 21-4N also is an iron-based austenitic precipitation-hardening stainless steel strengthened by carbon and nitrogen interstitial alloying elements. However, 21-4N is slightly superior in corrosion and elevated temperature strength properties. 21-4N is considered the premium iron-based valve alloy in leaded fuel applications.

- 21-4N-Nb-W, a modification to 21-4N, also is an iron-based austenitic precipitation-hardening stainless steel strengthened by carbon and nitrogen interstitial alloying elements. With additional tungsten (W), niobium (Nb), and tantalum (Ta), its elevated temperature strength properties are further improved. It is used primarily in light-, medium-, and heavy-duty engine applications.

- 23-8N also is an iron-based austenitic precipitation-hardening stainless steel designed for use in unleaded gasoline and diesel exhaust valve applications. This alloy possesses good high-temperature strength and provides good resistance to oxidation and sulfidation attack. Although not recommended for lead oxide environments, 23-8N is a suitable diesel exhaust valve alloy.

- Ni 30 is a precipitation-hardenable, iron-based hybrid austenitic superalloy. There are coarse gamma-prime phases or γ' precipitated after high-temperature exposure due to titanium and aluminum elements. This alloy provides better mechanical properties, including high-temperature fatigue strength, than iron-based austenitic alloys listed in this group. This alloy is used for automotive exhaust valve applications when iron-based austenitic alloys cannot meet the application requirements, but the application does not warrant the use of superalloys.

3.2.4.2.3 Exhaust Valve Alloys—Superalloys

The nickel-based superalloys are a class of specialty alloys that are employed in exhaust valve applications when the performance requirements cannot be met by the standard precipitation-hardening stainless steels. Their fatigue strength, oxidation, and leaded-fuel corrosion resistance are far superior to the precipitation-hardening stainless steels. But in high-temperature reducing environments, their sulfidation corrosion resistance is relatively poor. The microstructure of super-alloys consists of fine Ni_3Al particles dispersed in an fcc nickel-rich solid solution. These alloys are remarkably resistant to particle coarsening, owing to a low matrix-particle surface energy. Solid-solution hardening also is used to strengthen the matrix, and inert particles (e.g., carbides) may be dispersed on the material grain boundaries.

Inconel 751 is an age-hardenable nickel-based superalloy used extensively in heavy-duty exhaust valve applications requiring fatigue strength and corrosion resistance that are superior to austenitic exhaust alloys. This alloy is recommended in leaded fuel environments where 21-4N and 21-2N have failed because of lead oxide corrosion and lower fatigue strength. Heat treatment for Inconel 751 is solution treating and then aging. In many cases, double or triple aging is applied.

Nimonic 80A is a nickel-based superalloy similar to Inconel 751 in composition and performance. The higher chromium in this alloy may result in some improved performance over Inconel 751 in rich fuel applications. The alloy enjoys more extensive use in Europe. In the United States, it is used primarily for aircraft and tank valves.

Nimonic 90 is a cobalt-bearing, nickel-based superalloy that possesses excellent high-temperature strength. This alloy has seen limited applications in military tank valves.

Pyromet 31 is a nickel-based superalloy that basically is equivalent to Inconel 751 except that it is higher in chromium. This increased amount of chromium could improve the sulfidation resistance of the alloy in some exhaust valve applications.

Wasploy is a nickel-based superalloy that basically is equivalent to Nimonic 90 with high cobalt and chromium.

3.2.4.2.4 Hardfacing Alloys

Valve seat hardfacing alloys generally possess a combination of properties that make them suitable for maintaining a seal between the valve face and seat insert, preventing the occurrence of indentation, wear, and burning that might occur with an unprotected valve seat. These properties include high hot hardness, wear resistance, and corrosion resistance at operating temperatures.

Hardfacing alloys consist of a mixture of carbides or intermetallic compounds in a soft matrix. As a first approximation, the hard carbide or intermetallic compound particles provide the resistance to wear, and the tough matrix serves the role of binding the relatively brittle carbides. Both the matrix and hard phases must contain sufficient chromium to make the alloy corrosion and oxidation resistant. Because the matrix phase loses its strength more rapidly with an increase in temperature, it must be solid solution strengthened for elevated temperature service (hence, the tungsten and/or molybdenum additions).

The hardfacing alloys can be classified into four main groups: (1) the cobalt-based Co-Cr-C-W alloys (Stellite); (2) the cobalt-based Co-Mo-Cr alloys (Tribaloy) containing a closed-packed intermetallic compound (Laves phase); (3) the nickel-based Ni-C-W or Mo or Co alloys; and (4) the iron-based Fe-Cr-Ni-C-Mo alloys. These alloys provide a full range of hardness and corrosion resistance against essentially all corrosive environments. Iron- and nickel-based alloys have been introduced as cost-effective substitutes in selected applications. The following are typical alloys for valve seat hardfacing:

- Stellite 6 alloy is the most widely used cobalt-based hardfacing alloy for imparting abrasion, impact, and corrosion resistance to edges of earth-moving equipment, pump shafts, and cutting tools, as well as providing high-temperature properties suitable for valve facings mainly in diesel engines. With the low carbon (1.2%) and tungsten content, the alloy forms a hypo-eutectic dendritic structure in a matrix of chromium carbide ($Cr_{23}C_6$) or network-type carbides, and cobalt solid solution, on solidification during welding, which determines its properties. Its structure and properties are not affected by subsequent heat treatments. The alloy is weldable on all valve materials using both oxyacetylene and plasma-transferred arc (PTA) hardfacing processes.

- Stellite 1 alloy has higher carbon content (2.4%) than Stellite 6, resulting in a hyper-eutectic structure dispersed with chromium and tungsten carbide (M_7C_3) particles, which are hard and could grow to a large size, thus resulting in higher hardness.

- Stellite F also is a cobalt-based alloy that has a high nickel content to stabilize the austenitic structure and to prevent phase changes during engine operation; therefore, it is more suitable for exhaust applications.

- Eatonite 6 alloy is an iron-based alloy with Cr/Mo carbide dispersed in an Fe-Cr-Ni dendrite matrix. The microstructure for this alloy consists of iron-based dendrites and interdendrite lamella of chromium-rich carbides and iron phase that are similar in volume fraction to Stellite 6.

- VMS 585 also is an iron-based high-carbon austenitic stainless steel with Cr/Mo carbide dispersed in an Fe-Cr-Ni dendrite matrix.

- Eatonite 5 is a nickel-based alloy with Cr-Mo carbide dispersed in the Ni-Cr matrix. The microstructure of the nickel-based hardfacing consists of a high volume fraction of interdendrite carbides.

- Tribaloy (T 400 and T 800) is a cobalt-based Laves-type alloy. In this material, molybdenum and silicon are added at levels in excess of their solubility limit, with the objective of inducing the precipitation of the hard (and corrosion-resistant) Laves phase (an intermetallic compound $(Co, Mo)_2Si$). Carbon is held as low as possible in this alloy to discourage carbide formation. The microstructure of alloy T 400 includes the primary Laves phase, in a matrix consisting of areas of lamellar eutectic mixture plus solid solution. The differences between T 400 and T 800 are that T 400 has a finer eutectic mixture, a smaller volume fraction of primary Laves phase, and a smaller proportion of Laves-free cobalt solid solution. The Laves phase containing intermetallic types of alloys is unique, in that the Laves phase is thermally stable up to 790°C. Therefore, it retains its hardness at high temperatures. This behavior makes the Laves phase alloys possess even better wear resistance at high temperatures than the carbide type of Stellite-type alloys. The high Mo content also makes them highly corrosion resistant in seawaters and certain reducing acids.

3.2.4.3 Valve Alloy Heat Treatment

Performance characteristics of a valve alloy rely on the alloy microstructure, a cross product of composition and heat treatment. Alloy selection depends on stresses, corrosive agents, and temperatures encountered in service, as well as the economic and durability objectives. Heat treatments used to improve the mechanical properties of valves depend on the specific alloy, economics, and the level of properties desired. They can be general or selective, and it is not uncommon for a single valve to be subjected to two or more heat treatments in different locations.

3.2.4.3.1 Harden and Temper—Martensitic Alloys

Plain carbon and low-alloy and high-alloy martensitic steels (such as SAE 1547, SAE 8645, and Sil 1) are used primarily for intake valves. Extreme-duty martensitic steels generally have the highest carbon and alloy content to resist wear, to resist seat face indentation by deposits, and to provide increased strength. Elements such as chromium and silicon are added when increased oxidation or corrosion resistance is needed. Manganese and nickel are added as strengthening agents. Occasionally, refractory elements, such as molybdenum, tungsten, and vanadium, are used to enhance certain elevated-temperature properties.

Martensitic valves are most often quench hardened after being heated to more than the austenite temperature and tempered to hardness readings in the 25–45 HRc scale range. Hardening or quenching refers to the process of rapidly cooling metal parts from the austenitizing or solution-treating temperature, typically from within the range of 815–1050°C for steels. Stainless and high-alloy steels may be quenched to minimize the presence of grain boundary carbides or to improve the ferrite distribution. However, most steels (including carbon, low-alloy, and tool steels) are quenched to produce controlled amounts of martensite in the microstructure. Successful hardening usually means achieving the required microstructure, hardness, strength, or toughness while minimizing residual stress, distortion, and the possibility of cracking. Fundamentally, the objective of the quenching process is to cool steel from the austenitizing temperature quickly enough to form the desired microstructural phases, sometimes bainite but more often martensite. Quenching effectiveness depends on the steel composition, the type of quenchant or the quenchant use conditions, and the size of the parts being quenched. Although a more severe quench produces martensite to a greater depth (with a steel of given hardenability), it also increases the likelihood of distortion and cracking. Distortion is a result of warping, thermally induced deformation, and martensite formation. Warping is the result of non-uniform heating or non-uniform support of a part during heating. Thermal deformation is the result of non-uniform contraction during cooling. The expansion associated with martensite formation also induces stresses that cause distortion.

Tempering of steel is a process in which previously hardened steel usually is heated to a temperature below the lower critical temperature and then is cooled at a suitable rate, primarily to increase ductility and toughness, to relieve quenching stresses, and to ensure dimensional stability. The principal variables associated with tempering that affect the microstructure and the mechanical properties of a tempered steel include the tempering temperature, the time at temperature, the cooling rate from the tempering temperature, and the composition of the steel. In a steel quenched to a microstructure consisting essentially of martensite, the iron lattice is strained by the carbon atoms, producing the high hardness of quenched steels. Upon heating, the carbon atoms diffuse and react in a series of distinct steps that eventually form Fe_3C or an alloy carbide in a ferrite matrix of gradually decreasing stress level. During tempering, martensite decomposes into a mixture of ferrite and cementite, with a resultant decrease in volume as tempering temperature increases. The properties of the tempered steel are determined primarily by the size, shape, composition, and distribution of the carbides that form, with a relatively minor contribution from solid-solution hardening of the ferrite. These changes in microstructure usually decrease hardness, tensile strength, and yield strength but increase ductility and toughness. Typical tempering is done at temperatures of 175–705°C and for times ranging from thirty minutes to four hours.

Figure 3.20 shows the tempering curves for three martensitic valve alloys. Specimens are tempered for two hours at the various temperatures after being fully hardened. This is a compromise among good strength, adequate ductility, impact performance, and wear resistance along the valve stem. In some less-demanding applications, martensitic valves can be used in the annealed condition.

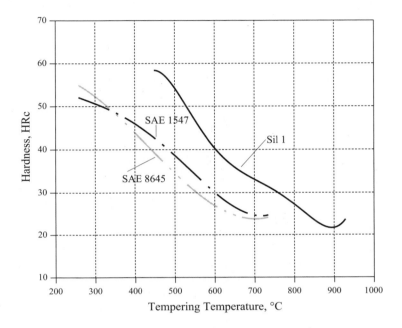

Figure 3.20 *Room-temperature hardness after tempering for two hours of fully hardened martensitic valve materials.*

Valve tips and seats often are hardened selectively to create high-hardness and wear-resistant surfaces. These surfaces are hardened selectively, generally to the greatest hardness practical for the alloy. Tip hardening may extend beyond the keeper groove to improve the fatigue strength and the wear resistance of the region.

3.2.4.3.2 Solution Treat and Age—Austenitic Alloys

Austenitic valve alloys are most often solution treated and aged to hardness readings in the Rockwell "C" scale range of 25–40 HRc. The selection of austenitic valve alloys is influenced strongly by economic considerations, as well as the mechanical, physical, and chemical attributes required to satisfy specific engine requirements. Mechanical properties are improved in austenitic valve alloys by precipitation hardening or strain hardening. The most common austenitic alloy hardening processes are high-temperature forging followed by aging heat treatment, high-temperature forging followed by a solution treatment and then aging heat treatment, or cold forming.

In lower-temperature applications, austenitic engine valves frequently are used in the forged and aged condition. More severe and higher-temperature service generally requires solution treatment followed by aging treatment. These operations produce hardness readings in the Rockwell "C" scale range of 20–40 HRc. The hardness developed depends on the capability of the individual alloy. These alloys develop fatigue, creep, wear, and seat face indentation resistance from the heat treatments.

Austenitic alloys have a face-centered cubic crystal structure termed austenite. The elements that stabilize the austenitic structure are carbon, manganese, nickel, copper, and nitrogen. Chromium, silicon, and sometimes aluminum are added for oxidation or corrosion resistance. Refractory elements such as molybdenum, niobium, tantalum, tungsten, and vanadium may be added for high-temperature strength. Typical austenitic valve alloys such as 21-4N, 21-2N, or 23-8N are solution treated at approximately 1050–1200°C for thirty to ninety minutes until carbides and nitrides are dissolved in solution. Then they are quenched into cold water to acquire a hardness of 22–35 HRc. Subsequently, age hardening for these alloys is performed above 700–850°C for one to sixteen hours for improved hardness, reduced residual stress, and dimensional stability. If the aging temperature is too high or the aging time is too long, this could lead to overaging, which would result in larger grain size and lower hardness.

Figure 3.21 illustrates the hardness-age response curve for austenitic valve materials. It shows different characteristics of hardness response to aging time, depending on pre-existing conditions or treatment. Different materials may show different characteristics, too.

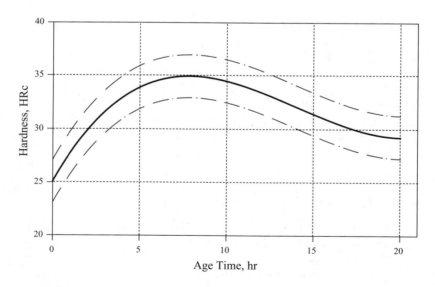

Figure 3.21 *Typical effects of aging on austenitic valve steels.*

3.2.4.3.3 Solution Treat and Age—Superalloys

For higher-stress and higher-temperature applications, a number of nickel-based superalloys are commonly used, such as Inconel 751, Nimonic 80A, Nimonic 90, and Pyromet 31. Iron-based austenitic valve steels are hardened by carbonitride precipitation. Nickel-based superalloys are hardened by precipitation of aluminum, nickel, niobium, tantalum, and titanium in the form of intermetallic compounds called gamma-prime phase or γ' consisting of $Ni_3(Al, Ti)$, precipitates. The aging characteristics of nickel-based superalloys are similar to those of austenitic steels, as shown in Figure 3.21.

Typical nickel-based valve alloys such as Inconel 751, Nimonic 80A, Nimonic 90, and Pyromet 31 are solution treated at approximately 1040–1150°C for fifteen to ninety minutes until intermetallic compounds are dissolved in solution. Then the alloys are air cooled to acquire a hardness of 22–35 HRc. Subsequently, age hardening for these alloys is performed above 700–850°C for one to four hours for improved hardness, reduced residual stress, and dimensional stability. A second aging sometimes is performed to further improve the hardness, and a third aging occasionally is applied. However, if the aging temperature is too high or the aging time is too long, this could lead to overaging that would result in a coarse γ' phase, which will deteriorate strength. Depending on the type of alloy, pre-existing condition, or treatment, different characteristics of hardness response to aging time may exist.

Typical heat treatments for various valve alloys are summarized in Table 3.7, and heat treatment parameters such as temperature, time, and cooling medium depend on the hardness/strength requirement.

TABLE 3.7
TYPICAL HEAT TREATMENT PARAMETERS
FOR VARIOUS VALVE ALLOYS

	Harden	Temper	Solution Treat	1st Aging	2nd Aging
Martensitic Steels	830–1050°C 15–30 min. Oil or air quench	175–750°C 30–240 min. Air cool	–	–	–
Austenitic Steels	–	–	1050–1200°C 30–90 min. Water quench	700–850°C 1–16 hours Air cool	–
Superalloys	–	–	1040–1150°C 15–90 min. Air quench	700–870°C 1–4 hours Air cool	650–750°C 1–4 hours Air cool

Notes:
- Tempering temperature and time depend on the hardness requirement. High tempering temperature and long tempering time usually result in low hardness.
- Aging temperature and time also depend on the hardness requirement. Aging usually increases hardness due to the precipitation hardening process.
- Nickel-based superalloys sometimes adopt a third aging process to optimize the alloy hardness.
- Occasionally, austenitic steel valves are used in the as-forged condition. These valves are being "aged" during engine operation.

3.2.4.4 Valve Alloy Properties

The properties of valve alloys depend on the alloy chemistry, processing, and heat treatment during manufacturing. The property data shown in the following are from various origins (Caird and Trela, Campo *et al.*, Giles, Goth, Hagiwara *et al.*, Jenkins and Larson, Jones, Kattus, Larson *et al.*, Narasimhan and Larson, Newton and Allen, SAE J775, Sato *et al.*, Schaefer *et al.*, TRW, Tunnecliffe and Jenkins, Umino *et al.*, and Wu *et al.*); therefore, their accuracy cannot be presumed. The data should be used only as a reference. Valve manufacturers generally can provide expected properties with specific chemistry and heat treatment when a design envelope is established.

3.2.4.4.1 Physical Properties

The critical physical properties of valve alloys are the coefficient of thermal expansion and thermal conductivity. Figures 3.22 and 3.23 show typical values for the coefficient of thermal expansion and thermal conductivity, respectively, as a function of temperature.

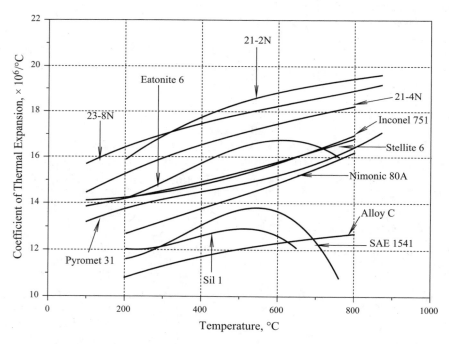

Figure 3.22 *Typical coefficient of thermal expansion for valve alloys.*

Austenitic alloys show higher coefficients of thermal expansion than nickel-based superalloys and iron-based martensitic steels. Both the coefficient of thermal expansion and thermal conductivity increase as the temperature increases for all valve alloys except SAE 1541H; the

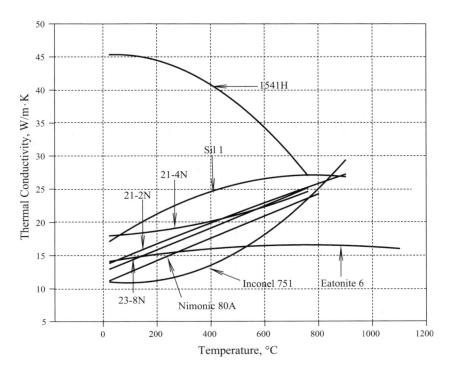

Figure 3.23 *Thermal conductivity of valve alloys.*

thermal conductivity of low-carbon steel SAE 1541H decreases as the temperature increases. The coefficient of thermal expansion of martensitic alloys, including SAE 1541H, Sil 1, and Eatonite 6, increases as the temperature increases to a point and then decreases as the temperature increases further.

3.2.4.4.2 Mechanical Properties

The crucial mechanical properties of valve alloys include tensile strength, yield strength, elongation, hot hardness, and hot fatigue at various temperatures.

Figures 3.24 and 3.25 show the ultimate tensile strength and yield strength, respectively, as a function of temperature for various valve alloys. At low temperatures (<350°C), low-alloy carbon steels have equivalent or better ultimate tensile and yield strength than high-alloy austenitic alloys. However, at higher temperatures (>500°C), austenitic alloys show higher tensile and yield strength than the martensitic steels. Nickel-based superalloys show consistently higher tensile and yield strength and less strength deteriorating as the temperature increases. Pyromet 31 stands out in ultimate tensile strength and yield strength among all valve alloys, including other nickel-based superalloys such as Inconel 751.

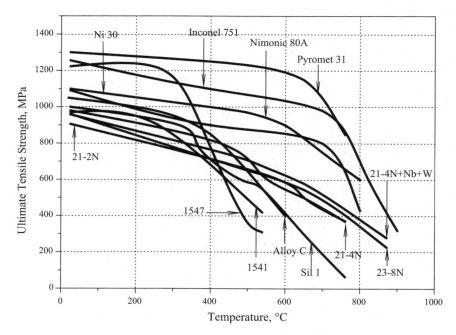

Figure 3.24 *Ultimate tensile strength at elevated temperatures for valve alloys.*

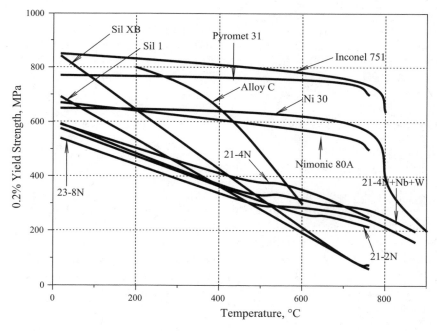

Figure 3.25 *Yield strength of valve alloys at elevated temperatures.*

Figures 3.26 and 3.27 show elongation and area reduction properties for various valve alloys at elevated temperatures. In general, fcc-structured materials such as all austenitic alloys and superalloys have much high plasticity than martensitic steels. Also, in general, plastic deformation reflected in elongation and area reduction increases as temperature increases. However, some abnormalities for individual alloys are caused by phase transformation, grain size growth, and dissolved carbide, which all affect the dislocation movement and influence plastic deformation. Note that both Nimonic 80A and Inconel 751 alloys have significant dips in elongation at 800°. Sil 1 also has a dip at 300°C in elongation, and austenitic alloys have similar dips as well at various temperatures. Although both elongation and area reduction measure the plasticity of a material, there is little correlation between the two sets of data shown in Figures 3.26 and 27.

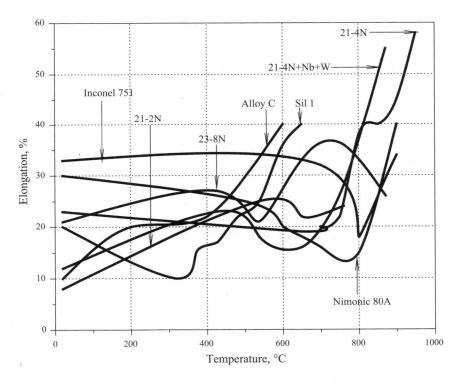

Figure 3.26 *Elongation of valve alloys at elevated temperatures.*

Hot hardness data are used most often to predict wear resistance of valve materials at elevated temperatures as well as at room temperature because interface temperatures can be very high due to friction, even at low operating temperatures. The compatibility of the pairing materials, however, can influence the wear outcome. Under no circumstances should hot hardness data substitute for wear test data in choosing the material for the wear resistance application, despite the fact that wear data are difficult to generate and are inconsistent because no standard test methodologies exist. Hardness is a measurement of only plastic deformation, and if the predominant

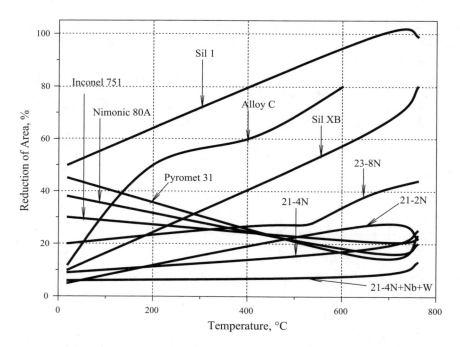

Figure 3.27 *Reduction in area of valve alloys at elevated temperatures.*

wear mechanism is a deformation-controlled one, the hot hardness data will be indicative of the wear resistance of the material. However, other types of wear mechanisms usually are involved in the tribological system. Therefore, hardness data alone are not enough to determine the valve material selection for wear resistance. Figures 3.28 and 29 show hot hardness data for various valve alloys and seat hardfacing alloys. The trend is that hardness decreases as temperature increases for all valve and hardfacing alloys. Martensitic valve alloys have higher hardness at low temperatures (<500°C) than do austenitic alloys and superalloys. However, superalloys have higher hardness than both martensitic and austenitic alloys at high temperatures (>550°C).

For hardfacing alloys, cobalt-based Tribaloy 400 has the highest hot hardness, followed by Stellite 1, the nickel-based alloy Eatonite, and iron-based Eatonite 6. The remainder of the hardfacing alloys shown in Figure 3.29 have close hot hardness characteristics.

3.2.4.4.3 Hot Fatigue Strength

Despite the varying conditions encountered in different internal combustion engines, hot fatigue strength can be defined singly as the most critical property for valve alloys among all properties. Hot fatigue strength data and corrosion-accelerated fatigue strength data of valve alloys are limited because of the lengthy testing required to generate meaningful data. The situation is improving, and more data are being generated and published in the literature. In lieu of these data, estimates of fatigue resistance also are established on the basis of elevated-temperature tensile, creep, and

Valvetrain Components

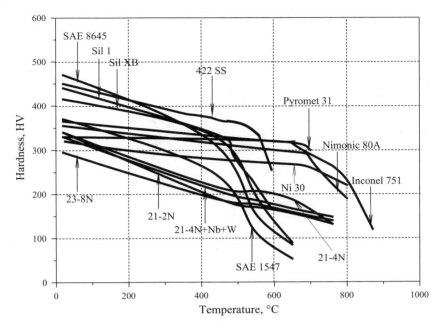

Figure 3.28 *Hot hardness of valve alloys.*

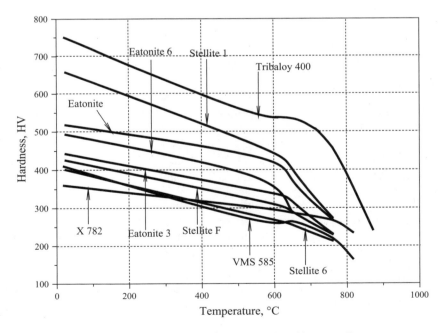

Figure 3.29 *Hot hardness of valve hardfacing alloys.*

stress rupture performances. Figure 3.30 shows the stress versus the number of cycles to failure (S-N) curves for some of the intake martensitic alloys, including Sil 1, Alloy C, SUH 11, and SAE 1547 tested at the 500°C typical intake valve application limit.

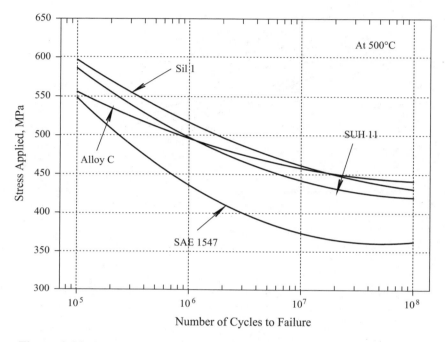

Figure 3.30 *S-N curves for representative intake valve materials at 500°C.*

Figure 3.31 shows S-N curves for the nickel-based superalloys and iron-based austenitic alloys at 800°C at the typical exhaust valve application limit. Nimonic 80A and Inconel 751 show the highest fatigue strength at this temperature, followed by Pyromet 31 and Ni 30, and then followed by the iron-based austenitic alloys 21-4N+W+Nb, 23-8N, 21-4N (SUH 35), and 21-2N. Again, different heat treatments may alter the results slightly.

Figure 3.32 shows the hot fatigue strength or limit for various valve alloys at various temperatures suspended at 100 million cycles. Fatigue strength is paramount in selecting the valve alloys. Figure 3.33 shows the effect of a typical safety factor on the failure frequency for reference. In valve design, the fatigue safety factor usually is chosen between 1.5 and 2. However, the challenge is to accurately estimate the stresses, considering that the valve is not seated properly under the worst scenario, that a corrosive environment results in reduced load bearing area, and that valvetrain dynamics may be poor.

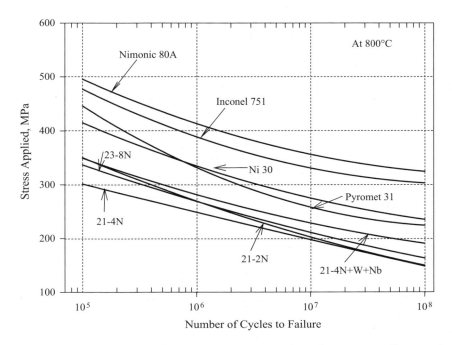

Figure 3.31 *S-N curves for representative iron-based austenitic alloys and nickel-based superalloys at 800°C.*

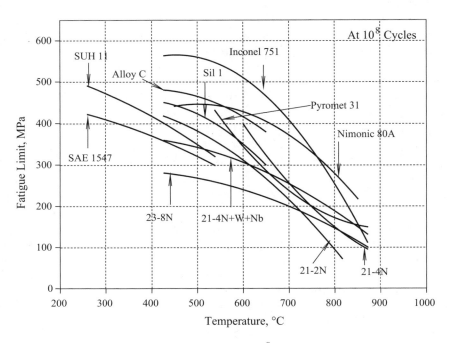

Figure 3.32 *Valve alloy fatigue strength at 10^8 cycles at various temperatures.*

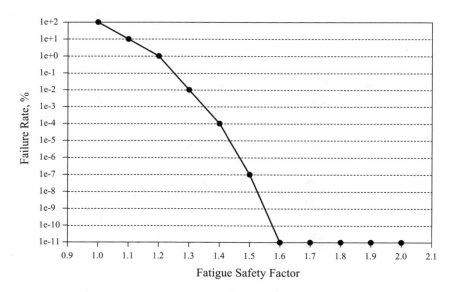

Figure 3.33 *Failure rate versus fatigue safety factor.*

3.2.4.4.4 Corrosion Resistance

Exhaust valve corrosion and burning have constituted a restriction to the durability of many internal combustion engines. The corrosive compounds formed by the combustion of gasoline and diesel fuels are complex. Due to the nature of higher operating temperatures in gasoline engines, oxidation is the primary corrosion form for gasoline engines, whereas diesel engine valves encounter more sulfidation corrosion due to higher sulfur in diesel fuel.

One method to obtain oxidation-resistant valve alloys is the selective oxidation approach, which consists of oxidizing essentially only one element in the valve alloy and relying on the oxide of this element for protection. For this approach to be feasible, the oxide that is formed must cover completely the surface of the alloy, and it must be an oxide through which diffusion of the oxygen takes place at comparatively slow rates. The elements having oxides that are sufficiently protective and that have affinities for oxygen sufficient for selective oxidation are aluminum, chromium, and silicon. It is necessary to emphasize that selective oxidation processes are affected by a number of factors that include alloy compositions, alloy surface conditions, gas environment, and cracking of the oxide scale. In alloy oxidation, the selective formation of specific oxide phases is governed primarily by the preferential reaction of those elements that form the most thermodynamically stable compounds. However, thermodynamics does not specify the scale-subscale morphologies, which are established principally by the availability of alloying elements and various environmental parameters such as concentrations, mobilities, and concentration gradients of the point defects in the oxides formed. Furthermore, thermodynamics will not predict the rate of the reactions; hence, some thermodynamically stable oxides may not exist due to the limiting rates of formation.

The oxidation resistance of valve alloys usually increases with chromium concentrations. The "critical" minimum chromium content to ensure the formation of a "protective" Cr_2O_3 scale is approximately 20–25% chromium. Although the protective Cr_2O_3 scale is growing, depletion of chromium occurs in the matrix. The danger of chromium depletion is the susceptibility of this diluted region to internal oxidation and NiO formation if a scale rupture should occur.

Figures 3.34 and 3.35 show oxidation resistance for various valve alloys at elevated temperatures. Austenitic valve steels are good only below 750°C, and nickel-based superalloys must be deployed when the valve operating temperature exceeds 800°C. In general, the higher the reactive element additions (aluminum, silicon, and rare earths) and the lower the refractory metal (Mo, W, Ta, Cb) content, the greater the oxidation resistance. Nickel-based superalloys have superior oxidation resistance, followed by austenitic alloys. Carbon steel, such as SAE 3140, has the lowest oxidation resistance among valve alloys.

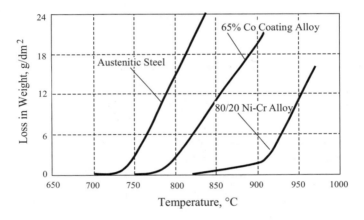

Figure 3.34 *Comparison of lead oxidation resistance of alloys at various temperatures (Mogford and Ball).*

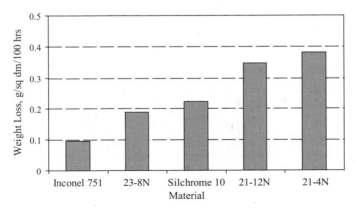

Figure 3.35 *Comparison of oxidation resistance of alloys at 816°C (Jenkins and Larson).*

Figure 3.36 compares the general valve alloy oxidation and corrosion resistance.

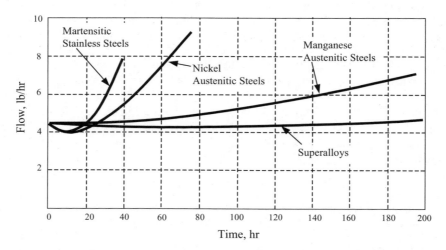

Figure 3.36 *Comparison of valve alloy corrosion resistance*
(Tunnecliffe and Jenkins).

Figure 3.37 shows the valve alloy sulfidation corrosion resistance ranking in the order of 21-12N, Tribaloy 400, 21-4N, Stellite 6, Eatonite 6, 23-8N, 21-4N+W+Nb, 21-2N, Eatonite 4, Pyromet 31, and Inconel 751. The tests were performed at 870°C for 80 hours by merging

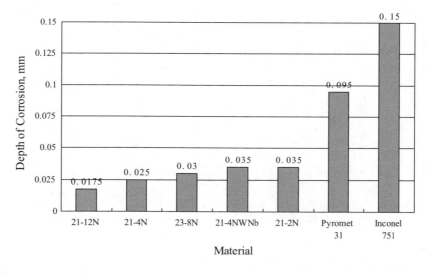

Figure 3.37 *Comparison of sulfidation corrosion resistance of valve alloys*
at 870°C for 80 hours (Ari-Gur et al.).

specimens in MgO crucibles containing a salt mixture of $10CaSO_4$, $6BaSO_4$, $2Na_2SO_4$, and 1 carbon. The drilled-hole method was used in the test, in which a hole 12.7 mm deep by 6.35 mm in diameter was drilled into the specimen, and salt mixture that had been ground into powder was filled into the hole.

The results indicate that nickel-based valve alloys have poorer sulfidation resistance than iron-based alloys, and nickel-based seat facings accelerate the corrosive attack on the base material in the faced region, compared to the unfaced portion of the valve. Nickel is important in the preservation of the oxide layer and protective Cr_2O_3 to less protective spinels. Nickel also restricts the spalling of the oxide layer because it reduces thermal stresses. The major drawback of nickel is its higher sulfur reactivity and sulfur diffusivity relative to iron and cobalt. Therefore, sulfur penetrates nickel-based alloys rapidly. Valves with cobalt- and iron-based seat facings exhibit similar good resistance, whereas nickel-based alloys shows poorly. The results correlated with the known high diffusion coefficient of sulfur into nickel-based alloys that allowed excessive penetration and damage.

Results indicate that for increased resistance of sulfidation attack, high levels of chromium and low levels of nickel are beneficial. The same statement may be true with the vanadate deposits. It has been shown that the hot corrosion resistance of nickel-based alloys generally is improved by the addition of chromium. Although it has been argued that the chromium oxide and not the chromium *per se* affords the sulfidation protection, a high chromium content generally favors the formation of the protective chromium oxide scale. It also can be argued that the chromium in iron-based alloys is needed for similar reasons. Nickel retards the transformation of protective chromium oxide scales to less protective spinels. Nickel also increases the spalling resistance of the oxide scale by reducing the differential thermal expansion between the alloy and oxide, thereby reducing the stresses at the alloy/matrix interface during cooling.

In summary, hot corrosion, including lead corrosion and sulfidation, is becoming less of an issue because of regulations that mandate the elimination or reduction of lead and sulfur in fuel. In a hot salt sulfidation environment, the austenitic steels are superior to the nickel-based alloys. In PbO environments, the reverse is true; the nickel-based alloys offer superior corrosion protection. The nickel-based alloys also have superior oxidation resistance in an air environment without corrodents. The problem of achieving and maintaining resistance to degradation induced by such environments is more difficult when the gas does not contain oxygen. Oxide phases generally are more suitable as barriers to exhibit high-temperature corrosion because oxide scales are more amendable to formation via selective oxidation. Likewise, oxide scales are more effective diffusion barriers to separate the alloy from the environment, and such scales can be made more adherent to the alloy substrates.

3.2.4.4.5 Wear Resistance

The action between two surfaces in contact is a difficult problem involving the statistical relationships among many interrelated variables. Among the parameters are the surface roughness, waviness, and stresses; prior history of machining; moduli of elasticity; friction (rolling and sliding); materials; lubrication; corrosion; and dynamic loads. The following categorization of wear adequately describes valvetrain wear:

1. **Removal of metal.** This is caused by (a) abrasives in the lubricant; (b) corrosion; (c) tearing away of high surface points; and (d) fatigue, which is commonly called pitting or spalling.

2. **Transfer of metal between surfaces, commonly called adhesion.** This occurs with sliding and produces fusion of contacting micropoints under high temperature, which sometimes also is called welding, scoring, galling, and scuffing.

3. **Displacement of metal by plastic flow, superficial wear, or smearing.** This occurs in all metals in contact under high pressure and often is called shear strain, radial flow, and plastic deformation.

In a valvetrain system, all three forms of wear may occur at the same time. Historically, valve wear has been assessed by the hardness of the alloy, facing, or coating. The harder the material, the more wear resistant, as reflected in Archard's equation

$$V = k\frac{SL}{3H} \tag{3.8}$$

where

V = wear volume
k = wear coefficient
S = sliding distance
L = normal load
H = hardness

However, many factors, including the environment, friction characteristics, and wear mechanisms, also can drastically change valve wear characteristics. Most importantly, wear on the valve seat, stem, or tip indisputably depends on the valve mating materials of insert, the guide and rocker arm, and so forth. Material that works well on the valve tip may not survive on the stem and seat. Coating material on the valve stems may not work on the seat and tip, and seat hardfacing material may not be applicable on the stem. The most important but least understood factor in all wear study is the effects of dissimilar metals on each other. Tests show that some combinations are compatible for long wear life, whereas others are poor.

The valve seat is subject to high impact and combustion stress during seating and combustion. The distance slid on the valve seat surface may be small; however, due to high contact stress and the toleration of seat recession, the challenge in selecting the compatible valve and insert material will always be there for valvetrain engineers. The implication of valve seat wear is that when it exceeds the preset lash, the valve will be held open and will not be able to seal the combustion chamber. The engine can lose its combustion pressure and thus its power. Furthermore, gas leakage could lead to valve burning or guttering. Several dominant wear mechanisms may be involved in any given application. The primary valve seat wear mechanisms are oxidation/corrosion, adhesion, and abrasion/shear strain. Figure 3.38 shows exhaust valve seat wear resistance ranking using a specially designed and built rig against a Sil XB insert at 538°C valve seat temperature.

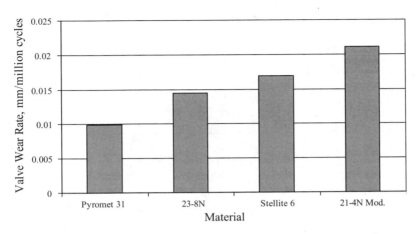

Figure 3.38 *Wear resistance ranking exhaust valve materials*
at 538°C (Zhao et al., 1997).

It cannot be emphasized enough that the wear resistance of one material is dependent on the compatibility with the counterpart material. As illustrated in Figure 3.39, a Sil 1 valve has the least valve wear or is ranked as the most resistant material when run against PL 7 among tests against other insert alloys. However, the total wear (valve wear plus insert wear) of the combination of the Sil 1 valve and PL 7 insert does not fare as well because of significant wear on the insert. Hardfacing, induction hardening, and salt bath nitriding have commonly been used to combat valve seat recession. Hardfacing materials include cobalt-based alloys such as Stellite 6, Stellite 1, Stellite F, and Tribaloy 400; nickel-based alloys such as Eatonite 2, Eatonite 3, and Eatonite 5; and iron-based alloys such as Eatonite 6. Cobalt-based alloys are preferred as hardfacing material in demanding applications. However, iron-based alloys can be as effective as cobalt- and nickel-based alloys if the mating insert is chosen properly.

Two criteria typically are used in evaluating stem and guide compatibility or scuffing characteristics: (1) the amount of wear from the stem and guide surface, and (2) the seizure characteristic. Both wear and seizure features are a function of the coefficient of friction; the higher coefficient of friction usually results in higher wear and earlier seizure. However, no direct correlation has been established in the literature. Typical valve stem and guide wear mechanisms are scuffing and adhesion under high-speed reciprocating movement that generates high-friction temperatures. Valve stems currently are either chrome plated or salt bath nitrided. However, due to environmental concerns related to chrome plating and salt bath nitriding, significant efforts have been made in the past in searching for alternative stem coatings. Thin coatings such as TiN, CrN, WC/C, diamond-like carbon (DLC) coating, and plasma ion nitriding all show promise in wear performance, in comparison with chrome plating and nitriding. However, one obstacle for these thin coatings to reach the market is the cost associated with the vacuum batch process.

Tip materials usually are hardenable martensitic steels. Sometimes, a piece of wafer can be welded onto the tip for wear resistance. Typical wafer materials are SAE 8645, and high-alloyed steels such as Sil 1 or 52100 sometimes are used. The predominant tip wear mechanism is pitting

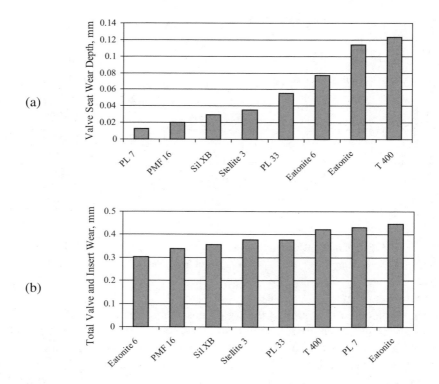

Figure 3.39 *Intake valve Sil 1 against various insert materials at 510°C: (a) valve seat wear ranking, and (b) total recession (valve wear plus insert wear) (Wang et al.).*

or contact fatigue due to a repeated high-contact stress under a boundary lubrication condition. This continual alternate flexing exceeds the fatigue limit of the material, so that the high spot flakes out, leaving a pit behind it. Sometimes, oil under high hydrostatic pressure is trapped in this hole, furthering the pit and forming a crack. Rough surfaces, high sliding velocities, corrosion, grinding cracks, thin layers of case hardening, high dynamic loads, and surface cracks from heat treatment accelerate this action. For applications where tip fatigue-pitting poses a challenge, the selection of harder material may not be the best solution because harder material usually results in higher contact stresses. Better lubrication for reduced friction and alternative design for less stress may provide the solution to the issue of tip pitting.

3.3 Cams

The function of a cam is to drive the valvetrain to open, and then to close the valve as rapidly as possible, with an overall valve event of predetermined duration. A precise and carefully controlled profile for the cam is of vital importance to the successful performance of the engine. The objective of the cam design is to obtain the most effective valve lift diagram possible with a given valvetrain system, considering the practical limits imposed on the design. These limits include the valvetrain dynamics or resonant frequency, the maximum allowable loads and stresses permissible in the valvetrain, and the practical cam manufacturing accuracy in production.

3.3.1 Cam Nomenclature and Design Considerations

The major parts of a camshaft include the quill, bearing journals, thrust faces, and cam lobe (Figure 3.40). The camshaft must be sufficiently stiff to minimize the deflection that affects the valvetrain dynamic characteristics. The quill size is dependent on the loads that the camshaft must support, which consist of (1) the sum of the individual loads (static and dynamic) of each of the camshaft lobes, and (2) the load transmitted by the camshaft drive due to the crankshaft torsions. The quill diameter must be less than the cam lobe base circle diameter to permit machining of the lobe and adequate lobe hardening.

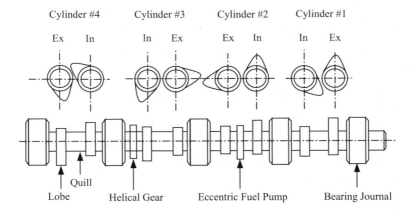

Figure 3.40 *Cam nomenclature and relative lobe position.*

The number of journals is determined by the configuration of the cylinder block or cylinder head. However, long spans should be avoided. Each lobe should be positioned at an equal distance from the journals for equal stiffness. The diameter of the journal must be greater than the lobe base circle diameter plus twice the maximum lift to facilitate assembly (i.e., so that it will go through the camshaft journal bearings). When bearing caps are used, the journal diameter can be reduced. The bearing journal width should be such that it prevents the edge of the journal from operating within the cam bore bearing surface, which will prevent the journal from digging into the bearing and blocking lubrication.

The cam lobe must open and close the valve properly, and the actual profile is composed of many parts, as shown in Figure 3.41:

- **Ramp**. The portion of the cam lobe event from zero lift (base circle) to the defined opening or closing point.

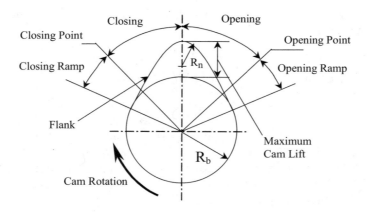

Figure 3.41 *Cam lobe nomenclature.*

- **Base circle radius (R_b).** The portion of the cam contour at zero lift (valve in the closed position).

- **Flank**. The portion of the cam lobe between the ramp and the nose radius.

- **Nose radius (R_n).** The instantaneous radius at the maximum lift point of the nose of the cam, which is tangent to the flank radii and the maximum lift point.

- **Inflection point**. The lift value on the lift curve where the acceleration instantaneously changes from positive to negative, or vice versa.

3.3.2 Cam Profile Characteristics

Three basic characteristics govern mixture flow, timing, and overall engine volumetric efficiency. These three characteristics involve the design of the cam lobe (profile) and commonly are referred to as lift, duration, and overlap. Each has a vital effect on the incoming fuel charge, as well as disposing of the exhaust gases after combustion has occurred. Therefore, precise cam profile design improves engine performance, fuel economy, and emissions.

The motion of a cam-operated valve can be described as follows. First, the valve is lifted from its seat and is accelerated to maximum velocity, at which point the cam follower contact is approximately halfway through the flank of the cam. From this point, the valve is decelerated from maximum to zero velocity at the top of the lobe (maximum lift point), with the acceleration reaching its maximum negative value. The decelerating force is provided by the valve spring, which opposes the movement of the valve in its opening motion. At the beginning of the closing motion from the top of the lobe, the valve commences on its return path under the accelerating force of the spring and reaches its maximum closing velocity. At this point, the valve is decelerated by the cam and finally is brought to rest.

3.3.2.1 Valve Lift Characteristics

The distance the valve travels off its seat is called valve lift. Figure 3.42(a) shows the valve lift curve as a function of cam rotation angle, and it normally is expressed in millimeters (mm). If the system is direct acting, the valve lift is the same as the cam lift. If there is a rocker arm, the valve lift equals the cam lift multiplied by the instantaneous rocker arm ratio. The valve lift is given by the mathematical relationship

$$y = f(\theta) \tag{3.9a}$$

where θ is the cam angle rotation in radians. However, because the cam rotates at a constant angular velocity, the displacement is

$$y = g(t) \quad \text{and} \quad \theta = \omega t \tag{3.9b}$$

where t is the time for the cam to rotate through angle θ (in seconds), and ω is the cam angular velocity in radians per second (rad/sec).

Equation 3.8a often is preferred because it is simpler to analyze and use.

3.3.2.2 Valve Lift Velocity

The rate of valve lift is the speed with which the valve opens. Velocity is the first derivative of the lift contour with respect to the cam angle (Figure 3.42(b)). Therefore, velocity is the slope of the displacement curve at angle θ or time t, in millimeters per second (mm/sec), and can be calculated from the lift or the acceleration values

$$v = \frac{dy}{dt} \tag{3.10a}$$

The cam profile usually is given as a function of angle θ instead of time. Therefore, the velocity with respect to time is

$$v = \frac{dy}{dt} = \frac{dy}{d\theta}\frac{d\theta}{dt} = \omega\frac{dy}{d\theta} \tag{3.10b}$$

The velocity of the opening ramp should remain constant, so that minute deflections in the valvetrain system (clearances, or slack, and manufacturing deviations) at the point of juncture at the base circle and the ramp can be taken up before the load (actual lift) is applied to the system. The velocity at the end of the closing side must be low to prevent a "hammering" noise as the valve approaches its seat and the system releases its flexibility. The velocity at the point of maximum lift must be zero; if it is not, the tappet will tend to lift off the cam contour and result in a hammering noise. It is possible to have many camshafts with the same net lift; however, with different opening contours, some will open faster and provide a higher volumetric efficiency than others.

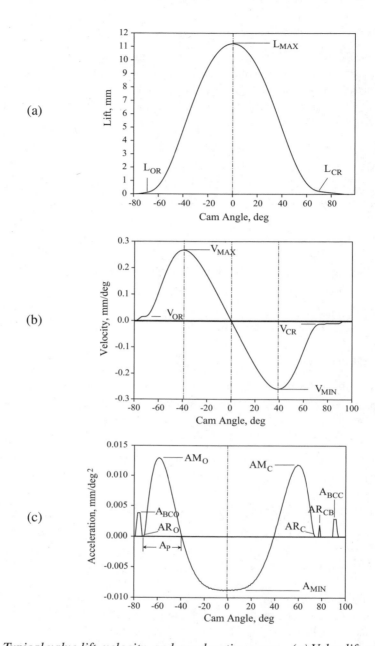

Figure 3.42 *Typical valve lift, velocity, and acceleration curve. (a) Valve lift curve, where L_{OR} is the lift at the opening ramp, L_{MAX} is the maximum lift, and L_{CR} is the lift at the closing ramp. (b) Velocity curve, where V_{OR} is the velocity at the opening ramp, V_{MAX} is the maximum velocity at the opening side, V_{MIN} is the minimum velocity at the closing side, and V_{CR} is the velocity at the closing ramp. (c) Acceleration curve, where AM_O is the maximum acceleration at the opening side, AM_C is the maximum acceleration at the closing side, A_{MIN} is the minimum acceleration, AR_O is the acceleration at the opening ramp, AR_C is the acceleration at the closing ramp, and A_P is the positive acceleration pulse width.*

3.3.2.3 Acceleration

Acceleration is the second derivative of the lift contour or the first derivative of the velocity with respect to the cam angle θ, or time t,

$$a = \frac{d^2y}{dt^2} = \frac{dv}{dt} \tag{3.11}$$

Acceleration is the slope of the velocity curve at angle θ mm/deg^2 or time t, mm/sec^2. Figure 3.42(c) shows the valve acceleration curve.

The shape and values of the acceleration curves are of critical concern for engine designers. From that information, analysis can be made for the shock, noise, wear, vibration, and general performance of a valvetrain system. For best action, the acceleration curve shall be smooth and have the smallest maximum values possible.

3.3.2.4 Effect on Pulse

Pulse, or jerk, is the instantaneous time rate of change of acceleration or the slope of the acceleration curve at angle θ, mm/deg^3 or time t, mm/sec^3,

$$p = \frac{d^3y}{dt^3} = \frac{da}{dt} \tag{3.12}$$

For high-speed actions, the maximum values of the pulse should not be too large. Vibrations then will be kept to a minimum.

3.3.3 Cam Profile Design

The fundamentals of cam profile design are discussed extensively by Turkish (1946), Rothbart, and others. The cam design starts from a specified valve event with an engine design speed based on the resonant frequency characteristics of the valvetrain system. The engine breathing is optimized by maximizing the area under the valve-lift diagram, with the valve lift determined from the maximum allowable force-stress limits of the system.

The basic curves of the valve displacement or lift curve are primarily of two families: (1) the simple polynomial, and (2) the trigonometric. The trigonometric curves are superior to the polynomial curves and give smoother action, easier layout, lower manufacturing cost, and less vibration, wear, stresses, noise, and torque, in addition to smaller cams. However, to provide an understanding of the basics of cam profiles, both types are discussed in the following paragraphs.

3.3.3.1 Simple Polynomial Curves

The displacement equations for simple polynomial curves are of the form

$$y = C\theta^n \tag{3.13}$$

where

    n = any number
    C = a constant

In this polynomial family, we have the following curves with integer powers:

- Straight line or constant velocity, n = 1
- Parabolic or constant acceleration, n = 2
- Cubic or constant pulse, n = 3

3.3.3.1.1 Straight-Line or Constant Velocity Curves

This curve of the polynomial family (n = 1) is the most simple. It has a straight-line displacement curve at a constant slope (Figure 3.43), giving the smallest length for a given rise of all the basic curves. We see that the valve lift is uniform, the velocity is constant, and the acceleration is zero during the rise. However, at the beginning and end of the valve lift, there is an impractical condition. That is, from the base circle (zero velocity) to a finite velocity, there is an instantaneous change in velocity, giving a theoretically infinite acceleration. This acceleration transmits a high shock throughout the valvetrain—the magnitude depends on its flexibility. In other words, a "bump" exists in the contour, which the follower could not follow.

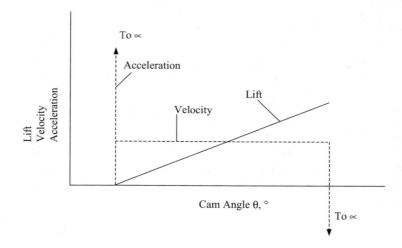

Figure 3.43 *Characteristics of a straight-line curve.*

3.3.3.1.2 Straight-Line Circular Arc Curves

To improve the poor condition of sharp bumps on straight-line cams, the junction at the ends of the valve lift can be smoothed. This often is achieved by employing circular arcs tangent to the straight line at both ends (Figure 3.44). The shorter the radius, the nearer is the approach to the undesirable condition of the straight-line curve. A longer radius produces a more gradual action at the beginning and at the end of the curve. Although such a curve is an improvement over the straight-line curve, it can be applied for low speeds only because large accelerations exist at the beginning and at the end of the stroke.

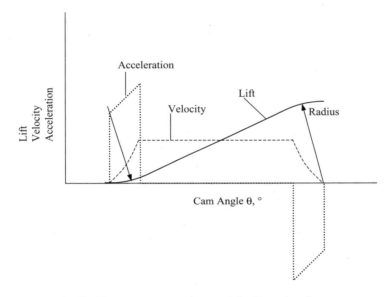

Figure 3.44 *Characteristics of a straight-line circular arc curve.*

3.3.3.1.3 Circular Arc Curves

This curve is composed of two circular arcs tangent to each other, and it is somewhat improved over the infinite acceleration of the straight-line curve. Although its acceleration is finite at all times, the curve gives large follower accelerations and excessive velocities. Therefore, it is used only for low speeds. Figure 3.45 shows the characteristics of the circular arc curve, which are

$$y = H - \left[H^2 - l^2 \right]^{1/2} \qquad (3.14)$$

where

H = radii of circular arcs, mm
R_P = radius of pitch circle, mm
l = developed length of cam for angle θ

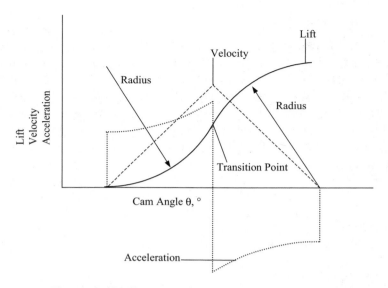

Figure 3.45 *Characteristics of a circular arc curve.*

For radial cams,

$$1 = R_p\theta$$

Therefore,

$$y = H - \left[H^2 - \left(R_p\theta\right)^2\right]^{1/2} \tag{3.15}$$

Differentiating gives velocity and acceleration

$$v = \frac{R_p^2\theta}{\left[H^2 - \left(R_p\theta\right)^2\right]^{1/2}} \tag{3.16}$$

$$a = \frac{R_p^2\left\{\left[H^2 - \left(R_p\theta\right)^2\right]^{1/2} + R_p\theta^2\right\}}{\left[H^2 - \left(R_p\theta\right)^2\right]^{3/2}} \tag{3.17}$$

3.3.3.1.4 *Parabolic Constant Acceleration Curves*

This curve of the polynomial family has the property of constant positive and negative accelerations (Figure 3.46). No other curves will produce a given motion from rest to rest in a given time with such a small maximum acceleration. With perfectly rigid members having no backlash or clearance in the system, the constant acceleration curve would give excellent performance. However, all members are somewhat elastic, and clearance or backlash always exists. The abrupt change of acceleration of the curve at the lift ends and the transition point produces noise, vibration, and wear, and requires a large spring size. Thus, the parabolic curve should be used only at moderate or low speeds. One reason for the popularity of this curve is the ease of determining the inertia forces, which are proportional to the constant accelerations.

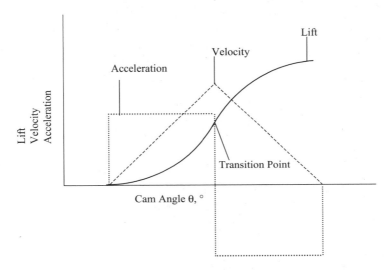

Figure 3.46 *Characteristics of a parabolic curve.*

The general lift equation for a continuous constant acceleration is

$$y = v_0 t + \frac{1}{2} a t^2 \tag{3.18a}$$

where v_0 is the initial velocity of the follower in millimeters per second (mm/sec). Substituting $t = \dfrac{\theta}{\omega}$ gives the displacement

$$y = v_0 \frac{\theta}{\omega} + \frac{1}{2} a \left(\frac{\theta}{\omega} \right)^2 \tag{3.18b}$$

Differentiating to give velocity,

$$v = \frac{dy}{dt} = v_0 + at = v_0 + a\left(\frac{\theta}{\omega}\right)$$
(3.19)

Often, the boundary condition is that at the initial point $\theta = 0$, the initial velocity $v_0 = 0$ gives

$$y = \frac{1}{2}at^2 = \frac{1}{2}a\left(\frac{\theta}{\omega}\right)^2$$
(3.18c)

The characteristics of the curve in terms of the cam angle θ, total rise h, and overall duration angle β are illustrated in Eqs. 3.19 through 3.21 inclusive. All values are to be measured from the initial point, where the cam angle θ is zero. At the transition point, $\theta = \frac{\beta}{2}$ and $y = \frac{h}{2}$.

Before transition, that is, $\theta \leq \frac{\beta}{2}$,

Lift:

$$y = 2h\left(\frac{\theta}{\beta}\right)^2$$
(3.19a)

Velocity:

$$v = \frac{dy}{dt} = \frac{4h\theta\omega}{\beta^2}$$
(3.20a)

Acceleration:

$$a = \frac{dv}{dt} = \frac{4h\omega^2}{\beta^2}$$
(3.21a)

After transition, $\theta > \frac{\beta}{2}$,

Lift:

$$y = h\left[1 - 2\left(1 - \frac{\theta}{\beta}\right)^2\right]$$
(3.19b)

Velocity:

$$v = \frac{dy}{dt} = \frac{4h\omega}{\beta}\left(1 - \frac{\theta}{\omega}\right)$$
(3.20b)

Acceleration:

$$a = \frac{dv}{dt} = -\frac{4h\omega^2}{\beta^2}$$ (3.21b)

3.3.3.1.5 Cubic or Constant Pulse Curves (I)

This curve of the·polynomial family has a triangular acceleration curve. It is a modification of the parabolic curve, eliminating the abrupt change in acceleration at the beginning and the end of the stroke. This has the advantage of reducing the vibration, shock, wear, and noise that occur at these points with the parabolic curve. Nevertheless, the cubic curve has the same poor infinite slope characteristic at the midpoint of the acceleration curve as that of the parabolic curve. In addition, it has the disadvantages of high maximum acceleration and large velocities, necessitating large cams and critical machining. Therefore, this curve is not practical except when combined with others.

The following equations and Figure 3.47 describe the constant pulse or cubic curve.

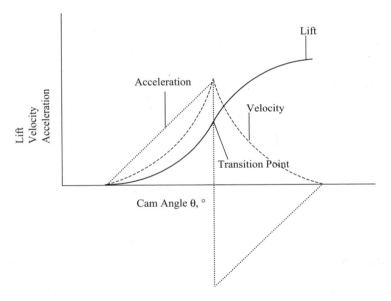

Figure 3.47 *Characteristics of the cubic curve I.*

Before the transition midpoint,

Lift:

$$y = 4h\left(\frac{\theta}{\beta}\right)^3 \qquad (3.22a)$$

Velocity:

$$v = \frac{12h\omega}{\beta}\left(\frac{\theta}{\beta}\right)^2 \qquad (3.23a)$$

Acceleration:

$$a = \frac{24h\omega^2}{\beta^2}\left(\frac{\theta}{\beta}\right) \qquad (3.24a)$$

Pulse:

$$p = \frac{24h\omega^3}{\beta^3} \qquad (3.25a)$$

After the transition midpoint,

Displacement:

$$y = h\left[1 - 4\left(1 - \frac{\theta}{\beta}\right)^3\right] \qquad (3.22b)$$

Velocity:

$$v = \frac{12h\omega}{\beta}\left(1 - \frac{\theta}{\beta}\right)^2 \qquad (3.23b)$$

Acceleration:

$$a = -\frac{24h\omega^2}{\beta^2}\left(1 - \frac{\theta}{\beta}\right) \qquad (3.24b)$$

Pulse:

$$p = \frac{24h\omega^3}{\beta^3} \qquad (3.25b)$$

3.3.3.1.6 Cubic or Constant Pulse Curves (II)

This curve is similar to the constant acceleration curve and the constant pulse curve I. However, it differs in that there is no abrupt change in acceleration at the transition point and that its acceleration is a continuous curve for the complete lift (Figure 3.48). Similar to the constant acceleration curve, it has the disadvantages of abrupt change in acceleration at the beginning and the end of the stroke. This cubic curve has characteristics similar to those of the simple harmonic motion curve. It is not used often but has advantages when used in combination with other curves.

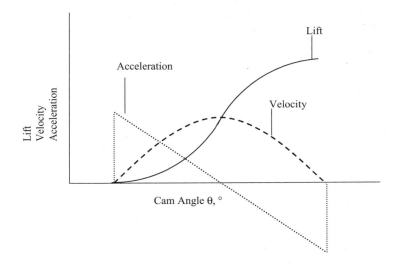

Figure 3.48 *Characteristics of the cubic curve II.*

Lift:

$$y = h\frac{\theta^2}{\beta^2}\left(3 - \frac{2\theta}{\beta}\right) \tag{3.26}$$

Velocity:

$$v = \frac{6h\omega\theta}{\beta^2}\left(1 - \frac{\theta}{\beta}\right) \tag{3.27}$$

Acceleration:

$$a = -\frac{6h\omega^2}{\beta^2}\left(1 - \frac{2\theta}{\beta}\right) \tag{3.28}$$

Pulse:

$$p = -\frac{12h\omega^3}{\beta^3}$$

(3.29)

3.3.3.2 Trigonometric Curves

3.3.3.2.1 Simple Harmonic Motion

This curve of the trigonometric family is one of the most popular curves primarily because of its simplicity in layout and understanding (Figure 3.49). It provides acceptable performance at moderate speeds. The resulting motion of the follower on such a cam is a simple harmonic movement similar to that of a swinging pendulum. The simple harmonic curve is a definite improvement over the previous curves. Shock is reduced so that it is no longer serious at moderate speeds.

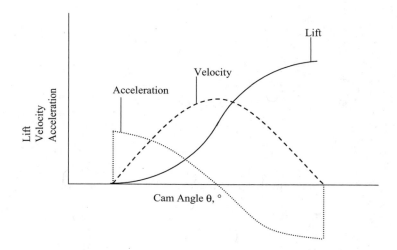

Figure 3.49 *Characteristics of a simple harmonic motion curve.*

The characteristics of the lift curve are a cosine curve plotted from

Lift:

$$y = \frac{h}{2}\left(1 - \cos\frac{\pi\theta}{\beta}\right)$$

(3.30)

Differentiating gives the velocity and acceleration as

Velocity:

$$v = \frac{dy}{dt} = \frac{h\pi\omega}{2\beta} \sin \frac{\pi\theta}{\beta} \qquad (3.31)$$

Acceleration:

$$a = \frac{dv}{dt} = \frac{h}{2} \left(\frac{\pi\omega}{\beta} \right)^2 \cos \frac{\pi\theta}{\beta} \qquad (3.32)$$

The lift and velocity curves are smooth and continuous. However, at the ends of the lift, there is a sudden acceleration and discontinuity in the acceleration curve. This is undesirable for high-speed cams because noise, vibration, and wear can result.

3.3.3.2.2   Sine Acceleration Curves

To overcome the sudden acceleration and discontinuity in the acceleration curve, the sine acceleration curve or cycloidal curve is used for reducing vibration, wear, stress, noise, and shock (Figure 3.50). The equation for lift is

Lift:

$$y = \frac{h}{\pi} \left(\frac{\pi\theta}{\beta} - \frac{1}{2} \sin \frac{2\pi\theta}{\beta} \right) \qquad (3.33)$$

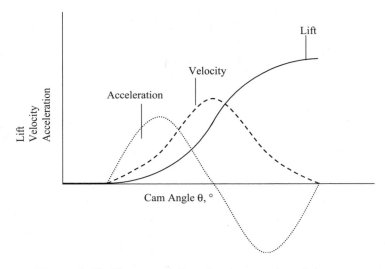

Figure 3.50 *Characteristics of a sine acceleration curve.*

Differentiating gives the velocity and acceleration as

Velocity:

$$v = \frac{dy}{dt} = \frac{h\omega}{\beta} \cos\left(1 - \cos\frac{2\pi\theta}{\beta}\right) \tag{3.34}$$

Acceleration:

$$a = \frac{dv}{dt} = \frac{2h\pi\omega^2}{\beta^2} \sin\frac{2\pi\theta}{\beta} \tag{3.35}$$

Comparing the curves most often used, the trigonometric ones (simple harmonic motion and cycloidal curves) give better overall performance than the basic polynomial family (straight-line, parabolic, and cubic curves). The advantages are smaller cams, lower follower side thrust, lower manufacturing cost, and easier layout and duplication.

In reality, valve profile design including duration (lift length), lift height, and ramp profile may not follow exactly the preceding equations. Several different equations may be required to satisfy the performance, fuel economy, and emissions requirements for ramp, duration, and lift of the cam profile design.

3.3.4 Considerations in Cam Profile Determination

The fundamental law of cam design requires continuity of all cam functions through acceleration. Thus, at minimum, boundary conditions for displacement, velocity, and acceleration are required that match the values of these functions with both adjacent cam segments. Small errors in the cam profile are found to increase considerably the dynamic loads and vibratory amplitudes. When designing cams, the mathematical expression defining the lift curve or the cam lobe profile need not be symmetrical for the opening and closing sides of the cam profile. The closing side is dependent on the required engine performance, specifically durability and noise, vibration, and harshness (NVH), for which the particular cam is being designed. The maximum possible acceleration will be dictated by the valvetrain weight and stiffness. Table 3.8 lists the critical parameters and their values for various types of valvetrain.

The problem of designing a suitable contour can be resolved by choosing a shape that induces the valve and its associated components to move in a specified manner so that the motion is fundamentally unaltered, notwithstanding valvetrain flexibility, throughout the entire speed range of the engine. The acceleration characteristics should be such that a spring of high natural frequency can be used so that the chosen harmonics of the cam motion cannot be excited in the speed range. Valve spring frequency is limited by spring load requirements so that it will not exceed cam contact stress limits. The sequence of the cam profile determination typically is as follows. First, choose the lift curve, then determine the cam base circle radius based on the follower type and size, and then design the ramp profile with acceptable dynamics and seating velocity. Next, design the nose radius with smooth connection with the lift curve and the permissible contact stress, and calculate and tabulate the radial distance to the cam profile versus

TABLE 3.8
CAM DESIGN PARAMETERS

Parameter		Valvetrain Type						Remarks
Symbol	Description	1H	1M	2	3	4	5	(1)
L_m	Max. cam lift	Objective: Establish maximum cam lift area with minimum possible cam lift and event						(2)
L_A	Cam lift area							
Event	Valve open to close							
LR_o	Ramp height open, mm	0.11	0.40	0.065	0.076	0.089	0.10	
LR_c	Ramp height close, mm	0.16	0.42	0.125	0.13	0.14	0.152	
$V_{max}(F)$	Max. cam velocity, mm/deg			NA	NA			(3)
$V_{max}(R)$	Max. cam velocity, mm/deg	NA	NA	0.32	0.19	0.173	0.168	(4)
VR_o	Open ramp velocity, mm/deg	0.014	0.018	0.018	0.011	0.0114	0.0114	
VR_c	Close ramp velocity, mm/deg	0.014	0.016	0.0145	0.01	0.01	0.01	
$Abco$	Open base cir. acc., mm/deg^2	0.010	0.005	0.005	0.0025	0.0025	0.0025	
$Abcc$	Close base cir. acc., mm/deg^2	0.010	0.004	0.0033	0.002	0.002	0.002	
AM_o	Max. acc. open side, mm/deg^2	0.0264	0.028	0.019	0.019	0.01	0.009	
AM_c	Max. acc. close side, mm/deg^2	0.0264	0.027	0.018	0.018	0.01	0.0084	
A_{min}	Min. acc., mm/deg^2	0.0073	0.0067	0.0076	0.0064	0.0064	0.0058	(5)
A_p	Positive acc. open pulse, deg	18	16	22	25	28	28	
L_{am}	Lift @ max. close acc., mm	0.27	0.60	0.48	0.51	0.56	0.58	
VS_{Pres}	Valve spring reserve rate, %	15	15	20	20	25	30	(6)
$RAMP_o$	Open ramp length, deg	9	25	8	8	11	11	
$RAMP_c$	Close ramp length, deg	14	30	13	18	19	22	
Fn	Valvetrain natural freq., Hz	1300+	1300+	1150	1000	830	500	
RBC	Cam base circle rad., mm	18	18	19		16	18	
	Tappet offset, mm	1	1	0	0	1.5	1.5	(7)
	Cam lobe taper, minute	1–4	0	0	0	7–11	7–11	(8)

Notes:

(1)   1M = Type 1 valvetrain with mechanical tappet; 1H = with hydraulic tappet.

(2)   Objective accomplished with highest cam acceleration. Stiffer valvetrain types allow higher acceleration.

(3)   Flat follower.

(4)   Roller follower.

(5)   Type 1 = Function of spring reserve and nose radius for contact stress.

(6)   At maximum operating speed.

(7)   Shown for non-roller cams. Roller cams have zero offset.

(8)   Shown for non-roller cams. Roller cams have zero taper. Hydraulic tappet has a crowned face. Mechanical tappet has a flat top shim.

the cam angle based on the base circle radius, lift, ramp, and nose profiles. Finally, verify the chosen cam profile having optimized velocity and acceleration characteristics.

3.3.4.1 Lift Curve Profiles

At the end of the ramp, the valve is accelerated by a curve of increasing slope. Many successful cams have been designed on the theory of constant acceleration during this period. However, even with a cam designed for constant acceleration, the valve will not follow such a curve because such action calls for the instantaneous transmission of a sudden change in load. Resilience in the valvetrain will modify the valve motion and force it to follow with a sinusoidal acceleration curve. In view of these facts, it would seem logical to design for a sinusoidal acceleration curve in the beginning.

At the end of the acceleration curve, the valve spring must decelerate the valve to zero velocity when the valve reaches maximum lift, and then accelerate it downward, where the load again is transferred to the cam for deceleration to the ramp velocity.

Because the valve-operating mechanism, including its supporting structure, is composed of elastic bodies, the actual motion of the valve will be the designed motion plus the motion that is due to the elasticity of the train. A basic problem in cam and valvetrain design is to achieve an actual valve motion as close to the theoretical rigid-body motion as possible. At low speeds, the valve follows the designed motion closely because the forces of cam acceleration and deceleration are small and are applied over relatively long periods of time. These forces increase as the acceleration increases; therefore, as the acceleration reaches a point, false motion becomes apparent. As the acceleration increases further, the amplitude of this motion may reach a point where the follower leaves the cam at one or more points. Increased spring force may reduce false motion, in particular the seating velocity and the degree of bounce. Thus, high spring force would seem desirable, within the limits imposed by space for the spring, valvetrain friction loss, and wear on the valvetrain parts. Some degree of false motion is always present at high engine speeds. It can be tolerated only to the point where there is danger of failure or excessive wear of the mechanism. The usual limit is set by the seating velocity of the valves—when the valve seating velocity increases to the point where it seats so quickly that it rebounds, or "bounces," sometimes several times. The danger of failure increases with increasing seating velocity, which increases tensile stress in the valve. Obviously, one principal objective of valvetrain design should be to minimize false motion.

Duration, or valve event length (i.e., how long the valve is off the seat), is related to crankshaft degrees (or degrees of crankshaft rotation) and occurs simultaneously with the lifting of the valve. The primary function of valve duration is to allow the incoming gases sufficient time to fill the cylinder with a fresh charge of fuel and air, and more time to discharge the combustion chamber of burned gases.

During high-speed engine operation, the in-rushing air/fuel mixture has a packing force (kinetic energy). If the design takes advantage of this packing force by holding the intake valve open longer (duration), the volumetric efficiency and power output of the engine will improve. This is why high-performance camshafts are designed with a longer duration. This provides more valve off-seat time and improves high-revolutions-per-minute mixture flow by taking advantage of the in-rushing packing force of the air/fuel charge. For good performance, the following cam lifts are recommended in rocker arm follower engines:

- 5.6–5.8 mm for conservative
- 6.1–6.4 mm for average
- 7.1–7.4 mm for high performance

A long valve event is preferred for good dynamics. It allows higher toss speed and results in lower stresses for related components, as well as higher engine output at higher engine speeds. A short valve event is used for better fuel economy, good torque at relatively low engine speeds, and better driveability and idling due to the lower overlap tendency; however, there is a loss of high-end power due to less lift and area.

3.3.4.2 Base Circle and Cam Size

The base circle is the smallest circle drawn to the cam profile from the radial cam center. The minimum cam size basically is affected by three factors: (1) the pressure angle or the steepness of its profile, (2) the curvature or sharpness of the profile, and (3) the size of the camshaft. The minimum size is desirable because of space limitations, unbalance at high speeds, longer paths of follower movement, and correspondingly higher wear. Stresses and deflections are the controlling factors in establishing the size of the camshaft.

For a flat-faced follower as shown in Figure 3.51(a), the radial distance to the cam profile r is

$$r = \left[\left(R_b + y \right)^2 + q^2 \right]^{1/2} \qquad (3.36)$$

where

q = eccentricity of the point of contact from the cam center, $q = \dfrac{v}{\omega} = \dfrac{dy}{dt}\dfrac{1}{\omega}$

R_b = radius of the base circle

r = radial distance to the cam profile

$\tan \eta = \dfrac{q}{R_b + y}$

$\psi_c = \theta + \eta$ (on the rise period of the follower motion shown)

$\psi_c = \theta - \eta$ (on the fall period of the follower motion)

Figure 3.51(b) shows a roller follower without offset, and for every cam angle of rotation θ, there is a displacement y for the follower. This gives the pitch-curve radius

$$r = R_a + y$$

where R_a is the radius of the prime circle. The prime circle is the smallest circle drawn to the pitch curve from the cam center and is similar to the base circle when the centerline of the cam is not aligned with the follower moving direction and there is offset e. The prime cam lobe circle radius (R_a) can be determined when the maximum lift of the follower (L_{MAX}) and the eccentricity of the follower and cam center (e) are known,

$$R_a = \left(y_0^2 + e^2 \right)^{1/2} \qquad (3.37)$$

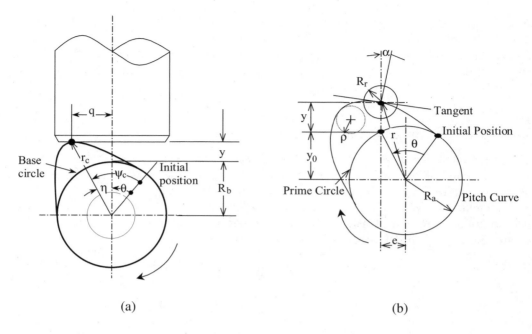

(a)   (b)

Figure 3.51 *Cam profile base circle determination: (a) flat follower,*
and (b) roller follower (Rothbart).

where y is the displacement of the follower and y_0 is the vertical displacement of the prime circle. The radius vector (r) from the center of rotation to the trace point at any angle of rotation is

$$r = \left(R_a^2 + 2y_0^2 2y_0y + y^2 \right)^{1/2} \tag{3.38}$$

Pressure angle (α) is the angle (at any point) between the normal to the pitch curve and the instantaneous direction of the follower motion. This angle is important in cam design because it represents the steepness of the cam profile, which, if too large, can affect the smoothness of the action.

3.3.4.3 Ramp Profiles

Ramps must be provided on the opening side of the cam contours for both the hydraulic lifters and the mechanical tappets. This opening ramp is needed to compensate for base circle runout and small deflections and to take up the slack in the valvetrain mechanism, including the hydraulic lifter, before the load is applied. It also ensures that minute manufacturing errors do not produce excessive accelerations where the base circle joins with the flank of the contour. Inadequate design of cam ramps can exacerbate vibration in the valvetrain system, causing higher levels of spring surge and noise.

Opening ramps required on cam contours for mechanical tappets are in excess of the values given here previously by the amount of clearance to control the impact velocity. This clearance is provided with mechanical tappets to permit thermal expansion of the valve, cylinder block, and head. Constant velocity ramps ahead of the actual lift event are blended into the lift function in the mechanical tappet designs. However, the presence of a hydraulic lash compensator that maintains a zero gap makes the job of the opening ramp different from that required with a mechanical system that does not have automatic lash adjusting. There is no particular need to maintain a constant velocity during the opening ramp because there is no impact event.

Closing ramps must be provided on the closing side of the cam contours to eliminate excessive impact when the valves are being seated. Although some seating impact occasionally is desired to induce valve rotation, excessive impact is detrimental to the life of valvetrain parts. A similar approach puts a constant velocity closing ramp there to control the closing impact velocity. The presence of a hydraulic lash adjuster does not eliminate the impact event completely because there is a need for an aggressive closure of the valve onto the seat to avoid blowby and valve burning. However, the valve closure cannot be so aggressive that it increases valve seat wear or NVH excessively. The closing ramp is always greater than the opening ramp because it is required to compensate for the cocking action of the valve in the guide, to allow for some valvetrain deflection caused by the closing inertia load and the spring load, to allow for hydraulic lifter leak-down when hydraulic lifters are used, and to allow for lash when mechanical tappets are used.

Typical ramp rates at valve closing are 6–19 μm/cam-degree for automotive and high-speed engines (13 maximum recommended for hydraulics), 13–25 μm/cam-degree for truck and low-speed engines, and 25–127 μm/cam-degree for diesel and aircraft engines. Higher and longer ramp results in better dynamics.

3.3.4.4 Cam Curvature

The pressure angle and cam size are directly related, that is, a limiting pressure angle will determine a certain size of cam. However, another condition may exist to preclude the use of a chosen cam size: the curvature of the cam may be too sharp. If this occurs, the follower may not follow the prescribed pitch curve, and the stresses on the cam profile may be prohibitive. The shape of a curve at any point depends on the rate of change of direction, called curvature. At each point of the curve, a tangent circle whose curvature is the same as that of the curve at that point exists. The radius of this circle is called the radius of curvature. Inadequate curvature of the cam profile is a frequent problem in cam design, and it is related to the acceleration of the follower. The minimum curvature or sharpness of a convex cam contour is dependent on the value of the maximum negative acceleration of the follower. Thus, the larger the maximum negative acceleration, the sharper must be the cam surface. Figure 3.52 shows that the positive profile has a convex lobe surface, whereas the negative radius of the curvature lobe has areas of concavity on its surface.

The control of either the curvature or acceleration will determine the limitation of the other, undercutting or jumping being the result of too high a negative acceleration. Undercutting is undesirable because of incorrect follower action, sharp cam profile, and high surface stresses. Therefore, the rule of thumb is that the minimum radius of curvature of the curve ($\rho_{k(min)}$) is

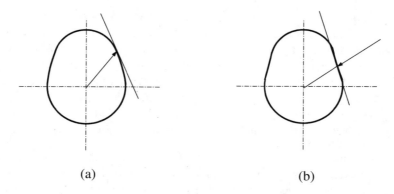

(a) (b)

Figure 3.52 *Negative radius of curvature in cam lobe profiles: (a) positive curvature, and (b) negative curvature.*

greater than the radius of the follower (R_r), or $\rho_{k(min)} > R_r$ (Figure 3.52). Generally, the smallest radius of curvature occurs at the point of maximum negative acceleration. The easiest solution to the problem of undercutting is to use a larger cam.

Because the pressure angle has only a minor effect on the flat-faced follower, a factor limiting the cam size of this follower is undercutting (Figure 3.53). Practically speaking, for flat-faced follower cam construction, there is no limitation on the curvature during the positive acceleration period. This requires that the distance u is positive below the center. The restriction occurs

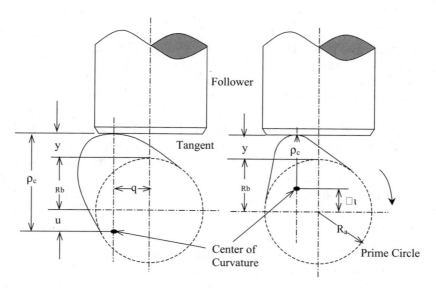

Figure 3.53 *Cam curvature determination.*

when the acceleration is negative, which gives u a minus value. To eliminate sharp corners and undercutting in the cam profile, the minimum radius of curvature shall be

$$\rho_c(\text{min.}) = R_b + (y + u) \underset{\text{max. neg.}}{>} 0 \qquad (3.39)$$

Because R_b and y are always positive, the undercutting ($\rho_c(\text{min.}) < 0$) is due to excessive negative acceleration. If small curvature or undercutting exists, then it is necessary to employ either a larger cam or a smaller maximum negative acceleration.

Often, the radius of curvature must be known to determine the contact stresses and lubricant film thickness between the cam and follower. The contact stress at the maximum stress point can be reduced exponentially by increasing the radius of curvature on the cam. Another way to reduce contact stress is to increase the cam width because contact stress is inversely proportional to cam width. The larger the radius of curvature, the smaller are the surface stresses. Thus, good practice requires that $\rho_{k(\text{min})}$ exceeds R_r by a reasonable margin of safety.

For camshafts with roller followers, high cam acceleration can introduce negative curvatures on the cam profile. To eliminate negative curvatures, the locus of points that result in negative curvature is identified. The maximum cam acceleration is limited to this locus, which prevents negative curvature from occurring.

Many cams now are generated from lift ordinates computed from a mathematical law incorporating the desired characteristics. Therefore, it is necessary to calculate the instantaneous radius of the cam curvature around the profile. At any cam angle, the instantaneous radius of curvature at that angle is given here (Neale, 1994).

For flat followers (tappets),

$$R_c = R_b + y + 3282.8y'' \qquad (3.40)$$

where

R_b = base circle radius, mm
y = cam lift at the desired angle, mm
y'' = cam acceleration at the chosen angle, mm/deg^2
R_c = radius of curvature, mm

For curved followers,

$$R_c = \left\{ \frac{\left[(R_b + R_F + y)^2 + V^2 \right]^{3/2}}{(R_b + R_F + y)^2 + 2V^2 - (R_b + R_F + y)A} \right\} - R_F \qquad (3.41)$$

where

R_F = follower radius, mm
V = follower velocity at the chosen angle, mm/rad, or $57.29 \times$ velocity, mm/deg
A = follower acceleration, mm/rad^2 = $3282.8 \times$ acceleration, mm/deg^2

The value for R_c will be positive for a convex cam flank and negative for a concave flank. It is common to assess cam/tappet designs on the basis of the maximum contact stress between the contacting cam and the tappet, with some consideration of the relative sliding velocity. This requires the determination of the loads acting between the cam and tappet throughout the lift period (at various speeds if the mechanism operates over a speed range), the instantaneous radius of curvature for the cam throughout the lift period, and the cam follower radius of curvature. Figure 3.54 shows the relationships among these various quantities for a typical automotive cam.

3.3.4.5 High-Speed Cam Profiles

In all respects, a valvetrain is an elastic and nonrigid system. The valve timing and event lengths control the fluid flow of gases having inertial and compressible properties. The fixing of these values for a cam design in any given engine system depends on the general subject of gas dynamics. In addition, there is inherent to the design of the engine elastic system the problem of the inertial dynamics of the mechanical components. In particular, consideration of the dynamic deflections in the valvetrain linkage and support points is vital to the design of the cam contour. This fact becomes more important as the valve loads increase with higher design engine speeds, and as the valvetrain stiffness-to-mass ratios decrease with the design effort toward lighter engine weights. The actuating forces in the valvetrain at high engine speeds in combination with inherent valvetrain flexibility result in a valve motion that bears no simple relationship to cam contour, rather a precise and carefully controlled contour given the considerations for the gas dynamics and inertial dynamics in the engine.

In a cam follower system, vibrations of some sort always are induced when high speeds, low stiffness, high mass, or resonance are encountered. To reduce vibrations in the cam follower system, it is suggested that the members from the driver to the follower end be made as rigid as possible.

Increasing the maximum cam acceleration results in an increased area under the lift curve, which improves engine performance characteristics. It also reduces cam-tappet maximum contact stresses due to the larger nose radius obtained from increasing the area under the lift curve. The maximum acceleration for a particular cam contour is dependent on the type of valvetrain. A system with few components, such as a Type I valvetrain or direct-acting OHC, can tolerate a higher maximum acceleration value compared to that of a Type V pushrod system (many components), which would require a lower permissible maximum acceleration. The valve spring force required to hold the valve open and to hold the tappet on the cam is in direct proportion to the mass of the system and its acceleration (force = mass × acceleration). Therefore, the maximum acceleration in the Type V pushrod system must be lower than that of the Type I direct-acting system to offset the greater number of components, the greater mass, and the higher flexibility. If not severe, higher cam harmonic resonance factors affecting overall valvetrain dynamics can

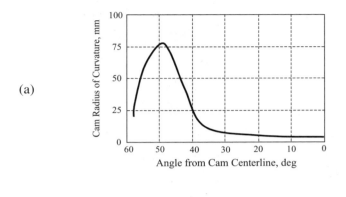

(a)

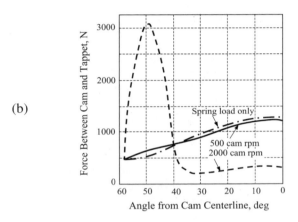

(b)

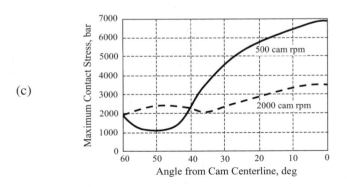

(c)

Figure 3.54 *Cam radius of curvature, contact force, and stress as a function of engine speed and cam angles: (a) instantaneous radius of curvature, (b) cam/tappet force, and (c) maximum contact stress (Neale, 1994).*

be compensated for by a high natural frequency valve spring design. Figure 3.55 shows typical cam profiles for various applications, including high-performance and racing applications.

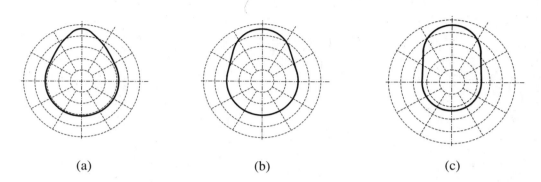

(a)   (b)   (c)

Figure 3.55 *Typical cam profiles for various applications: (a) normal application, (b) high-performance application, and (c) racing application.*

3.3.5 Lightweight Cam Design

Valvetrain weight, including cam weight, is a major factor affecting engine performance, valve-train dynamics, maximum engine speed, and friction/torque loss. Each of the following two cam designs and a combination of both have shown significant reductions in weight.

3.3.5.1 Base Lobe Weight Reduction

Figure 3.56 shows a cam design that reduces the cam weight by reducing material in the base circle of the cam lobe. By reducing material from the base circle of the cam lobe to achieve reduced cam weight, it also reduces the contact area at the base circle contact with the cam fol-lower. Thus, it inevitably increases the contact stress. However, because the spring load at the valve closed position is minimal compared to that at the valve lift position, it has room for the calculated contact area reduction or stress increase.

3.3.5.2 Hollow Camshaft

Hollow camshafts or composite camshafts can significantly reduce camshaft weight by 20–40% and cut machining costs. Typical cam tube wall thickness is 2–3 mm (Figure 3.57). This usually is accomplished by using a hollow steel tube through interference stress fitted to the cam lobes. The fitting can be accomplished by using cam lobe shrink fitting or camshaft tube mechanical expansion fitting. The principle of the shrink fit process is to heat the lobe, which in its cold state has a smaller inner diameter (I.D.) than the O.D. of the tube. Then the tube and lobe are assembled together and cooled until the lobe shrinks onto the tube. The camshaft tube mechani-cal expansion fit is accomplished by generating an interference stress between the tube and the

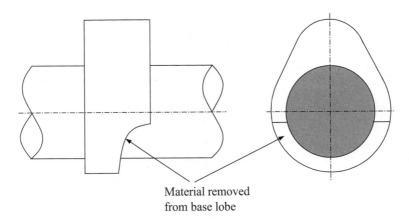

Material removed
from base lobe

Figure 3.56 Lightweight cam lobe design.

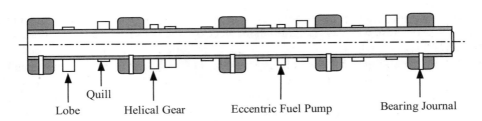

Lobe   Quill   Helical Gear   Eccentric Fuel Pump   Bearing Journal

Figure 3.57 Schematic of a composite camshaft (Thumuki et al.).

cam lobes through mechanical expansion of the tube with a mandrel. The shrinking technology allows the use of both wrought and sintered powder metallurgy cams, depending on the application. Precision sintered powder metal lobes allow complex lobe geometry, and the precise joining process eliminates the need for form grinding of the cams. In addition to weight reduction, the hollow tube can improve the engine lubrication system by providing a main passage of engine oil through the hollow tube. Thus, the oil delivery tube is eliminated. Another advantage of the hollow camshaft tube combined with a precision sintered cam lobe is the machining cost savings (20% compared to that for a cast iron camshaft) due to the near-net shape sintering process. Another advantage of assembled cam methodology is that the material choice for the camshaft, cam lobes, and auxiliary gears does not have to be compromised.

3.3.6 Cam Lobe Surface Finishes and Tolerances

With a maximum calculated contact stress on the order of 1 GPa or more at the cam lobe surface, it is imperative that the surfaces are finished properly and are as accurate as possible.

Roughness and inaccuracies that rupture the borderline oil film will produce relatively higher local stresses that may result in scuffing and rapid wear. Although early scuff and wear do not show up due to improved materials, the increased stress due to these factors will break the oil film and cause friction to increase, consequently resulting in spalling, flaking, or contact fatigue from high shear stresses.

Some cam and follower surface imperfections (deviations from the theoretical contour) often are difficult to control or determine. Depending on their magnitude, these surface irregularities may induce shock, noise, wear, and vibration. The vibrations produced are superimposed and may exceed all other kinds already discussed. Three kinds of surface imperfections may be evident: (1) errors, (2) waviness, and (3) roughness. The effects of these three types of imperfections may be compounded.

Errors, local in character, may be large or small or of any shape or duration. They can be produced by an incorrect setting of the milling cutter, a file scratch, or by holding the grinding wheel at a point on the cam for too long a time. On the other hand, waviness is a periodic imperfection. It is a uniform distribution of high and low points of longer duration than roughness. Waviness of a surface is the vertical distance between peaks and valleys of relatively long wavelengths. It may be produced by a milling cutter taking a larger feed in the continuous generation of a cam or by the increment locations of the milling cutter in which scallops are formed that can never be eliminated. Because waviness is periodic, its successive effect may be a vibratory reinforcement at certain speeds, which may seriously impede operation. Roughness is a random distribution of many small high and low points. Because they are random, any change is as likely to decrease previously existing vibrations as to increase them. The net effect is very low-level vibration (usually negligible) which, furthermore, is absorbed by damping due to residual oil on the surface. Roughness is a relatively short-wavelength irregularity. The type and degree of surface finish are important factors that affect cam performance, and they must be kept under tight control. The higher the contact stress, the more important is the surface finish. Rough or irregular surface finish increases the asperity or local contact stress and lowers the efficiency of lubrication. A high cam surface finish increases scuffing tendency, and a high follower finish increases pitting tendency. Imperfections in the cam profile usually produce high-frequency and low-amplitude vibrations.

Finishing of cams often is the most critical of the fabricating operations. In general, any hand operation is subject to error and is completely dependent on the operator's skill and experience. It is known that rough surfaces wear more than smooth parts, but the degree of smoothness may vary. With rough surfaces, the microstresses will be high because the contacting surface asperity area is low. Thus, failure generally originates at these high points. It has been stated that polishing of the automotive sliding follower to a maximum of 0.1 μm is used, whereas cam finishes as high as 1.27 μm are acceptable. Roller followers have been finished as smoothly as 0.1 μm.

Figure 3.58 shows a typical cam profile with a statement of the production tolerances. These tolerances are given as average and are typical of those presently used in industry. The tolerances used on any particular profile may be somewhat less demanding for an engine of relatively lesser performance or more rigid for an engine of higher performance.

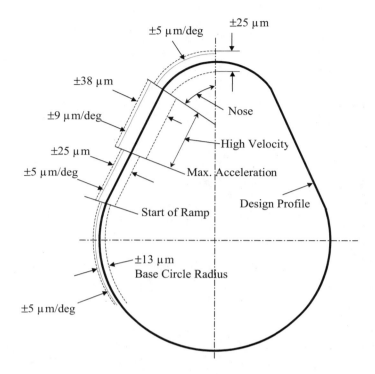

Figure 3.58 *Cam tolerance (Nourse).*

Typical cam profile tolerances should be on the order of ±5 μm for point-to-point consecutive degree positions on the profile with the tolerance at 180-degree positions on the order of ±50 μm. The tolerance consists of two kinds of statements. First, there is a maximum allowable deviation (or bandwidth) from which the profile "as made" can be permitted to depart from the design profile. Normally, this bandwidth changes with the location along the profile. The second kind of statement is a profile waviness specification. The intent of this tolerance is to limit the rate at which the profile "as made" can be permitted to deviate from the design values. Figure 3.59 shows typical dimensional tolerances for the hollow camshaft with a precision sintered powder metal cam lobe.

3.3.7 Cam Materials

As engine power and speed increase, the cam operating conditions become more arduous. An increase in power results in steeper cam ramps and higher valve lift, resulting in higher contact stress between the cam and follower. An increase in speed requires higher spring loads to offset the increased valvetrain inertia loads, and this results in higher contact stresses when the engine is running at low speeds. In addition, the cam and follower must operate at a higher relative velocity at the maximum engine speed.

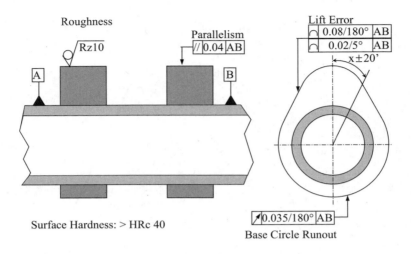

Roughness

Parallelism

Lift Error

Surface Hardness: > HRc 40

Base Circle Runout

Figure 3.59 *Typical tolerance precision of cam lobes (Müller and Kaiser).*

Selection of cam materials is intended to fulfill the requirements of smooth cam action with minimum wear over a long period of time. Three groups of materials typically are chosen as cam materials: (1) cast iron (including grey cast iron with chill, induction or flame hardened, nodular, and malleable iron), (2) hardenable steel, and (3) sintered powder metal (PM) alloys. The choice of materials for the camshaft depends largely on the design of the engine, the maximum contact stress between the cam and follower contact surfaces, and the compatibility of the cam and follower materials. Table 3.9 summarizes materials used by many of the world's passenger car manufacturers, and Table 3.10 lists the chemistries. In Europe, chilled cast iron is widely used as camshaft material; however, steel, nodular or spherical graphite (S.G.) iron, and malleable iron are favored for OHC engines. In North America, hardenable cast iron predominates; in Japan, chilled cast iron and chilled low-alloy cast iron are favored (Roberts and Tournier). The rule of thumb in choosing the cam and follower material is to use a chilled cast iron where the cam/follower stresses are high, and a low-alloy hardenable iron for low-stress applications. Camshafts for diesel engines are manufactured predominantly from steel due to the fact that cam contact stresses tend to be lower than in gasoline engines, and cam wear usually is not a problem. In addition, the possibility exists for oil pump drive gear problems because of the need for a strong camshaft if auxiliary equipment is to be driven from the camshaft. For OHC engines, reduction in valvetrain inertia leads to lower valve spring loads and a consequent reduction in contact stress. This will tend to give more latitude in the choice of materials, including nodular iron, malleable iron, or PM.

Hardenable steels include SAE 1010, 1020, 1039, 2515, 4340, 4615, 8520, and 52100. These steels typically are induction hardened to 50–60 HRc or case hardened/carburized to a case depth of ~1 mm with a surface hardness of 50–60 HRc.

TABLE 3.9
PASSENGER CAR CAM AND FOLLOWER
MATERIAL COMBINATIONS (JARRETT)

Manufacturer	Valvetrain	Cam Material	Follower Material	Comments
Alfa Romeo	OHC	C.C.I.	Steel	
Audi	PR	Steel	C.C.I.	
BMW	OHC	C.C.I.	C.C.I	
Austin Morris	OHC	C.C.I.	C.C.I.	Follower phosphated
Jaguar	OHC	C.C.I.	C.C.I.	
Rover	OHC	C.C.I.	C.C.I.	
	PR	Flame hardened alloy iron	Hardenable C.I.	Hydraulic followers
Triumph	OHC	C.C.I.	C.C.I.	
Simca (Chrysler)	PR	H.C.I. and C.C.I.	C.C.I.	
Talbot (Chrysler)	PR	C.C.I.	C.C.I.	
Citroën	PR	Steel	C.C.I.	
Daimler-Benz	OHC	I.H. malleable iron	I.H. low-alloy steel	Follower chrome plated
Fiat	PR	S.G. iron, steel, and C.C.I.	C.C.I. & hardenable C.I.	Some cams and followers phosphated
Ford (Europe)	OHC	I.H. low-alloy C.I.	Nitrided nodular iron	Cams and followers phosphated
	PR	I.H.C.I. (K iron)	Chilled & I.H.C.I.	Some cams and followers phosphated
Lancia	OHC	Steel	Steel	
Lotus	OHC	C.C.I.	Nitrided steel	
Opel	OHC	S.G. iron	C.C.I.	
Peugeot	OHC	C.C.I.	C.C.I.	
	PR	C.C.I.	C.C.I.	
Renault	PR	C.C.I.	Steel	
Rolls-Royce	PR	C.C.I.	C.C.I.	Follower phosphated
Vauxhall	OHC and PR	C.C.I.	C.C.I.	
Volkswagen	OHC and PR	Hardenable C.I.	Hardenable C.I.	H.C.I. high-phosphorus Steadite material
		C.C.I.	C.C.I.	
Volvo	PR	I.H. alloy C.I.	C.C.I.	Follower ferrox treated
Datsun (Nissan)	PR	C.C.I.	C.C.I.	
Honda	OHC	I.H.M.I.	C.C.I.	
Toyota	OHC	C.C.I.	Cr plated steel & phosphated C.C.I.	Some cams soft nitrided
	PR	Soft nitrided and C.C.I.	Chilled alloy C.I.	Some followers soft nitrided
Chrysler	PR	I.H. alloy C.I.	I.H. alloy C.I.	
Ford	OHC	I.H. alloy C.I.	I.H. alloy C.I.	Cams and followers phosphated
	OHC and PR	I.H. alloy C.I.	I.H. alloy C.I.	Cams phosphated

Notes: C.C.I. = Chilled cast iron; C.I. = Cast iron; H.C.I. = Hardenable cast iron; I.H. = Induction hardened; I.H.M.I. = Induction hardened malleable iron; OHC = Overhead camshafts; PR = Pushrod operated valves.

TABLE 3.10
CAM LOBE CAST IRON CHEMISTRIES (JARRETT)

	C	Si	Mn	S	P	Ni	Cr	Mo
Chilled Cast Iron	3.10–3.25	1.90–2.10	0.60–0.70	0.10 max.	0.10 max.	–	0.50–0.80	
	3.10–3.40	2.00–2.40	0.50–0.80	0.10 max.	0.20 max.	0.15–0.25	0.80–1.00	0.15–0.25
	3.60 max.	2.30 max.	0.60–1.20	0.12 max.	0.30 max.	0.50 max.	0.15 max.	
Hardenable Cast Iron	3.40–3.60	2.20–2.40	0.80–1.10	0.15 max.	0.20 max.	0.40 max.	0.90–1.15	0.40–0.45
	3.10–3.30	2.10–2.40	0.70–0.90	–	0.20 max.	0.40–0.70	1.00–1.25	0.50–0.70
	3.00	2.50	0.80	0.10 Ti	0.70	–	0.30	0.50 Cu
	3.20–3.40	2.25–2.50	0.75–0.85	–	–	0.27–0.30	0.90–1.05	0.45–0.50
Carburized Steel	0.1–0.18	0.05–0.35	0.6–1.0	<0.07	<0.05	–	–	–

Cast irons include alloyed chilled grey cast iron, hardenable grey cast iron, hardened ductile (nodular) cast iron, and hardened malleable iron. For chilled camshafts, chilling is achieved by casting against iron chills, placed around the noses of the cams, resulting in a hardened surface on the cam nose in the range of 45–55 HRc.

Powder metal sintered alloys often are iron-based hardenable alloy steels. Iron-based Fe-Cr, Fe-Mo, and Fe-Ni often are used with copper for high density and carbon for hardenability. An example of this alloy is 5% Cr, 1% Mo, 0.5% P, 2% Cu, 2.5% C, and <2% others, with iron for balance.

Other materials not mentioned here may be acceptable substitutes. Safe values for contact stress (Hertzian stress) depend on a number of factors, such as the combination of materials in use, heat treatment, surface treatment, and quality of lubrication. Figure 3.60 gives allowable contact stress for iron and steel components of various hardnesses. These values can be applied only if lubrication conditions are good.

Cam material requirements depend on the type of application, particularly when considerations other than cam/follower compatibility are involved. For example, many camshafts have an integral gear driving the distributor and oil pump, and the requirements of a material to give satisfactory performance of this gear can conflict with cam/follower compatibility. A compromise on material selection often is necessary.

3.4 Valvetrain Lash Compensators

3.4.1 Lash and Mechanical Lash Adjusters

Lash is the gap or space in a valve actuating system and is used to compensate for the variation that occurs during manufacturing, assembly, wear, and thermal expansion, so that proper valve sealing and valve timing can be accomplished. Figure 3.61 illustrates examples of lash in a valvetrain system. Without any lash and without an automatic compensation mechanism built into the valvetrain system, the valve could be held open and lose its sealing function. The variation in valvetrain lash has an adverse effect in power output, fuel economy, and the emissions

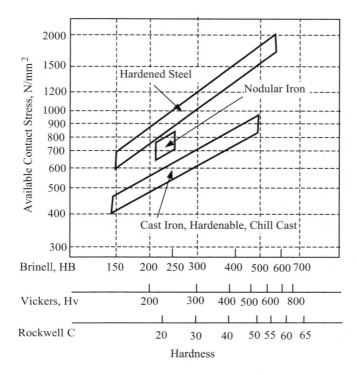

Figure 3.60 *Typical allowable contact stresses under good lubrication conditions (Neale).*

of an engine. Insufficient lash will not allow the valve to close at the base circle cam position and will lead to the condition causing valve sealing failure. Excessive lash can lead to excessive seating velocities and valve closing noise. This condition, if extreme, will cause valvetrain component breakage.

There are two types of lash compensation methods: (1) mechanical (or manual), and (2) hydraulic (or automatic). The latter (hydraulic) is an automatic compensation mechanism and has seen increased application. In a valvetrain system without automatic hydraulic compensation, a shim usually is used in the direct-acting or Type I valvetrain to achieve the required lash needed for the system; typically, it is 0.20–0.75 mm. In other types of valvetrains, manually adjusted systems usually are built into the rocker arm for lash adjustment. Typically, lash is measured with the valve closed and the cam at the base circle position. One advantage of mechanical valvetrains is less valvetrain mass, which leads to high engine speed potential; another is that the mechanical tappets do not require an oil supply. The disadvantages include initial lash setting at assembly plants for each valvetrain of every engine, and service adjustments for changes in lash caused by component wear. Lash changes with respect to engine speed and load, and it results in valvetrain noise, increased valve seat recession, and changed valve timing. The greater valve timing variation can cause unpredictable overlaps, HC levels, vacuum levels, idle quality, and torque.

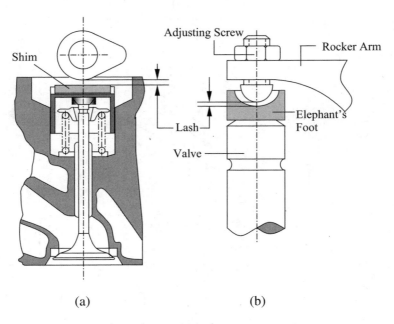

(a)   (b)

Figure 3.61 *Valve lash schematic.*

3.4.2 Hydraulic Lash Compensators

The hydraulic lash compensator, interchangeable with the hydraulic bucket or adjuster or lifter, quietly transfers the rotary motion of the cam into the reciprocating motion required for poppet valve actuation while maintaining zero lash. Currently, most modern internal combustion engines adopt some kind of automatic lash compensation mechanism. The hydraulic valve lash compensator has been developed, not with the idea of maintaining the lash at constant, but with the idea of maintaining the gap at zero. At the same time, it automatically compensates for mechanical and thermal variations. The advantages of the hydraulic lifter over the mechanical tappet are that no assembly lash adjustments are required; no service lash adjustments are required; there is no valvetrain noise resulting from lash; and valve timing is more accurate, which yields better emissions, fuel consumption, vacuum levels, and torque. The disadvantages are that an oil supply to hydraulic lifters is required, and hydraulic lifters have more mass added to the valvetrain; thus, maximum engine speeds are lower. Figure 3.62 summarizes the typical hydraulic lash compensation devices for different types of valvetrain applications and the nomenclature of its components. Although many designs exist other than those listed here, the principle of lash control by controlled leakage past the plunger and replenishment via a check valve is the same for all types of hydraulic lifters.

The hydraulic lash compensation automatically eliminates all lashes in the valvetrain. This zero-lash condition ensures that the valve begins to open at precisely the same time during each valve event. A small amount of lash is present after the valve seats to ensure good valve seating, but this lash is eliminated by the lifter before the valve is lifted again.

Valvetrain Components

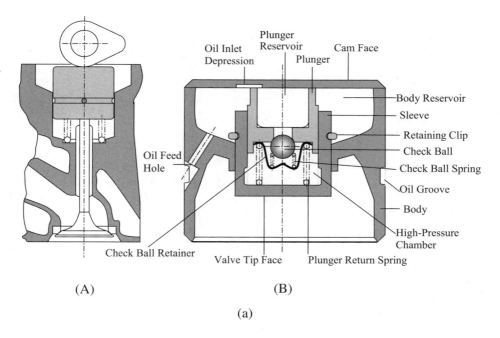

(A)   (B)

(a)

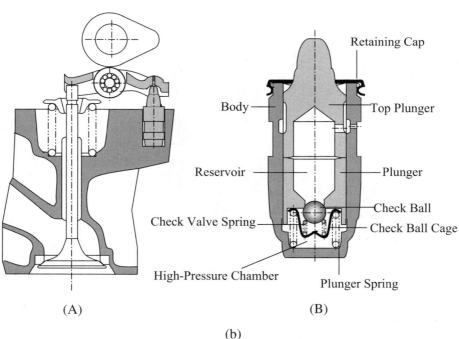

(A)   (B)

(b)

Figure 3.62 *Hydraulic lash compensator application (A) and component nomenclature (B): (a) Direct-acting bucket lifter, OHC, Type I valvetrain. (b) Lash adjuster, end-pivot rocker arm, OHC, Type II valvetrain.*

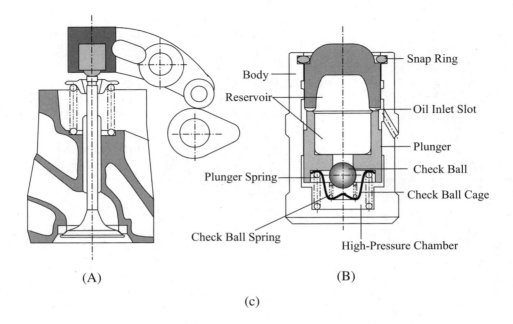

(A)   (B)

(c)

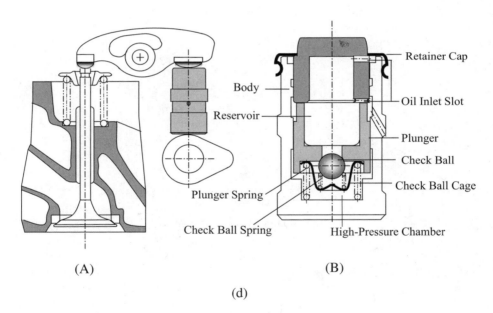

(A)   (B)

(d)

Figure 3.62 *(Continued) (c) Capsule, center-pivot rocker arm, OHC, Type III.*
(d) Flat lifter, center-pivot rocker arm, OHC, Type IV.

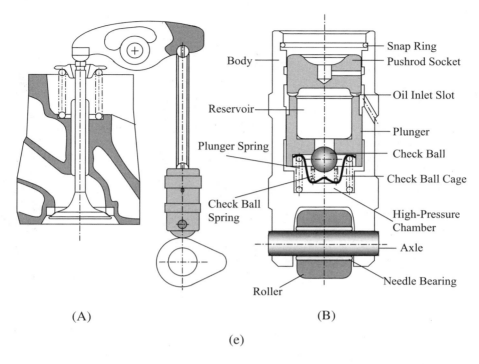

Body
Reservoir
Plunger Spring
Check Ball Spring
Roller
(A)

Snap Ring
Pushrod Socket
Oil Inlet Slot
Plunger
Check Ball
Check Ball Cage
High-Pressure Chamber
Axle
Needle Bearing
(B)

(e)

Figure 3.62 *(Continued) (e) Roller lifter with pushrod, center-pivot rocker arm, OHV, Type V valvetrain.*

3.4.3 Hydraulic Lash Compensation Mechanism

Figure 3.63 illustrates the working mechanism of hydraulic lash compensation, using the Type V pushrod valvetrain configuration for explanation; however, the principle applies to all configurations. At the end of the base circle and prior to the lift cycle, the plunger has finished upward recovery and is at the zero lash position. The check valve ball is seated and therefore has sealed the high-pressure chamber, as shown in Figure 3.63(a). When the valve lift cycle begins, axial valve spring force is applied through the pushrod via the rocker arm to the socket and leak-down plunger. This force increases the high-pressure chamber pressure because the oil is nearly incompressible, the lifter acts as a solid link, and the cam motion is transmitted to the pushrod. High pressure causes a controlled leak of fluid from the high-pressure chamber through the leak-down clearance. Figure 3.63(b) shows the leakage path. This controlled leak (leak-down) reduces the high-pressure chamber volume and allows the leak-down plunger to move downward into the body a controlled amount (approximately 0.025 mm). After the cam returns to the base circle and the valve is seated, the plunger recovers from approximately 0.025-mm lash to zero lash.

After the valve event is complete and the valve is closed, the leak-down plunger moves upward to the zero lash position for the next lift cycle. The high-pressure chamber then is replenished

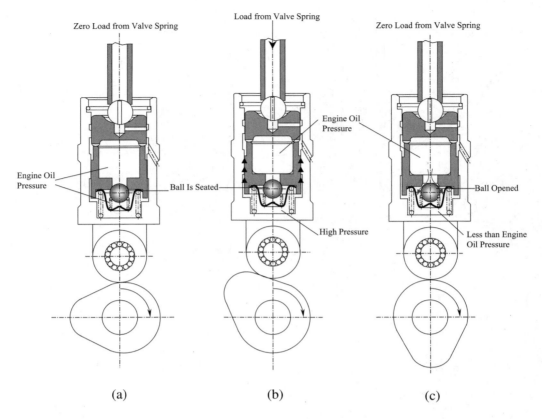

Figure 3.63 *Hydraulic lash compensation mechanism: (a) cam just before the lift cycle begins, (b) cam during the lift, and (c) cam on the base circle (Abell).*

with oil, as illustrated in Figure 3.63(c). The plunger spring force pushes on the bottom of the leak-down plunger, which causes an increase in the high-pressure chamber volume. This increase in volume drops the high-pressure chamber pressure to less than that of the reservoir. As the pressure drops low enough to overcome the check ball spring force, the check ball opens off its seat. The increasing volume in the high-pressure chamber then draws in the amount of fluid required to allow the leak-down plunger to return upward to the zero lash position. When the leak-down plunger reaches the zero lash position, the pressure differential between the high-pressure chamber and the reservoir is diminished. The check ball spring then can force the check ball to its seat before the beginning of the next event.

To improve hydraulic lifter performance, the socket is deliberately manufactured as a separate piece. This prevents pushrod side forces from disturbing the function of the plunger. With the separate socket, the plunger spring holds the plunger against the socket while the oil pressure and plunger spring force combine to push the socket upward to remove lash. An additional function of many hydraulic lifters in Type V valvetrains is to meter oil into the pushrod through the socket to provide lubrication to other areas of the valvetrain.

When the engine valve is closed and no load is applied to the lifter (generally the cam base circle), the resultant force of the plunger spring and the oil pressure in the reservoir removes all lash in the valvetrain. When a load is applied to the lifter at the beginning of the valve event, a slight amount of lifter collapse (up to 0.025 mm) can occur while the check valve is closing, prior to sealing the incompressible column of oil in the high-pressure chamber. The lift motion of the cam then is transmitted to the valve through this incompressible volume of oil. The oil that is squeezed past the plunger during the valve event results in an additional lift loss at valve closing (0.025–0.075 mm, depending on the application). The lift loss at opening (due to leakage by the plunger) must be anticipated in the cam design and must be included in the opening and closing ramps, respectively.

The oil reservoir acts as a sump to prevent the introduction of air into the high-pressure chamber, which would result in valve closing noise during engine startup. When the engine is shut down, many of the lifters will be on the lift portion of the cam. The valve spring load eventually will cause these lifters to collapse to the bottom of their travel. When the engine is restarted, these lifters must remove this temporary valve lash and return to their normal operating positions. This normally takes one or two engine revolutions before the engine oil pressure will be established again in the lifters; under cold conditions, it takes longer.

To reduce cold start emissions, cold start spark retarding is used in the beginning of engine operation. It means that engine combustion initially puts a lot of heat in the exhaust for fast heat-up of the catalytic converter. This heat creates an immediate increase of the valve head diameter, and as a consequence, the valve is pulled downward due to the increase in head diameter. All hydraulic lash adjusters have the ability to immediately compensate for lash because the check ball is, in fact, a one-way valve. It allows oil to easily enter the high-pressure chamber but closes oil flow out from the chamber. Immediately after that, the valve continues to heat up, and the valve insert grows in diameter. This creates the gap. (In normal hot-oil operating conditions, this would be compensated for by a hydraulic lash adjuster.)

To close the valve, the oil from the high-pressure chamber of the hydraulic lash adjuster must leak out through the leak-down surface. If the hydraulic lash adjuster is still cold thick oil, the hydraulic lifter cannot collapse and allow the valve to close. This creates conditions where no compression occurs in the cylinder, and misfiring occurs. To prevent misfiring caused by the hydraulic lash adjuster and to guarantee valve closing in all oil viscosity conditions, a normally open lash adjuster can be used (Edelmayer *et al.*). As opposed to a normally closed lash adjuster, the check ball remains open during the base circle steady state. At the start of lift, quick downward movement of the plunger causes the check ball to close and the plunger to move slowly downward during the lift controlled by the leak-down. The check ball opens after the valve closes and the plunger moves upward (Figure 3.64).

3.4.4 Hydraulic Lifter Design Considerations

Before the specific features of any particular hydraulic lifter design can be addressed, several considerations must be examined within the confines of the engine objectives to determine which valvetrain type or hydraulic lifter is the best design for that engine.

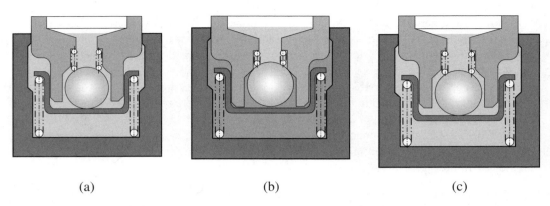

(a)   (b)   (c)

Figure 3.64 Normally open hydraulic lifter mechanism: (a) base circle with the ball open, (b) the ball closed during lift, and (c) the ball open after the valve is seated.

3.4.4.1 Engine Design Considerations

Hydraulic valvetrains have the same engine design requirements as mechanical valvetrain engines, with the additional requirements of supplying pressurized oil to the hydraulic component, a new cam and cam ramp design, and, of course, the new hydraulic unit. The engine design criteria of performance, high-speed capability, durability, installation, and cost dictate the selection of the type of valvetrain. To achieve a projected engine performance, the valve event should be flexible and repeatable, with frictional losses minimal. High-speed capability requires a stiff and lightweight valvetrain, including the hydraulic lifter. Engine durability points toward lower stresses, better material selection, lower friction, good dynamic behavior, and lubrication. Packaging, versatility, and cost also weigh tremendously in choosing a specific valvetrain. After going through the process of choosing the type of valvetrain to meet the engine design criteria, the choice of the hydraulic lifter Type I is readily determined. For example, bucket lifters typically are used for a Type I direct-acting valvetrain, a lash adjuster for a Type II finger follower valvetrain, and a barrel or tappet lifter for a Type V pushrod-type valvetrain.

3.4.4.2 Dry Lash Considerations

In an assembled engine with the cam on the base circle and the compensator fully collapsed, the clearance between the compensator and its mating part is called "dry lash." It is important to realize that dry lash is measured at different locations in different valvetrains. This can be translated into clearance at the compensator by considering the rocker ratio as appropriate.

Plunger travel inside the hydraulic lifter is the maximum distance the internal components can move between the fully extended and bottomed positions. This travel represents the total amount of dimensional variations that could be accommodated by a hydraulic lifter design. Dry lash determines where the plunger will be located during engine operation. It generally is practical to have dry lash equal to approximately half of the plunger travel, thus locating the internal components midway in their bore. This value may be biased if the valvetrain is expected to grow more in one direction than another throughout the life of the engine (i.e., valve seat wear). In production, dry lash varies from engine to engine as a result of the tolerance stackup of all valvetrain components and support castings. It is important to realize these tolerances and to not allow the dry lash to approach either extreme of plunger travel.

Another important consideration when calculating dry lash is the capacity of the oil reservoir of the lifter. When an engine has been stopped, some of the valves would have been in the open position. These valves would maintain a load on the lifter, causing it to eventually leak-down completely. When the engine is restarted, the lifter must take up this lash, which can be as much as the dry lash. Because the lifter responds in the first cam cycle long before it receives pressurized engine oil, it must have sufficient reserve capacity to prevent it from ingesting air. The reservoir should have a capacity equal to the area of the high-pressure chamber times the maximum dry lash plus some additional margin. Providing adequate internal oil can ensure that the lifter will not gulp air and will guarantee quiet engine cold starts.

3.4.4.3 Leak-Down Considerations

A hydraulic lifter operates on the incompressibility of oil and the controlled leakage of oil from the high-pressure chamber. The controlled passage of oil through the small clearance between the plunger and the body during a lift event is called leak-down. Leak-down is a critical aspect of lifter performance to allow it to make cycle-to-cycle adjustments to changes in length within the valvetrain. The controlled leakage is maintained by select-fitting of the plunger and body to diametrical clearances typically of the magnitude of 0.013 mm. The leak-down rate is specified as the time required for the plunger to travel between two gage points under a given load with a specific fluid at a specific temperature. The leak-down rate typically is expressed as the time (in seconds) that is required to displace a plunger to a given distance with a given fluid viscosity and a constant force. The carefully controlled radial clearance between the plunger and the body in conjunction with the length of the leak path (called leak-down land) controls how quickly oil is bled from the high-pressure chamber and consequently for how much length adjustment it may compensate at the end of one cycle. The purpose of leak-down is to compensate for changes in the plunger operating height in the downward or plunger collapsed direction.

3.4.4.4 Oil Supply Considerations

Design considerations must be given to the incorporation of an oil gallery and oil feed passages from the oil filter to the hydraulic units, as well as the size of the drainback passages to the oil sump. Special oil flow considerations should be given to those designs that have significantly different expansion properties between the lifter and the lifter bore.

The desirable minimum oil pressure is approximately 1 bar with a new engine during hot idle. This minimum pressure will help the hydraulic unit purge any air that may have entered. Some applications see the supply pressure in the range of 0.2–0.35 bar. The maximum oil pressure is limited by the minimum valve-closed spring load that can be tolerated. The reaction of the oil pressure on the internal components of the hydraulic unit tends to counteract the valve spring force holding the valve closed. The following equation is used to determine the total force created by a hydraulic lifter:

$$\text{Total Force} = \text{Plunger Area} \times \text{Oil Pressure} + \text{Plunger Spring Force}$$

The plunger area ranges from 100–200 mm^2, and maximum oil pressures below 5 bar are acceptable. The typical plunger spring force is approximately 20 N, and the total force created by a hydraulic lifter is in the range of 70–120 N. In a valvetrain with rocker arms, the lifter force must be adjusted by the rocker ratio to determine the force at the valve. If cold oil pressures are excessively high, the valves consequently could be held off the seat because the separation force in the lash adjuster when transmitted through the rocker ratio is greater than the valve-closed spring load. In such cases, modification of the conventional size lash adjuster to reduce the plunger area thus reduces the force and makes the system acceptable. Another solution is to add a restrictor in the supply gallery to the lash adjusters to keep the oil pressure down to an acceptable level. The oil supplied to the hydraulic lash compensating mechanism must be free of air. Oil pump bypass valves that route excess oil back to the pump inlet and not to the sump are desirable to minimize oil foaming and to provide air-free oil to the hydraulic lash compensator.

As a further aid to prevent cold startup valve noise, it is recommended that the hydraulic lifter oil supply galleries be positioned so that it is difficult for oil to drain out when the engine is shut off. All vents to the atmosphere should be avoided.

For good lifter performance, it is essential to minimize particulate contamination in the oil. Full flow oil filtration also is recommended to ensure a clean oil supply to the lifter.

3.4.4.5 Oil Aeration Considerations

Aerated oil is an oil that has become compressible due to the air or gases mixed within it. The proper function of hydraulic lifters depends on the incompressibility of oil. The oil trapped in the high-pressure chamber of the lifter causes the lifter to act as a solid link in the valvetrain and to effectively transfer the cam motion to the valve lift. Oil compressibility occurs only if aerated oil is present, causing lifter collapse and deterioration of the valve lift event. In addition, oil aeration that results in a sponginess effect of the hydraulic lifter function could lead to valvetrain noise and potential dynamic problems at high speeds, which will result in valvetrain failures. With the increasing use of smaller high-speed engines, particular attention should be paid to engine oil aeration and its detrimental effect on hydraulic lifter function. Restriction of the oil flow into the lash adjuster has been observed to cause lash and sponginess when air is present in the oil. The lash adjusting mechanism must be a purging design to eliminate the possibility of air entrapment.

There are three types of oil aeration, and each type differs in the degree of its effect on lifter performance. Excessive foaming in an engine can be detrimental to engine performance—not only to the hydraulic lifters but to all lubricated bearings and surfaces. Entrained air is the most critical type of aeration. The small bubbles of air are dispersed evenly throughout the oil, and they travel with the oil through the oil passages and into the lifters. The air bubbles are compressible; therefore, the aerated oil is compressible. Dissolved air is not critical because it does not have a significant effect on oil compressibility. However, dissolved air can come out of solution and form bubbles of entrained air. Most oil contains some amount of dissolved air. If a sample of oil is placed in a vacuum, small bubbles can be seen forming as the air comes out of solution.

Typical causes of oil aeration are attributed to the foam (inhibiting oil additives are broken down), the agitation of oil in the crankcase and/or overheat, a sump oil level below the minimum, the admission of air/gas to the suction side of the oil pump, and the oil pump bypass to the sump instead of the pump inlet.

Common symptoms of oil aeration are spongy lifters, noisy valve seating (due to the elimination of opening and closing ramps caused by partial lifter collapse), poor engine performance (rough running, decreased power output, increased exhaust emissions), possible valve seat recession, and potential valve breakage. The most direct effect of air in the oil is to allow the lifter to collapse more than is desired during engine operation. The result is that the valve seats early and possibly seats hard because it seats before the cam closing ramp is reached. This early valve closing is indicated by valve noise (especially audible at low engine speeds), possible rough running, and possible valve breakage. Perhaps the most common symptom of aerated oil is in valve noise during starting. Previous high-speed running of the engine could generate entrained air, and the low oil pressure during starting would make this aerated oil especially compressible.

The presence of aeration can be confirmed in several ways. In addition to the direct measurement of aeration, the following methods are useful:

1. Determine the valve opening time relative to the camshaft position by means of strain gages or displacement transducers. The valve will open later under conditions of significant aeration.

2. Measure the lifter plunger motion relative to the lifter body. Under conditions of aeration, the plunger will sink hundreds of micrometers rapidly before the valve opens. This method is easiest on an end-pivot rocker arm valvetrain in which the lash adjuster body is stationary.

3. Stop the engine and depress a lifter when it is on the base circle. The lifter should feel solid; aeration will be indicated by an instant and measurable sinkage of the plunger when the force is applied.

Figure 3.65 shows the effect of engine speed on oil aeration inside a hydraulic lifter. The average air fraction decreases as the engine speed increases.

3.4.4.6 Cam Design Considerations

Valvetrain loads and drive requirements are comparable for mechanical and hydraulic lifter systems, and additional design changes typically are not required. The primary difference between mechanical and hydraulic camshafts is the smaller ramp height requirement for a hydraulic system. Because the manufacturer's objectives for engine performance, emissions, and fuel consumption are affected by the shape and placement of the valve events, a new camshaft with a hydraulic lifter, as with any new cam design, can be optimized for achieving desired engine performance.

Camshaft ramp requirements are established primarily to account for valvetrain deflections, camshaft runout, and up to 0.1-mm closing ramp lift loss that can occur during the valve event.

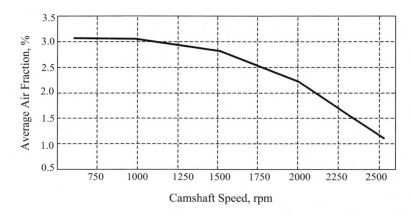

Figure 3.65 *Average air fraction in the lifter versus camshaft speed (Zhao, 1999).*

Variations in valvetrain stiffness can reduce or amplify the ramp height requirements, with rigid valvetrains obviously requiring less ramp height than the more flexible valvetrain designs. Because the hydraulic lash compensators automatically make cycle-to-cycle adjustments, maintaining a zero lash condition, their required ramp heights are inherently less than those of a mechanical design. Because opening ramp requirements are not as critical as those for closing, it may be possible to design a cam with very small or no opening ramp in applications having a rigid valvetrain. Because of this added flexibility in design, hydraulic ramps may be useful for optimizing the valve timing points, especially the area of valve overlap. Table 3.11 shows typical values for ramp heights and velocities.

TABLE 3.11
RAMP HEIGHT AND VELOCITY COMPARISON FOR
MECHANICAL AND HYDRAULIC LIFTERS

		Valve Opening		Valve Closing	
		Intake	Exhaust	Intake	Exhaust
Ramp Height (mm)	Mechanical	0.25–0.50	0.25–0.64	0.25–0.50	0.25–0.64
	Hydraulic	0–0.15	0–0.20	0.10–0.25	0.10–0.25
Ramp Velocity (mm/cam-deg)	Mechanical	0.015–0.025		0.011–0.020	
	Hydraulic	0.010–0.015		0.010–0.013	

Valvetrain deflection is an important consideration not only for ramp height requirements, but also for determining allowable runout of the base circle. Camshaft base circle runout tolerances usually can be the same as those on mechanical camshafts; however, they could become critical with rigid valvetrain designs such as the Type I OHC direct-acting valvetrain. With the hydraulic unit acting essentially as a solid link in the valvetrain, base circle runout effectively will induce valvetrain deflections, and based on the system stiffness, forces will be generated. These forces must be kept below the valve-closed spring load to prevent it from opening at the base circle. Typical maximum permissible base circle runout for the Type I OHC direct-acting valvetrain is approximately 0.015 mm, and approximately 0.075 mm for the Type V pushrod-type valvetrain.

The maximum valve lift and duration of the valve event are regulated by the engine design parameters as well as engine performance requirements. Engine performance is affected by the complete engine breathing system, and the cam must supply a valve motion that is in harmony with this system. When compared to a cam design for a mechanical valvetrain, the cam lift (including ramps) for a hydraulic cam can be reduced by the difference in ramp heights between the mechanical and hydraulic designs. The shape of the lift curve between the opening and closing ramps and its resulting velocities and accelerations can remain the same as a mechanical cam design, or they may be changed as desired. For the same O.D., the direct-acting hydraulic bucket design has a larger effective diameter than its mechanical counterpart using shims at the cam face. This allows the hydraulic cam design to have higher velocities and therefore more lift area of valve opening. This feature may be utilized to improve engine performance with this type of valvetrain.

3.4.4.7 Valvetrain Dynamics Considerations

Hydraulic lash compensators have a considerable influence on the natural frequencies of the valvetrain due to the low stiffness of the high-pressure chamber. At low engine speeds, an inaccuracy in the radius of the cam base circle can reduce the preload of the valve spring. A loss in solid contact between valvetrain components can produce check-valve opening during the engine valve event. This results in an "intermittent collapse" of the hydraulic lash adjuster caused by an undue oil flow through the check-valve bore. At high engine speeds, oscillations of the check-valve due to camshaft deflections on the cam base circle result in a delayed check-valve closing. This causes a "precollapse" of the hydraulic lash adjuster.

Valvetrain dynamics generally are considered "bad" when the valve significantly deviates from its intended motion. This false motion is associated with severe valvetrain loading and unloading, and if severe enough, it can generate destructive forces. Although a mechanical valvetrain can tolerate periods of bad dynamics, a hydraulic valve lifter usually can respond quickly enough to the periods of unloading (system lash) and eventually create a pump-up condition that prevents the valve from seating. This pump-up condition is not malfunction of the lifter itself but merely the reaction of the lifter to dynamic behavior in the valvetrain. In any valvetrain, there are limitations to the maximum speed of operation before the dynamic conditions become unstable, resulting in a separation of the valvetrain components. Factors that govern this maximum speed are the stiffness of the valvetrain, its resonant frequencies, the weights of the moving components, the valve spring design, and the cam design. During engine operation, any unloading of the valvetrain during the valve event as the result of exceeding the limitation of any of the controlling factors will present lash to the hydraulic lifter. If the lifter can respond quickly enough (usually it can under severe conditions), it will adjust to remove this new lash and gradually "pump up," preventing valve seating and resulting in a loss in engine power. Therefore, the lifter may function as an overspeed protector to help prevent damage from occurring as the result of over-speeding the valvetrain under power. When jumping separations occur in a valvetrain and the lash adjustment mechanism responds quickly enough to adjust for them before these separations vanish, the leak-down plunger operates above its proper zero lash position and holds the valve open on the base circle.

When dynamic problems are encountered, all areas of the valvetrain can be examined for potential improvement. Generally, any modifications to reduce effective valvetrain weight and increase system rigidity are beneficial. Camshaft profile improvements also are important due to the role of the profile in creating the forces and exciting the resonant frequencies in the valvetrain. The harmonic content of the cam profile, its acceleration levels, and the length of the positive acceleration interval all have been considered as important design factors relating to valvetrain dynamics. Also, the amount of excess spring load over what is required to maintain contact between components during cam deceleration (i.e., the spring load margin) at maximum engine speed is critical. The necessary spring load margin for each type of valvetrain is dependent on the rigidity of the valvetrain and can vary for different designs. Typically, 10% spring load margin is required for a Type I direct-acting valvetrain, 30% for a Type V pushrod-type valvetrain, and 20% for other types of valvetrains.

Normally, valve spring load prevents excessive upward movement of the hydraulic lifter plunger beyond that required to eliminate valvetrain clearance. But at high engine speeds, hydraulic force can overcome valve spring pressure, raising (or pumping up) the plunger from its normal position. The now-higher plunger location prevents the valve from closing fully, so the engine appears as if it is experiencing valve float. Of course, true valve float also causes lifter pump-up. By causing the lifter to bounce on the cam lobe, valve float creates extra valvetrain clearance. The self-compensating hydraulic lifter tries to remove the clearance from the system, preventing the valve from closing fully. The only solution to real valve float is higher-load valve springs.

Sometimes it is possible to run an engine to higher speeds with mechanical tappets than with hydraulic lifters, but the difference represents operating the valvetrain in a region of undesirable dynamic response. Valvetrain jumping usually is associated with insufficient valve spring load. In general terms, pushrod engine speeds are limited by valvetrain resonant frequencies, valve spring design, or cam design. In contrast, most OHC engine designs have high valvetrain frequencies (800 Hz or greater) and would be expected to have valve spring design or cam design as the limiting factor.

Another mishap for a hydraulic lifter is a "pump-down" or collapse, which is too slow of a response time for the plunger to recover upward to its zero lash position during the base circle portion of the cam event. The check ball does not return to its seat before the next valve event occurs, and the check valve ball restricts fluid flow from the reservoir to the high-pressure chamber. The magnitude of pump-down is a function of engine speed, that is, the hydraulic lifter collapses more as the engine speed increases. As Figure 3.66 shows, at idle speed, there is 70-µm lifter collapse or pump-down. As engine speed increases, there is 90-µm pump-down at medium engine speed. At maximum engine speed, the pump-down is 200 µm. The consequences of pump-down are hard valve seating, noise, and valvetrain component breakage.

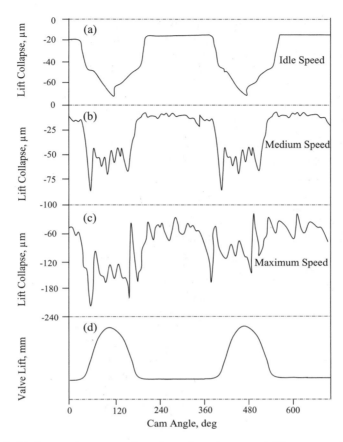

Figure 3.66 *Pump-down as a function of engine speed (Kreuter and Maas).*

Typically, the stiffness of the hydraulic lifter decreases as the engine speed increases; therefore, it contributes to the decrease in valvetrain natural frequency at higher engine speeds.

3.4.4.8 Seating Velocity Considerations

Leak-down time has considerable effect on valve seating velocity. A slow leak-down time results in a milder valve closing velocity due to the cushion effect of hydraulic fluid leaked from the high-pressure chamber. Fast leak-down leads to higher seating velocity, especially at higher engine speeds.

3.4.4.9 Lift Loss Considerations

Valve lift loss is experienced by using hydraulic lifters because of aeration and leak-down during lifting. The amount of lift loss is proportional to leak-down time and engine speeds.

Obviously, the greater the leak-down time, the less lift loss results. Lift loss also is reduced when the engine speed increases; however, when the engine speed exceeds 3000 rpm, the rate change in lift loss diminishes. Lift loss can be expected from valvetrain deflection and fluid compression as well.

3.4.5 Dimensions, Tolerances, and Surface Finishes

In valvetrains in which the cam contacts the lifter directly (e.g., Types I, IV, and V valvetrains), the lifters are cylindrical and reciprocate in a bore. These lifters are free to rotate and are designed to do so, except that roller lifters are used. The requirements of lifter dimension and surface finish are the most stringent among all engine components. The tolerance and surface finish information shown in the following sections are only for reference purposes. Lifter manufacturers can provide more accurate tolerance and surface finish data, based on the engine performance requirements and manufacturing capabilities.

3.4.5.1 Mechanical Lifters

Lifters for use in this design tend to have large diameters because the absence of a rocker ratio results in high-lift and high-velocity camshafts. These lifters normally have a flat face and an offset centerline (typically 1 mm) to ensure lifter rotation. A larger gage diameter should be used to specify flatness; a typical example would be flat to 0.018-mm crown at 30 mm. Dry lash is measured at each cam base circle to the lifter face contact point. The most commonly used mechanical lifters are shimless and shimmed bucket lifters for direct-acting Type I valvetrain applications and mushroom-type tappets for Types IV and V valvetrains.

3.4.5.1.1 Shimless Buckets

A typical application for shimless buckets is in direct-acting Type I valvetrains. Figure 3.67(a) shows the critical dimensional tolerances and surface finish requirements for a shimless bucket. Representative material used for the bucket is low-alloy steel such as $16MnCr_5$. The surface usually is carbonitrided or nitrided to achieve a surface hardness of 670–910 HV300. The bucket cam surface is manganese phosphate coated to reduce friction and running-in wear. Typically, a 0.01 ± 0.005-mm crown is added to the tappet cam face to avoid any concave profile. Even those lifters designed as "flat" have some small crown height to ensure that production tolerances never result in a concave lifter face. Those designs that use a radius to induce rotation have a considerably larger crown. Crown usually is measured as the height of the lifter at its centerline relative to its height at a gage diameter. A typical specification for a flat follower would be flat to 0.0125-mm crown at 20 mm, whereas a typical spherical-faced lifter would have 0.048/0.071-mm crown at 20-mm gage diameter. Followers that have a spherical face radius are always used with camshafts that have a taper on their lobes. This combination results in the offset of the point of contact used to induce rotation of the follower. Offset values typically are 1.0–3.5 mm.

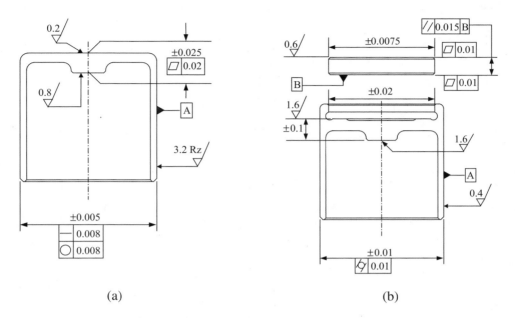

(a)   (b)

Figure 3.67 *Mechanical bucket tolerance and surface finish: (a) shimless bucket, and (b) shimmed bucket.*

3.4.5.1.2 Shimmed Buckets

A typical application of shimmed buckets also is in direct-acting Type I valvetrains. Figure 3.67(b) shows the typical dimensional tolerances and surface finish requirements for a shimmed bucket. Representative material used for the shim typically is low-alloy steel such as $15Cr_3$. The shim surface usually is carbonitrided or nitrided to achieve a surface hardness of 670–910 HV300, or 62 minimum HRc. The shim surface is manganese phosphate coated to a thickness of 0.002–0.008 mm. The lifter body material can be aluminum alloy for reduced valvetrain weight. Toyota uses a shimmed bucket lifter in its V8 engine, using SAE 4118 as the shim and 4T12 as the lifter body (Ezaki *et al.*)

3.4.5.1.3 Tappets

Figure 3.68 shows typical tappet dimensional tolerances and surface finish requirements. Material used for tappets usually is chilled cast iron with 50 HRc minimum on the stem and 55 HRc minimum on the face taken 1/4 mm off center of the face.

3.4.5.2 Hydraulic Lifters

Hydraulic lifters have very tight and precise tolerance control, especially plunger-to-lifter body clearance, because it affects the leak-down rate and thus engine performance. There are several types of hydraulic lifters, depending on the application and type of valvetrain.

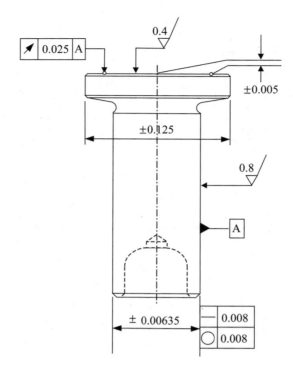

Figure 3.68 *Mechanical tappet tolerances and surface finish requirements.*

3.4.5.2.1 Flat Barrel Lifters

The two valvetrain types, Types IV and Type V, use similar lifters. The difference is that the pushrod lifter has a socket to accommodate the spherical pushrod end, whereas the Type IV lifter may have a flat button on which the rocker rides. Figure 3.69 shows typical dimensional tolerances and surface finish requirements of a flat barrel lifter. The tolerance on a lifter body diameter is ±0.0064 mm with a 0.3-μm surface finish. The lifter face can be finished to an 0.8-μm finish.

The lifter bore should be reamed with a ±0.038-mm tolerance on the diameter to a size that will result in a 0.0254–0.0762-mm diametrical clearance with the lifter body. The bore should be designed such that the ratio of the supported length of the lifter divided by the O.D. is 1.5 mm minimum (1.7 mm preferred). The oil gallery should be located such that a 5.08-mm minimum oil seal (6.35 mm preferred) will occur in any position of the cam event to prevent excessive loss in the oil pressure due to leakage around the lifter body. It is possible to design the oil groove on the lifter body to register with the oil supply only during part of the cam event. This intermittent registration has been used successfully in some engines (shutoff at partial lift) to extend the last good engine speed by making the lifter less sensitive to dynamic unloading of the valvetrain. Many engines use the lifter O.D. oil grooves as part of the oil gallery to pass supply oil from lifter to lifter, in which case full registration is required.

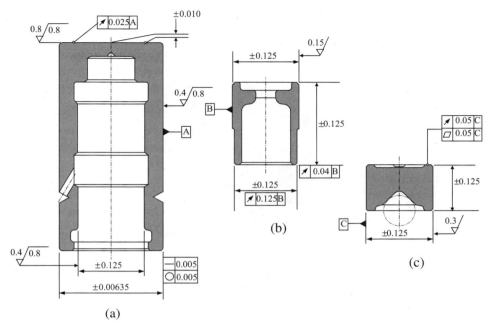

Figure 3.69 *Hydraulic lifter tolerance and surface finish: (a) body, (b) plunger, and (c) socket.*

The statistical stackup of tolerances in a component should meet the overall tolerances shown on an assembly. In manufacturing, the lifter body and plunger/socket are sorted by fitting, based on the clearance requirement for leak-down.

3.4.5.2.2 Roller Barrel Lifters

Incorporation of a roller element to follow the cam motion requires packaging modifications from that of a flat-faced follower. The overall length of the unit will increase by the O.D. of the roller plus clearance. Typical roller O.D. sizes are in the range of 17.75–25.5 mm. A means to keep the roller in alignment with the cam must be employed. Current methods include mating flats on the O.D. of the lifter body with slots on a plate that is secured to the boss of the lifter bore and guiding the sides of the roller via slots in the lifter bore.

Current roller applications utilize steel cams that are carburized or induction hardened. Alignment of the roller onto the cam is critical to the life of the cam and roller. Crowning of the roller (0.75–2.25 meters) has been found to reduce the tight alignment tolerances with a noncrowned roller. However, crowning also increases contact stresses; therefore, care must be taken to keep stresses within recommended limits. The tolerances and surface finish requirements are similar for the roller lifter to those of a flat-faced lifter, except for the roller bearing (Figure 3.70). Lash compensator length is measured from the face or bottom surface of the body to a socket gage

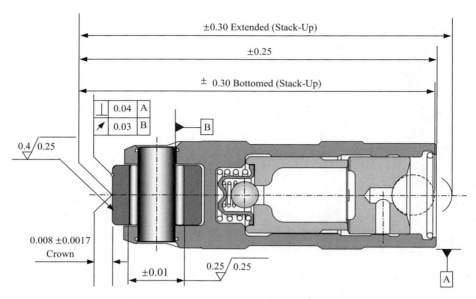

Figure 3.70 *Roller lifter tolerances and surface finish requirements.*

ball, a rocker arm foot center, or all pivot gage diameters at both the plunger fully bottomed and extended positions. The maximum (bottomed) and minimum (extended) lengths of the lash compensator usually are specified, establishing the operating envelope or usable travel of that lash compensator design. Statistical methods are used in determining these lengths, encompassing lash compensator manufacturing tolerance stackups. Each individual lash compensator is a unique assembly of components.

3.4.5.2.3 Lash Adjusters

Because no actual lifter-cam contact occurs in this design, the diameter can be kept to a minimum to conserve space. Current production lash adjusters have a cast iron body for low cost. The "mini" design offers a substantial reduction in space and weight. Dry lash is measured at each cam base circle to follower contact point, with rocker ratio adjustments made to determine the resultant value at the adjuster.

The bore design parameters for the other valvetrain types also apply to this design but with some modifications. Because this hydraulic device is installed into a blind bore, it is necessary to provide a vent hole that intersects with the bottom of the bore to aid assembly. Processing of the bore usually results in a step or radius at its bottom, resulting from the finish ream operation. This is accounted for in the design of the bottom portion of the lash adjuster body by incorporating an O.D. relief to provide clearance for the step. Figure 3.71 shows typical requirements of length, tolerance, and surface finish of a lash adjuster. The relationships of component lengths are

$$L_B = L_P + L_S + L_L \quad \text{and} \quad L_E = L_b + L_P - L_{E'}$$

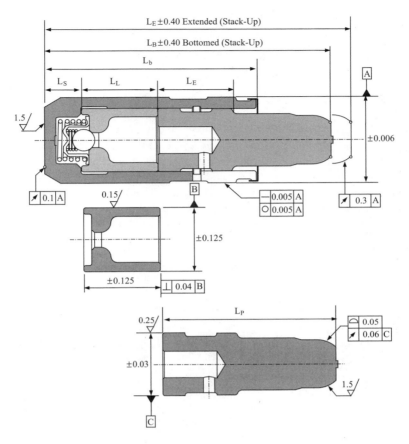

Figure 3.71 *Lash adjuster lengths, tolerances, and surface finish requirements.*

where

L_B = maximum bottomed length
L_P = maximum ball plunger length
L_S = maximum seat-to-face thickness
L_L = maximum leak-down plunger length
L_E = minimum extended length
L_b = minimum body length
$L_{E'}$ = maximum ball plunger engagement

A so-called "gothic" socket axis has several notable differences when compared to a spherical socket used with a spherical ball pivot:

1. There is no contact at the edge of an axially drilled hole.
2. The axial position of the contact line may be designated.

3. The socket is less likely to slide upward on the pivot due to side loads.
4. The socket may be more easily cold-formed.
5. There is less sensitivity to ball diameter tolerances and surface finish requirements.

3.4.6 Materials and Stresses

A hydraulic valvetrain is exposed to the same stresses and wear conditions as a mechanical valvetrain, with the exception of some additional loading on the camshaft base circle. In gasoline engines, base circle loading does not create any problems that require special considerations; consequently, material requirements for hydraulic valvetrains are identical to those of mechanical valvetrains. In some diesel engine designs, the increase in base circle loading could aggravate cam and follower wear problems, requiring some changes to minimize its effect. Table 3.12 lists some of the more common cam and follower material combinations for gasoline and diesel engine applications.

TABLE 3.12
PERMISSIBLE CONTACT STRESSES AT
THE CAM/FOLLOWER INTERFACE

Material Combination (Cam/Follower)	Follower Type	Stress Level, MPa
Hardened Cast Iron/Hardened Cast Iron or Chilled Cast Iron/Steel	Rotating follower (gasoline)	1240
	Rotating follower (diesel)	825
	Nonrotating follower	690
Steel/Steel	Rotating follower	1725

3.4.6.1 Contact Stresses

Surface contact stress is a key criterion for determining the longevity of the cam/follower interface, and stress levels should be maintained within certain guidelines. Table 3.12 identifies acceptable stress levels for various applications.

Under ideal contact, flat tappet surfaces will have a larger contact area and thus lower contact pressure or stress compared to spherical tappet surfaces (Figure 3.72). However, when a camshaft is deflected under high load, the tappet faces with a spherical radius do not lead to much change in contact stress. The stress can rise considerably in the flat tappet case, thus causing the cam to cut into the tappet face and leading to premature cam and follower failure.

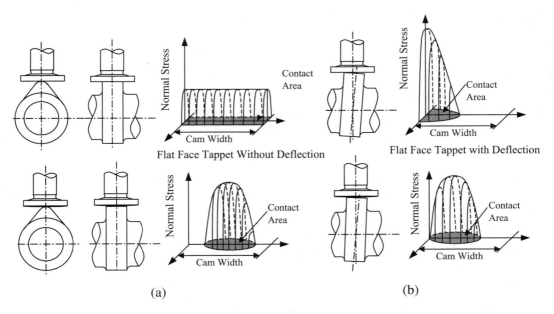

Figure 3.72 Stress distribution in tappet faces: (a) spherical face tappet without deflection, and (b) spherical face tappet with deflection (Turkish, 1946).

3.4.6.2 Lifter Materials

Typical materials used for the lifter body are hardenable steel, hardenable alloy cast iron, and chilled cast iron. Carbon steels typically are used for other lifter components, including the plunger, snap ring, pushrod seat, return spring, check ball, ball cage, and ball spring. Components with tribological contact (e.g., pushrod seat, plunger, check ball, ball cage) usually are case hardened for wear resistance. Other materials and treatments also are used frequently: carbonitrided SAE 1522 and 1215 for the ball plunger, carburized SAE 9310 or 8620 or hardened SAE 52100, chilled alloy cast iron, hardened steel, and nodular iron, phosphated for the roller. Carburized or hardened steels such as SAE 4100 and 5100 may be used. Hardened and tempered cast iron and steel components may be phosphated for uniform break-in. Either one or both mating components may be phosphate treated. Chilled iron components may be ferroxed or steam tempered for uniform break-in. The use of surface treatments such as nitriding and chrome plating has been reported. Hardened and tempered steels may be surface hardened by carburizing or carbonitriding.

In addition to stress, cam/follower material compatibility (including consideration for surface treatment), lubrication (including the nature of the lubricant chemistry), pre-lube at assembly, interface oiling, interface film thickness or elastohydrodynamic lubrication factors, and valvetrain dynamics generally are considered to be of prime importance for achieving satisfactory cam/follower life.

3.5 Seat Inserts

Valve seat inserts are installed in internal combustion engine heads to provide a seating surface for the valves. The insert seat surface is exposed to harsh environments that include high contact stresses and temperatures, as well as corrosive gases and other compounds. Metal-to-metal contact with little lubrication is common at the valve and insert seat interface. Inserts are used for two fundamental reasons. First, and most important, insert alloys are more wear resistant and heat resistant than the cylinder head material (mostly cast irons and aluminum alloys), making them better suited for the harsh environments encountered during valve seating. Thus, engine durability can be improved. Second, the use of inserts permits easier field repair or rebuild of engines.

3.5.1 Nomenclature

Figure 3.73 shows the most common type of insert and counterbore in use today, along with the related nomenclature.

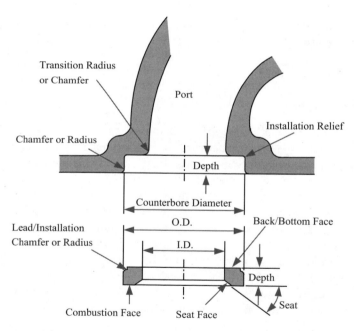

Figure 3.73 *Typical valve seat insert and counterbore nomenclature.*

3.5.2 Insert Design Considerations

3.5.2.1 Insert Installation and Retention

The valve seat insert is a cylindrical, ring-shaped component that usually is pressed into a head counterbore. The most common installation practice is to retain the insert through an interference fit between the insert and the mating counterbore. In special applications, the valve seat insert is held in place by mechanical means, such as threads or retaining rings. Inserts also may be cast, welded, or clad in place. Note that the installation relief radius in the counterbore must be of a smaller radius than the insert lead/installation chamfer or radius on the insert, so that the insert can seat at the bottom of the counterbore. In addition to retaining the insert, the interference fit promotes efficient heat transfer from the insert to the mating material, to the cylinder head, and ultimately to the engine coolant (or cooling fins in the case of an air-cooled engine). To accomplish these functions, the insert must be round, perpendicular to the bottom face, essentially free of O.D. size variations and/or taper, and sufficiently smooth to facilitate intimate counterbore contact. Also realize that the mating counterbore must have similar characteristics. The typical tolerance for interference fit between a valve seat insert and its mating counterbore is 0.05–0.13 mm. The actual amount used depends on the insert size and the materials of both the insert and the cylinder head.

The insert may be pressed or hammered into the counterbore. Pressing is preferred because it reduces the possibility of insert and/or counterbore damage. Both the insert and the cylinder head may be at room temperature during installation; however, the inserts commonly are cooled to reduce their O.D. size, thus reducing the pressing force required. Additional caution must be employed when ferritic or martensitic insert alloys are cooled cryogenically because they will be extremely brittle at those low temperatures. The interference in cast iron heads is 0.019 mm per millimeter of O.D., and the interference in an aluminum alloy cylinder head is 0.025 mm per millimeter of O.D. For heavy-duty applications, many inserts are prefinished. To facilitate the fitting of these rings, they normally are shrunk in liquid nitrogen to below 180°C and then readily are pressed into position.

In addition to the press-fit and shrunk-fit method of securing inserts in their recessed bores, extra precautions sometimes are taken for the retention of the inserts, such as a rolled-edge fit or mechanical lock design as illustrated in Figure 3.74.

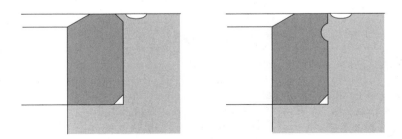

Figure 3.74 Mechanical lock designs using a rolled-edge fit.

Figure 3.75 shows other types of enhanced insert retention mechanisms. The flexible or skirted insert is used in some air-cooled engine applications. The clearance between the flange diameter and the cylinder head allows slight movement of the cylinder head to occur without distorting the seating surface. Insert retention is accomplished by the interference fit between the skirt diameter and the head. This puts the insert head interface farther away from the port in a relatively cooler area, thus aiding retention. Insert retention also is aided by a nominal 1/4-degree reverse taper on the skirt O.D.

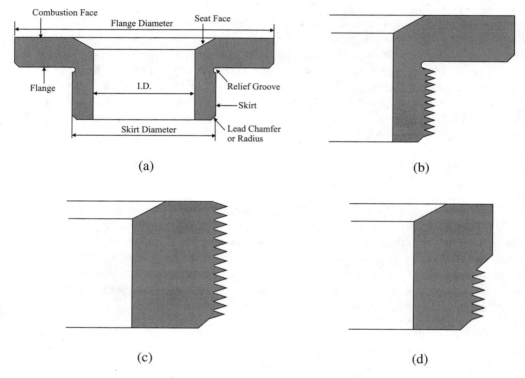

Figure 3.75 *Skirt and screw insert: (a) flexible or skirted, (b) skirt with screw, (c) full screw, and (d) partial screw.*

3.5.2.2 Dimensions

Proper component design and appropriate material selection are two keys to valve seat insert success. Insert design must address the obvious functional requirement of valve sealing and wear resistance, plus a number of secondary requirements such as installation and retention. The dimension of the insert must be such that the insert is rigid enough to absorb the continuous hammering and to provide a sufficiently large mass to dissipate the heat from the valve head to the cylinder head coolant system. To meet these requirements, the radial thickness of the insert wall should be at least 0.10–0.14 times the throat diameter (I.D.), and the external diameter (O.D.) of the insert should be within 1.2–1.3 times the throat diameter. A typical height for an insert is 0.15–0.25 times the throat diameter. Figure 3.76 graphically shows the relationship of the insert wall thickness and O.D. It is recommended that the aspect ratio, which is the insert height (h) divided by the radial wall thickness (T), be 1.7–2.5 (Dooley *et al.*).

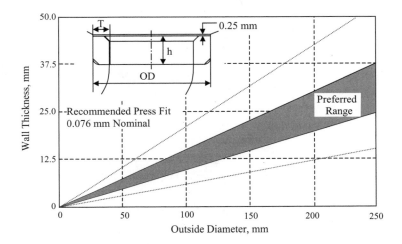

Figure 3.76 *Relationship of an insert O.D., wall thickness, and depth (Newton and Allen).*

3.5.2.3 Tolerances

For reference, Figure 3.77 shows typical tolerances and surface finishes of a valve seat insert.

3.5.2.4 Seat Face Width and Angle

The seat face is the working surface of the insert. The choice of seat width and angle depends primarily on two conflicting characteristics: sealing versus wear. Both sealing and wear depend on the seat angle, the friction coefficient at the interface, the combustion temperature and pressure, the seating velocity, deposit formation, the valve and insert material selection, and others. Figure 3.78 compares seating load versus seating angle at various assumed coefficients of friction. This graphic shows that large seat angles and low coefficients of friction result in high normal forces and pressures, which in turn increase the wedging or nesting between the valve and the insert seat. Large seat angles also may aid in crushing any combustion deposits. The increased seat stresses with large angles tend to be a drawback in terms of wear rate, particularly with nonlubricated conditions such as those encountered with "dry" fuels such as natural gas. Shallow seat angles reduce seating stresses and can reduce seat wear, but at the risk of guttering due to increased deposit buildup and incomplete seating. The two most common angles utilized are

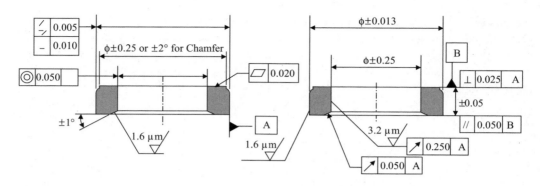

Figure 3.77 *Typical valve seat insert tolerances and surface finish requirements.*

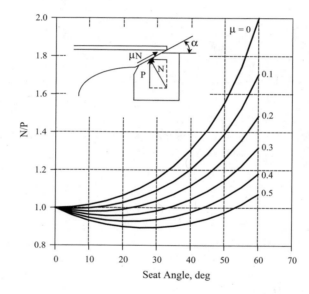

Figure 3.78 *Seating pressure as a function of seat angle*

and friction, $\dfrac{N}{P} = \dfrac{1}{\cos\alpha + \mu\sin\alpha}$ *(Giles).*

45° and 30°; however, 20° seating angles are becoming more common, particularly on intakes in nonlubricated diesel fuel conditions with high combustion loads.

In gasoline engine applications, the combustion pressure is relatively low; therefore, the primary consideration is proper sealing. If there is any misalignment of the insert to the guide, or any mechanical (assembly) distortion or thermal distortion, valve and insert seat runout can lead to leakage. The preferred choice will be a narrow seat width and a great seat angle. However, in diesel engine applications, the combustion pressure is much higher. Thus, seat wear is a major concern, and a wide seat width and a small seat angle are the typical choices. It is intuitive that a wide seat width with a large contact area results in small contact stress. For the insert to function properly, the insert seat must be concentric with the valve seat face. To accomplish this, it is common to machine the final insert seat in place, that is, after both the insert and the valve guide have been installed in the cylinder head. Either turning or grinding can be used to accomplish this, and the choice will depend on the available equipment and the machinability of the insert material. Prefinished insert seats also are available, but the tolerance requirement is more stringent. Gasoline engine passenger car inserts are machined mostly in the cylinder head, whereas heavy-duty diesel engine inserts frequently are prefinished.

3.5.2.5 Lead Chamfer/Radius and Outside Diameter Break

The purpose of the insert lead chamfer or radius is to guide the insert into the counterbore during insertion. Figure 3.79 shows some insert O.D. design options. For cast iron cylinder heads, a 45° chamfer with a slight break at the chamfer/O.D. intersection generally is adequate. A slight secondary "break" also is commonly added where the chamfer meets the O.D. With aluminum heads, however, care must be taken to prevent the harder insert from scraping or shaving the softer aluminum counterbore. If a radius is employed, it should be well blended with the O.D. of the insert. Lead chamfers for aluminum head applications should be a shallow angle. It is important that the insert fully bottoms out during installation, and the tolerance stackups of both the insert and counterbore should be reviewed to ensure that this occurs. After installation, the O.D. of the insert is in intimate contact with the counterbore. This contact serves two purposes. First, it provides an interface over which heat is conducted from the insert to the cylinder head. Second, the frictional forces set up by the interference fit retain the insert in the counterbore. The thickness of a valve seat insert must be sufficiently large to adequately perform the functions.

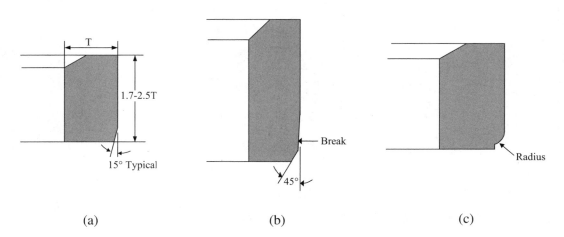

Figure 3.79 *Insert O.D. design options: (a) typical depth/thickness ratio, (b) chamfer design, and (c) radius design.*

3.5.2.6 Inside Diameter

The use of the insert I.D. configuration as produced by casting or sintering is the most economical option available. The two concerns that must be satisfied are that the as-produced dimensions and surface finish are acceptable. In the case of the dimensions, one key criterion is that of the concentricity of the I.D. to the O.D. Because the insert seating surface will be machined concentric with the insert O.D., the I.D. to the O.D. concentricity and runout will affect seat width variation. If the as-cast or as-sintered concentricity is off too much, machining of the I.D. will be required to reduce seat width variations. Another alternative is to machine a slight taper between the seat and the I.D. to control the seat width variation. A taper also can be used to adjust the size of the seat width if desired. The use of increased seat width sometimes has been successful in reducing seat wear. A profiled or venture-shaped I.D. may be used for flow optimization. Again, the most economical method to accomplish this is in the as-cast or sintered state, if feasible. Machining an I.D. contour is expensive because it must be done either with form tools or by single-point computer numerically controlled (CNC) profiling.

3.5.2.7 Swirl-Type Inserts

To improve combustion efficiency and reduce emissions on certain engine families, it is desirable to create a swirl motion of the intake flow. This motion may be accomplished in a number of ways: by porting changes, by incorporating a masked intake valve, or by installing a separate swirl plate or baffle (usually a stainless steel stamping) below the intake valve seat insert. Another possibility is to make the swirl plate integral with the valve seat insert (Figure 3.80). This can be incorporated in either a P/M insert or in a casting.

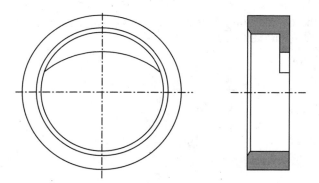

Figure 3.80 *Swirl-type insert (Dooley et al.).*

3.5.2.8 Interference Angle, Overhang, and Underhang

It is common to make the insert seat angle 0.5–1.0° larger than that of the valve seat face angle so that the interference angle (α) is in the range of 0.5–1.5°. This interference angle causes the valve to seat on the combustion chamber side of the insert seat with higher localized pressure and improved sealing. This can be particularly useful during initial engine operation until the valve and the insert have had sufficient time to wear in together. The valve head O.D. should overhang and underhang the finished seat by approximately 0.5–1.0 mm. The valve seat angle versus the insert angle should be tolerance stacked to provide line contact at the seat O.D. Under no circumstances should the tolerance stack of the angles produce a negative interference angle or a crevice opening toward the combustion chamber. For best results, finish machining of the insert seats should be done with three angles. The top cut and throat cut will provide uniform seat contact widths. Uniform seats promote uniform valve head cooling and more predictable wear characteristics. Contact widths are in the range of 1.00–1.50 mm on intake valves and 1.50–2.00 mm on exhaust valves for gasoline engine applications. For diesel engines, the contact widths should increase accordingly due to higher combustion pressure and potential seat wear problems.

3.5.3 Insert Materials

Valve seat insert materials are designed to provide wear resistance in an environment subject to corrosion at moderate to high temperatures of approximately 200–540°C. They also must have a thermal expansion rate similar to that of the engine head counterbore at operating temperatures to minimize retention problems, and they must be produced at a reasonable cost. Furthermore, the inserts must be compatible with the valves with which they are coupled, so that the two components will exceed the warranty service requirements of the engine. Valve seat inserts usually are metallic rings and are manufactured using cast, wrought, or sintered PM processes. Wear resistance generally is related to the volume fraction of hard phases such as martensite, carbides, nitrides, and oxides. The major alloy categories in each of these groups follow, with a brief metallurgical description of each category.

3.5.3.1 Powder Metal Valve Seat Inserts

Powder metal (PM) inserts are used widely in automotive engines. The PM process has long been recognized as a cost-effective manufacturing method because of near-net shape production, the versatility in raw material selection, post-sintering treatments, and labor and energy savings. The PM process typically consists of blending the raw metal powders, plus any necessary lubricants or other additives, filling a die with the blended powder, pressing the powder together in the die to form a "green" compact, and then sintering at high temperatures to bond the powder particles together. Figure 3.81 shows a typical PM processing flow-chart, including optional operations when desired. The base powders can be prealloyed, partially prealloyed, or not prealloyed. Prealloyed powder is believed to give better wear resistance but poor compressibility. Tool steel powders are preferred in seat insert applications. The type and amount of solid lubricant/additive easily can be blended in as desired. The parts can be heat treated, such as by hardening, carburizing, or steam treatment, to the desired properties.

A number of special operations may be utilized with PM valve seat inserts, as described in the following sections.

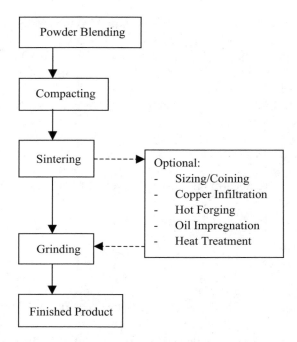

Figure 3.81 *Typical processing flow of a PM insert.*

3.5.3.1.1 Powder Forging

Powder forging is a secondary operation that can be employed to increase the density of PM inserts and subsequently improve their mechanical properties. The process consists of fabricating a preform by conventional press and sintering and then forging that preform to minimize contained voids.

3.5.3.1.2 Liquid Phase Sintering

Liquid phase sintering also has been used to achieve a high-density valve seat insert. Here, the powder blend and the sintering temperature are specially selected so that a liquid begins to form in the compact during sintering. This liquid phase leads to both rapid and a high degree of densification. Considerable dimensional changes may occur in the compact during liquid phase sintering. These changes may require the machining of geometric features that normally would be pressed only in the green compact.

3.5.3.1.3 Infiltration

Another method of filling the pores remaining in a PM insert is to infiltrate the solid compact with a liquid. This differs from liquid phase sintering, in that the source of the liquid is external for infiltration methods. Capillary action draws the infiltrating liquid into the compact. In practice, a second compact of the infiltrating material is pressed into a ring shape. This infiltrating ring is placed either below or on top of the insert compact, and the two-piece assembly is run through a furnace to produce the infiltrated valve seat insert.

Lead has been employed as an infiltrant in several insert applications. The purpose of the lead is not to enhance strength but to act as a lubricating medium. In essence, it replaces the lead compounds that previously were found in gasoline fuel. The use of lead raises serious concerns regarding insert manufacture in terms of the potential for lead in the manufacturing environment and the potential waste disposal concerns.

Infiltration with copper is popular for PM parts and has been used successfully for a number of PM valve seat inserts. There are a number of advantages with copper infiltration. Unlike lead, copper improves the mechanical properties of the insert. Size control of the insert compact also is improved. The presence of copper will increase the thermal conductivity of the insert. As discussed previously, this increase can lower both the insert and valve operating temperatures. Machinability likewise is improved.

3.5.3.1.4 Solid Lubricants

As discussed, infiltration with materials such as lead and copper can provide a degree of lubricity to the valve seat insert. Additives also can be mixed into the powder blend to accomplish the same function and may improve machinability, too. Common solid lubricants include MnS, MoS_2, CuS_2, CaF_2, and graphite.

3.5.3.1.5 Dispersed Hard Phase

One technique available to the PM process is to add a hard phase material to the powder blend. This results in a dispersal of the hard phase throughout the microstructure. The hard phase is bound with the matrix and transmits loading pressures to that matrix. This essentially is similar to that of many cast alloys, where a dispersion of hard primary carbides is supported by a tougher alloy matrix. Various metallic alloys have been used as the dispersed hard phase in PM insert alloys such as the Co-Mo-Cr-Si Laves phase alloy and the Co-Cr-W-C hardfacing alloy. Non-metallics, such as ceramics, and cerments also have been employed.

Powder metal offers the potential advantage of pressing in many of the geometric features of an insert, such as the lead chamfer/radius, the I.D., and a rough seat. This process offers a reduction in machining requirements, in that the sintered parts must be ground only to size on their faces and O.D., which involves fast machining operations. On the downside, PM parts normally contain some amount of residual porosity that may limit their ability to function in certain applications.

Typical microstructures of PM seat inserts are listed in SAE J1692. Martensitic or tempered martensitic structure is the primary structure for an intake insert. Steam treatment may be applied in some applications and results in an oxide layer at pores for increased wear performance. An exhaust seat insert has hard particles such as carbides or Fe-Mo dispersed in the martensite and austenitic matrix as enforcement, for wear resistance.

Heavy-duty valve seat inserts for diesel and gasoline engines usually are hardened and tempered. Small carbides together with solid lubricants are embedded in the matrix, and pores are infiltrated substantially with copper.

High-performance exhaust and intake valve seat insert materials for diesel and gasoline engines, liquefied petroleum gas (LPG), compressed natural gas (CNG), and flex fuel applications have a hard phase embedded in a martensitic/bainitic matrix with evenly distributed solid lubricants.

3.5.3.2 Cast Inserts

Grey iron finds some use for light-duty applications in both the as-cast and heat-treated conditions. The presence of graphite in the microstructure improves machinability and provides a measure of lubricity for the mating valve. Heat-treated ductile iron has seen limited duty as a mate for titanium valves. The high volume fraction of primary carbides present in alloyed white cast irons makes these alloys well suited for insert applications, particularly for intake use. Austenitic cast iron has a high thermal expansion coefficient and is used for aluminum head intake applications. This iron is alloyed heavily with nickel and copper to stabilize the austenitic phase. It is modified from normal austenitic cast iron, in that the casting is designed not to have free graphite. Carbon exists in the form of carbide for wear resistance.

Cast tool steels (usually of the molybdenum variety) and martensitic stainless steels commonly are employed for heavy-duty intake and moderate-duty exhaust applications. The microstructure of these alloys typically consists of a matrix of tempered martensite surrounded by interdendritic alloy carbides.

The iron-based alloys described here have operating temperature limits in the vicinity of 430–600°C. For temperatures higher than this, nickel- or cobalt-based wear-resistant alloys can be employed. The microstructures of the nonferrous alloys normally consist of primary carbides, interdendritic carbides, and either a nickel- or cobalt-based matrix. Nickel-based alloys are much less expensive than the cobalt-based alloys. Nickel-based alloys have a relatively low compressive yield strength that must be factored into the insert design by means of a higher aspect ratio for retention. The high alloy content of both nickel and cobalt alloys significantly reduces their thermal conductivity. Cobalt-based alloys, such as Tribaloy, utilize a hard corrosion-resistant intermetallic Laves compound in their microstructure rather than carbides. They are one of the premiere heavy-duty intake insert materials. Laves phase alloys are expensive, however, due to both the inherent cost of the constituent alloying elements and the difficulty in machining the alloy.

In general, automotive gasoline engines use PM inserts and sometimes cast iron inserts, whereas heavy-duty diesel, CNG, and other dry fuel engines adopt cast alloy inserts. Table 3.13 lists the typical insert material chemistries that are shown in SAE J1692.

TABLE 3.13
TYPICAL AVERAGE CHEMICAL COMPOSITIONS
(IRON BALANCE UNLESS SPECIFIED) (SAE J1692)

Trade Name	C	Mn	Si	Cr	Ni	Co	W	Cu	Mo	Others
Bleistahl Produktions GmbH										
AR 16 FS	0.7/1.1			3.5/4.5			5.5/6.5		6.0/7.0	V 1.5/2.5, S 1.0/1.4
CN 024 FS	1.5/2.5	0.2/0.6			2.5/3.5					S 0.3/0.6
CN 10CU	0.8/1.2				0.3/0.7			8/12		S 0.2/0.5
COMO 12	0.6/1.1				1.0/2.0	10/12			1.0/2.0	Pb 0.5/1.5
COMO 12 07 Cr	0.8/1.2	0.3/0.7		2.0/3.0	1.0/2.0	9/11			1.5/2.5	S 0.3/0.7, Pb 0.5/1.5
COMO 12 FS	0.6/1.0	0.3/0.7			1.0/2.0				2.0/3.0	S 0.2/0.7, Pb 0.5/1.5
COMO 406 FS	0.8/1.2	0.4/0.8			1.0/2.0	3.0/5.0			2.5/3.5	S 0.4/0.6, Mo 2.5/3.5
FCN 335 15	0.5/1.0				2.5/3.5				0.7/1.1	S 0.4/0.8, Pb 2.0/3.5
FCR 2 05 FS	0.2/0.6	0.3/0.7		1.5/2.5	1.8/2.8					S 0.1/0.4
MS 20 CU	0.4/0.8			0.5/1.5	2.5/3.5		0.4/0.6	14/20	2.5/3.5	S 0.3/0.5
Engineered Sintered Components, Inc.										
E 554 M	1.3/2.0	0.5/2.0		2.0/7.0	4.0/7.0			1.5/2.0	5.0/9.0	N .05/0.3, S 0.2/0.6
E 554 M Cul	1.0/1.5	0.4/2.0		1.7/6.0	3.6/6.5			10/20	4.0/8.5	N .04/0.3, S 0.2/0.6
EMS 541	0.7/1.1	3.0/4.7		8.0/11	0.8/1.5			1.5/2.5		N 0.1/0.3, S 0.1/0.3
EMS 542	0.6/0.9	0.2/0.5			0.4/0.6			1.5/2.5	0.7/0.9	
EMS 544	0.7/1.1	3.0/5.0		8.0/11	0.8/1.8			1.5/2.5	0.3/0.5	S .04/.15, N 0.13/0.30
EMS 554	1.3/2.0	1.0/2.7		4.0/6.5	1.6/2.4			2.5/4.0	0.3/0.5	S .05/.35, N 0.13/0.30, Mg 0.1/0.2
V 625	0.6/1.5			6.0/10				1.0/2.0	2.0/4.0	V 0.5/2.5
V 625 Cul	0.6/1.5			4.0/9.5				10/20	1.8/3.8	V 0.4/2.5
V 626	0.6/1.5			2.5/4.0			4.0/6.0	1.0/3.0	4.0/6.0	V 1.0/3.0
V 626 Cul	0.6/1.4			2.5/4.0			4.0/6.0	10/20	4.0/6.0	V 1.0/3.0
Federal-Mogul Sintered Products										
PMF-15	0.6/1.4	<0.75		3.5/7.5			4.5/7.5		3.5/7.5	V 1.5/3.5
PMF-16	0.6/1.4	<0.75		3.5/7.5					3.5/7.5	
PMF-2	0.5/1.4		<0.5		<0.75			1.8/3.8	<1.0	
PMF-22	0.6/0.9	<0.6		1.5/2.5			2.8/3.8		3.5/6.5	V 0.5/1.5
PMF-22D	0.8/1.2	<0.75		1.0/2.2			2.0/3.0		2.5/4.5	V 0.4/1.0
PMF-25	0.6/1.1	<0.6		2.0/4.0			4.0/6.0		4.0/7.0	V 0.5/2.5
PMF-27	0.8/1.2			2.0/3.5			5.0/6.5		3.5/6.0	V 1.5/2.5
PMF-3	0.5/1.4	<0.5			<0.75			1.8/3.8	<1.0	
PMF-33	0.7/1.1							0.5/1.5		
PMF-36	0.9/1.4			2.5/5.0		1.5/3.0	7.0/9.0	8.0/12	4.0/6.0	V 2.0/4.0
PMF-4	0.3/0.9			12/15						
Fuji Ozecks										
FMS 601	0.7/1.4			1.5/2.5						
FMS 604	0.8/1.2	0.4/1.0	0.1/0.3	2.0/3.0			0.6/1.5		0.4/1.0	
FMS 608	0.8/1.2			1.0/2.0	0.3/0.7	0.8/2.0				
FMS 610	0.8/1.2			1.5/2.5	1.5/2.5			Cu	0.4/1.0	(Sintering—hot forge)
FMS 612	0.7/1.3			3.0/5.0	0.5/2.0			Cu		S
FMS 613 base	0.2/1.0									Double layer
FMS 613 seat	0.8/1.2			1.0/2.0	0.3/0.7				6.0/8.0	Double layer
FMS 614	0.7/1.3			1.0/3.0	1.0/2.0			Cu		S
FMS 615	0.7/1.3			1.0/3.0	0.5/2.0	3.0/6.0			5.0/8.0	Nb (Sintering—hot forge)

TABLE 3.13 (*Continued*)

Trade Name	C	Mn	Si	Cr	Ni	Co	W	Cu	Mo	Others
Hitachi P/M Co., Ltd.										
EH-2	0.7/1.1			2.0/4.0					3.5/6.0	V 1.5/2.5
EH-8	0.6/1.1		0.1/0.3	1.0/2.0	0.5/1.0	5.0/10			1.5/4.0	V 0.1/0.2
EH-9	0.6/1.1			1.0/4.0	0.3/3.0		0.5/2.0	0.5/2.0	2.5/8.0	
EH-10	0.6/1.1		0.1/0.5	2.0/4.0	0.5/2.0	10/15			4.0/7.0	V 0.1/0.2
EH-11	0.4/0.8		0.1/0.5	0.5/2.0	0.5/1.0	10/15			3.0/7.0	Pb 13/23
Imperial Piston Ring										
V513	0.9							2.0	0.6	Other <1.0
V514	1.3			5.0					0.5	Other <1.0
V521	0.9			1.5				12	1.5	Other <1.0
V532	1.1			4.0	1.5	6.0			4.0	Other <1.0
V536	1.2			5.0	2.0	8.0			5.0	Other <1.0
V537	1.2			5.0	2.0	7.0			8.0	Other <1.0
V556	1.0			3.5	1.5	5.5		12	3.5	Other <1.0
V557	1.2			6.5	2.0	6.5		12	6.5	Other <1.0
V567	1.1			4.5	2.0	6.5		12	4.5	Other <1.0
V571	1.2			6.5	2.0	6.5		12	6.5	Other <1.0
V573	1.1			6.5	2.0	6.5		12	6.5	Other <10
Keystone Carbon										
VMS 622	1.6/2.2	0.2/0.5		0.1/0.3	0.7/0.9	0.3/0.5		<0.5	0.4/0.7	
VMS 643	0.8/1.2	<0.5	<0.5	1.5/2.5	0.1/0.4		2.3/3.5	8.0/10	7.0/9.0	
VMS 644	0.7/0.9	0.5/0.9	<0.5		0.3/0.6			1.8/2.3	0.7/0.9	
LE Jones Company										
AMS 5700	0.4/0.5	<1.0	0.3/0.8	12/15	12/15				<0.5	
AMS 5710	0.8/0.9	0.2/0.6	1.9/2.6	19/21	1.0/1.6					
J10	<0.08		2.2/2.6	7.5/8.5		BAL			27/30	Fe+Ni <3.0
J100	2.0/2.8		<1.0	27/31	BAL	9.0/11	14/16			Fe <8.0
J101H	<3.0	1.0/1.5	1.0/2.8	1.8/3.0	14/18			5.5/7.5		
J120	1.2/1.5	0.3/0.6	0.3/0.6	3.5/4.3	<1.0		5.0/6.0	<0.25	6.0/7.0	S <0.1
J120V	1.2/1.5	0.3/0.6	0.3/0.6	3.5/4.3	<1.0		5.0/6.0	<0.25	6.0/7.0	V 1.3/1.7
J122	2.5/3.0	0.5/0.8	1.5/2.5	2.8/3.3					4.0/5.0	
J125	1.5/1.8	0.2/0.6	1.9/2.6	19/21	1.0/1.6					S <0.1, P <0.15
J3	2.3/2.6	<1.0	<1.0	29/32	<3.0	BAL	11/14			Fe <3.0
J589	2.9/3.4	<0.5	0.5/1.5	16/19					16/18	V 1.7/2.1
J6	0.9/1.4	<1.0	<1.5	27/31	<3.0	BAL	3.5/5.5	<1.5		Fe <3.0
J70	2.4/2.8		<1.25	27/32		BAL	15/19	<1.50	1.8/2.8	Fe <9.0
J96	2.0/2.8		<150	27/31		BAL	14/16			Fe <8.0
JP200	0.4/1.1	<0.9			<1.0			1.0/3.0	0.2/1.5	S <1.5
JP21	0.5/1.2			2.0/4.5			3.0/6.0	10/25	2.5/6.5	V 1.0/4.0
JP300	0.5/1.5	<0.8		2.5/7.0			4.0/10		3.0/10	V 0.5/4.0
JP350	0.5/1.2			1.0/4.0			2.0/5.0		3.0/6.0	V <2.0
Mitsubishi Material										
FH-15M	1.0			0.6	1.0					S <0.9
FH-15MH	1.0			1.5	1.0			3.0		S <0.8
TC-1	1.0			1.8						
W-15F	0.7/1.3			1.0/3.0	0.5/2.0	3.0/6.0			5.0/8.0	Nb 0.5/1.0
W-16F	0.7/1.3			1.0/3.0	0.5/2.0	3.0/6.0			12/15	Nb 0.5/1.0
W-17F	0.7/1.3		0.3/0.8	1.7/4.0	0.5/2.0	9.0/14			8.0/13	Nb 0.5/1.0
W-20	1.0			1.5	0.5	1.0			7.0	

TABLE 3.13 (*Continued*)

Trade Name	C	Mn	Si	Cr	Ni	Co	W	Cu	Mo	Others
Mitsubishi Material *(continued)*										
W-22	1.0			1.5	0.5	1.0			12	S <0.1
W-22CS-7	0.5/1.2		0.1/0.6	1.6/3.2	0.5/1.2	7.0/13		10/19	6/11	Nb 0.2/0.7
W-22DC	0.9			1.3	0.4	0.8		16	10	S <0.1
W-23CD-53C	0.2/0.8			1.3/3.6				10/20		Other <3.0
W-23CD-23C	0.5/1.7		0.1/0.8	3.3/7.8	0.4/1.4	7.6/16		10/20	2.1/5.1	
W-23D-23	0.7/1.8		0.2/0.8	4.2/8.6	0.5/1.5	9.6/18			2.6/5.6	
W-23D-53	0.3/0.8			1.7/4.0						Other <3.0
W-3	1.0			1.8					7.0	
W-6	1.0			1.5	0.5			3.0	5.0	
W-7	0.7/1.3			1.0/2.0	1.0/3.0				4.0/6.0	
MWP Pleuco GmbH										
PL 12M	1.8/2.3	<0.6	0.8/1.2	12/14	<0.5				2.0/2.5	
PL 33M	1.8/2.3	<0.6	1.8/2.3	33/35	<0.5				2.0/2.5	
PL 476	1.2/1.6	<0.7	<1.0	3.5/4.5			5.0/7.0		6.0/8.0	S<0.15
PL 7N	3.1/3.5	0.4/0.8	1.8/2.4	0.5/0.8	0.8/1.2				1.0/1.3	P 0.4/0.6
PLCoCr30W15	2.0/2.3	<1.0	<0.4	28/32	<3.0		10/15		BAL	B<1.0
PL Ni40	2.0/2.4	1.0/1.3	1.5/1.9	11/13	38/42		0.4/0.6		5.0/7.0	
PL 12 V4000	1.8/2.3	1.0	0.8/1.3	12/14	<0.5				2.0/2.5	
PL 33/20	1.8/2.3	<1.0	1.8/2.1	33/35	<0.5				2.0/2.5	
PL 500	1.0/2.0	<1.5	<1.5	5.0/10					9.0/15	V 2.0/4.0
PL 7 V400	3.1/3.5	0.6/0.8	1.8/2.2	0.5/0.8	0.8/1.2				1.0/1.3	P 0.4/0.6
PLS 085	0.8/1.3				1.3/2.5				0.8/1.2	S 0.2/0.4
PLS 100	1.0/1.4							8.0/10		
PLS 250	0.6/1.1	1.5/2.2		2.5/3.5			4.0/5.5	15/25	3.5/4.5	S 0.6/0.9
PLS 300	0.3/0.6			1.4/1.7	2.0/3.0			18/20	4.5/5.5	
PLS 301	0.3/0.6			0.5/0.8	3.0/3.5			18/20	2.4/3.0	
PLS 340	<1.0			0.5/1.5	2.0/4.0			15/20	3.0/4.0	S <0.5
PLS 345	<1.0			1.0/3.0	2.0/4.0			15/20	5.0/7.0	S <0.8
PLS 350	0.4/0.7	1.5/1.9		1.2/2.0	2.2/2.8			16/22	4.5/5.5	S 0.6/0.9
PLS 380	<1.0	<2.0		<4.0	<2.5			15/25	8.0/11	S <1.0
Nippon Piston Ring										
PB1	0.8/1.6			7.0/10	1.0/3.0	7.0/12	2.0/4.5		0.3/0.8	
PB18	1.0/1.6			5.5/8.0	1.0/3.0	5.0/8.0	1.5/3.0	11/18	0.3/0.8	
PB1E	0.8/1.6			7.0/10	1.0/3.0	7.0/13	2.0/4.5		0.3/0.8	
PB21	1.1/1.6			7.0/10	1.0/3.0	2.0/6.0	3.0/4.5			
PB37	0.9/1.5			3.0/5.0	1.0/3.0	0.5/3.5	0.5/2.0	6.5/11	0.3/0.8	
PB37A	0.9/1.5			3.0/5.0	1.3/3.0	0.5/3.5	0.5/2.0		0.3/0.8	
PB38	0.9/1.5			2.5/4.5	1.0/3.0	0.5/3.5	0.3/1.4	11/18	0.3/0.8	
PB42	1.2/1.6			8.5/11	1.0/2.5	7.5/9.5	2.5/4.0		0.4/0.8	
PB43	1.0/1.5			5.0/7.5	0.8/2.3	6.0/8.0	1.8/3.3	4.5/14		
PB47W	1.46	0.43	0.11	9.0	2.0	8.0	3.0			
PB50	1.0/1.5			5.0/7.5	0.8/2.3	6.0/16	1.5/3.0	8.0/16		
PB55W	0.6/1.1	<0.3	<1.0		3.0/4.0	6.5/7.5		2.5/3.5	7.0/9.0	
PB6	0.9/2.0								0.4/0.8	
PB6B	0.9/1.5							2.0/6.0	0.4/0.8	
PB6S	0.9/1.5							2.0/3.0	0.4/0.8	
PB7E	1.0/1.6			5.5/8.0	1.0/3.0	5.0/8.0	1.5/3.0	3.0/5.0	0.3/0.8	
PX190	0.6/1.5			5.9/9.0	0.8/2.7	5.9/11	1.7/4.1		0.3/0.8	

TABLE 3.13 (*Continued*)

Trade Name	C	Mn	Si	Cr	Ni	Co	W	Cu	Mo	Others
Nippon Powder Metal										
PMZ1026	0.47	0.18	0.12		1.81	10	0.73		6.06	
TVA24	0.7/1.2					7.0/10			4.0/6.5	Pb 10/22
TVA28	0.6/1.2					2.0/3.0			4.0/6.5	Pb 10/22
EH-4	0.5/1.0			1.5/3.5					0.2/0.4	Pb 10/20, Other <1.0
EH-10	0.6/1.1			2.0/4.0	0.5/1.0	10/15			4.0/7.0	V 0.1/0.2, Other <1.0
EH-11	0.6/1.1			2.0/4.0	0.5/1.0	10/15			4.0/7.0	V 0.1/0.2
EH-11H	0.6/1.0		0.1/0.5	0.5/2.0	0.5/2.0	10/15			3.0/7.0	Pb 8.0/18
EH-2	0.7/1.1			2.0/4.0					0.2/0.4	Other <1.0
EH-5	0.5/1.0				1.0/2.0	3.5/7.5			1.0/2.0	Pb 10/20, Other <1.0
EH-5H	0.5/1.0				1.0/2.0	3.5/7.5			1.0/2.0	Pb 5/15, Other <1.0
EH-6	0.7/1.1			1.0/2.0	0.5/1.0	2.0/4.0			0.6/1.2	V 0.1/0.2, Other <1.0
EH-7	0.6/1.1			1.0/2.0	0.5/1.0	3.5/8.0			1.5/4.0	V 0.1/0.2, Other <1.0
EH-8	0.6/1.1			1.0/2.0	0.5/1.0	5.0/10			1.5/4.0	V 0.1/0.2, Other <1.0
EH-9	0.6/1.0			1.0/4.0	0.5/3.0				2.5/8.0	Other <1.0
Riken										
SRF V2	0.7/2.0			0.7/1.3	1.0/1.5				0.3/0.5	
SRF V3	0.7/2.0			3.0/5.0	0.5/3.0	0.2/1.0			3.5/6.0	V 0.05/0.4
SRF V4	0.5/1.5			5.0/8.0		0.5/2.0	1.5/4.0		0.1/0.4	
SRF V5	0.7/2.0			1.0/5.0		0.5/2.0	1.0/3.0		2.0/6.0	
SRF V8	0.7/2.0			3.0/8.0	1.0/3.0	6.0/10	1.0/2.5			
SRF V9	1.1/2.0			9.0/14					0.5/1.5	Ti 5.0/10
SRF V4WFQ	0.5/1.5			5.0/8.0		0.5/2.0	1.5/4.0		0.1/0.4	V 0.05/0.4, Other <5.0
SRF V9	1.1/2.5		<1.0	9.0/14					0.5/1.5	Ti 5.0/10, V <2.0, Other <5.0
SRF V12A	0.8/1.8				8.0/12				8.0/20	Other <5.0
SRF V12B	1.2/1.5		0.2/2.0		2.0/12				6.0/18	Other <5.0
SRF V15B	0.6/1.0			0.5/1.5	4.0/8.0				10/14	Solid lubricant 3.0/7.0
SRF V23	0.5/1.5			2.0/4.0					0.1/2.0	S <0.4, Other <5.0
SRF V25	0.7/2.0			0.7/1.3	1.0/3.0			2.0/4.0	0.1/0.5	Other <5.0
SRF V3	0.7/2.0			3.0/5.0	0.5/3.0	0.2/1.0			3.5/6.0	Other <5.0, S 0.1/0.4
SRF V31	0.7/2.0			1.4/3.4	0.5/3.0	0.1/0.7			2.0/4.5	Other <5.0
SRF V32	0.7/2.0			1.0/3.0	0.5/3.0	0.1/0.7			2.0/4.5	Other <5.0
SRF V41	0.7/1.8			8.0/11		0.8/1.8	2.0/5.0		0.7/1.4	V 0.7/1.4, Other <5.0
SRF V5K	0.5/1.5			1.0/5.0		0.5/2.0	1.0/3.0	10/20	2.0/6.0	Other <5.0
SRF V5W	0.5/1.5			1.0/5.0		0.5/2.0	1.0/3.0	<3.0	2.0/6.0	Other <5.0
SRF V7W	0.5/1.5			5.0/8.0		0.5/2.0	1.5/4.5		0.1/0.4	V 0.05/0.4, Other <5.0
SRF V8	0.7/2.0			3.0/8.0	1.0/3.0	6.0/10	1.0/2.5			Other <5.0
SRF V83	0.7/1.8			8.0/11		0.8/1.8	2.0/5.0		0.7/1.4	V 0.7/1.4, Other <5.0
SEC										
SVS3	0.5/1.5			4.0/6.0	1.5/2.5	2.0/4.0			0.2/0.6	Other <2.0, Sinter + hot forge
SVS37	0.5/1.5			0.5/1.5	1.5/2.5	4.0/6.0			4.0/7.0	Other <2.0
SVS37F	0.5/1.5			0.5/1.5	1.5/2.5	4.0/6.0			4.0/7.0	Other <2.0, Sinter + hot forge
SVS37I	0.5/1.5			0.5/1.5	1.5/2.5	3.0/6.0		9.0/13	4.0/7.0	Other <2.0 Other <5.0
SVS55	0.5/1.5			1.5/4.5	1.5/2.5				4.0/7.0	Other <2.0
SVS55F	0.8/2.0			1.4/4.5	1.5/2.5				4.0/7.0	Other <2.0
SVS55FH	0.8/2.0			1.4/4.5	1.5/2.5				4.0/7.0	Other <2.0
SVS55I	0.8/2.0			1.4/4.1	1.5/2.5			9.0/13	3.0/6.0	Other <2.0

TABLE 3.13 (*Continued*)

Trade Name	C	Mn	Si	Cr	Ni	Co	W	Cu	Mo	Others
Riken (continued)										
SVS55IH	0.8/2.0			1.4/4.1	1.5/2.5			9.0/13	3.0/6.0	
SVS55IH520	1.1			2.2	1.7			13	5.3	Ca 0.4
SVS5B	0.5/1.5			4.0/6.0	1.5/2.5					Other <2.0
SVS61IH	1.1			5.0	0.4	1.4	2.5	13	1.5	
SVS710	2.0/5.0			15/19						Other <3.0
SVS75	1.0/4.0			13/17						Other <3.0
SVS75FH	2.2			15						Ca 1.0
Winsert										
W100	2.3/2.7	<1.0	<1.0	29/32	<3.0	BAL	11/14			Fe <3.0
W150	<0.08		2.2/2.6	7.5/8.5	<3.0	BAL			27/30	
W230	1.8/25		<1.5	27/31	BAL		<1.0		7.0/9.0	Fe <25
W240	2.0/2.8		<1.0	27/31	46/52		14/16			Fe <8.0
W260	2.0/2.8		<1.0	27/31	37/41	9.0/11	14/16			Fe <8.0
W280	1.3/1.5		<1.0	25/28	BAL	9.0/11	9.0/11		9.0/11	Fe 11/14
W50	2.5/3.0	0.5/0.8	1.5/2.5	2.8/3.3					4.0/5.0	
W60	<3.0	1.0/1.5	1.0/2.8	1.8/2.5	14/18					
W70	1.2/1.5	0.3/0.6	0.3/0.6	3.5/4.3	<1.0		5.0/6.0		6.0/7.0	S <0.1
W70V	1.2/1.5	0.3/0.6	0.3/0.6	3.5/4.3	<1.0		5.0/6.0		6.0/7.0	S <0.1, V 1.4/1.7
W76	1.2/1.5	0.3/0.6	0.4/0.8	4.0/5.0			5.0/6.0		6.0/7.0	V 1.4/1.7
W90	1.3/1.8	0.2/0.6	1.9/2.6	19/21	1.0/1.6					S <0.1, P <0.15
W93	2.0/2.5	<1.0	<1.5	14/18	1.0/2.0	5.5/7.5			10/14	V 1.3/1.7

Notes: BAL = balance; S = sulfur.

3.5.3.3 Other Insert Materials

Wrought alloys are used for valve seat inserts to a limited extent. The starting material may be either a round bar or an extruded hollow tube, if available. Flexible design inserts used in aircraft engines typically are made of either wrought AMS 5700, a modified austenitic stainless steel, or AMS 5710, a modified martensitic stainless steel. Wrought material is a good choice for low-volume work to avoid pattern or die charges. One other application for wrought alloys is that of temperature-check inserts used to evaluate insert operating temperatures. These utilize techniques similar to those employed for temperature-check valves. Inserts are machined from a wrought, fully hardened steel. These inserts then are run in the engine for a predetermined time and set of conditions. The degree of temper softening they undergo is compared to previously established temperature/hardness response curves to estimate operating temperatures at various points within the insert.

Ceramic inserts possess a number of properties that make them attractive for valve seat inserts—in particular, the potential for improved adhesive and abrasive wear resistance. Ceramics have been demonstrated successfully in the application of engine valve seat inserts in reducing valve and insert seat wear. However, the low reliability of ceramics and the high cost of machining remain the major concerns in utilizing ceramics as engine components. In addition, the inherent brittleness of ceramics raises reliability concerns about their use on a widespread production basis, even though inserts are subjected primarily to compressive stresses. Another concern is thermal expansion mismatch with the cylinder head. The solutions to these problems may lie in the ability to develop tougher and more consistent ceramics inserts, as well as more efficient and cost-effective machining methods.

Table 3.13 also shows the chemical compositions of valve seat inserts from various manufacturers.

Insert material selection for a specific engine depends on the fuel type, the valve seat material used, and the expected temperature, load design, and revolutions-per-minute design of the engine, along with the expected service life of the engine. Table 3.14 shows examples of insert and valve material combinations.

TABLE 3.14
EXAMPLES OF INSERT AND VALVE COMBINATIONS
(STRONG AND LIANG)

Fuel Type	Intake		Exhaust	
	Valve	Insert	Valve	Insert
Gasoline	Sil 1 (HNV3) 1541 (NV1)	Sil XB (W90)	21-2N (EV12)	W70V W90
Diesel	Sil 1 21-4N (EV8)	Sil XB (W90) W70 (Tool steel)	Inconel 751 (HEV3) 21-4N+W+Nb	Eatonite 2 (W240) Eatonite (W260)
Dry Fuels	Stellite 6 (VF2) Stellite 1 (VF6) Eatonite 6 VMS 585 (VF11)	Stellite 3 (W100) W77T6 T400 W10	Stellite 6 (VF2) Stellite 1 (VF6) Eatonite 6 VMS 585 (VF11)	Stellite 3 (W100) W77T6 T400 W10

Note: Dry fuel includes compressed natural gas, methane, methanol, etc.

3.5.3.4 Alternatives to Press-Fit Inserts

In press-fit seat inserts, an insulating layer of air exists between the valve seat insert and the cylinder head counterbore. Thus, the rate of thermal conductivity is low, and the thermal load on the valve material becomes more severe. To address this problem, the valve seat is metallurgically bonded directly to the aluminum alloy by means of a laser cladding method. As a result, thermal conductivity between the valve seat and the cylinder head is improved, and the valve seat insert can be made smaller because there is no concern about it coming loose and falling off (Figure 3.82).

As a result of the improved thermal conductivity, the temperature of the combustion chamber wall is lowered by a maximum of 20°C, the engine becomes more resistant to knocking, and the temperatures of the valve and the valve seat insert are reduced by 30–50°C.

The powdered cladding material must be appropriate in terms of both the cladding properties required for an aluminum alloy (e.g., adhesion, toughness) and the characteristics necessary for a valve seat insert (e.g., thermal conductivity, high temperature, wear resistance, lubricity).

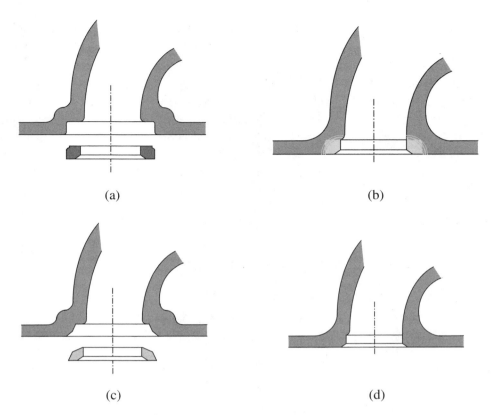

Figure 3.82 *Alternatives to press-fit seat inserts: (a) press-fit seat insert, (b) laser-clad seat, (c) welded or bonded seat, and (d) cast-in seat.*

Table 3.15 shows examples of cladding powder alloys.

TABLE 3.15
LASER CLADDING POWDER CHEMISTRIES (WT%)

Source	Cu	Ni	Si	Fe	Co	V	Mo	Cr	Al	P
Kano *et al.*	Balance	12	3	1.5	10	2	—	—	1	—
	Balance	14	3	1.5	—	2	—	2	1	0.5
Kawasaki *et al.*	Balance	16	3.0	6	8	—	7	1.5	—	—
	Balance	18.5	3	8	—	—	—	—	—	1.3B

Copper is selected as the base material because it has high thermal conductivity, forms relatively stable compounds when joined with other materials, and is commercially available at a low cost. Then nickel, which dissolves completely into a solid solution in the copper matrix when the two metals are alloyed, is added to improve heat resistance and toughness. Next, silicon is added to facilitate destruction of the oxide film on the aluminum alloy so that the cladding material will bond strongly to the aluminum.

Because the valve seat insert can be made smaller, the intake and exhaust valves can be made larger, making it possible to improve fuel economy, torque, and output from the medium-low to high-speed ranges. Figure 3.83 compares the effect of a press-fitted valve seat insert and the laser-clad seat on the torque and power output. It also established that the durability and reliability are more than twice as good as those of the press-fitting method.

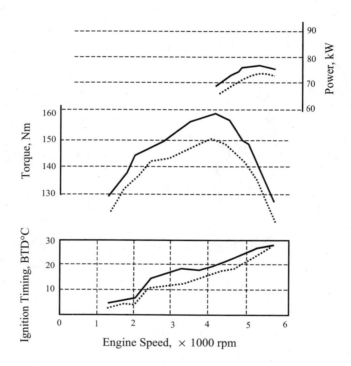

Figure 3.83 *Effects of laser-clad valve seat (solid lines) on engine performance (Sato et al., 1999).*

The defects that may occur after the cladding process can be divided roughly into three types: (1) pinhole, (2) crack, and (3) scale. Pinhole defects occur on the seat surface after the cladding is machined to the finished product. Pinholes may be caused by gas that is trapped inside the casting raw material, or residual oil or water in the pores. Cracks are defects that occur on the seat surface. They occur because toughness is compromised when the cladding layer is diluted excessively by the aluminum alloy. Scale is the residue of cladding bead surface that remains after the final machining of the seat. It is caused mainly by an unstable powder supply.

To address the same problem of a thermal barrel in a press-fit insert, the valve seat insert can be metallurgically bonded to the aluminum alloy directly by means of resistance welding under high pressure or directly cast in the aluminum head. Materials used for both welded and cast-in inserts are similar to those of press-fit inserts. The thickness of seat inserts in the cast-in or welded case does not have to be as thick as the press-fit insert because of the strong metallurgical bond between the seats and matrix.

3.5.4 Insert Material Properties

To perform satisfactorily, a valve seat insert material must possess various specific properties. These properties include compressive yield strength, hot hardness, thermal expansion, elevated temperature stability, thermal conductivity, machinability, corrosion resistance, and wear resistance. Properties data presented in this section serve only as references; they are averaged from various sources. Because material properties are influenced significantly by their chemistries and heat treatment, insert suppliers should be able to provide more accurate material property data pertaining to the specific heat/batch of the insert.

3.5.4.1 Physical Properties

Inserts must be dimensionally and metallurgically stable over the temperature region in which they will operate. Relief of residual stresses or phase changes, such as the transformation of retained austenite, may result in a loss of retention. Normal manufacturing practice is to subject inserts made of a susceptible material to a heat-treat cycle at a temperature higher than that seen in service. Also note that some alloys or phases are inherently unstable, and their use must be restricted only to intake applications.

Density control in PM inserts is extremely critical because it affects compressive yield strength and the ability of the material to resist indentation and material flow. The densities being attained depend on the types of powders and limitations, such as thin-walled punches. Heavy-duty inserts that demand superior density requirements (i.e., >97% of theoretical density) must rely on expensive processing such as liquid phase sintering and PM hot forging. However, the dimensional control of these processes is not as precise as the conventional press and sintering technique, thus negating the near-net processing advantage typically enjoyed by PM fabrication (Figure 3.84).

Ideally, the rate of thermal expansion for the insert material should be similar to that of the cylinder head. With cast iron heads, the expansion coefficients of the metallic inserts usually will match or exceed those of the iron. The results normally are satisfactory. However, aluminum heads have a thermal expansion coefficient (e.g., 20+μm/m/K) that can be much higher than that of the insert. For the exhaust insert, this normally is not a problem because the head flux will cause the insert to run significantly hotter than the surrounding aluminum and thus be retained by differential thermal expansion. However, this is not necessarily true for the intake insert, particularly for aluminum air-cooled engines. In these situations, the insert and surrounding aluminum may be of similar temperatures, causing a loss of interference with increasing temperature and potentially leading to insert drop-out at high temperatures. The use of a high

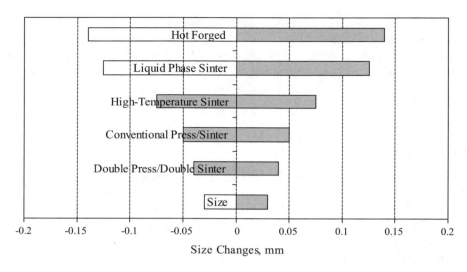

Figure 3.84 *Effect of processing on dimensional tolerance.*

thermal expansion insert material, such as an austenitic alloy, can be of assistance in these cases. Figure 3.85 shows the coefficient of thermal expansion as a function of temperature for various insert materials.

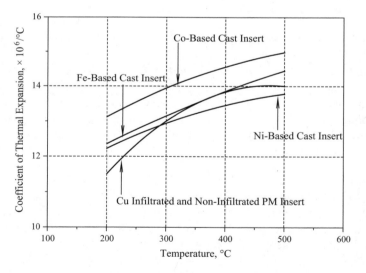

Figure 3.85 *Coefficient of thermal expansion for various insert materials.*

Various researchers have estimated that most of the heat absorbed by the valve flows out the valve seating face. The valve seat insert must provide the medium for this heat flow from the valve to the cylinder head coolant; thus, the thermal conductivity of the insert is important. The model predicts that doubling the insert thermal conductivity from 20 to 40 W/Mk would reduce both the maximum insert temperature by 50°C and the maximum valve temperature by 30°C. Valve seat inserts are not as efficient in conducting heat away as an integral seat would be. The insert material often is highly alloyed and possesses an inherently lower thermal conductivity. In addition, the interface between the insert and the counterbore presents an impediment to the heat transfer. As a result, valves generally will run hotter with inserts than with integral seats. Although high insert thermal conductivity usually is desired, the opposite is needed in low-heat-rejection engines. Here, inserts with insulating properties are needed. Ceramics can meet this need. Figure 3.86 compares the thermal conductivity of nickel- and cobalt-based insert alloys with a cast tool steel alloy.

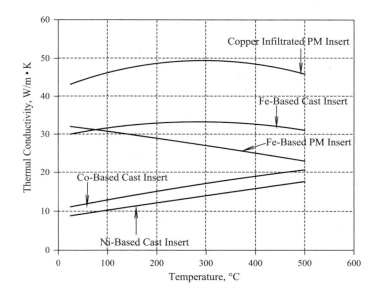

Figure 3.86 *Thermal conductivities of various insert materials (Dooley et al.).*

3.5.4.2 Mechanical Properties

Yield strength is an indicator of insert retention. The interference fit used to retain most inserts in the cylinder head subjects the insert to compressive loading. In addition, temperature gradients, particularly for exhaust inserts, can add a significant amount of thermal compressive stress. Plastic deformation can occur in an insert with a low compressive yield strength, which results in a reduction in the interference fit. In extreme cases, catastrophic total loss of retention or insert "drop-out" may occur. It is not unusual to observe some amount of plastic flow on the combustion chamber side of an exhaust insert because of high thermal gradients. Sufficient compressive yield strength is required throughout the temperature range to which the insert will be subjected.

Correlation of yield strength and hardness has been examined, and the following relation has been developed to relate yield strength to hardness data (i.e., hardness is approximately three times the yield strength). Correlation of yield strength and hardness depends on the strengthening mechanism of the material. For many metals and alloys, a reasonably accurate correlation between hardness and yield strength has been found. However, it must be emphasized that these are empirically based relationships; therefore, testing still may be warranted to confirm a correlation of tensile strength and hardness for a particular material.

Ultimate tensile strength (UTS) is the maximum stress a material can sustain without fracture; however, the radial crush test usually is used to quantify the insert strength in practice. Most insert materials, including cast and PM inserts, are brittle in nature and contain a high percentage of carbides for wear resistance. The radial crush strength is a good gage to ensure the insert does not break during insertion. SAE J1692 provides tensile and yield strength data for various insert materials.

Hot hardness is a measure of plastic deformation or indentation resistance of material at elevated temperatures. Hot hardness has been proven correlating to wear resistance at elevated temperatures and is one of the key requirements for valve seat inserts in the absence of actual wear data. It also is a good indicator of elevated temperature strength and can be used to qualitatively compare several potential material candidates. Sharp drops in hot hardness indicate an operating temperature limitation above which a significant degradation in mechanical properties can be expected. Hardness generally relates to chemical compositions, processing, and microstructure. However, high hardness at room temperature does not necessarily ensure high hardness at operating temperatures. In PM inserts, a distinction must be made between microhardness (a measure of hardness on a particular phase of the material) and macrohardness (bulk or apparent hardness that also is affected by porosity). Figure 3.87 shows hot hardness data for various insert materials.

3.5.5 Wear Resistance

This is the prime property for which a valve seat insert material is selected. Unfortunately, it also is one of the most difficult to accurately quantify and predict because many factors contribute to insert wear, including insert materials, design, cylinder head design, and operating conditions.

Valve seat insert materials consist of a base material containing a solid solution of elements such as chromium, silicon, molybdenum, tungsten, and others. They also contain wear-resistant carbides or hard Laves phases, and solid lubricant for PM inserts. The carbides impart various levels of wear resistance, depending on the type of carbides and the volume fraction percentage of carbides. Iron carbides are the least wear resistant and break down quickly at temperatures as low as 482°C. Chromium forms more stable carbides; molybdenum and tungsten M_6C refractory carbides attain greater hardness and are more thermally stable. The vanadium, titanium, tantalum, and niobium MC-type carbides provide the highest level of stability and wear resistance.

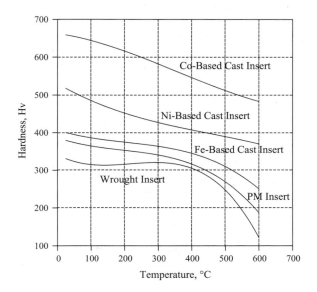

Figure 3.87 *Hot hardness for typical insert materials (Dooley et al.).*

The size, shape, distribution, and volume of carbides have a major impact on the wear life of the insert. The solid lubricant is used extensively in PM inserts for reduction of friction and thus wear. Types of solid lubricants being added to powder or impregnated into the porosity include CaF_2, MoS_2, MnS, graphite, mica, silicates, BN, Cu_2S, Cu_2O, Pb, and FeO/Fe_3O_4. The benefit of solid lubricants in inserts is not as significant as in guides; in addition, they reduce the density and the load-bearing capability. It is advised that the total amount of solid lubricant should be limited to less than 2% wt. The base material, often referred to as the matrix, is equally important. The excellent wear resistance of the hard carbides can be utilized only when the matrix is strong enough to support these carbides. This is achieved through solution and precipitation hardening of an alloy. The ideal matrix also will have the least tendency to form bonding with the valve alloy at complex contact pressure and high-temperature conditions. The ability to form a tenacious, continuous oxide film is another important factor.

From an insert design standpoint, a reduction in seat angle will reduce the tangential force to the insert surface, which consequently reduces insert wear. It is common practice to reduce the seat angle when converting a diesel or gasoline engine to dry natural gas fuel to reduce the severity of wear. An increase in seat width will reduce the load per area or contact pressure and leads to a reduction in wear. A balance must be reached between seat wear and sealing, and reduced pressure may be detrimental to sealing. The concentricity of the insert to the guide and the seat runout also are critical in minimizing insert wear. The large offset and runout will lead to non-uniform contact and less contact area, and thus high contact stress, and can result in severe wear. The interference angle between the valve and insert is another factor that can affect insert wear as well as sealing. A positive 0.5–1° interference angle is recommended, so that under combustion pressure, valve "oil canning" ensures that the whole insert seat is in contact with the valve seat and lands the least contact pressure. Valves also seal better when under only spring load due to the relatively high contact pressure of line contact. A zero or negative interference angle should be avoided, recognizing its detrimental effect on wear and sealing.

The operating environment of an insert is complex and involves the mating valve and seat insert materials, elevated temperatures, combustion deposits, and complex dynamic loading. Although the designed valve seating velocity is very low by proper closing ramp design in the cam, seating velocity increases exponentially as engine speed increases. The valve cannot follow the cam profile closely due to the weak spring rate. However, too strong a spring can cause power loss due to the friction force that the cam needs to open the valves by overcoming the spring forces. Valve bouncing and valvetrain vibration are phenomena associated with high-speed engine operation, and they cause accelerated seat insert wear. The presence of oxide films on the valve and/or insert seat can markedly affect wear rate.

Improper cylinder head design, including the valve arrangement and the port and coolant jacket design, can lead to non-uniform cooling of inserts, resulting in distortion of the head and insert. A distorted head and insert can cause non-uniform seat contact and high local contact stress, consequently leading to severe wear.

Standard room-temperature wear tests, such as pin-on-disk and rubber wheel abrasion, have been used by a number of researchers as a starting point for materials characterization. The logical extension is to then run these wear tests at the elevated temperatures at which the insert operates. Others have developed test apparatus in an effort to simulate engine operation. Ultimately, the final test must be done in the engine.

3.5.6 Machinability

For a commercially viable valve seat insert, the insert material must be able to be machined (i.e., turned and/or ground) to the desired geometry by the manufacturer, or by the engine builder if the insert seat surface is generated after installation. However, the property for which an insert material is selected—wear resistance—typically makes these materials difficult to machine. Fortunately, advances in cutting tool and grinding technology have made it practical for manufacturers to machine these materials on the production bases. A prefinished insert is an option for engine builders to avoid the aggravation of machining the insert after installation. However, the prefinished insert may pose the problem of alignment or concentricity to the guide, which may adversely affect the sealing function of the valve. Figure 3.88 shows factors that affect the machinability of valve seat inserts.

3.5.7 Corrosion Resistance

Inserts are exposed to the effects of oxidation and corrosive gases in addition to elevated temperatures. Generally, chromium is added to metallic insert materials to improve corrosion resistance via the formation of a protective oxide film. The chromium oxide film generally is strong, thin,

Valvetrain Components

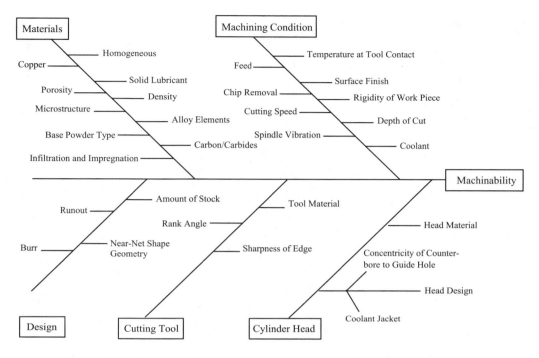

Figure 3.88 *Factors affecting valve seat insert machinability (Rodrigues).*

and tenacious. Materials having inadequate oxidation resistance form thick layers of scale at high temperatures, which are easily broken away. At high temperatures, however, corrodents such as sulfur and vanadium pentoxide may remove this film and cause more rapid attack. The bench tests that have been used with valve materials also can be used to screen insert materials for high-temperature corrosion. One method is to embed the sample into a crucible containing the desired corrodents, heat both to an elevated temperature, and subsequently measure the weight loss per unit area. Other researchers have used inorganic acid baths to simulate service, and others have subjected samples to high-temperature gas mixtures to measure resistance. Because corrosion of inserts has rarely been an issue, little data are available in the literature.

3.6 Valve Guides

Valve guides are installed in internal combustion engine heads as axial bearings for the valve stem, and they hold the valve seat coaxial to the insert seat and guide the valve actuation. They also serve as heat sinks to cool the valves. In cast iron cylinder heads, the guide often is an integral part of the cylinder head; however, with aluminum alloys, a ferrous guide is used. Valve guides typically are hollow metallic cylinders. Other materials such as ceramic and ceramic-coated metals have been used on an experimental basis.

3.6.1 Valve Guide Types and Nomenclature

Figure 3.89 shows a typical valve guide installation, illustrating the relationship of the valve guide to other components.

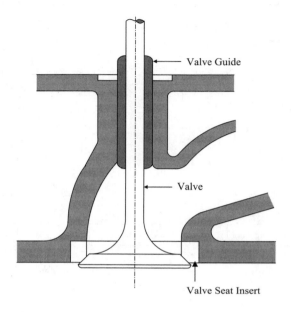

Figure 3.89 *Typical valve guide relationship with other components.*

Figure 3.90 shows the most common types of valve guides and their nomenclature. The most common installation practice is to retain the valve guide through interference fit between the guide and the mating counterbore. In addition to retaining the guide, the fit promotes efficient heat transfer from the guide to the mating counterbore material. Smooth bore guides are most common in passenger car operations (Figure 3.90(a)). For heavy-duty engines, helical grooves often are machined, as shown in Figure 3.90(b), into the valve guide bore for oil retention. However, heavy-duty engine manufacturers are abandoning this practice because of emissions concerns. Figure 3.90(c) shows a guide with a retaining flange that can be used for location and retention.

3.6.2 Valve Guide Design

3.6.2.1 Tolerances

Figure 3.91 lists typical valve guide dimension tolerances and surface finishes for passenger car applications. The diametric clearance between the valve stem and the valve guide typically is

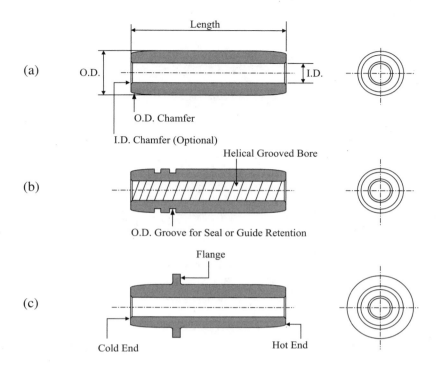

Figure 3.90 *Typical valve guide configurations: (a) smooth bore guides for passenger car engines, (b) helical grooves machined for heavy-duty engines, and (c) a guide with a retaining flange for location and retention.*

0.02–0.05 mm for intake applications and as much as 0.04–0.07 mm for exhaust applications. In general, 0.005–0.010 mm larger clearance is designed for the exhaust side due to different thermal expansion coefficients of the mating materials operating at higher temperatures. The exact clearance is dependent on design considerations such as material compatibility, valve stem thermal expansion, lubrication, emissions, and noise requirements. A smaller stem-to-guide clearance promotes improved stem cooling and minimizes the amount of valve cocking during the lift event. However, if excessive eccentricity of the guide and insert exists, a smaller stem-to-guide clearance can lead to poor valve sealing.

3.6.2.2 Guide Dimension Relationships

The length of the valve guide should be as long as space permits to provide adequate support for the valve and to transfer heat away from the valve. This reduces the contact stresses arising from the cocking action imparted to the valve from side loading. The hot end of the guide should be located as close as possible to the valve head without affecting port restriction. It is good practice to avoid exhaust valve guide protrusion into the hot exhaust gas stream, because this will greatly reduce its ability to serve as a heat sink and increase scuffing tendencies. A typical valve guide

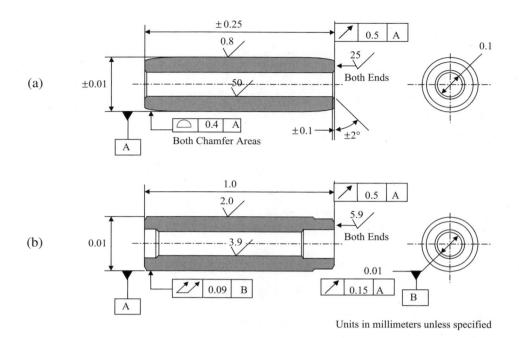

Figure 3.91 *Typical valve guide tolerances and surface finish requirements: (a) I.D. not finished for passenger car applications, and (b) prefinished for heavy-duty diesel applications.*

length for automotive applications varies from 30–50 mm for a Type I direct-acting valvetrain, and from 40–55 mm for other types of valvetrains with rocker arms. Table 3.16 shows typical relationships of the valve guide to the valve, representative of industry averages.

The valve guide length should be as long as the cylinder head and port shape will allow. A good rule of thumb for the length of a guide is seven times the stem diameter. A guide of this length should provide adequate piloting and wear surface to withstand the side loads imparted by the rocker arm or cam follower. In engine applications where no side loading or scrubbing occurs on the valve tip, guide lengths can be shorter. Figure 3.92 shows the typical relationship of the valve guide length and valve guide I.D. and wall thickness, based on the function and tooling limitations. Another rule of thumb is the guide length should be 20 times the wall thickness. For a PM guide, regardless of length, 1.8 mm is the minimum wall thickness that will allow proper powder flow for compaction.

The valve guide and cylinder head bore are interference fit. The recommended interference fit between a valve guide and its mating counterbore is 0.02–0.08 mm. It breaks down to 0.04–0.08 mm and 0.02–0.05 mm for aluminum and cast iron heads, respectively, for PM valve guides. The surface roughness on the O.D. can affect the press-fit assembly of the guide into the engine head. The grinding process required to conform to the strict O.D. specifications normally produces

TABLE 3.16
TYPICAL RELATIONSHIPS OF VALVE GUIDE TO VALVE

	Ratio	Types II–V with Rocker Arm	Type I Direct Acting
	$\dfrac{A-B}{B+R}$	>1.8	>1.3
	$\dfrac{D}{B+R}$	>1.0	>0.95
	$\dfrac{A-R}{C}$	>1.6	>1.4
	$\dfrac{A}{B+Lv}$	>1.3	>1.05
	$\dfrac{D}{B+Lv}$	>0.9	>0.75
	$\dfrac{A}{C-Lv}$	>2.5	>1.75

a surface finish well within the requirements. An extremely coarse or smooth surface may alter the press-fit load and initiate some noise. Valve guides are installed into the counterbore such that the guide can fit directly against the full length of the metal counterbore or so the center of the guide is exposed to coolant. Guides usually are inserted by the following methods:

1. The guide is pressed in at ambient temperatures.

2. The guide is heated with the cylinder head.

3. The guide is chilled with liquid nitrogen, called cryogenic insertion.

Be careful not to crack or deform the guide during installation. Particular care must be taken if the guide bore is not machined after installation. Also, start the guide squarely and press without interruption until it is flush with the counterbore in the cylinder head port surface or reaches its final location. If the guide has a retaining flange, it can be held in place by the valvetrain. The valve guides generally should not protrude into the exhaust port, because this tends to raise the operating temperature of the exhaust valve head. The norm has been to insert the guide at room temperature. Figure 3.93 shows the press-fit load and shrinkage amount of the valve guide I.D. versus press-fit stock. The press-fit load also is a function of the contact area between the head and guide; therefore, these figures will vary according to the guide and head design.

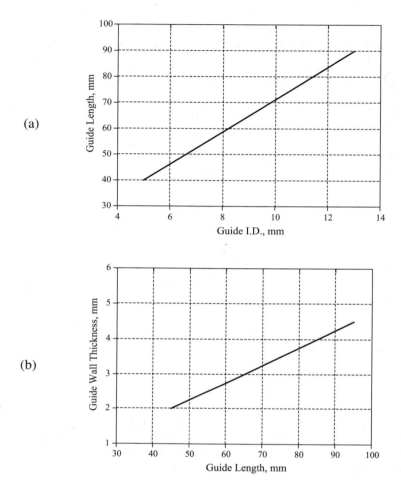

Figure 3.92 *Typical relationship of valve guide length, I.D., and wall thickness:*
(a) guide length versus I.D., and (b) guide wall thickness versus length.

3.6.3 Guide Materials

Valve guides are manufactured using cast, wrought, or sintered PM processes. Wear resistance generally is related to the bulk hardness and volume fractions of the soft and hard phases present, as well as to lubrication. Valve guides may be heat treated or coined for wear resistance. Table 3.17 shows typical chemical compositions of the guides.

Valve guides most frequently are made from pearlitic grey cast iron. The microstructure of these guides should consist primarily of Types A and B graphite, sizes 4 to 7. Types D and E graphite should be held to a minimum. The matrix structure after etching shall consist of a pearlitic matrix with a maximum of 5% free ferrite. The phosphide constituent should be distributed uniformly as a noncontinuous network. Applications requiring greater wear resistance can be obtained by

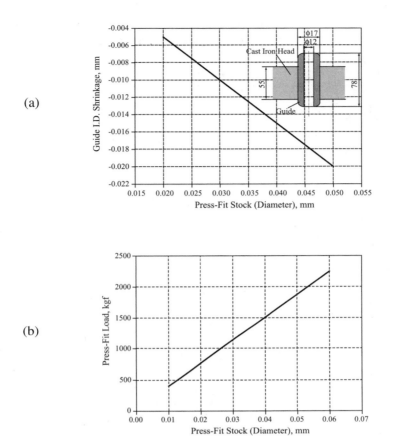

Figure 3.93 *Press-fit load and shrinkage amount: (a) guide I.D. shrinkage, and (b) press-fit load (Fundabashi et al.).*

higher alloy contents providing uniformly dispersed phosphide and/or carbide constituents and by providing a harder matrix consisting of tempered martensite and/or bainite.

Copper-based wrought alloys are used occasionally for high-performance applications to promote more efficient heat transfer from the guide to the mating counterbore. However, these materials generally have inadequate long-term wear durability.

Sintered PM alloys have been developed and are widely used for automotive engine valve guides. One unique feature of powder metallurgy is the ability to form compounds that normally are incompatible through conventional metallurgical processes. Powder metal valve guide compositions are Fe-C-graphite alloys. Alloying additions may include copper, lead, tin, phosphorus, molybdenum, or nickel. A solid lubricant often is added to improve wear resistance. The microstructure consists of free graphite, pores, and a hard phase (such as Fe-P-C) in a lamellar pearlitic matrix. Sintered valve guides can be oil impregnated prior to installation to lubricate the valve guide/valve stem interface, providing additional wear resistance particularly during the

TABLE 3.17
TYPICAL VALVE GUIDE CHEMISTRIES (SAE J1682)

Trade Name	C	Mn	Si	Ni	P	S	Cu	Mo	Sn	Others[1]
Bleistahl Produktions GmbH										
BLV 013 D2	0.5/1.0	0.7/1.3	0.2/0.7				11/16		0.6/1.3	
BLV 073 C8	0.4/0.8	0.4/0.6				0.4/0.7	11/16	0.4/0.8	0.3/0.7	
Engineered Sintered Components, Inc.										
EMS 543	0.5/1.0	0.5/1.0				0.1/0.5	3.5/5.5			Mg 0.2/0.7
V602	1.0/2.0	0.3/1.0				0.1/0.5	4.0/10			Mg 0.1/0.5
V603	1.5/2.5	0.3/1.0			0.2/0.5	0.1/0.5	4.0/7.0			Mg 0.1/0.5
V604	0.5/1.0					0.2/0.7		5.0/10		Cr 10/20, Co 8.0/13
Federal-Mogul Sintered Products										
PMF-10	0.6/1.0			0.2/0.5			1.5/3.5	1.5/3.5		Solid lubricant <3.0
PMF-11	0.6/1.0						1.0/4.0	1.0/3.0		Solid lubricant <3.0
PMF-6B	0.5/1.4		<0.5					1.0/4.5		Solid lubricant
Hitachi P/M Co., Ltd.										
EB-4	1.5/2.5				0.1/1.0		2.0/6.0		0.2/1.0	
HY-LIFT Sealed Power Tech.										
JP-406	1.8/2.2				0.2/0.3		4.0/5.0			Other <2.0
SPVG10	3.1/3.8	0.5/0.9	2.2/2.9		<0.2	<0.15				
SPVG12	3.1/3.6	0.6/0.9	2.2/2.8		<0.2	<0.15				
SPVG14	3.0/3.6	0.6/0.9	1.9/2.6	0.2/0.8	<0.25	<0.15	<0.6	0.2/0.4		Cr 0.2/0.8
SPVG15	3.0/3.5	0.6/1.2	1.8/2.6		0.3/0.6	<0.12				Cr <0.25
SPVG16	3.0/3.5	0.7/1.0	2.0/2.7	0.4/0.7	<0.2	<0.1		0.3/0.7		Cr 0.8/1.3
MWP Pleuco GmbH										
PL 105	3.2/3.5	0.3/0.8	1.8/2.4			<0.12				
PL 106	3.2/3.5	0.6/0.9	1.8/2.2		0.7/0.9	<0.12				Cr <0.2
PL 108	3.2/3.5	0.6/0.8	1.8/2.2		0.4/0.6	<0.12	0.5/0.7	0.4/0.5		Cr <0.2
PL 33 10	1.0/1.4	<0.6	1.8/2.1	<0.5	<0.06	<0.04		2.0/2.5		Cr 33/35
PLS 103	1.8/2.2	<0.3	<0.3		0.4/0.6					
PLS 105	0.8/1.2	0.4/0.6	<0.3	<0.2		0.1/0.2	1.8/2.2			
PLS 115	0.8/1.0	0.4/0.6		<0.2		0.1/0.3	1.8/2.2			
PLS 116	1.3/1.8	0.4/0.6				0.1/0.3	1.7/2.2			
PLS 120	1.8/2.2				0.2/0.5		3.5/4.5		0.4/0.6	
PLS 126	0.6/1.3	<0.8				<0.4	<3.0			

Note: (1) Iron as balance. Solid lubricants are proprietary in nature and are added to improve machinability.

break-in period. Microporosity may be higher in the center of the valve guide length than at the ends. After sintering, the valve guides usually are impregnated with oil (typically ISO-VG-47 equivalent to SAE 20) containing a rust preventive additive for the purpose of enhancing the machinability of the material.

3.6.4 Guide Material Properties

The required properties for a valve guide are heat resistance, mechanical strength, wear and seizure resistance, and machinability.

3.6.4.1 Physical and Mechanical Properties

Table 3.18 shows typical physical and mechanical properties of various guide materials.

TABLE 3.18
VALVE GUIDE PROPERTIES (SAE J1682)

Trade Name	Density (g/cm^3)	E[1] (GPa)	T. Conduct.[2] (W/m°C)	Co. T.E.[3] (µm/m°C)	0.2%YS (MPa)	UTS (MPa)	Hardness[4] (HB)	Process
Bleistahl Produktions GmbH								
BLV 013 D2	6.4	110	33	14		350	120–200	P/M
BLV 073 C8	6.4	120	26	15		490	120–200	P/M
Engineered Sintered Components, Inc.								
EMS 543	>6.5	110	21	9.2	310	330	110–160	P/M
V602	6.5	120	20	9.1	300	320	120–185	P/M
V603	6.5	125	20	9.0	300	310	120–185	P/M
V604	6.5		23	9.6		300	135–210	P/M
Federal-Mogul Sintered Products								
PMF-10	6.5/6.9	117	38	13	540		120–245	P/M
PMF-11	6.3/6.8	109	29	13	415		90–160	P/M
PMF-6B	>6.3	67				350	135–245	P/M
Hitachi P/M Co., Ltd.								
EB-4	6.5	101	29	7.4	280	340	120–160	P/M
HY-LIFT Sealed Power Tech.								
JP-406	>6.2						90–120	P/M
SPVG10	7.1					250	185–245	Cast
SPVG12	7.1					250	200–245	Cast
SPVG14	7.1					250	215–280	Cast
SPVG15	7.1					250	200–280	Cast
SPVG16	7.1					250	210–270	Cast
MWP Pleuco GmbH								
PL 105	7.1	110	50	10		250	200–260	Cast
PL 106	7.3	120	50	10		250	210–270	Cast
PL 108		130	50	10		330	265–410	Cast
PL 33 10	7.8	180	19	10–11		700	310–350	Cast
PLS 103	>6.8	100	42	13	250	300	210–270	P/M
PLS 105		80	42	13	250	300	130–190	P/M
PLS 115	6.2	100	27			300	>100	P/M
PLS 116	6.2	80	27	10–13		250	>100	P/M
PLS 120	6.2	90	27	10–13		250	>105	P/M
PLS 126	6.2	90	27	13–15		300	>100	P/M

Notes:
(1) Elastic modulus.
(2) Thermal conductivity extrapolated to room temperature.
(3) Coefficient of thermal expansion.
(4) Apparent hardness.

3.6.4.2 Wear and Seizure Resistance

Many factors contribute to guide wear and stem/guide seizure characteristics, including guide materials and design, cylinder head design, and operating conditions. However, due to the different functional requirement and tribological characteristics, the degree of influence on the wear and seizure characteristics of the stem and guide varies.

Frictional characteristics are more prominent at the stem and guide interface due to the high-speed reciprocating motion. Surface temperature can be many times that of the guide bulk temperature due to friction heat at the interface. Therefore, lubricant is fed to the stem and guide interface metered through the stem seal. A typical metering rate is 0.01–0.1 cc/10 hours. However, less oil consumption and better emissions control are being mandated, with a trend that the oil metering rate will continue to be reduced. Eventually, frictional behavior will rely only on the improved tribological characteristics of the guide and stem surface materials with little lubricant replenishing.

Open pores in PM guides are beneficial to feed the impregnated oil to the stem/guide interface during operation. Synthetic temperature oils often are used as impregnating oil. However, pores formed in the PM process usually are unavoidable and are detrimental to mechanical properties. Solid lubricants are used more extensively in PM guides than in inserts for reduction of friction and have the propensity for wear and seizure. Types of solid lubricants being added to powder or being impregnated into the porosity include CaF_2, MoS_2, MnS, graphite, mica, talc, silicates, Cu_2S, Cu_2O, and Pb. The benefit of solid lubricants in guides is more significant than in inserts, because of the less stringent requirement of the density and load-bearing capability. The total amount of solid lubricants can be higher than 2% wt. The base material, often referred to as the matrix, is equally important. The excellent wear resistance of the hard carbides can be utilized only when the matrix is strong enough to support these carbides. Pearlite is a preferred microstructure for guide matrices and provides a compromise between good wear resistance and machinability.

The concentricity, straightness, and cylindricity of the guide I.D. bore and the seat runout are critical to guide wear. The large offset and/or runout will lead to non-uniform contact and thus high side load on the guide, resulting in severe wear. The length-to-diameter ratio is another factor that affects guide wear and seizure. The guide surface finish can affect guide wear, but not as much as the effect of stem surface finish. Most stem surfaces are chromium plated or nitrided and have very high hardness. A rough stem surface finish can act as a file to the guide and can cause severe abrasive wear.

Low friction at the stem/guide interface is most desirable and can be achieved by continuously fed oil metered through the stem seal. Temperature has the most detrimental effect on lubricant oil; most oil decomposes when temperatures exceed 500°C. Although the guide bulk temperature may not exceed the oil decomposition temperature, the surface asperity temperature can be very high if the surface is not properly lubricated at high engine speeds. High friction at the stem/guide interface due to poor lubrication and high sliding speed can lead to accelerated guide wear or possible seizure.

Improper head design including the valve arrangement and the port and coolant jacket design can lead to non-uniform cooling of inserts, resulting in distortion of the head. A distorted head causes non-uniform contact and high local contact stress and, consequently, severe wear.

3.6.4.3 Machinability

Good machinability characteristics are essential for the established tool life for valve guide machining operations in engine assembly plants. The factors that affect guide machinability are

Valvetrain Components

similar to those that affect insert machinability. They include the guide material, the I.D. stock amount to be removed, the tool design and materials, and machining conditions (Figure 3.94).

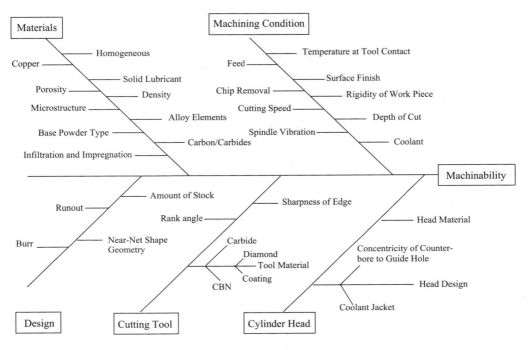

Figure 3.94 *Factors affecting valve guide machinability (Rodrigues).*

Guide materials including the matrix material, volume percentage of carbides and pores, the amount of solid lubricant, and impregnated liquid oil influence the guide machinability. Pearlite structure is the preferred matrix; martensite will be hard, and ferrite or austenite will be too gummy for tooling. If not impregnated with oil or not infiltrated with copper, pores can be detrimental to tool life. The microstructural variation should be kept to a minimum for consistent tool life. Figure 3.95 shows the effect of hardness of the guide material on tool wear. Using a carbide tool, the harder guide material results in more tool wear. However, the cubic boron nitride (CBN) tool shows less sensitivity to guide material hardness.

Tool material and design obviously affect guide machinability as well. Valve guide machining involves reaming the hole using a carbide reamer, following insertion of the valve guide into the cylinder head. Although various styles of reamers are in use, a TiN-coated spiral-shaped multifluted design carbide reamer has been shown to be effective in reaming guides. Machining conditions such as the cutting speed, feeding rate, reaming stock per pass, machining fluid, and the type and rigidity of the reaming machine also influence the quality and efficiency of guide machining. Long guides with small I.D.s can result in premature reamer breakage.

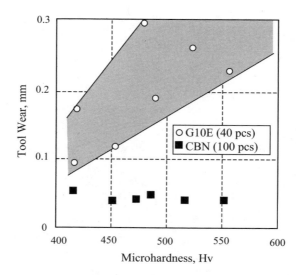

Figure 3.95 *Effect of microhardness on tool wear (Rodrigues).*

Near-net shape processing of the PM guide and accurate estimation of the guide I.D. dimension change due to press fit can minimize the amount of stock for machining. The I.D. dimension of the pre-reamed guide, which determines the machining stock, is determined based on the cumulative error of the cylinder head positioning against the reamer, the housing position of the head, and the concentricity of the I.D. to O.D. of the valve guides. A commonly specified reaming stock is 0.3–0.8 mm as measured on the diameter. It is advantageous to minimize the reaming stock to obtain longer life of the reaming tool.

3.7 Rocker Arms

3.7.1 Rocker Arm Configurations

The rocker arm is a device required to transfer the rotating motion imparted by the camshaft lobe to the reciprocating motion of the intake and exhaust valves. Valvetrain Types II through V use some form of rocker arm follower. Type II uses an end-pivot rocker arm, and the arm can contain either a radius follower or a roller follower. Type III uses a center-pivot rocker arm, Type IV uses a center-pivot rocker arm with a tappet, and Type V uses a fulcrum-mounted rocker arm for cam-in-block engines. However, for convenience of discussion, the typical rocker arm configurations can be divided into two groups based on the pivot and cam locations: (1) end pivot with the cam above the rocker arm (Figure 3.96), and (2) center pivot with the cam below the rocker arm (Figure 3.97).

End-pivot rocker arms usually have the cam located above the arm and have a relatively higher rocker ratio compared to that of the center-pivot rocker arm. The rocker ratio is the ratio of the valve lift divided by the cam lift, or the distance from the rocker-pivot center to the valve centerline

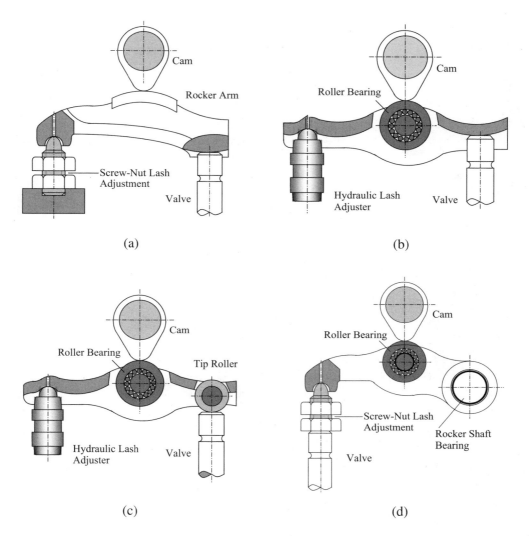

Figure 3.96 _Typical end-pivot rocker arms, Type II valvetrain. Note the cam is above the arm. (a) End-pivot rocker arm with sliding contact at the cam face (mechanical lash adjuster), (b) end-pivot rocker arm with rolling contact at the cam face (hydraulic lash adjuster), (c) end-pivot rocker arm with dual rollers at the cam and tip face (hydraulic lash adjuster), and (d) end-pivot shaft rocker arm with rolling contact at the cam face (mechanical lash adjuster)._

at the valve tip divided by the distance from the rocker-pivot center to the cam/follower contact center. Therefore, for the same amount of valve lift, the cam lobe design in the end-pivot rocker arm is less aggressive. However, any machining deviation of cam contour will be exacerbated at the valve lift profile, thus affecting the velocity and acceleration or valvetrain dynamics.

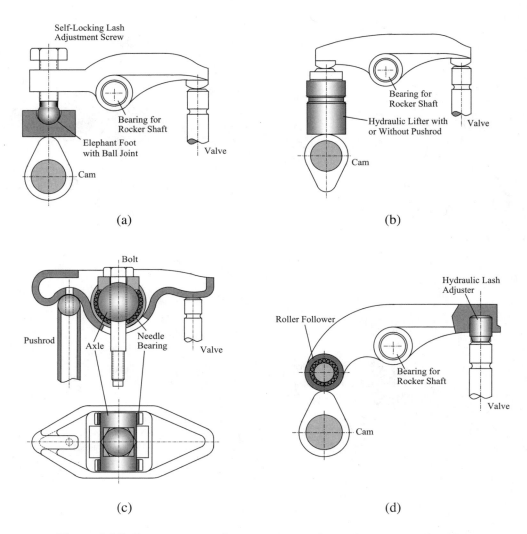

Figure 3.97 *Center-pivot rocker arm, cam under rocker arm, applicable to Types III, IV, and V valvetrains. (a) Center-pivot flat follower with mechanical lash adjuster, Types III, IV, or V; (b) flat-faced hydraulic lifter, Type IV; (c) pushrod with a roller bearing fulcrum, Type V; and (d) roller arm with hydraulic lifter, Type III.*

Figure 3.96(a) shows a typical end-pivot rocker arm with a sliding contact at the cam face. This type of rocker arm is simple and inexpensive. The cam contact pressure usually is low, but tangential stress at the cam face may be high, depending on the coefficient of friction at the interface. The side load exerting on the valve usually is high. Figure 3.96(b) shows a typical end-pivot rocker arm with rolling contact at the cam face. This is the most common type of rocker arm used in current passenger car applications. The rolling contact at the cam face reduces friction and is more efficient. However, depending on the size of the roller and the relative position of the roller and cam, the larger pressure angle can result in a great side load on the valve and lash adjuster. Figure 3.96(c) shows a dual-roller rocker arm that can significantly reduce the friction and tangential stress at the tip contact (i.e., contact pressure may be higher due to the small radius of the roller), thus reducing tip wear and valve side load in addition to the benefits of low friction with a center roller. Figure 3.96(d) shows the end-pivot rocker arm with the rocker shaft as the pivot. The side load on the valve is relatively low and depends on the friction at the tip interface. The stiffness may be enhanced, too. However, the rocker arm shaft may add more weight to the overall engine weight.

Center-pivot rocker arms usually have the cam located below the arm and have a relatively low rocker ratio compared to that of the end-pivot rocker arms. Therefore, for the same amount of valve lift, the cam lobe design is slightly more aggressive but is less aggressive than the direct-acting cam contour. Depending on the cam position, a pushrod may have to be used between the cam and the rocker arm if the cam is in the engine block.

Figure 3.97(a) shows a center-pivot rocker arm, OHC with a mechanical lash adjuster. If a hydraulic tappet is used on the cam side and a cam is located in the head, it is designated as a Type IV valvetrain. as shown in Figure 3.97(b). The rocker arm configuration is the same, but if the cam is located in the engine block, then it is a Type V valvetrain, either with mechanical or hydraulic lash adjuster, as shown in Figure 3.97(c). Figure 3.97(d) shows a center-pivot roller follower rocker arm with a hydraulic lash capsule (a mechanical lash capsule also can be used) at the valve tip, and it is a Type III valvetrain. There can be many derivatives of end-pivot and center-pivot rocker arms.

3.7.2 Design Guidelines

Rocker arm design requires a combination of analytical techniques and experience. It also involves the choice of materials and the manufacturing method. Design validation is accomplished by structural analysis through the finite element method and actual testing.

3.7.2.1 Geometry and Layout

The design of the rocker arm begins with a layout of the valvetrain system geometry. This results in a stable system that will simplify the design and validation of the rocker arm. The following factors, including the rocker ratio, the camshaft base circle radius, and the roller diameter (for a roller rocker arm), are the most critical parameters in rocker arm design.

The rocker ratio is not a constant value but changes due to the change of contact locations over the lift cycle. The rocker ratio is expressed as $RR = \dfrac{R_1}{R_2}$, where R_1 is the distance from the lash adjuster pivot to the valve tip contact point, and R_2 is the distance from the lash adjacent to the pivot to the cam/follower contact point.

Significant variation in the rocker ratio can occur over the cam event, depending on the valvetrain system geometry, as discussed in Chapter 2. For an end-pivot rocker arm such as in a Type II valvetrain system, the rocker contact ratio is the dimension from the lash adjuster contact to the valve tip contact point divided by the lash adjuster contact to the camshaft contact point. Because both the valve tip contact and the cam follower contact change throughout the cam event, significant variation in the ratio will occur. For a center-pivot rocker arm such as in a Type V valvetrain system, the lift ratio is the dimension from the rocker arm fulcrum pivot point to the valve tip contact point divided by the rocker arm pushrod socket contact to the rocker arm fulcrum pivot point. The only variation in ratio occurs due to valve tip movement because the socket-to-fulcrum dimension is unchanged. Rocker ratios for center-pivot rocker arms range from approximately 1.2 to 1.9; for end-pivot rocker arms, they range from 1.5 to 2.0. They should not exceed 2.1. Higher ratios cause a loss of rigidity and allow greater magnification of the effect of deviations in cam manufacturing. To ensure undistorted transfer of the cam motion to the valve, it is essential to design the arm and its supporting structure for maximum stiffness.

With a fixed cam profile, higher-ratio rocker arms open the valve faster and higher and provide greater lift area, thus achieving higher engine performance. The downside is that any manufacturing error in the cam profile could be magnified on the valve lift. To achieve the same valve lift height and area without changing the rocker ratio, the cam profile must be increased. Such a cam lobe would be aggressive and would require much heavier springs to keep the lifter from flying off the lobe. Aggressive lobes also will add more side stress on the lifters/bores and could cause lifter/bore failure. Typically, the optimization of the end-pivot rocker arm in a Type II valvetrain system can have approximately 20% more valve lift area than the center-pivot rocker arm in a Type V system when considering different design parameters such as rocker ratio, valvetrain weights, and natural frequency limitations. A larger rocker ratio tends to lead to greater valve tip scrubbing.

The primary emphasis on the pressure angle is the distribution of forces in the cam follower or rocker arm. Thus, the maximum value is of concern. The maximum pressure angle establishes the cam size, torque, loads, accelerations, wear life, and other pertinent factors. The cam base circle radius and roller diameter, as well as the relative position of the two, can affect the pressure angle. In turn, this can affect the pivot side loading. The pressure angle of the cam/roller rocker arm is a function of the angle of the centerlines of the cam/roller and roller/pivot, cam size and profile, and the roller size and percentage of lift. From a design standpoint, to minimize side thrust on the lash adjuster and valve stem, the angle of the centerlines of the cam/roller and roller/pivot is recommended as $\angle ABC = 90°$ when the valve is halfway open, or $95–105°$ when the valve is in the closed position (Figure 3.98). Aggressive cam profiles and a large ratio of cam/roller size usually cause greater side load. The pressure angle can be reduced by using a larger radius (i.e., either a larger cam or a larger roller).

3.7.2.2 Dimensions and Tolerances

Figure 3.99 shows typical dimensions and tolerances of rocker arms for automotive applications.

3.7.2.3 Stiffness

The overall valvetrain stiffness depends on the component with the lowest stiffness. Generally, in Types II–IV valvetrains, the arm is the component with the minimum stiffness. In Type V, the

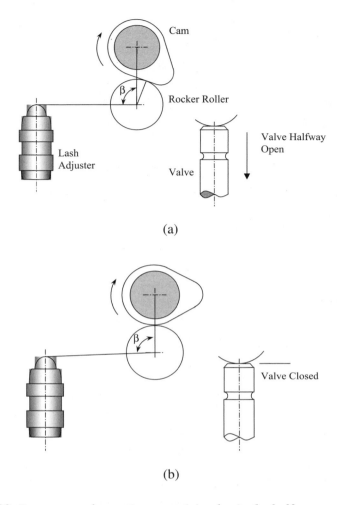

(a)

(b)

Figure 3.98 *Pressure angle requirement: (a) valve in the halfway open position,*
β = 90°, and (b) valve in the closed position, β = 95–105°.

rocker arm and the pushrod typically have limiting stiffness values. Stiffness also is a function of overall rocker arm length, cross-section area, weight, rocker ratio, and so forth. A longer arm, a smaller cross-section area, a heavy valvetrain, or a larger rocker ratio usually results in lower stiffness. High valvetrain stiffness results in a large valvetrain system natural frequency that allows a more aggressive camshaft profile design and a high-performance engine. System natural frequency can be calculated from the equation

$$f = \frac{1}{2\pi} \sqrt{\frac{k}{m}}$$

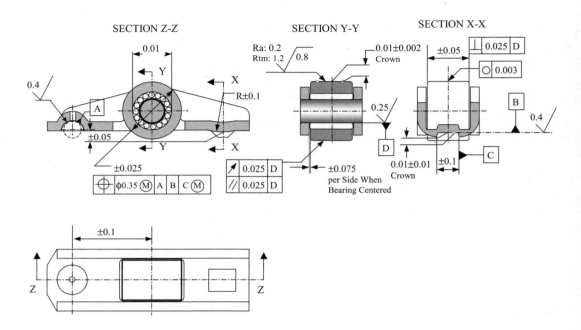

Figure 3.99 *Example of dimensional tolerances and surface finish requirements for an end-pivot roller rocker arm in a Type II application.*

where k is the valvetrain system overall spring rate (N/mm), and m is the valvetrain system mass. Overall, valvetrain stiffness is obtained by dividing the maximum test load by the maximum deflection at the valve side with the lifter evacuated of oil and spaced with washers to the nominal installed position. A Type II valvetrain system using a pivoted shaft rocker arm can result in a stiffer rocker mechanism, compared to a stamped or cast fulcrum pivoted arm system with a lash adjuster.

3.7.2.4 Friction Characteristics

The main advantage of rolling contact over sliding contact is that there is much less friction under the same loading condition. In theory, there is no friction between two perfectly smooth bodies rolling on each other. However, in practice, there is resistance to moving due to slip because both bodies are elastically deformed under load. The lowest frictional force is found in materials that have a high elastic modulus due to the minimized elastic deformation. The tradeoff is an increased stress due to a smaller contact area. Other factors leading to increased slip include that on the opening flank, the roller must accelerate against the static friction forces between the roller and its pin. Slip occurs when there is deviation between the actual roller speed and theoretical values. Misalignment between the roller axis and the camshaft axis also adds slip. A reduced roller pin and needle diameter contributes to lower friction torque. Although roller followers reduce friction and frictional losses, they increase contact stress between the camshaft and follower. Follower arm/camshaft contact stress increases with a smaller cam base circle radius, smaller roller follower radius, and increased rocker contact ratio, as with a smaller roller pin and needle diameter. Despite clear advantages, roller followers have added complexity in function and manufacturing.

Rocker arms with a roller can reduce valvetrain friction torque loss up to 70%, depending on the engine speed and type of roller bearing. Figure 3.100 shows an example of valvetrain torque reduction with a needle bearing roller rocker arm, compared to a sliding rocker arm. At a cruise engine speed of 2000 rpm (1000-rpm camshaft speed), 60% of the valvetrain torque reduction is observed with the roller rocker arm.

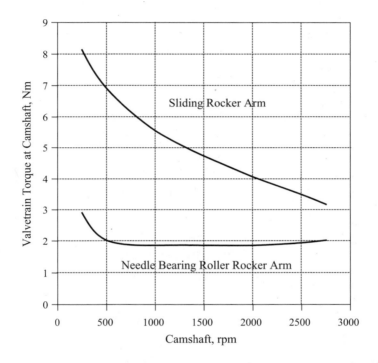

Figure 3.100 *Comparison of flat and roller rocker arms in friction reduction, Type II valvetrain (Armstrong and Buuck).*

3.7.3 Rocker Arm Materials

Typical materials used in the construction of rocker arms include SAE 1008, 1010, 4140, and 8620 or 16MnCr 5, and often are carburized or carbonitrided with a case depth of approximately 0.375 mm and a surface hardness of 90 HR_{15N}. Other materials such as 15-5 PH stainless steels also are used. Lightweight materials such as aluminum alloys are crucial to efficient operation, but strength plays a major function in longevity. Materials for the roller bearing including the outer roller, needle, and axle are SAE 52100 or equivalent, with surface hardness hardened to HRc 60 minimum for the roller and needle and HRc 55 minimum for the axle.

3.7.4 Variable Actuation Rocker Arms

Several automotive manufacturers use rocker arms to achieve variable valve actuation (VVA), including variable valve timing and lift. As described in Chapter 2, the Honda VTEC and i-VTEC are typical examples of using rocker arms to achieve variable valve lift. Figure 3.101 illustrates the mechanism of the VTEC. At low speeds, the primary and secondary rocker arms are not connected to the mid-rocker arm but are driven separately by cam lobes A and B at different timing and lift. The lift of the secondary cam lobe is small, so that one intake valve opens slightly (one-valve control). Although the mid-rocker arm is following the center cam lobe with the lost motion assembly, it has no effect on the opening and closing of the valves in the low revolutions-per-minute range. When driving at high speeds, the timing piston moves in the direction shown by the arrow in Figure 3.101(b). As a result, the primary, secondary, and mid-rocker arms are linked by two synchronizing pistons, and the three rocker arms move as a single unit. In this state, all rocker arms are driven by cam lobe C, opening and closing the valves at the valve timing and valve lift set for high operation.

Another example of VVA using a rocker arm is the Toyota VVT-i system shown in Chapter 2. This Toyota system uses a single rocker arm follower to actuate the valves. It also has two cam lobes acting on that rocker arm follower. The lobes have different profiles: one with a longer valve opening duration profile for high speeds, and another with a shorter valve opening duration profile for low speeds. At low speeds, the slow cam actuates the rocker arm follower via a roller bearing (to reduce friction). The high-speed cam does not have any effect on the rocker follower because there is sufficient spacing beneath this hydraulic tappet. When the speed has increased to the threshold point, the sliding pin is pushed by hydraulic pressure to fill the spacing. The high-speed cam becomes effective. Note that the fast cam provides a longer valve opening duration, while the sliding pin adds valve lift.

Figure 3.102 shows other types of end-pivot rocker arms with dual lift or deactivation mechanisms appearing in the literature. The rocker arm design is a two-piece arm system, with latching by means of a movable outer arm. The outer arm is located at the lash adjuster by a slotted pinpoint. The slot allows the outer arm to move perpendicular to the axis of the camshaft to perform the latching function. The inner arm carries the roller and axle assembly and is located at the lash adjuster end by a rotational pinned joint. The outer arm is located by the outer arm latch surface near the valve tip. A spring force keeps the inner arm in contact with the cam surface when the system is in the unlatched mode. The same spring also provides the necessary torque to prevent the lash adjuster from pumping up and thus maintains a minimum latching clearance between the inner and outer arms. A second spring holds the outer arm in the latched position until an external actuation force unlatches the system.

Valvetrain Components

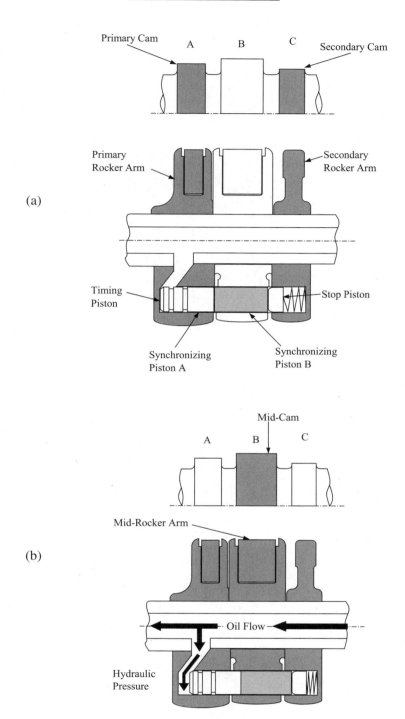

Figure 3.101 *Honda VTEC rocker arm: (a) at low speed, and (b) at high speed.*

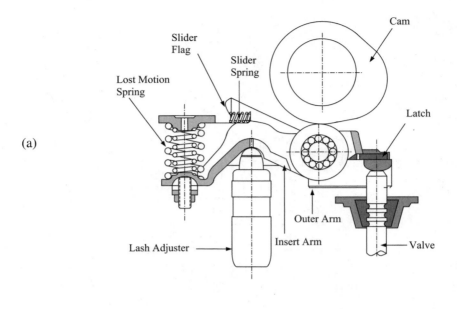

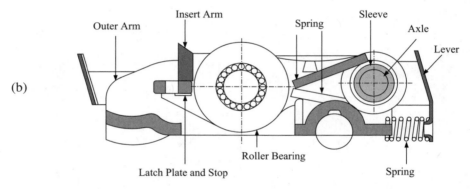

Figure 3.102 *Eaton's variable displacement rocker arms: (a) dual-lift mechanism, and (b) dual-lift rocker arm cross section (Buuck and Hampton).*

3.8 Valve Springs

3.8.1 Introduction

Valve springs are helical compression springs. They are used to follow precisely the corresponding cam-profile motion during each cycle to keep the cam and transfer elements in contact at all times. During the acceleration period, the cam exerts a positive accelerating force on the follower. The magnitude of this force is equal to the product of the mass of the valvetrain (m) and the acceleration (a) at the speed of operation, that is,

$$f = ma$$

The actual force on the camshaft will exceed this value by the amount of the spring load and the gas pressure on the head of the valve at the beginning of the lift period. When the follower is on the deceleration portion of the cam contour, it must be forced to follow the prescribed path by the use of a spring.

Valve springs must operate the engine valves at different speeds (i.e., 600 to more than 7000 rpm crankshaft speed). The movement is restricted in the axial direction of the valve, and the ends must be ground for proper seating. Also, the I.D. is restricted by the cylinder head boss diameter. The loads also change from valve closed to valve open. The environment temperature ranges from ambient to 150°C. The valve spring is restricted on one end by a boss on the cylinder head and on the other end by a retainer. The wire hardness must be compatible with the retainer. If aluminum is used for the cylinder head, a spring seat washer must be used between the spring and the head. There are two basic considerations for durability: (1) the fatigue strength of the valve spring should survive the normal life of an operating vehicle, and (2) it should not lose more than a specified load-carrying capacity during its lifetime. Failure is defined as breakage or load loss exceeding a maximum specified percentage of the original valve open load during any test. Figure 3.103 shows valve spring nomenclature and formulae. The spring index is the ratio of the mean diameter to the wire diameter. The preferred index range is 4–12. Springs with high indexes may tangle and require individual packaging. Springs with indexes lower than 4 are difficult to form.

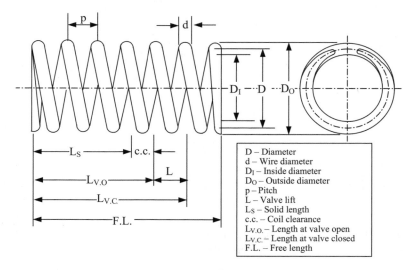

Figure 3.103 *Valve spring nomenclature and formulae: solid height* $L_C = (N_t - 0.25)d$; *free length* $L_f = pN_a + 2d$; *total coils* $N_t = N_a + 2N_i$ (N_i = *inactive coils*); *pitch* $p = \dfrac{L_f - 2d}{N_a}$; *active coils* $N_a = \dfrac{L_f - 2d}{p}$ *(Turkish, 1951).*

–277–

3.8.1.1 Spring Rate

Linear rate springs initially were used for low-operating-speed applications where lower spring natural frequencies were acceptable and damping of spring surging was not necessary. The spring rate (k) for helical compression springs is defined as the change in load per unit deflection and is expressed as

$$k = \frac{F}{d}$$

where F is the applied load and d is the deflection. This equation is valid when the pitch angle is less than 15 or when the deflection per turn is less than $\dfrac{D}{4}$. The total deflection curve for linear compression springs essentially is a straight line up to the elastic limit, provided that the amount of active material is constant. The initial spring rate and the rate as the spring approaches solid often deviate from the average calculated rate. When it is necessary to specify a rate, that rate should be specified between two test heights that lie within 15–85% of the full deflection range (Figure 3.104(a)). Figure 3.104(b) shows an example of valve spring performance where L_r is the free length, L_1 is the installed height or the height at valve closed, L_2 is the height at valve open, F_1 is the spring load at L_1, F_2 is the spring load at L_2, and L_C is the solid height.

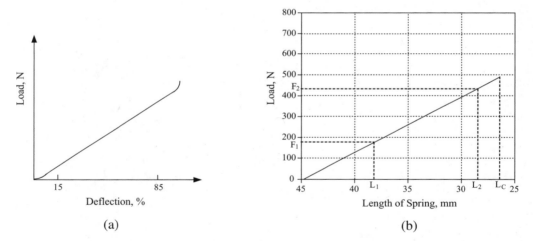

Figure 3.104 *Valve spring rate: (a) load versus deflection for a valve spring, and (b) example of spring performance.*

3.8.1.2 Spring Types and Configurations

Although the most prevalent form of valve springs is a straight cylindrical spring made from round wire, many other forms can be used. Conical, beehive, barrel, or cylindrical forms are available, with or without any variable spacing between the coils. Such configurations are used to reduce solid height, buckling, or surging, or to produce nonlinear load deflection characteristics.

Figure 3.105 shows some typical valve spring and damper designs. The objectives of the various designs with various pitches and dual springs with interference are to achieve the required spring rate and to avoid spring surge. Sometimes, combinations such as dual helical springs with interference, dual conical springs with interference, dual conical springs with friction damper and interference, and beehive springs with variable pitch and/or damper are used in applications to achieve the objectives (Figure 3.106). Straight cylindrical springs are designs with one O.D. or I.D. throughout the length of the spring. Straight cylindrical springs of round wire are the most common design, but they can be linear, variable rate, or various wire cross sections such as round or oval.

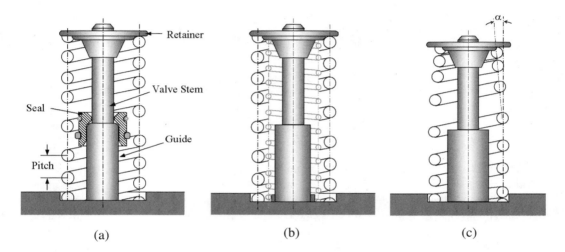

(a)   (b)   (c)

Figure 3.105 *Typical valve spring configurations: (a) variable pitch, (b) dual springs with interference, and (c) beehive spring with variable pitch.*

Conical and beehive springs have variable diameters and are used in applications that require a low solid height, increased lateral stability, or resistance to surging. Conical springs can be designed such that each coil nests wholly or partially into an adjacent coil. The rate for these springs usually increases with deflection because the number of active coils decreases progressively as the spring approaches solid.

3.8.1.3 Wire Cross Sections

Most valve springs use commercially available round wires. From a design or package standpoint, some drawbacks of round wire are higher solid and installed heights, heavier weight, and lower natural frequency than their oval counterparts. From a manufacturing standpoint, round wire is easier to process and coil, as well as to detect seams and inclusions with eddy current testing.

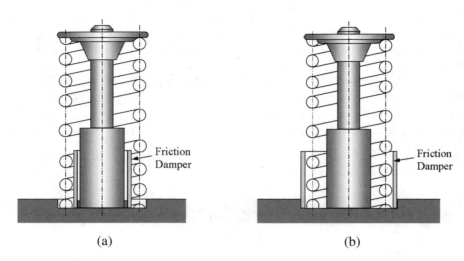

(a)   (b)

Figure 3.106 *Valve spring friction dampers: (a) internal friction damper, and (b) external friction damper.*

Oval wire cross sections typically have a nonround or egg-shaped cross section on the I.D. and a round cross section on the O.D. Figure 3.107 shows a typical design. Some advantages of the oval wire are that the solid height and installed height can be decreased, the weight can be decreased, and the natural frequency can be increased versus comparable round wire springs. Also, the maximum stress moves away from the I.D. (major axis) for round wire toward r_1 and r_3 for oval, which allows a higher design stress. The oval wire also is available with chrome-silicon material that theoretically can be designed to a higher stress level. Disadvantages of oval wire are maintaining the cross section during material processing, preventing twisting during coiling, difficulty in detecting seams and inclusions with eddy current equipment, and higher cost than round wire.

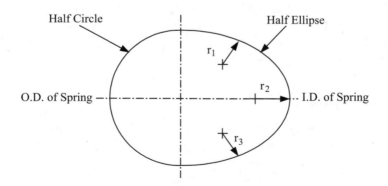

Figure 3.107 *Typical oval wire cross section.*

3.8.2 Design Considerations

Initial parameters given on valve spring designs are valve installed height, valve open height, installed load, outside or internal coil diameter allowed, and the desired valve spring theoretical natural frequency. It is desirable to design a spring to the maximum stress allowed by the selected wire material. This allows the designer either to increase the spring rate by using a larger valve open load (if allowed by the camshaft contact stresses) or to decrease the wire diameter size (maintaining the prescribed spring rate) and thereby reduce the number of coils and the installed height by means of the decreased solid height. The latter is advantageous in new engine applications for improved valvetrain dynamics. This allows a reduction in the spring mass and a reduction in the valve mass through a decrease in the valve stem height due to the lower valve spring installed height.

3.8.2.1 Space Limitations

It is valuable to determine the maximum volume of spring material that can be used within a given space. Figure 3.108 shows "space efficiency factors" V/V_S as a function of spring index, where V is the volume of active spring material, and V_S is the volume of the cylindrical space in which the spring is contained when solid. For an engine valve spring, the curve that subjects it to fatigue loading with corrected stresses is more applicable than the static loading.

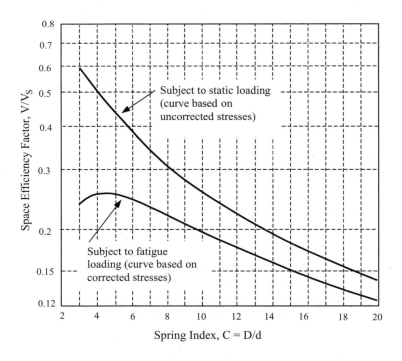

Figure 3.108 *Space limitation (SAE HS 795).*

3.8.2.2 Dimensions and Tolerances

Free length is the overall spring length in the free or unloaded position. Pitch is the distance between the centers of adjacent coils and is related to the free length and the number of coils. Figure 3.109 shows some spring terminology and tolerances for a helical compression valve spring. Solid height is the length of a spring with all coils closed. For ground springs, solid height is the number of total coils minus one-fourth coil multiplied by the wire diameter. This formula agrees well with measurements made on production springs. A spring should be designed so that the coils do not touch or clash in service because such contact will induce wear, fretting corrosion, and failure. Therefore, valve springs must have at least 1.5 mm of additional spring travel available at maximum valve lift.

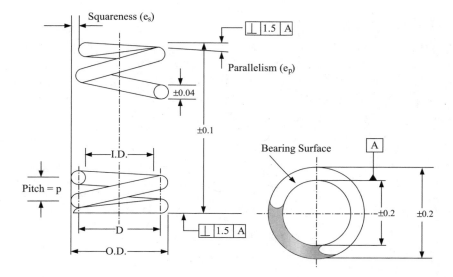

Figure 3.109 *Dimensional terminology and typical tolerances for helical compression valve springs.*

Squareness (e_s) of helical compression valve springs can be measured by standing a sample spring on end on a horizontal flat plate and bringing the spring against a straight edge at right angles to the plate. The spring is rotated to produce a maximum out-of-square dimension. Normally, squared and ground springs are square with 2° when measured in the free position. Squareness should be checked at both ends. Specifying squareness or parallelism in the free position does not assure squareness or parallelism under load. Parallelism (e_p) refers to the relationship of the ground ends. It is determined by placing a spring on a flat plate and measuring the maximum difference in free length around the spring circumference.

Coil diameter tolerance may be specified on either the inside or outside coil diameter. Figure 3.110 gives typical coil diameter tolerances for wire diameters in the range of 0.30–9.50. Round the

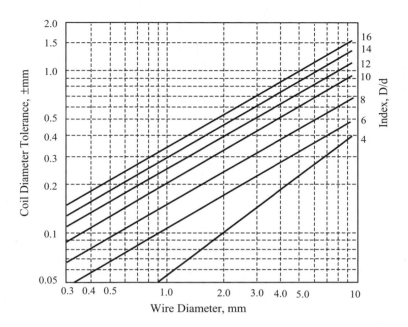

Figure 3.110 *Coil diameter tolerance for wire diameter*
of 0.30–9.50 mm (SAE HS 795).

index to the nearest whole number, and interpolate when the rounded value is an odd number. Use a tolerance of 0.30-mm wire diameter when the wire diameter is less than 0.30 mm.

3.8.2.3 Types of Ends

The types of ends available are plain ends, plain ends that are ground, squared ends, and squared ends that are ground (Figure 3.111). To improve squareness and reduce buckling during operation, a bearing surface of at least 270° is required. Valve springs have ends that are square and ground. Typically, the end coils must not project over the body of the spring and must have a minimum tip thickness of 0.9 mm.

3.8.2.4 Spring Coils

End coils or dead coils should not be less than 1-1/16-mm coils at each end. The number of end coils should be as close to the minimum as possible for two reasons: (1) economics (i.e., less wire means less cost), and (2) minimum weight to reciprocate at the retainer end of the spring. The end coils are determined at the free length position. The same number of end coils typically is used at each end of the spring, so that orientation of the springs at the assembly is not needed. If the number of end coils is different, the greater number of coils should go to the cylinder head end. Over-travel between the valve open positions to the solid height is to be 1.5–2.5 mm and depends on stress at the solid height.

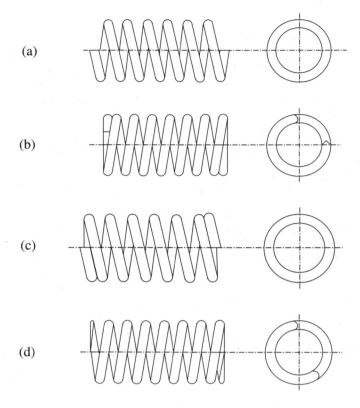

Figure 3.111 *Types of ends for helical compression springs: (a) plain ends coiled right-hand, (b) squared and ground ends coiled left-hand, (c) squared or closed ends coiled right-hand, and (d) plain ends, ground and coiled left-hand (Maker).*

The number of active coils equals the total number of coils less those that are inactive at each end. For plain ground ends, the number of inactive coils depends on the wire diameter and the pitch of the spring. The number of active coils in a spring with plain, unclosed ends ground would approximately equal the number of turns of the wire untouched by grinding. One inactive coil on each end should be allowed in springs with squared ends or squared and ground ends.

3.8.2.5 Direction of Coiling

A helical compression spring can be either left- or right-hand coiled. If the index finger of the right hand can be bent to simulate the direction of coil, so that the fingernail and coil tip are approximately at the same angular position, the spring is right-hand wound (Figure 3.112). If the index finger of the left hand simulates the coil direction, the spring is left-hand wound. If the direction of coiling is not specified, the springs may be coiled in either direction. Nested springs with small diametrical clearances should be coiled in opposite directions.

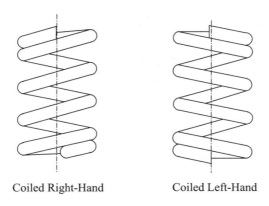

Coiled Right-Hand   Coiled Left-Hand

Figure 3.112 *Valve spring coil directions.*

3.8.2.6 Hysteresis

Hysteresis is the loss of mechanical energy under the cyclic loading and unloading of a spring. It results from frictional losses in the spring support system due to a tendency of the ends to rotate as the spring is compressed. Hysteresis for compression springs is low, and the contribution due to internal friction in the spring material generally is negligible.

3.8.3 Spring Performance

3.8.3.1 Load and Stress

The minimum load (i.e., the load with the valve closed) should be high enough to hold the valve firmly on its seat during the period when the valve is closed. In engines that are throttled, the exhaust valve must remain closed at the highest cylinder vacuum. In supercharged engines, the intake valve must not be opened by the highest expected manifold pressure. The dynamic requirements are set by the required engine operating speeds and may be calculated to the first approximation from the cam-lift curve, using rigid-body assumptions. The true jumping speed (i.e., the speed at which the follower leaves the cam at some point in the lift curve) will always be lower than the theoretical jumping speed.

In cyclic applications, the load-carrying ability of a spring is limited by the material fatigue strength. To select the optimum stress level, it is necessary to consider the operating environment, expected life, stress range, frequency of operation, speed of operation, and permissible levels of stress relaxation. Because maximum stress is at the wire surface, any surface defects (such as pits or seams) severely reduce fatigue life. Shot peening improves fatigue life and minimizes the harmful effect of surface defects, but it does not totally remove them.

The wire in a helical compression spring is stressed in torsion. Torsional stress is expressed as (Taylor)

$$\sigma = \frac{8FC^3}{\pi D^2}(Y) = \frac{\Delta}{N}\frac{G}{\pi dC^2}(Y) \qquad (3.42)$$

$$Y = \frac{4C-1}{4C-4} + \frac{0.615}{C} \qquad (3.43)$$

$$\delta = \frac{\Delta}{N} = \frac{8PC^4}{GD} = \frac{\pi dC^2}{GY}\sigma \qquad (3.44)$$

$$k = \frac{P}{\Delta} = \frac{GD}{8C^4N} \qquad (3.45)$$

where

σ = maximum shear stress in the spring
δ = axial deflection of the spring, per coil
Δ = total deflection
k = spring stiffness
P = applied load
C = spring index = $\frac{D}{d}$
N = number of active coils
G = shear modulus of the material (81 GPa for typical spring steel)
D = pitch diameter of the coil
d = wire diameter
Y = stress correction factor for the helical spring

The spring rate for conical springs is calculated by considering the spring as many springs in a series. The spring rate for each turn or fraction of a turn is calculated by Eq. 3.42. The rate for the complete spring is determined by summing the individual rates of a series of springs per the relationship

$$\frac{1}{k} = \frac{1}{k_1} + \frac{1}{k_2} + \frac{1}{k_3} + ... + \frac{1}{k_n} \qquad (3.46)$$

The mean diameter of the largest active coil at load is used to calculate the highest stress at a given load. Some additional benefits of variable diameter springs are reduced spring mass plus the retainer mass at the smaller end.

Bending stresses are present but can be ignored except when the pitch angle is greater than 15 and the deflection of each coil is greater than $\frac{D}{4}$. Under elastic conditions, torsional stress is not uniform around the wire cross section due to coil curvature and a direct shear load. Maximum stress occurs at the inner surfaces of the spring and is computed using a stress correction factor (Y). In some circumstances after yielding occurs, the resultant stresses are distributed more uniformly around the cross section. Then a stress correction factor Y_W, which accounts for only the direct shear component, is used. The preferred value for the spring index (C) is 4–12 (Figure 3.113), where Y is calculated from Eq. 3.43 for 2% setpoint or fatigue, and $Y_W = 1 + \dfrac{0.5}{C}$ for springs with set removed.

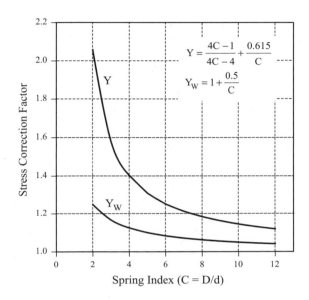

$$Y = \frac{4C - 1}{4C - 4} + \frac{0.615}{C}$$

$$Y_W = 1 + \frac{0.5}{C}$$

Stress Correction Factor

Spring Index (C = D/d)

Figure 3.113 *Stress correction factors for round wire helical compression springs.*

For linear rate springs, when the deflection is known, the loads are determined by multiplying the deflection by the spring rate. When the stress is known or assumed, the loads are determined from Eq. 3.42.

For variable rate springs, the procedure used to determine the loads of the variable rate springs is complex. In this case, the load deflection curve is approximated by a series of short chords. The spring rate is calculated for each chord and is multiplied by the deflection to obtain the load. The load is then added to that calculated for the next chord. The process is repeated until the load has been calculated for the desired value of the deflection. The loads should be specified at a test height. Because the load deflection curve often is not linear at very low loads or at loads near solid, the loads should be specified at test heights of 15–85% of the full deflection range.

When springs are designed to work at maximum stress, allowance should be made for the effect on stress of the specified type of end. For example, in a compression spring with squared, or squared and ground, ends, the number of active coils will vary throughout the entire deflection of the spring, and this results in a loading rate and stress that differ from those originally calculated.

This is especially important in springs having fewer than seven coils. The number of active coils changes because part of the active coil adjacent to the squared end closes down solid and becomes inactive as the spring approaches the fully compressed condition.

3.8.3.2 Buckling

Compression springs that have lengths greater than four times the spring diameter can buckle. Figure 3.114 shows critical buckling conditions for axially loaded springs with squared and ground ends.

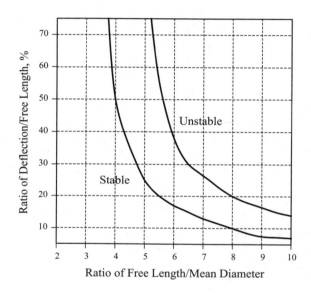

Figure 3.114 *Critical spring buckling condition curves (SAE HS 795).*

3.8.3.3 Spring Relaxation

Spring relaxation occurs when the spring—which could be operating at average temperatures of approximately 150°C or higher—under cyclic conditions of stress plastically deflects so that it does not recover its original free length when an external load is released. Therefore, over a period of time, the force exerted by the spring is reduced. The amount of this relaxation, or set, generally increases as the stress and/or temperature increases. Normally, the amount of set also depends on the duration at the elevated temperature; the set is greater for long periods of time than for short ones. Figure 3.115 shows spring relaxation for two types of valve springs.

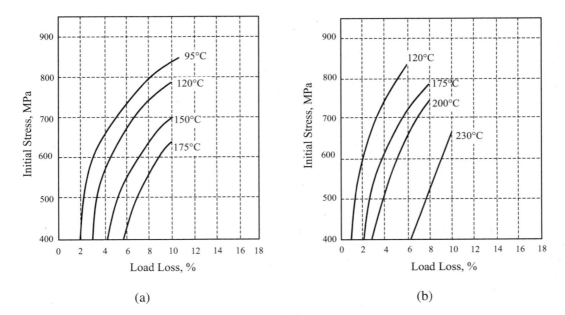

Figure 3.115 _Spring relaxation tested for 100 hours at indicated temperatures:_
(a) SAE J351 carbon valve spring wire, and (b) SAE J132 chrome vanadium
valve spring wire (SAE HS 795).

To avoid undesirable load loss or to enhance relaxation resistance, valve springs frequently are heat set. By subjecting the springs during manufacturing to a greater combination of stress, temperature, and time than encountered during normal operation, the primary creep of the spring can be removed. A residual stress can be built up in the spring so that it will resist further load loss during normal operating conditions. The amount of improvement in the relaxation process of a spring during heat set depends on the material, the spring stress, and the working temperature.

Presetting is an operation performed during manufacturing in which the valve spring is compressed beyond the yield point of the material. The yielding of the surface layers of the wire, which occurs during presetting, produces beneficial residual stresses, thus increasing the elastic limit of the spring and thereby reducing the amount of settling or sag in subsequent service. The spring initially is coiled to a free length greater than the final desired free length and then is compressed to a point where the spring may be at or near the solid length (Figure 3.116, point B). On releasing the load, a permanent set OC is produced, while the spring behaves nearly elastically along line CB.

When presetting on helical compression springs at some elevated temperature, the operation known as hot pressing, heat setting, or warm setting may be employed to produce greater resistance to relaxation than cold setting. Table 3.19 shows some typical results of heat set on valve material.

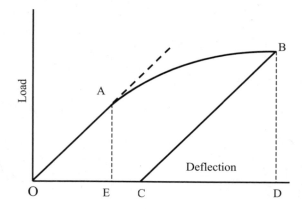

Figure 3.116 *Typical load-deflection diagram of a helical spring during presetting (SAE HS 795).*

TABLE 3.19
VALVE SPRING LOAD LOSS EVALUATION

Material	Valve Spring Condition	% Relaxation After 10 × 10^6 Cycles at 110°C
Carbon Wire	Not Heat Set	7.0%
	Heat Set	2.7%
Chrome Vanadium Wire	Not Heat Set	3.0%
	Heat Set	0.1%

In heat set parts, negative relaxation can occur at low stress, which is acceptable because a minimum load may exist throughout the spring life. Chrome silicon is more resistant to relaxation; however, shot peening degrades relaxation resistance.

3.8.3.4 Spring Natural Frequency or Surge

The basic function of a valve spring is to keep the cam and follower in contact. However, at high speeds, a phenomenon occurs that may seriously reduce the effective force of the spring and allow the follower to leave the cam, even though considerable surplus spring force was provided. This is called spring surge or intercoil vibration. Resonance is the condition when the natural frequency of vibration of the spring synchronizes with disturbing vibrations or periodic force generated by the action of the cam against its follower. If resonance occurs, a torsional wave is transmitted through the wire up and down the spring at the natural frequency of the spring. The coils at the stationary end of the spring close together completely or almost completely during the initial stages of valve lift due to their inertia, and successive coils away from this end close to a lesser extent. During the final stages of valve opening, the coils farthest from the cam follower close the most. A collapse of spring resistance so that some of the coils temporarily lose their pitch and move closer together sweeps or surges from one end of the spring to the other. When the spring surging takes place, the natural closing action of the coil spring will be out of control; therefore, the valve movement will no longer correspond to the designed cam profile rise, dwell, and fall.

Spring natural frequency can be calculated as

$$m = \frac{\pi d^2}{4} 2\pi DN\rho, \qquad \text{and} \qquad k = \frac{GD}{8C^4 N}$$

$$(3.47)$$

$$f = \frac{1}{2\pi}\sqrt{\frac{k}{m}} = \frac{1}{\pi}\sqrt{\frac{G}{2\rho}} \frac{d}{ND^2}$$

where

    N = number of coils
    d = wire diameter
    D = mean radius of the spring
    ρ = density of the wire material
    G = shear modulus

Figure 3.117 plots the first-mode natural frequencies of intercoil vibration as a function of stress range and lift. Increasing stress, increasing stress range, and decreasing lift will increase the spring natural frequency.

Figure 3.118 shows the relationship of the number of active coils to natural frequency with a given lift and maximum stress.

Fundamentally, the lower the harmonic number, the higher are the vibratory amplitudes. Therefore, the natural frequency of the spring should be high enough so that if resonance occurs, it will be with higher harmonic numbers, and vibratory amplitudes will be kept to a minimum. Figure 3.119 shows harmonic coefficients for a typical cam contour. In this instance, the harmonic number is the ratio of the spring natural frequency to the cam speed. This number indicates that if the spring natural frequency is higher than five times the camshaft speed, the spring vibration should be small. In practice, the harmonic number should be 11 or higher. However, a ratio as low as 9 may be used with good follower dynamics and a smooth acceleration curve. The maximum allowable harmonic amplitude for valve lift is 0.075 mm, based on experience.

Another method for controlling surging is by designing springs that have variable pitches. A variable pitch spring is one whose natural frequency varies along the length of the spring. Because there is no unique resonant spring frequency, the effects of surge are minimized. A variable rate spring may be constructed by closing the coils at the stationary end of the spring and continuously varying the distance between successive spring coils. Note that the spring

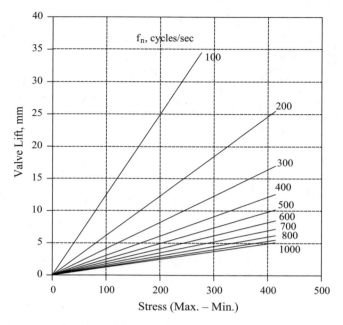

Figure 3.117 *Natural frequency of helical springs, Y = 1.15,
σ = stress at a round wire surface (Taylor).*

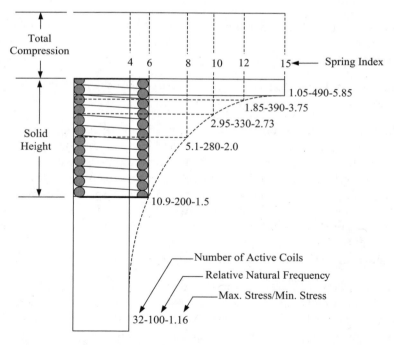

Figure 3.118 *Helical springs of equal stiffness and equal maximum stress
under the same maximum load (Taylor).*

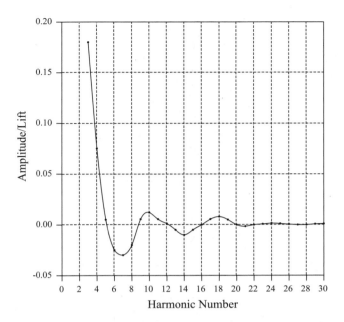

Figure 3.119 *Harmonic analysis of a typical engine cam: A = amplitude/lift; n = order of harmonic number based on camshaft speed; and a = 0.48 at n = 1 (Taylor).*

coils will not be stressed evenly. The spread coils, being deflected under load a greater distance, will be more highly stressed.

Variable pitch springs are used to achieve a variable rate, as shown in Figure 3.120. Load and deflection follow the relationship

$$P_n = k_1 f_1 + k_2(f_2 - f_1) + ... + k_n(f_n - f_{n-1})$$

As the valve opens, the coils will be progressively compressed solid, starting from the close-pitch cylinder-head end; hence, the number of active coils is reduced. Conversely, as the valve closes, the number of active coils increases. This variation of the number of active coils during opening and closing of the valve produces a variable spring rate and a constantly changing natural frequency that helps to minimize surging and spring resonance.

3.8.3.5 Valve Spring Damping

In addition to increasing the spring frequency, altering the harmonics of the cam, and using variable rate springs, another method of controlling undesirable spring surge is to introduce mechanical spring dampers or friction dampers. This can be accomplished by inserting a coiled flat (crowned) spring or a metal cylinder inside the valve spring, or by a metal sleeve on the O.D. of the spring.

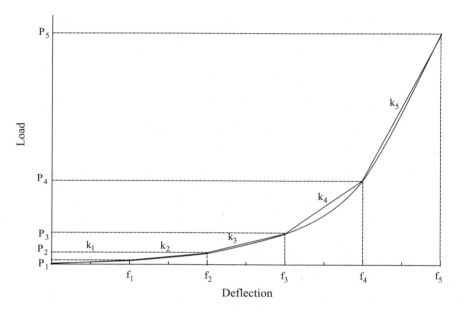

Figure 3.120 *Load deflection curve for a variable rate spring.*

The damper is effective for low-amplitude surge; however, it is least effective for high-amplitude resonant conditions. Surge damping for such a system will become less effective with time because wear of the surfaces will cause a decrease in friction. This wear also may cause stress concentrations when dampers are used on the inside surface of the spring, which is the maximum stressed area in a round wire helical spring, thereby reducing the spring life. It is believed that the O.D. damper will not be as greatly affected by the reduction in friction because the outer damper is a spring that grips the valve spring O.D. with more force per area than the I.D. The O.D. spring damper assembly consists of a stationary damper that is wrapped around the O.D. of the valve spring and is retained to the spring by interference between the spring O.D. and the damper I.D.

Spring stresses are considerably lower on the O.D. than on the inside. The O.D. spring dampers avoid the direct contact at the I.D. of the spring, which has the maximum shear stresses, and avoid galling at the spring I.D., which can initiate a stress riser that could lead to spring failure.

Another advantage is that at high engine speeds, the damper seat restrains the valve spring from excessively bouncing and/or rotating in clockwise and counterclockwise directions, thereby controlling valve rotation. The type of damping effect between the valve spring resonant amplitudes can be induced by camshaft harmonics and/or valvetrain excessive masses. Excessive valvetrain masses can reduce the valvetrain natural frequency and consequently can lead to early valvetrain surge. The valve spring natural frequency will not be affected by the damper because it is independent of friction force.

3.8.4 Valve Spring Materials

3.8.4.1 Characteristics

Valve springs can be made from plain high-carbon steels, low-alloy chromium-vanadium steels, chromium-silicon steels, stainless steels, and high-temperature alloys. Table 3.20 shows the chemistries and mechanical and special properties of various spring grades.

TABLE 3.20
COMMON VALVE SPRING MATERIALS (MAKER)

Materials (ASTM)	Composition (%)	Min. Tensile Strength (MPa)	Elastic Modulus (GPa)	Shear Modulus (GPa)	Hardness (HRc)	Max. Allowable Temperature (°C)	Special Properties
High-Carbon Steel							
A227 Hard Drawn	C 0.45-0.85 Mn 0.30–1.30	1010–1950	210	80	31–52	120	For average stress application.
A229 Oil Tempered	C 0.55–0.85 Mn 0.30–1.20	1140–2020	210	80	42–55	120	For general purpose.
A230 Carbon, Valve Spring Quality	C 0.60–0.75 Mn 0.60–0.90	1480–1650	210	80	45–49	120	HT before fabrication. Good surface condition.
Alloy Steel							
A231, A232 Cr-V	C 0.48–0.53 Cr 0.80–1.10 V 0.15 min.	1310–2070	210	80	41–55	220	HT before fabrication. For shock loads and moderate temperature.
A401 Cr-Si	C 0.51–0.59 Cr 0.60–0.80 Si 1.20–1.60	1620–2070	210	80	48–55	245	HT before fabrication. For shock loads and moderate temperature.
Stainless Steel							
A313 SS302 (18–8)	Cr 17–19 Ni 8–10	860–2240	190	69	35–45	290	For corrosion and heat resistance.
A313 SS631	Cr 16–18 Ni 6.5–7.75 Al 0.75–1.50	1620–2310	200	76	38–57	340	High strength, for heat and corrosion resistance.
High-Temperature Alloys							
Inconel 718	Ni 52.5 Cr 18.6 Fe 18.5	1450–1720	200	77	45–50	590	Good corrosion resistance at elevated temperature.
A-286	Fe 53 Ni 26 Cr 15	1100–1380	200	72	35–42	510	Good corrosion resistance at elevated temperature.

Carbon steels are easier to manufacture, are the least costly of the valve spring quality wires, and are not as notch sensitive (i.e., susceptible to fracture due to surface imperfections) as low-alloy steels. However, the alloy steels have good fatigue properties and show less propensity of suffering from relaxation.

The chromium-vanadium and chromium-silicon valve springs are more efficient in terms of mass and spring rate. The chromium-vanadium steel contains 0.48–0.53% carbon, 0.2% silicon,

0.6% manganese, 0.8–1.0% chromium, and a minimum of 0.15% vanadium. The steel wire of the valve spring quality (A232) is superior to the same quality of carbon steel wire. It is supplied in the softened state, and it is surface ground and then coiled to form. Finally, it is heat treated by hardening and tempering. Springs of chromium-silicon steel wire (A401) can be used at temperatures as high as 230°C.

Cold-drawn Type 302 stainless steel spring wire (A313) is high in heat resistance and corrosion resistance.

Carbon steel wire of valve spring quality, and chromium-vanadium steel wire of both spring and valve spring quality, can be supplied in the annealed condition. This will permit severe forming of springs with a low spring index and will permit sharper bends in the end hooks. Although a sharp bend is never desired in any spring, it sometimes is unavoidable. Springs made from annealed wire can be quenched and tempered to spring hardness after they have been formed. However, without careful control of processing, such springs will have greater variations in dimensions and hardness. This method of making springs usually is used only for springs with special requirements, such as severe forming, or for small quantities because springs made by this method may have less uniform properties than those of springs made from pretempered wire. In addition, they have a higher cost.

Static stresses represent spring stresses encountered at low engine speed before stress amplification occurs. Table 3.21 shows theoretical maximum allowable valve spring static stresses.

TABLE 3.21
THEORETICAL MAXIMUM ALLOWABLE
VALVE SPRING STATIC STRESS

	Carbon Steel Wire	Cr-V Alloy Wire	Cr-Si Alloy Wire
Stress @ Valve Open	790 MPa	860 MPa	900 MPa
Stress @ Solid Height	860 MPa	965 MPa	1035 MPa

The selection of round wire for cold-wound springs is based on the minimum tensile strength for each wire size and grade (Figure 3.121). Maximum tensile strength generally is approximately 200 MPa above the minimum tensile strength. For helical compression springs, the design stress of torsion typically is taken as 75% of the minimum tensile strength.

Compression valve springs, cold wound and cold formed from pretempered high-carbon spring wire, should always be stress relieved to remove residual stresses produced in coiling. Stress relieving affects the tensile strength and elastic limit, particularly for hard-drawn spring wire. Strength is increased by heating in the range of 230–260°C. Oil-tempered spring wire, except for the chrome-silicon (Cr-Si) grade, shows little change in either tensile strength or elastic limit after stress relieving below 315°C. Both properties then drop because of temper softening. Wire of Cr-Si steel temper softens only above approximately 425°C.

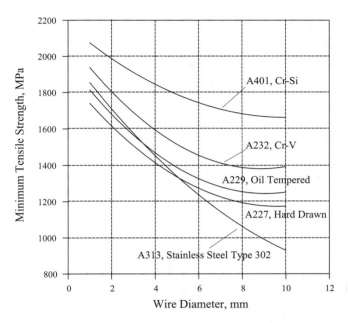

Figure 3.121 *Minimum tensile strength of steel spring wire (Maker).*

3.8.4.2 Residual Stress and Shot Peening

Residual compressive stresses are induced into the wire outer surface by shot peening and significantly improve life under repetitive high-stress conditions (i.e., fatigue life). It is important that full coverage of the spring surface be obtained, particularly in the high-stress I.D. of the spring. Essentially, the shot peening process consists of imparting round particles of hardened steel shot at high velocity against the wire surface. Two types of shot peening process are available. One type (the batch process) consists of placing the valve springs into a machine that holds and tumbles the springs. A second type (the in-line process) consists of passing springs longitudinally (i.e., end to end) through a shot blast. As the springs pass through the shot pattern, they are rotated to ensure that all surfaces (I.D. and O.D.) are well covered.

The shot peening process is dependent on the shot size, velocity, and coverage. Because shot peening is a mechanical working process, it can be affected by heat. If springs are heated sufficiently after shot peening, the benefits will be reduced and the endurance limit will drop, eventually to that of the wire as originally received (Figure 3.122). Shot peening begins to lose its effectiveness at temperatures around 250°C, with complete loss at temperatures near 430°C on steel springs. This indicates that at these temperatures, residual stresses are considerably reduced, and the benefits of presetting will be reduced correspondingly.

Fatigue tests on shot peened springs have demonstrated that fatigue life improvement is so significant that the springs must be shot peened to operate acceptably in internal combustion engines

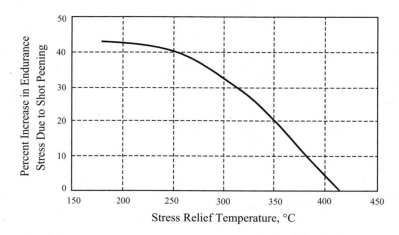

Figure 3.122 *Reduction in effectiveness of shot peening by subsequent heating or stress relieving (SAE HS 795).*

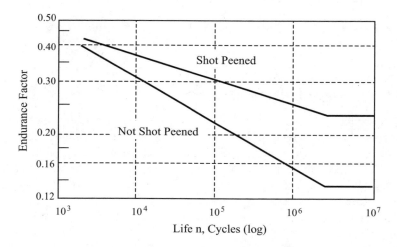

Figure 3.123 *Typical S-N curves for helical springs (SAE HS 795).*

(Figure 3.123). The endurance factor $K_E = \dfrac{S_E}{S_U}$ is normalized to represent the ratio of the peak alternating stress as a fraction of the minimum ultimate tensile strength of the material.

It also has been demonstrated that excessive shot peening produces a spring that is more susceptible to local relaxation. Consequently, the shot peening process must be optimized and controlled to ensure that acceptable valve spring performance will occur.

3.9 Valve Stem Seals

3.9.1 Introduction

The interface of the valve stem and guide must be lubricated adequately to prevent valve stem scuffing. However, excessive lubrication may adversely affect engine emissions and carbon buildup on the valves. In extreme cases, it can lead to valve sticking. If too little lubricant enters the interface of the stem and guide, then the valves will be insufficiently lubricated, and scuffing or seizure can occur. The basic function of a valve seal is to meter the amount of oil flowing down the valve guide. The primary objective for valve seal design must be to provide adequate oil flow to control the valve guide and stem wear, with acceptable oil economy and minimum friction increase without adversely affecting emissions. An optimum lubrication condition is required to minimize valve guide wear while providing acceptable oil economy and emissions requirements. A typical oil metering rate down the guide by the stem seal is in the range of 0.01–0.2 cc/10 hours. The collateral benefits of a good seal design also allow sufficient valve guide lubrication to maintain valvetrain hash and tick noise at an acceptable level, to minimize valve throat carbon deposits through adequate valve guide oil flow, and to control exhaust smoke at startup and high-vacuum conditions.

3.9.2 Design Considerations

Valve stem seals evolved when oil economy and emissions became more important and mandated that oil leaking down the guide must be controlled more closely. In valve stem seal design, the minimum oil flow variability from guide to guide at all engine operating conditions, low cost, and ease and speed of assembly must be considered.

A stem-mounted valve stem seal following stem actuating provides a better sealing capability; however, the primary objective of the stem seal is to meter oil through the stem/guide interface to provide acceptable stem/guide friction and wear. The current trend in valve seal design is to use guide-mounted valve seals that are attached to the top of the guide and that remain stationary during valve motion. Figure 3.124 shows a typical valve stem seal design schematic and its nomenclature.

Gravity and pressure differentials combine to force oil downward between the valve stem and the valve guide. For the intake guide, there usually is a significant pressure difference between the overhead area and the intake manifold where the vacuum condition exists when the engine is running. This large pressure differential forces the oil downward past the intake valve/guide and is the reason for the great sensitivity of oil economy to intake seal length on valve stem-mounted seals. For the guide-mounted valve seal, this condition is minimized because the valve seal is spring loaded at the guide seal lip by the garter spring and at the valve guide boss by the retaining band, thus providing positive sealing.

For exhaust valves, a pressure difference also can exist across the guide, but from a different phenomenon. When the exhaust valve is open, combustion gas exits from the cylinder. With sudden closure of the exhaust valve, the supply of hot gas ceases abruptly. When this occurs, the inertia of the gas leaving the port can create a low-pressure area in the port. There also is a partial pressure effect at the lower end of the valve guide because the exhaust gas flowing

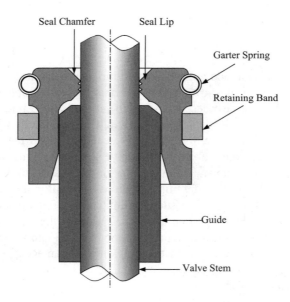

Figure 3.124 *Typical valve stem seal schematic and nomenclature.*

along the bottom guide has, by nature, a reduced static pressure. The combination of these two effects can produce a pressure differential across the exhaust valve guide, even though the total pressure in the manifold is greater than atmospheric pressure. The exhaust guide can contribute to oil consumption, although less significantly than the intake. However, worn exhaust valve seals and guides can greatly increase the exhaust valve guide contribution to oil consumption and exhaust emissions.

3.9.2.1 Valve Stem Seal Configurations

The shape of the sealing surface of the guide-mounted seal contacting the valve stem appears to be the primary parameter affecting oil metering and thus valve guide lubrication. Undoubtedly, many different guide-mounted seal configurations are possible, which, with proper development and tests, could meet engine objectives. The basic designs for valve stem seals currently are a multi-lip design and a single-lip design. Figure 3.125 shows schematics of these designs, comparing them to a design without a lip. A multi-lip design is characterized by a block lip with a number of grooves or spirals and a metal insert to increase and stabilize radial force.

Figure 3.126 shows the total amount of oil leakage as a function of duration hours for three types of seal designs: (1) controlled type, (2) wet type, and (3) dry type. The controlled oil leakage lip-type valve stem seals show the linear relationship between leakage and test duration. The controlled-type seals maintained the consistent leakage rate of 0.2 cm^3/10 hours throughout the test duration of 1000 hours. On the other hand, dry- and wet-type design seals

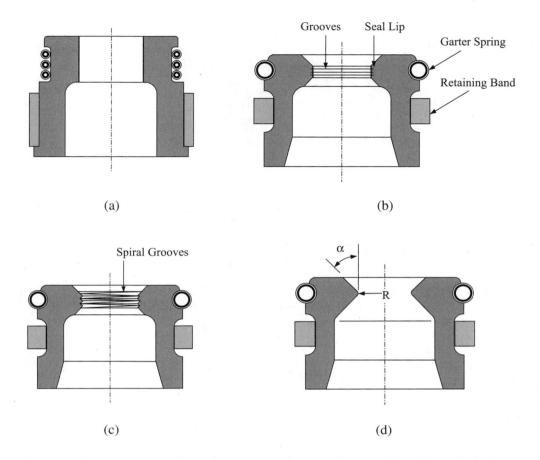

Figure 3.125 *Valve stem seal configurations and nomenclature: (a) without-lip design, (b) annular groove or multiple-lip dry design, (c) spiral-groove wet design, and (d) single-lip design.*

show the inconsistent condition of leakage due to the nature of the seal design with their multi-step grooves (dry type) or spiral grooves (wet type).

3.9.2.2 Parameters Affecting Oil Metering Rate

The lip seal design restricts oil to a minimum level. The amount of oil leakage depends on the application condition, the seal design, and the contact condition at the seal and stem surface. Figure 3.127 shows a fishbone cause-and-effect diagram for oil leakage through the stem seal. The oil leakage is controlled by the lip configuration and the tension of the garter spring. With the multi-lip design of the valve stem seal, an adjustment of the seal to a required oil metering rate can be achieved easily by varying the number of grooves in the lip. Because of the wide contact area with the valve stem, only little deformation occurs due to pressure in the port area.

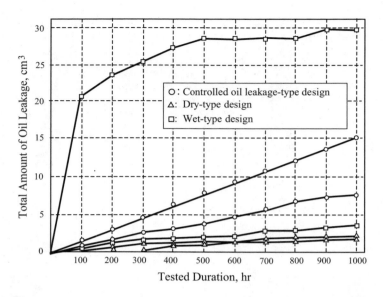

Figure 3.126 *Oil leakage for various types of seals as a function of test hours (Matsushima et al.).*

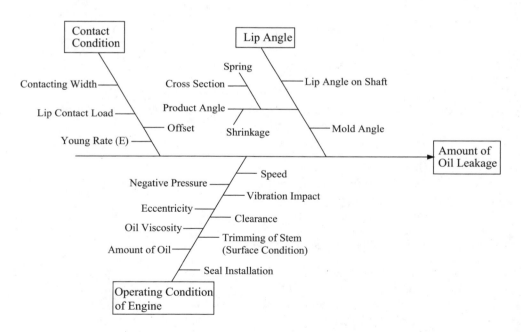

Figure 3.127 *Fishbone cause-and-effect diagram for oil leakage through a stem seal (Matsushima et al.).*

The seal chamfer in contacting the stem acts as a reservoir to collect oil and provides a constant supply for metering down the valve guide through the seal threads. The theory of this type of seal is that the action of the valve stem past the oil source creates a hydrodynamic pressure by pulling the oil into the wedge-shaped zone. The resulting effect introduces oil into and down the seal threads at a controlled consistent rate, providing adequate guide lubrication and acceptable oil consumption. Variation in thread pitch, thread form, chamfer angle, and depth can be made to provide the required lubrication for each application. The thread must break out into the chamfer for proper operation. Figure 3.128 shows the metering rate as a function of cam speeds for various types of seal designs.

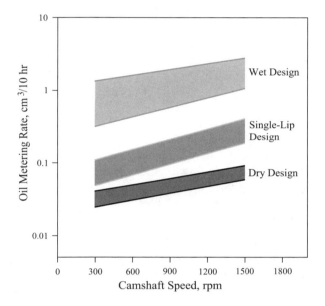

Figure 3.128 *The effect of cam speed and types of seals on oil metering rate (Matsushima et al.).*

Increased negative pressure (vacuum) increases the amount of oil leakage for both the dry and wet types, but especially the wet type. Increased oil temperature (from 80 to 120°C) also increases the amount of oil leakage for both the dry and wet types, although it will not affect the single-lip design. Higher lubricant viscosity reduces the amount of oil leakage.

3.9.2.3 Single-Lip Designs

The block lip or multiple-lip seal is capable of drastically reducing oil flow down the valve stem, but greater potential to meet more stringent requirements can be seen in a single-lip seal. A single-lip design provides the opportunity to define the oil metering rate more sensitively because the relevant parameters are infinitely variable. A garter spring can be fitted to compensate for the loss of radial force of the elastomer compound. There is only a single lip to be controlled on functional parameters. Oil metering tests show that the oil metering rate of the single-lip seal is lower and, more importantly, is much more constant than that of the multiple-lip seal (Table 3.22).

TABLE 3.22
VARIATION OF THE OIL METERING RATE
(NETZER AND MAUS)

Seal	Average Oil Metering (%)	Standard Deviation
Single-Lip Seal	100	0.12
Multiple-Lip Seal	6200	51

The oil metering rate of valve stem seals must be adjusted accurately to engine requirements to avoid oil deposits on the valves or measurable effects on exhaust emissions. The critical parameters on the sealing lip that influence the oil flow for a single-lip seal include the lip chamfer angle (α), lip radius (R), and radial force. Figure 3.129 shows that the oil metering rate decreases as the chamfer angle and lip radial force increase, and that the oil metering rate increases as the lip radius increases. The recommended design parameters for a single-lip seal are 35–55° lip angle, <0.4 lip radius, and 0.3–1.0 N/mm^2 specific radial force.

Figure 3.130 shows the effect of offset (ε) between the centerline of the garter spring and the lip radius on oil leakage. The oil leakage is minimal when the spring centerline to the contact point offset is 0.4 mm.

3.9.2.4 Effect of Emissions on Oil Metering Rate

The ideal oil metering rate lies between the compromise of the two contradictory requirements of effective lubrication and minimal emissions. Although the requirements for each engine differ according to the design parameters and operating conditions, generally they are within the range 0.1–1.0 mg per valve per hour. Figure 3.131 shows the effect of some parameters on emissions. A 10–30% reduction in carbon monoxide (CO), nitrogen oxide (NO$_x$), and hydrocarbon (HC) emissions using a single-lip seal design in a four-cylinder, four-valves-per-cylinder engine comparison test was recorded.

Hydrocarbons are primarily a result of the incomplete combustion of fuel and oil. Therefore, with more oil entering the exhaust ports, an increase of HC emissions is to be expected. Oil entering the combustion chamber via the intake port will be partially burned and therefore will increase HC emissions. Because the oil flow level of the standard seal is already low, a significant improvement with even less oil was not to be expected. However, the seals providing high

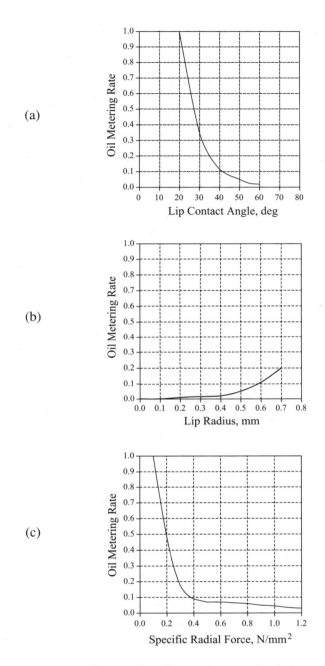

Figure 3.129 _Key parameters affecting the oil metering rate for a single-lip seal design: (a) lip contact angle, (b) lip radius, and (c) lip radial force (Netzer and Maus)._

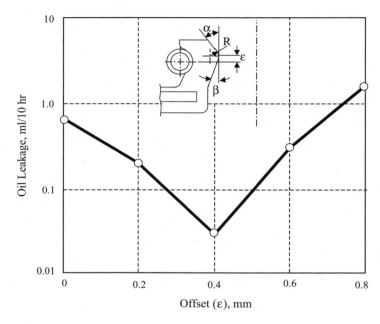

Figure 3.130 *The effect of offset on oil leakage (Mori et al.).*

oil flow have a negative impact on HC emissions. Damaged seals or no seals on the intake or exhaust result in even higher values (Figure 3.131(b)). The better performance of the valve stem seals with a high oil metering rate in comparison to the damaged seals basically results from the more constant oil flow at differing port pressures.

The amount of NO_x emissions basically depends on the combustion peak temperature and the amount of oxygen behind the flame propagation front in the chamber. Although seals were not expected to have a major impact on these combustion parameters, the standard valve stem seal shows a slight advantage over seals providing lower or higher oil flow (Figure 3.131(c)).

Carbon monoxide emissions depend primarily on the mass of available oxygen. With more oil, higher CO emissions behind the catalyst were measured (Figure 3.131(d)).

The exhaust positions (due to higher operating temperatures and backpressure conditions) require a seal with more leakage than intake plus a heavy chrome plate on the exhaust valve stem to prevent scuffing and galling. This presents an assembly complexity problem of keeping intake and exhaust seals separated after engine assembly. To reduce the complexity and oil consumption, both the intake and exhaust stems can be heavy-chrome plated with a common valve stem seal between the intake and exhaust of the lowest leakage class.

Valvetrain Components

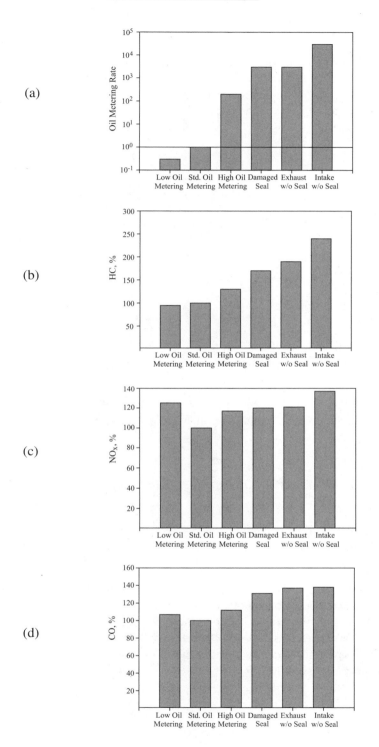

Figure 3.131 *The effect of oil metering rate on exhaust emissions: (a) oil metering rate, (b) HC emissions, (c) NO_x emissions, and (d) CO emissions (Netzer et al.).*

3.9.2.5 Integral Spring Seat Seal

Figure 3.132 shows a guide-mounted seal design with a molded or pressed-in integral spring seat (for aluminum cylinder heads). Special attention must be paid to this design and how it relates the cylinder head top of the guide to the spring seat dimension. The spring seat portion of the seal assembly must always rest on the cylinder head spring seat (before bottoming on the top of the valve guide) but still have enough of the guide for support and retention.

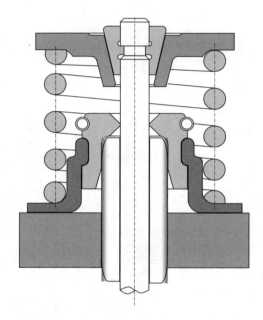

Figure 3.132 *Valve stem seal with integral spring seat.*

3.9.2.6 Dimensions and Tolerances

Although the seal dimensions and tolerances are important to the oil metering rate, the dimensions and tolerances of the pertinent matching components such as the valve stem and guide are critical in the successful control of a consistent oil metering rate and durability of the seal. Tolerance in the valve stem design is important in sealing or metering the oil. The valve seal also must be designed such that it does not interfere with the function of other components. The following and Figure 3.133 show the relationship of the seal to other components and tolerance guidelines (Chicago Rawhide):

Guide O.D. to I.D. runout:	0.25 mm (concentricity of the guide O.D. to the guide I.D. is 0.20±0.05 mm)
Seal lip radius R:	0.55±0.05 mm
Stem surface roughness:	$R_z \leq 3$ μm / $R_{max} = 4$ μm

Seal chamfer angle and depth α: $45 \pm 2° \times 1$ mm

Seal I.D. and stem O.D. interference: $\phi_1 \pm 0.05$ mm

Guide O.D. and seal membrane interference: $\phi_2 \pm 0.8$ mm

Guide lead-in chamfer β: $30 \pm 2° \times 1.5$ mm

Guide surface finish R_z: 15 μm (R_{max} 3–12 μm)

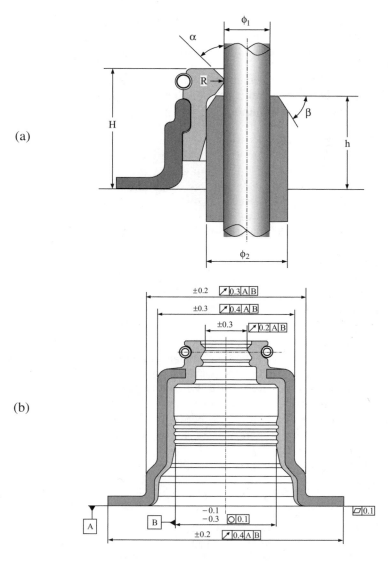

Figure 3.133 *Dimensions and tolerances of a seal and its pertinent matching components: (a) seat relationship to other components, and (b) dimensional tolerances.*

When the valve is fully open, the gap between the top of the valve stem seal and the bottom of the spring retainer should exceed 1.0 mm.

3.9.3 Seal Materials

Depending on the application and the design of the seal, the material used may be nitrile, poly-acrylate, fluoroelastomer (Viton or FKM rubber), silicone, or polytetrafluoroethylene (nylon or Teflon PTFE). Nitrile is the most commonly used sealing material and is one of the least expensive materials. It has been used for many years in umbrella or deflector-type seals for Type V pushrod engines. The temperature range for nitrile is –40°C to 110°C. It can withstand an intermittent operating temperature of up to 120°C, which usually is adequate for intake valve seals but not for exhaust valve seals. It has good abrasion resistance.

A step up from nitrile is polyacrylate. Polyacrylate has a temperature range of –40°C to 175°C and is well suited for use with extreme pressure lubricants. It has a higher resistance to oxidation but should not be used with water or at temperatures below –40°C.

For high- or low-temperature ranges and low-friction applications, silicone seals can operate at temperatures ranging from –73°C to 163°C. The high lubricant absorbency of silicone minimizes friction and wear. However, it has poor compatibility with oxidized oils, some extreme pressure additives, and abrasive contaminants. Silicone should not be used in dry-running applications.

Nylon is a hard material with a temperature rating of –40°C to 150°C. Nylon is impervious to oil, but it can melt if the engine overheats. Fluoroelastomer (Viton and FKM rubber) seals cost roughly ten times as much as nitrile but have a temperature range of –35°C to 230°C, making them some of the best high-temperature seals available. Viton has good flexibility similar to nitrile, which means it can handle some runout between the valve stem and guide. It also is considered to be a more durable material than silicone. Likewise, Viton has better wear resistance than most other seal materials, making it a good choice for applications where long-term durability is critical. Fluoroelastomer material possesses the flexibility, wear resistance, and necessary low friction to provide a stable oil metering rate for an engine lifetime.

The highest-rated positive seal material is Teflon, with a range of –70°C to 260°C, and it has the highest abrasion wear resistance and lowest friction. Similar to nylon, Teflon is a hard material; therefore, it cannot handle as much runout between the stem and guide as more flexible seal materials can handle. Teflon is the most expensive seal material.

Figure 3.134 shows applicable temperature ranges and the abrasive wear resistance ranking for various stem seals.

3.10 Keys

Valve keys also are known as valve locks, keepers, or cotters and are located in the recessed valve stem keeper groove area. The function of the valve key is primarily to secure the valve stem to the valve spring retainer through the wedge pressure exerted by the valve spring load.

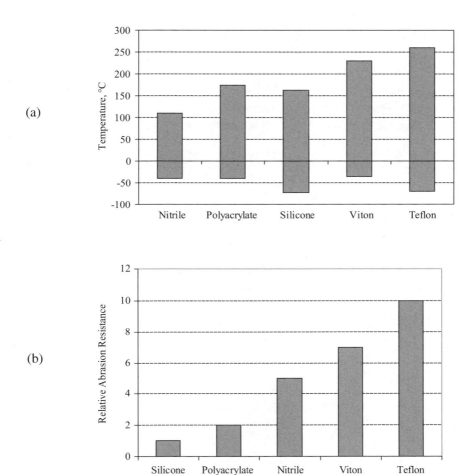

Figure 3.134 *(a) Seal temperature rating and (b) relative abrasion resistance ranking for various stem seals (Chicago Rawhide).*

The basic valve key designs currently used are single-bead, three-bead, and four-bead designs, as illustrated in Figure 3.135.

3.10.1 Single-Bead Keys

Single-bead locking keys are designed to lock the valve stem to the valve spring retainer through the wedge pressure exerted by the valve spring. These types of keys do not butt against each other (a gap exists between the keys at both ends, 180° apart), resulting in a circumferential contact with the valve stem. The force is transmitted through friction created by tightly fitting it around the valve stem. The single-bead key can be used with a one-piece retainer when little

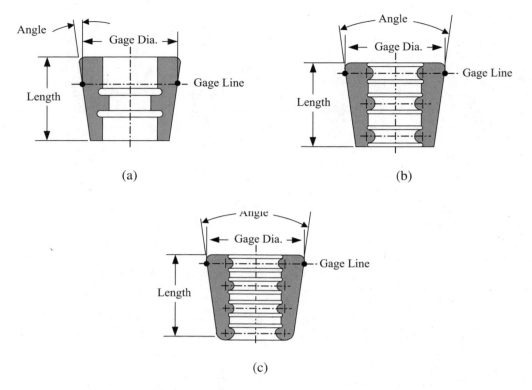

Figure 3.135 *Valve key designs: (a) single-bead key, (b) three-bead key,
and (c) four-bead key.*

or no valve rotation is required. A single-bead key is the only type of key used with two-piece retainers and rotocoils. Figure 3.135(a) illustrates a typical single-bead valve key design.

3.10.2 Multiple-Bead Keys

In addition to the primary function described in the preceding section for single-bead keys, multiple-bead valve keys are designed mainly to improve valve rotation capabilities over the single-bead key/retainer configuration. Both the three- and four-bead valve key designs, which each use a one-piece retainer, provide rotation similar to that of the single-bead two-piece retainer combination. The difference is that the multiple-bead keys butt against each other and do not grip the stem, providing the low-friction interface required for the valve to rotate. Both halves of the key are half circles that form a cone that wedges into the retainer because of the valve spring load. A diametrical clearance is maintained between the valve stem and the keys.

Multiple-bead key designs currently are being used in high-speed engines for both intake and exhaust valves. Figures 3.135(b) and 3.135(c) illustrate three-bead and four-bead key designs, respectively. However, because of the clearance between the beads and grooves in the valve, multiple-bead keys are more susceptible to pound-in at sustained high-speed operation. The valve rotation allowed by the three- and four-bead keys is influenced by groove dimensions such as the location and the groove/bead radius, because these control the amount of clearance between the bead and the groove.

3.10.3 Dimensions and Tolerances

Figure 3.136 shows a typical dimensional tolerance of a single-bead key. A typical wedge angle α ranges from 10–15°, the angularity of the key O.D. ∠ is 0.03, and the gap between the two halves δ is approximately 0.3 mm.

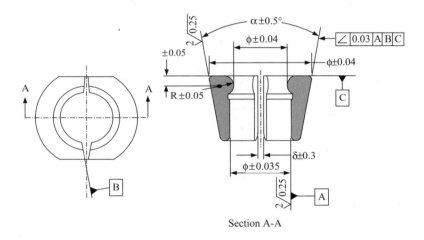

Section A-A

Figure 3.136 *Valve key dimensional tolerance requirements.*

3.10.4 Key Materials

Typical key materials are hardenable steel or low-carbon steel such as SAE 1010. These materials can be machined but often are cold formed and then are carbonized or carbonitrided with a case depth of 0.2 ± 0.1 mm. Surface hardness for carbonitrided or carbonized steel typically is 58 HRc minimum. Hardness for hardened steel ranges from 40–50 HRc.

3.11 Retainers

A valve spring retainer is fixed to the movable end of the valve to maintain a working spring height. The primary function of a valve spring retainer is to secure the valve stem through the valve keys and the upward pressure exerted by the valve spring load. Figure 3.137 illustrates

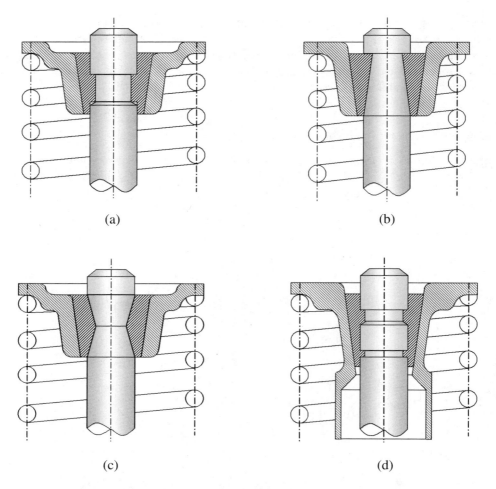

Figure 3.137 *Various configurations of valve spring retainer assemblies: (a) the design used most often in automobile engines; (b) through (d) other designs.*

various valve spring retainer assembly designs. Figure 3.137(a) describes the design most widely used in automobile engines.

3.11.1 One-Piece Retainers

There are two types of valve spring retainer designs: one piece and two piece. The one-piece retainer may have either single-bead or multiple-bead valve locking keys, depending on the application and amount of valve rotation required. The single-bead design consists of non-butting valve keys that grip the valve stem, thus locking the retainer and keys to the valve under valve spring load. The single-bead design prevents relative motion between the retainer and the valve. For this particular configuration, rotation of the retainer is restricted by a large moment arm (R_{MA}) existing from the center of the valve outward to where the retainer seats on the valve spring top coil (Figure 3.138(a)). This results in negligible valve rotation at low engine speeds. At engine speeds greater than 3000 rpm, the inherent helical coil feature of the valve spring will impart some rotation to the valve in the coil direction of the helix. This phenomenon occurs when slippage is induced between the top coil of the valve spring and the bearing surface of the retainer due to engine vibration. The one-piece retainer with the single-bead configuration is considered because of its strength, weight, and cost. However, the magnitude, direction, and even certainty of rotation for the one-piece retainer are not predictable. Compared to other types of retainer/key designs, this configuration results in the least amount of valve rotation.

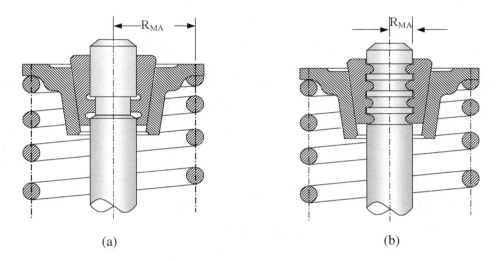

(a)   (b)

Figure 3.138 One-piece retainer designs: (a) single-bead valve keys, and (b) multiple-bead valve keys.

The multiple-bead design consists of either three- or four-bead valve keys. The keys butt against each other, leaving a small diametrical clearance between the valve stem and the keys. The large moment arm, which restricted the rotation of the one-piece retainer with non-butting single-bead locks, has been reduced for this design configuration. The moment arm extends from the valve centerline to the line of contact between the valve key beads and the valve stem key grooves (Figure 3.138(b)). This condition improves valve rotation capabilities over the one-piece retainer/single-bead keys. For this configuration, the valve rotation is produced primarily by means of engine vibration and valve seating as a result of the clearance existing between the keys and the valve stem. Currently, most high-speed engines use one-piece retainers with either three- or four-bead butting valve keys. New trends are one-piece retainers with three-bead keys for valve rotation. This retainer is well suited for conical and beehive valve springs that have a smaller opening at the retainer end plus smaller-diameter valve stems that require a smaller cone opening in the retainer.

3.11.2 Two-Piece Retainers

The two-piece retainer (McPherson Rotator) assembly consists of a spring retainer cap and a sleeve (Figure 3.139). A small diametrical clearance is designed between the sleeve and the retainer cap. When assembled in an engine under valve spring load, the sleeve and keys are locked to the valve by way of the retainer cap. In the case of a two-piece retainer, the moment arm restricting the rotation stretches from the valve centerline to the line of contact between the sleeve and the bottom of the retainer (Figure 3.139(a)). When comparing the two-piece to the one-piece retainer with single-bead keys, valve rotation capabilities are improved due to the reduced moment arm that exists for this case. For this configuration, valve rotation results primarily from a combination of engine vibration, valve seating, and valve spring helix, and by way of a low-friction interface that exists within the retainer, permitting occasional oscillation. The retainer may remain stationary while the valve stem, keys, and sleeve rotate as a unit within the retainer. Little rotation is induced in the valve at low engine speeds, but as the engine speed increases, valve rotation is improved. Two-piece retainers are not widely used, mainly because of their weight, cost, and tendency to break at the inside bottom corner of the radius. A similar function, in terms of valve rotation, can be accomplished at reduced weight and with less complexity by using a one-piece retainer with either three- or four-bead keys.

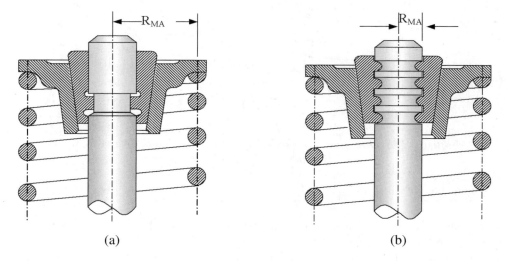

(a)   (b)

Figure 3.139 *Two-piece spring retainer: (a) two-piece positive retainer, and (b) two-piece spherical retainer.*

Helical compression valve springs have nonsymmetrical ground surface area at the ends, inactive coils, an inexact number of active coils, variation in the distance between active coils, variation in the distance between the inactive coils and the first active coil at each end, and an unequal number of inactive coils on the ends. These characteristics plus the direction of coiling and nontrue perpendicularity of the spring ends to the axial center of the spring can actuate non-axial forces through the ends of the spring, thus promoting valve rotation. A spherical sleeve inserted between the retainer and key can self adjust by reducing or eliminating non-axial forces caused by the valve spring (Figure 3.139(b)). The spherical spring retainer sleeve also may help in alleviating the non-axial force or side load from the rocker arm.

3.11.3 Dimensions and Tolerances

Figure 3.140 shows typical valve spring retainer dimensions and tolerances.

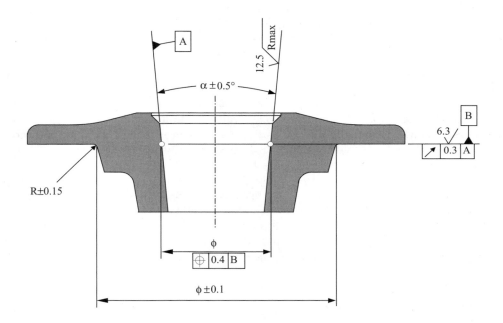

Figure 3.140 *Typical spring retainer dimensions and tolerances.*

3.11.4 Retainer Materials

Typical spring retainer materials are hardenable steels or low-carbon steels such as SAE 1010, 1018, or 1215. Low-carbon steels usually are machined but often are cold formed and then are carbonized or carbonitrided with a case depth of 0.2 ± 0.1 mm. Surface hardness for case-hardened steel typically is 55 HRc minimum, with a core hardness in the range of 30–45 HRc.

3.12 Other Components

Other components, such as valve rotators, pushrods, and valve bridges, also are used in some engines.

3.12.1 Valve Rotators

The major benefit of rotation of a valve is that if a minor amount of leakage between the valve and its seat exists, the valve rotation reduces the chances of combustion pressure drop or valve guttering. The leakage at the valve seat creates hot spots on the valve, and if those hot areas can be moved to a nonleaking location, this will allow the hot spot to cool and therefore extend the useful service life of the valve. Deposits can build up on the valve seat face and then later break away to cause leakage or valve burning. Valve rotation can minimize deposit buildup and can result in better valve sealing and improved valve life. The excellent contact between the valve seat and the insert improves heat transfer through the seat and reduces the valve head temperature. Furthermore, the improved valve sealing assists in maintaining idle vacuum and engine smoothness. An exhaust valve most often is subjected to non-uniform temperature distribution around the head of the valve. Rotation of the valve equally exposes all parts of the valve head to hot spots that exist in the combustion chamber. Thus, more uniform valve rim temperatures are obtained. Rotation results in less valve head distortion, more uniform valve seat contact, and less blowby at the valve seat. In addition, the rotation of the valve distributes the valve stem wear more uniformly and helps carry lubrication to the thrust surfaces in the valve guide. There are conflicting opinions about the effect of valve rotation on valve seat and insert wear. Some believe that it is detrimental because parting agents such as oxide film on the seat surface can prevent direct metal-to-metal contact and thus avoid adhesive wear. Valve rotation prevents such a film from forming.

Two types of valve rotation mechanisms are designed to cause rotation of the valve, that is, nonpositive and positive rotation. A number of ways exist to free or partially free the valve from the restrictions of the keys and the rocker arm pad that limit its ability to rotate. The positive valve rotation mechanism generally is located in the spring retainer or is used as a spring seat. These typically are referred to as positive rotators.

3.12.1.1 Nonpositive Valve Rotators

Nonpositive valve rotators enable the valve to revolve at random when it is opening and closing. Three common methods are used. The first approach uses loose-fitting keys with three- or four-groove keys, whereas the second approach adopts a thimble rotator cap that fits over the valve stem tip (Figure 3.141). The third approach applies a two-piece retainer or split spring retainer to achieve valve rotation. The valve rotates primarily while it is open or not in contact with the seat position when using nonpositive rotators. The rotation direction and speed depend on the type of valvetrain, the number of valve keeper grooves, the engine speed, the spring coil direction, and the direction of motion of the valve.

The three- and four-groove valves use keys that butt against one another when installed, and their I.D. is slightly larger than the valve O.D. in the groove area, as shown in Figure 3.141(a). This permits the valve to be free of the restriction of the keys and allows engine vibration to encourage rotation or oscillation. The nonpositive rotators start to work when the engine speed exceeds 3000 rpm.

The conventional valve is secured by a valve spring retainer, and the spring always maintains pressure between the valve-stem end groove and the tapered keys that secure the retainer. As

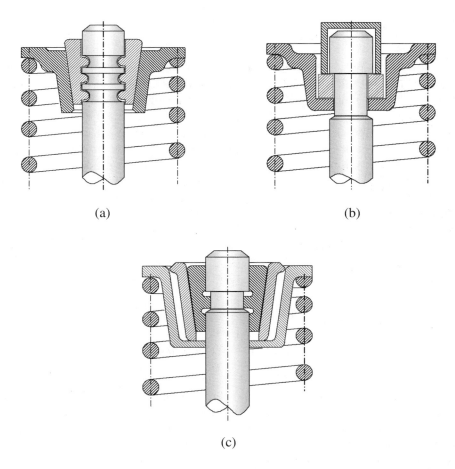

(a)　　　　　　　　　　　(b)

(c)

Figure 3.141 *Nonpositive rotators: (a) multiple-bead loose fit, (b) rotor cap, and (c) two-piece retainer (Tunnecliffe and Jenkins).*

a consequence, the valve usually does not turn around relative to its seating. The free-valve or rotator cap type completely isolates the valve from the friction of the contacting components (Figure 3.141(b)) and allows engine vibration to rotate or oscillate the valve. A steel cap fits over the end of the valve stem and rests on two semicircular keys that fit into the valve stem groove below them. The valve spring presses the retainer against these keys and keeps the valve shut. However, when the valve is to be opened, the cap is pressed downward and in turn bears against the two keys, and then moves the valve spring and retainer downward. Because the spring pressure now is taken by the cap, the valve is freed from spring pressure. It still is moved downward because the closed end of the cap then abuts the valve-tip end, but it is free to turn. There is no automatic provision for turning the valve relative to its seating, but it has been found in practice that the valve does tend to rotate in operation. The clearance between the tip of the valve and the bottom of the tip cup ranges from 25–50 μm and must be held accurately to obtain satisfactory rotation.

Split spring retainers are designed to allow the valve, keys, and one part of a two-piece retainer to rotate (Figure 3.141(c)). They also rely on engine vibration to produce rotation.

It has been observed that the valve rotates in one direction first and then in the reverse direction when a certain engine speed is exceeded. At low engine speeds, the valve rotates in the spring-winding direction during valve lifting when the spring tangential force (momentum) exceeds the friction force (momentum) at the valve interfaces. The valve typically rotates clockwise with a left-hand spring and counter-clockwise with a right-hand spring. However, at high engine speeds, the friction resistance becomes smaller during closing than during lifting because of separation or near-separation at the valve tip. Therefore, the valve starts to rotate toward the least-resistant direction, that is, the valve rotates counterclockwise with a left-hand spring and clockwise with a right-hand spring during spring unwinding. This critical reverse rotation speed depends on the natural frequency of the valvetrain. Most reverse rotation phenomena that are observed occur in valvetrains with rocker arms, which have lower natural frequencies than do direct-acting Type I valvetrains. Valves in a direct-acting valvetrain will rotate in the reverse also when the engine operates at high speeds.

3.12.1.2 Positive Valve Rotators

This type of retainer imparts a positive increment of valve rotation. Positive rotation means that the valve is forced to rotate instead of depending on engine vibration or spring motion. Generally, the positive rotator is used on exhaust valves to meet emissions requirements and in engine applications demanding sustained load requirements at elevated operating temperatures, such as heavy trucks, taxis, and industrial engines. Figure 3.142 shows examples of positive valve rotators. The operation of the rotocoil including the coil-spring type and the spring-ball type is as follows. When the valve is in the closed position, the valve spring force is applied to the retainer collar and onto the Belleville spring washer. At the closed position, the spring washer is cupped inward. In this position, the spring washer force is greater than the installed valve spring load. Thus, the load goes from the retainer collar to the spring washer and into the retainer body through the valve keepers to the valve stem. As the valve opens, the valve spring load increases. The increased load acting against the spring washer forces it to flatten and causes the coils of the coil spring to lay over. This, in turn, forces the retainer body to rotate. This type of rotator advances the valve approximately 1° per cycle. Valve spring surges must be controlled over the range of engine revolutions per minute to prevent rotocoil breakage and/or excessive valve rotation.

In a Type I direct-acting valvetrain, a taper on the cam lobe and a certain amount of offset between the tappet and cam lobe are believed to assist valve rotation. Positive rotators typically are used in heavy-duty engines but typically are not used in passenger car engines.

3.12.2 Pushrods

The pushrod is a strut whose function is to relay the to-and-fro cam follower movement to one end of the pivoting rocker arm for Type V valvetrains. Both ends of the pushrod form part of a pair of semispherical ball-and-socket joints to permit the rod to tilt slightly and revolve when the

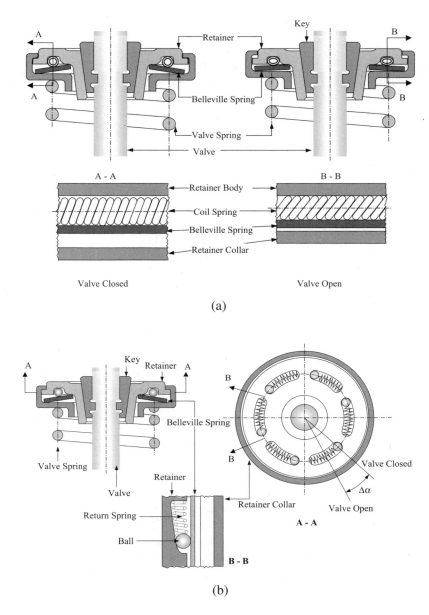

Figure 3.142 *A positive valve rotator: (a) coil spring type, and (b) spring-ball-slope type.*

rocker arm oscillates about its pivots (Figure 3.143). The bottom of the rod is convex and fits in a matching recess in the follower, while the top of the rod is expanded to support a concave recess seat that will locate with the adjustable tappet screw on the end of the rocker arm. For

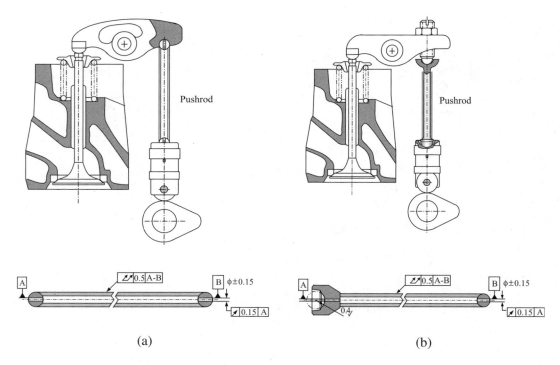

Figure 3.143 *Pushrod application schematics, dimensions, and tolerances: (a) sphere-sphere hollow pushrod, and (b) sphere-socket solid pushrod.*

medium-sized engines, the pushrod may be solid; however, for large engines, hollow rods that have hardened, shaped end pieces that are forced into the tubing are preferred.

Pushrod tubes usually are made from carbon steel such as SAE 1018 or carbon-manganese steel, with a popular composition being 0.35% carbon, 0.2% silicon, 1.5% manganese, and the balance iron. The heat treatment given is to quench and harden the rods from a temperature of 840–870°C and then temper between 550–660°C. This should produce a hardness of 220 to 280 Brinell number. Ball material used at the pushrod ends typically is SAE 52100.

3.12.3 Valve Bridges

A valve bridge is the component that allows a pair of intake or exhaust valves to be actuated by a single rocker arm. Figure 3.144 shows a valvetrain layout with a valve bridge and the valve bridge dimensions and tolerances.

Typical material used for the valve bridge is hardened and tempered, carburized SAE 8620, HR_{15N} 89/92, effective case 0.35 ± 0.15 mm. Powdered metal parts have been used as an effective material alternative to wrought steel for the valve bridge (Haugen and Wynthein).

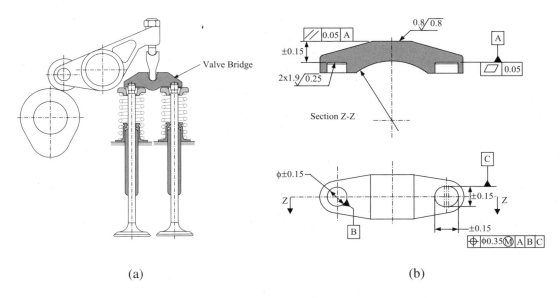

(a)   (b)

Figure 3.144 *Valve bridge schematic and dimensional tolerances: (a) schematic, and (b) dimensions and tolerances.*

3.12.4 Crosshead Rocker Arms

The crosshead rocker arm is a valvetrain component that functions as a combination of a valve bridge and a rocker arm. Figure 3.145 illustrates a schematic and typical dimensions and tolerances for a crosshead rocker arm.

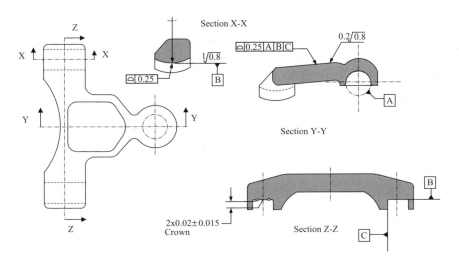

Figure 3.145 *Crosshead rocker arm schematic and tolerances.*

Typical materials used for the crosshead rocker arm are the same as those used in the valve bridge, that is, harden and tempered, carburized SAE 8620, HR_{15N} 89/92, effective case 0.35 ± 0.15 mm.

3.13 References

Abell, R., "The Operation and Application of Hydraulic Valve Lifters," SAE Paper No. 690347, Society of Automotive Engineers, Warrendale, PA, 1969.

Annand, W. and Roe, G., "Gas Flow in the Internal Combustion Engine," Foulis, Yeovil, 1974.

Aoi, K., Nomura, K., and Mtsuzaka, H., "Optimization of Multi-Valve, Four Cycle Engine Design—The Benefit of Five-Valve Technology," SAE Paper No. 860032, Society of Automotive Engineers, Warrendale, PA, 1986.

Ari-Gur, P., Noble, V., Narasimhan, S., and Larson, J., "Hot Corrosion Studies of Automotive Exhaust Valves," SAE Paper No. 910633, Society of Automotive Engineers, Warrendale, PA, 1991.

Armstrong, W. and Buuck, B., "Valve Gear Energy Consumption: Effect of Design and Operation Parameters," SAE Paper No. 810787, Society of Automotive Engineers, Warrendale, PA, 1981.

Buuck, B. and Hampton, K., "Engine Trends and Valvetrain Systems for Improved Performance, Fuel Economy, and Emission," *Proceedings of the International Symposium on Valvetrain System Design and Materials*, edited by Bolton, H.A. and Larson, J.M, ASM International, Materials Park, OH, 1997.

Caird, S. and Trela, D., "High-Temperature Corrosion-Fatigue Test Method for Exhaust Valve Alloys," SAE Paper No. 810033, Society of Automotive Engineers, Warrendale, PA, 1981.

Campo, E., Quaranta, S., and Pleragostini, F., "High-Temperature Mechanical Properties of SAE EV11 Engine Valve Steel," SAE Paper No. 821085, Society of Automotive Engineers, Warrendale, PA, 1982.

Chicago Rawhide, *Seal Handbook*, Catalog 457010, Elgin, IL, 1999.

Dooley, D., Trudeau, T., and Bancroft, D., "Materials and Design Aspects of Modern Valve Seat Inserts," *Proceedings of the International Symposium on Valvetrain System Design and Materials*, edited by Bolton, H.A. and Larson, J.M., ASM International, Materials Park, OH, 1997.

Edelmayer, T., Hillerand, G., and Paulson, R., "Hydraulic Lash Adjuster and Biased Normally Open Check Valve System," U.S. Patent 5,758,613, 1998.

Etchells, E., Thomson, R., Robinson, G., and Malone, G., "The Interrelationship of Design, Lubrication, and Metallurgy in Cam and Tappet Performance," SAE Paper No. 472, Society of Automotive Engineers, Warrendale, PA, 1955.

Ezaki, S., Masuda, M., Fujita, H., Hayashi, S., Terashima, Y., and Motosugi, K., "Aluminum Valve Lifter for Toyota New V8 Engine," SAE Paper No. 900450, Society of Automotive Engineers, Warrendale, PA, 1990.

Funabashi, N., Endo, H., and Goto, G., "U.S.–Japan PM Valve Guide History and Technology," *Proceedings of the International Symposium on Valvetrain System Design and Materials*, edited by Bolton, H.A. and Larson, J.M., ASM International, Materials Park, OH, 1997.

Giles, W., "Fundamentals of Valve Design and Material and Selection," SAE Paper No. 660471, Society of Automotive Engineers, Warrendale, PA, 1966.

Goth, G., "Recent Developments in Hardfacing Alloys for Internal Combustion Engine Valves," SAE Paper No. 831287, Society of Automotive Engineers, Warrendale, PA, 1983.

Hagiwara, Y., Ishida, M., Oka, T., Watanabe, R., and Sato, K., "Development of Nickel-Base Superalloy for Exhaust Valves," SAE Paper No. 910429, Society of Automotive Engineers, Warrendale, PA, 1991.

Haugen, D. and Wynthein, P., "Use of Powdered Metal for a Valve Bridge in a 4-Valve Heavy-Duty Diesel Engine," SAE Paper No. 980330, Society of Automotive Engineers, Warrendale, PA, 1980.

Jarrett, M., "Material Considerations for Automobile Camshafts," SAE Paper No. 710545, Society of Automotive Engineers, Warrendale, PA, 1971.

Jenkins, L. and Larson, J., "The Development of a New Austenitic Stainless Steel Exhaust Valve Material," SAE Paper No. 780245, Society of Automotive Engineers, Warrendale, PA, 1978.

Jones, D., "Fatigue Behavior of Exhaust Valve Alloys," SAE Paper No. 800315, Society of Automotive Engineers, Warrendale, PA, 1980.

Kano, M., Suzuki, K., Matsuyama, H., Sato, S., Yamaguchi, M., Ninomiya, R., and Nakahara, Y., "New Copper Alloy Powder for Laser-Clad Valve Seat Used in Aluminum Cylinder Heads," SAE Paper No. 2000-01-0396, Society of Automotive Engineers, Warrendale, PA, 2000.

Kattus, J., *Aerospace Structural Metals Handbook*, Belfour Stulen, Watertown, MA, 1981.

Kawasaki, M., Takase, K., Kato, S., Takagi, S., and Sugimoto, H., "Development of Engine Valve Seats Directly Deposited onto Aluminum Cylinder Head by Laser Cladding Process," SAE Paper No. 920571, Society of Automotive Engineers, Warrendale, PA, 1992.

Kreuter, P. and Maas, G., "Influence of Hydraulic Valve Lash Adjusters on the Dynamic Behavior of Valvetrains," SAE Paper No. 870086, Society of Automotive Engineers, Warrendale, PA, 1987.

Larson, J., Jenkins, L., Narasimhan, S., and Belmore, J., *Engine Valves—Design and Material Evolution, New Materials and Manufacturing Process—ICE Vol. 1*, edited by Bailey, J., American Society of Mechanical Engineers, New York, 1986.

Lilly, L., *Diesel Engine Reference Book*, Butterworths, London, 1984.

Maker, J., *Steel Springs, ASM Metals Handbook, 9th Edition, Vol. 1, Properties and Selection: Irons, Steels, and High-Performance Alloys*, ASM International, Materials Park, OH, 1990.

Matsushima, A., Hatsuzawa, H., Yamamoto, Y., and Iida, S., "Comments on Valve Stem Seals for Engine Application in Small Size Passenger Cars," SAE Paper No. 850331, Society of Automotive Engineers, Warrendale, PA, 1985.

Mogford, R. and Ball, F., *Exhaust Valve Life*, Inst. of Mech. Engrs., Vol. 5, 1955–1956.

Mori, G., Umeki, T., Ueno, Y., and Ohishi, T., "Engine Valve Stem Seal, Oil Leakage Control Technology and Performance," SAE Paper No. 960208, Society of Automotive Engineers, Warrendale, PA, 1996.

Müller, H. and Kaiser, A., "Composite Camshaft—Avoid Lobe Grinding Using Precision PM Lobes," SAE Paper No. 970001, Society of Automotive Engineers, Warrendale, PA, 1997.

Narasimhan, S. and Larson, J., "Valve Gear Wear and Materials," SAE Paper No. 851297, Society of Automotive Engineers, Warrendale, PA, 1985.

Neale, M., ed., *Drives and Seals, A Tribology Handbook*, Society of Automotive Engineers, Warrendale, PA, 1994.

Netzer, J. and Maus, K., "Improvements of Valve Stem Seals to Meet Future Emission Requirements," SAE Paper No. 980581, Society of Automotive Engineers, Warrendale, PA, 1998.

Netzer, J., Piasecki, T., and Deussen, N., "The Influence of the Valve Stem Seal Oil Metering Rate on Exhaust Emissions," SAE Paper No. 2000-01-0683, Society of Automotive Engineers, Warrendale, PA, 2000.

Newton, J. and Allen, C., "Valve Gear Fundamentals for the Large-Engine Designer," *ASME Transactions,* Vol. 76, No. 2, American Society of Mechanical Engineers, New York, 1954.

Nourse, J., "Recent Developments in Cam Profile Measurement and Evaluation," SAE Paper No. 964A, Society of Automotive Engineers, Warrendale, PA, 1965.

Roberts, D. and Tournier, M., "The Influence of Engine Oil Formulation on the Prevention of Valvetrain Wear in Modern European Passenger Cars," SAE Paper No. 810328, Society of Automotive Engineers, Warrendale, PA, 1981.

Rodrigues, H., "Sintered Valve Seat Inserts and Valve Guides: Factors Affecting Design, Performance and Machinability," *Proceedings of the International Symposium on Valvetrain System Design and Materials*, edited by Bolton, H.A. and Larson, J.M., ASM International, Materials Park, OH, 1997.

Rothbart, H., *Cams—Design, Dynamics, and Accuracy*, John Wiley and Sons, New York, 1956.

SAE HS 795, *Manual on Design and Application of Helical and Spiral Springs*, Society of Automotive Engineers, Warrendale, PA, 1990.

SAE Standard J775, "Engine Poppet Valve Information Report," Society of Automotive Engineers, Warrendale, PA, Revision 1993.

SAE Standard J1682, "Valve Guide Information Report," Society of Automotive Engineers, Warrendale, PA, 2001 Revision.

SAE Standard J1692, "Valve Seat Insert Information Report," Society of Automotive Engineers, Warrendale, PA, 2001 Revision.

Sato, A., Kawasake, M., Nunokawa, S., Masuda, Y., and Yamamoto, Y., "Development of Mass Production Technology for Laser-Clad Valve Seat," *Powertrain International*, 1999, p. 38.

Sato, K., Saka, T., Ohno, T., Kageyama, K., Sato, K., Noda, T., and Okabe, M., "Development of Low-Nickel Superalloys for Exhaust Valves," SAE Paper No. 980703, Society of Automotive Engineers, Warrendale, PA, 1998.

Sato, K., Takagi, Y., and Saka, T., "The Progress of Valvetrain Design and Exhaust Valve Material Research for Automobiles," *Proceedings of the International Symposium on Valvetrain System Design and Materials*, edited by Bolton, H.A. and Larson, J.M., ASM International, Materials Park, OH, 1997.

Schaefer, S., Larson, J., Jenkins, L., and Wang, Y., "Evolution of Heavy Duty Engine Valves—Materials and Design," *Proceedings of the International Symposium on Valvetrain System Design and Materials*, edited by Bolton, H.A. and Larson, J.M, ASM International, Materials Park, OH, 1997.

Strong, G. and Liang, X., "A Review of Valve Seat Insert Material Properties Required for Success," *Proceedings of the International Symposium on Valvetrain System Design and Materials*, edited by Bolton, H.A. and Larson, J.M., ASM International, Materials Park, OH, 1997.

Taylor, C., *The Internal-Combustion Engine in Theory and Practice, Vol. II: Combustion, Fuels, Materials, Design*, The M.I.T. Press, Cambridge, MA, 1968.

Thumuki, C., Ueda, K., Nakamura, H., Kondo, K., and Suganuma, T., "Development of Sintered Integral Camshaft," SAE Paper No. 830254, Society of Automotive Engineers, Warrendale, PA, 1983.

TRW Valve Division Handbook for Internal Combustion Engine Valves, TRW, Cleveland, OH, 1987.

Tunnecliffe, T. and Jenkins, L., "Why Valves Succeed," SAE Paper No. 249B, Society of Automotive Engineers, Warrendale, PA, 1960.

Turkish, M., "The Relationship of Valve Spring Design to Valve Gear Dynamics and Hydraulic Lifter Pump-Up," *SAE Transactions,* Vol. 61, Society of Automotive Engineers, Warrendale, PA, 1951, p. 707.

Turkish, M., *Valve Gear Design, Eaton Manufacturing Company*, Eaton Manufacturing Company, Detroit, MI, 1946.

Umino, S., Hamada, A., Kenmoku, T., and Nishizawa, Y., "New Fe-Base Exhaust Valve Material for Higher Heat Resistance," SAE Paper No. 980704, Society of Automotive Engineers, Warrendale, PA, 1998.

Wang, Y., Narasimhan, S., Larson, J., and Schaefer, S., "Wear and Wear Mechanism Simulation of Heavy-Duty Engine Intake Valve and Seat Inserts," *Journal of Materials Engineering and Performance*, Vol. 7, No. 1, 1998, p. 53.

Wang, Z., Fang, X., He, W., Sun, Z., Wu, M., Wu, G., Wu, P., Chen, Y., Yao, W., Lu, M., and Lu, H., *Handbook of Machinery Engineering Material Property Data*, Machinery Industry Publisher, Beijing, China, 1994.

Zhao, R., Barber, G.C., Wang, Y.S., and Larson, J.E., "Wear Mechanism Analysis of Engine Exhaust Valve Seats with a Laboratory Simulator," *STLE Tribology Transanctions*, Vol. 40, No. 2, 1997, p. 209.

Zhao, Y., Tong, K., and Lu, J., "Determination of Aeration of Oil in High Pressure Chamber of Hydraulic Lash Adjuster in Valvetrain," SAE Paper No. 1999-01-0646, Society of Automotive Engineers, Warrendale, PA, 1999.

Chapter 4

Valvetrain Testing

4.1 Introduction

The successful application of materials for valvetrain components relies on the ability of the material to meet design and service requirements and to be fabricated to the proper dimensions. The capability of a material to meet these requirements is determined by the mechanical and physical properties of the material. In this chapter, typical relationships between metallurgical features (such as crystal structures and microstructures) and the mechanical behavior of metals first are introduced. Then, typical mechanical properties and testing standards are presented, including tension, compression, bend, shear testing, hardness testing, impact toughness testing, fracture mechanics, fatigue testing, and tribological testing, including friction, wear, and surface testing. Next, many specialized valvetrain test methods for component testing and nondestructive testing are discussed. In component testing, one or more components are tested in a bench to simulate the engine environment to validate on a preliminary basis the design and/or material concept. Finally, examples of valvetrain system testing are presented. Validation of the system design is the ultimate goal. In the system or in engine testing, the valvetrain is validated as a system from the standpoint of valvetrain dynamics, performance, durability, and reliability.

4.2 Materials and Testing

In selecting a material for valvetrain component applications, one must take into account properties such as formability, machinability, mechanical durability, thermal properties, chemical stability, electrical behavior, and the important factor of cost. However, the first and foremost property that comes to mind usually is mechanical strength. Mechanical properties, the primary focus of this chapter, are described as the relationships among forces (or stresses) acting on a material and the resistance of the material to deformation (e.g., strain) and fracture. Mechanical properties include elastic modulus, yield strength, elastic and plastic deformation (e.g., elongation), hardness, fatigue resistance, and fracture toughness. Mechanical properties are highly dependent on the type of crystal structure (i.e., the arrangement of atoms), microstructure (e.g., grain size, phase distribution, second phase content), elemental composition (e.g., alloying element content, impurity level), and defects.

The following discussions are designed to briefly introduce typical relationships among metallurgical features (such as crystal structures and microstructures) and the mechanical properties of metals. Relationships between metal structure and performance make mechanical property determination important for valvetrain applications, for failure analysis and prevention, and for material development for advanced applications.

4.2.1 Material Behavior

To understand the origin of material properties, one must understand material structure at the atomic level. Two things are especially important in influencing material properties: (1) atomic forces, and (2) crystal structure. The atomic forces hold atoms together (the interatomic bonds) and act similarly to little springs, linking one atom to the next in the solid state. The crystal structure is the way in which atoms pack together (the atomic structure), because this determines how many little springs are present per unit volume, as well as the angle at which they are pulled.

4.2.1.1 Atomic Forces

The various ways in which atoms can be bound together involve primary and secondary bonds. Primary bonds include ionic, covalent, or metallic bonds, all of which are relatively strong. (They generally melt between 1000 and 4000K.) Secondary bonds include van der Waals and hydrogen bonds, which both are relatively weak. (They melt between 100 and 500K.)

Metals and alloys are held together entirely by primary bonds, specifically, metallic and covalent bonds. In a solid metal, the highest-energy electrons tend to leave the parent atoms (which become ions) and combine to form a "sea" of freely wandering electrons that are not attached to any ion in particular. The energy and force curves for interatomic metallic bonding are illustrated in Figure 4.1 and are described as

$$U = -\frac{A}{r^m} + \frac{B}{r^n} \quad (m<n) \tag{4.1}$$

where r is the interatomic distance, U is the potential energy, and A, B, m, and n are constants.

The interatomic forces shown in Figure 4.1(a) are the derivative of the energy equation (Eq. 4.1), that is, $F = \frac{dU}{dr}$. When a material is stressed, the attractive forces between the atoms resist the stress and keep the materials from deforming and pulling apart. If the nuclei of two atoms are forced into intimate contact, their positive charges repel each other until a balance is reached between the attractive forces that pull the atoms together and the repulsive forces that hold the nuclei apart. Both the attractive and repulsive forces increase as the atoms come into closer proximity, but the latter operate at much closer range. At the distance a-a' in Figure 4.1, the two forces are equal. This becomes the equilibrium distance between the two atoms, because energy must be supplied to either increase or decrease their interatomic spacing.

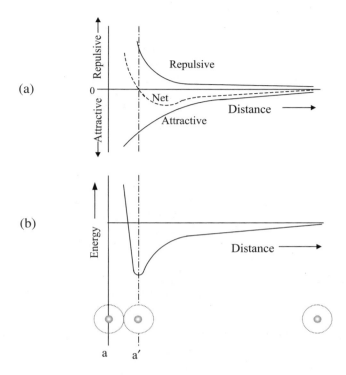

Figure 4.1 *Interatomic force and energy. (a) Attractive forces pull two atoms together until the repulsive forces equal the attractive forces. (b) At that distance, a-a′, the atoms are most stable because additional energy is required either to press them closer together or to pull them apart.*

4.2.1.2 Crystal Structure

Metals and many nonmetallic solids are crystalline, that is, the constituent atoms are arranged in a pattern that repeats itself periodically in three dimensions. The basic building block of the crystal lattice is the unit cell. The atoms are considered as hard spheres that vary in size from element to element. From the hard sphere model, the parameters of the unit cell can be described directly in terms of the radius of the sphere, r. In the three-dimensional drawings shown in Figure 4.2, the diagrams illustrate a unit cell of the crystal structures, with the nuclei shown as small circles/spheres. The large dotted-line circles represent the full hard sphere sizes of the atoms. The crystal structure can be described or constructed by stacking identical unit cells face to face in perfect alignment in three directions. The intersection of the lines through the centers of the atoms is called a space lattice. Every point of a space lattice has identical surroundings. By placing a motif unit of one or more atoms at every lattice site, the regular structure of a perfect crystal is obtained. The planes, directions, and point sites in a lattice are described by reference to the unit cell and the three principal axes, x, y, and z (Figure 4.2). The

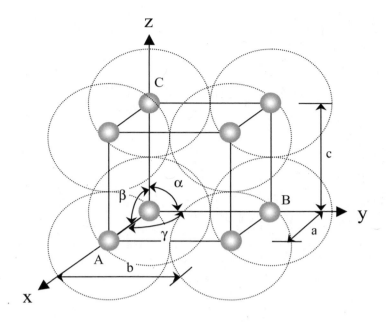

Figure 4.2 *Unit cell of the crystal structure.*

cell dimensions OA = a, OB = b, and OC = c are the lattice parameters. These, along with the angles $\angle BOC = \alpha$, $\angle COA = \beta$, and $\angle AOB = \gamma$, completely define the size and shape of the cell. In cubic crystals, a = b = c and $\alpha = \beta = \gamma = 90°$.

Most metals or alloys fall into three groups of crystal structures: (1) body-centered cubic (bcc), (1) face-centered cubic (fcc), and (3) hexagonal close-packed (hcp) structures. This is illustrated in the unit cell structure shown in Figure 4.3.

There are six groups of lattice systems (in order of increasing symmetry): (1) triclinic, (2) mono-clinic, (3) orthorhombic, (4) tetragonal, (5) hexagonal, and (6) cubic. They can be classified further into fourteen different types (or Bravais lattices), as shown in Table 4.1.

Many common metals (e.g., copper, silver, gold, aluminum, and nickel) and their alloys have fcc structures. In an Fe-C system or in most steels, austenite or γ-Fe is an fcc structure. Iron (ferrite or α-Fe), molybdenum, tantalum, vanadium, chromium, tungsten, niobium, sodium, and potassium are bcc metals. Metals such as beryllium, titanium, zirconium, magnesium, cobalt, zinc, and cadmium have hcp structures. Iron carbides, Fe_3C, are hcp structures with carbon atoms in the interstitials.

4.2.1.3 Deformation of Metals

In a uniaxial tension test, stress generally varies linearly with strain in the initial stages of deforma-tion. This deformation is regarded as elastic because the material will return to its original shape

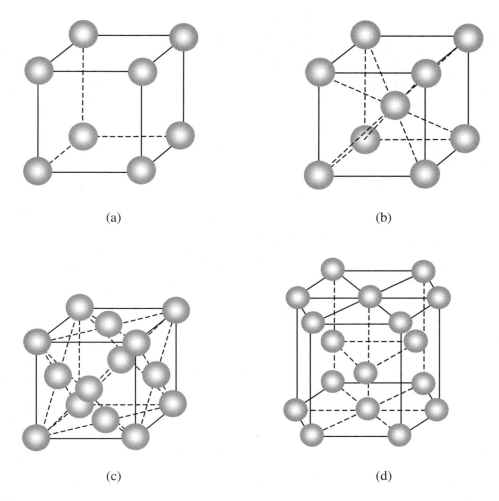

Figure 4.3 *Typical metal crystal structures: (a) simple cubic, (b) body-centered cubic (bcc), (c) face-centered cubic (fcc), and (d) hexagonal close-packed (hcp).*

when the applied stress is removed. However, if the sample is not unloaded and deformation continues, the stress-versus-strain curve becomes nonlinear. A permanent elongation occurs that will not be recovered after unloading of the specimen, and it is called plastic deformation. The stress at which plastic deformation starts is called the yield strength. The tensile yield strength of most alloys is on the order of 10^2–10^3 MPa (135–480 MPa for low-carbon steels, 200–480 MPa for aluminum alloys, and 1200–1650 MPa for high-strength steels).

To understand the different deformation modes, the structure of a metal must be considered. Elastic deformation can be conceptualized by considering the bonds between individual atoms as springs. In the same way that a spring constant relates the force to the applied displacement (i.e., $F = kx$), the elastic modulus (E) relates the tensile stress to strain (i.e., $\sigma = E\varepsilon$) and is simply

TABLE 4.1
THE SIX CRYSTALLOGRAPHIC SYSTEMS (MYERS)

System	Number of Lattices in System	Nature of Unit Cell Axes and Angles	Lengths and Angles to Be Specified	Illustration of Examples
Triclinic	1	$a \neq b \neq c$ $\alpha \neq \beta \neq \gamma$	a, b, c α, β, γ	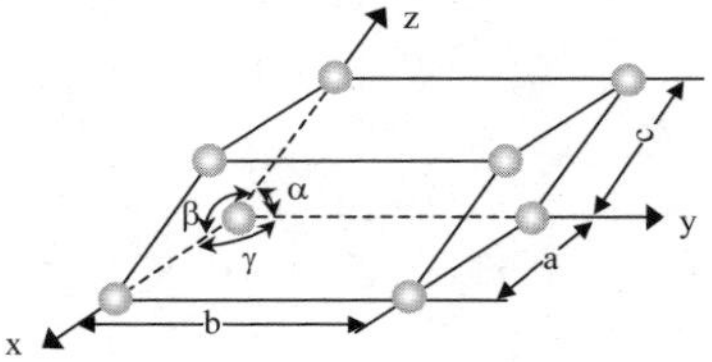
Monoclinic	2	$a \neq b \neq c$ $\alpha = \gamma = 90° \neq \beta$	a, b, c β	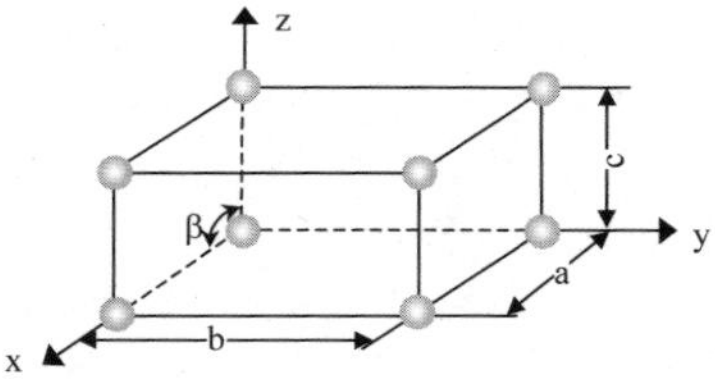
Orthorhombic	4	$a \neq b \neq c$ $\alpha = \beta = \gamma = 90°$	a, b, c	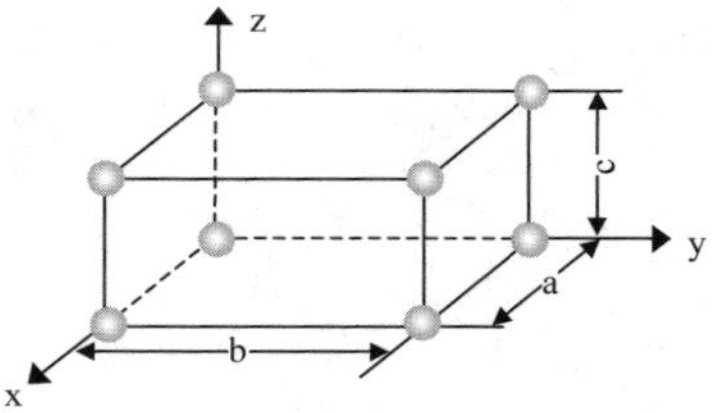
Tetragonal	2	$a = b \neq c$ $\alpha = \beta = \gamma = 90°$	a, c	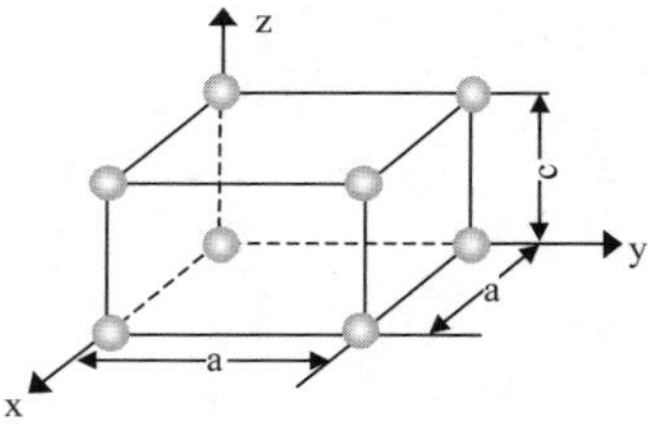
Hexagonal	2	$a = b \neq c$ $\alpha = \beta = 90°$ $\gamma = 120°$	a, c	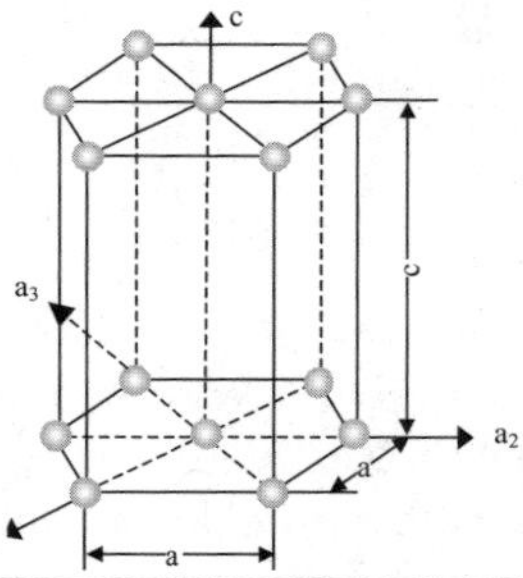

TABLE 4.1 *(Continued)*

System	Number of Lattices in System	Nature of Unit Cell Axes and Angles	Lengths and Angles to Be Specified	Illustration of Examples
Cubic	3	$a = b = c$ $\alpha = \beta = \gamma = 90°$	A	

the slope of the linear portion of the tensile stress versus the tensile strain curve produced in the tension test. Differences in the measured elastic moduli for different metals therefore can be rationalized in part by the differences in the atomic bonds between individual atoms within the crystal lattice.

Plastic deformation results in a permanent change of shape, meaning that after the load is removed, the metal will not return to its original dimensions. This implies a permanent displacement of atoms within the crystal lattice. If a perfect crystal is assumed, this deformation could occur only by breaking all of the bonds at once between two planes of atoms and then sliding one row (or plane) of atoms over another. Based on calculations using the theoretical bond strengths, this process would result in yield strengths on the order of 10^4–10^5 MPa (Hull and Bacon). These strengths are much greater than those typically observed in actual metals (10^2 MPa); therefore, deformation must occur via a different method.

Even under the most ideal crystal growth conditions, metals are not crystallographically perfect; instead, the lattice may contain many imperfections. The imperfections include point, line, surface, or volume defects, and they disturb locally the regular arrangement of the atoms. Their presence can modify the properties of crystalline solids significantly. The line defects called dislocations will be the primary concern and focus of this discussion.

Dislocation theories were remarkably successful in reconciling theoretical and experimental values of the applied shear stress required to plastically deform a single crystal. In a perfect crystal (i.e., in the absence of dislocations), the sliding of one plane past an adjacent plane is a rigid cooperative movement of all the atoms from one position of perfect registry to another. It is assumed that there is a periodic shearing force required to move the top row of atoms across the bottom row, which is given by the sinusoidal relation (Hull and Bacon)

$$\tau = \frac{Gb}{2\pi a} \sin \frac{2\pi x}{b} \tag{4.2}$$

where

τ = applied shear stress
G = shear modulus
b = spacing between the atoms in the direction of the shear stress
a = spacing of the rows of atoms
x = translation of the two rows away from the low-energy equilibrium position

The right side of Eq. 4.2 is periodic in b and reduces to Hooke's law for small strain, $\frac{x}{a}$, that is, in the small-strain limit,

$$\sin \frac{2\pi x}{b} = \frac{2\pi x}{b}$$

The maximum value of τ then is the theoretical critical shear stress

$$\tau_{th} = \frac{b}{a} \frac{G}{2\pi} \tag{4.3}$$

Because $b \cong a$, the theoretical shear strength is only a small fraction of the shear modulus. Using more realistic expressions for force as a function of shear displacement, values of $\tau_{th} \cong \frac{G}{30}$ have been obtained. Although these are approximate calculations, they show that τ_{th} is many orders of magnitude greater than the observed values (10^{-4}–10^{-8} G) of the resolved shear stresses for slip measured in real, well-annealed crystals. This striking difference between prediction and experiment was accounted for by the presence of dislocations, which move easily. When they move, the crystal deforms; the stress needed to move them is the yield strength. Dislocations are the carriers of deformation.

In general, dislocations entail the stepwise movement of the dislocation across the crystal lattice, as opposed to the displacement of one entire plane over another. This means that only one set of bonds is broken at a time, as opposed to an entire plane.

Dislocation movement or slip produces plastic strain or deformation in crystalline solids. It can be envisaged as sliding or successive displacement of one plane of atoms over another on so-called slip planes. Further deformation occurs either by more movement on existing slip planes or by the formation of new slip planes. Figure 4.4 shows how the atoms rearrange as the dislocation moves through the crystal and that when one dislocation moves entirely through a crystal, the lower part is displaced under the upper by the distance b, which is called the Burgers vector.

The slip plane normally is the plane with the highest density of atoms, and the direction of slip is the direction in the slip plane in which the atoms are most closely spaced, because such planes and directions are more widely separated than are others. Slip results in the formation of steps on the surface of the crystal. These are readily detected if the surface is polished carefully before plastic deformation.

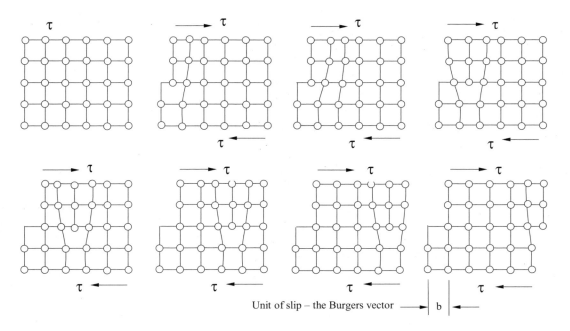

Figure 4.4 *Illustration of an edge dislocation as it moves through a crystal.*

Motion within a slip system is governed by the critical resolved shear stress (τ_c). As Figure 4.5 shows schematically, a single crystal is being deformed in tension by an applied force F along the axis of the cylindrical crystal. If the cross-sectional area is A, the tensile stress parallel to F is

$$\sigma = \frac{F}{A}$$

The force has a component $F \times \cos\lambda$ in the slip direction, where λ is the angle between F and the slip direction. This force acts over the slip surface, which has an area $\frac{A}{\cos\phi}$, where ϕ is the angle between F and the normal to the slip plane. Thus, the shear stress τ is (Hull and Bacon)

$$\tau = \frac{F}{A}\cos\phi\cos\lambda \tag{4.4}$$

If F_c is the tensile force required to start slip, the corresponding value of the shear stress τ_c is called the critical resolved shear stress for slip.

In polycrystalline metals, plastic flow typically does not occur at a constant stress. In contrast, an increased stress must be applied to produce additional deformation. This trend can be rationalized by considering the motion, interaction, and multiplication of dislocations. As plastic flow continues, the number of dislocations increases, typically in a parabolic fashion.

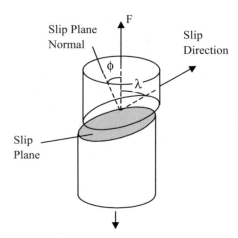

Figure 4.5 *Illustration of the geometry of slip in crystalline materials (Hull and Bacon).*

These dislocations begin to interact with each other and with interfaces such as grain boundaries. When a dislocation encounters a grain boundary, motion usually is halted. Although direct transmission to the neighboring grain may occur, dislocations more typically start to build at the grain boundary, and dislocation tangles may be created. As this buildup continues, a back stress develops that opposes the motion of additional dislocations, giving rise to work hardening.

4.2.1.4 Strengthening of Metals

4.2.1.4.1 Alloy Stress-Strain Behavior

Plastic deformation occurs by the glide of dislocations; hence, the critical shear stress for the onset of plastic deformation is the stress required to move dislocations. This usually is measured by a tensile test in which the specimen is elongated at a constant rate, and the load on the specimen is measured simultaneously with the extension. Figure 4.6 shows some typical stress-strain curves. The curve shown in Figure 4.6(a) represents a ductile material that undergoes extensive plastic deformation before fracture at F, and the curve shown in Figure 4.6(b) represents a brittle material that exhibits little plasticity. The dislocations shown in Figure 4.6(b) are either too low in density or too immobile to allow the specimen strain to match the elongation imposed by the testing machine. The curves show a linear region OE in which the specimen deforms elastically, that is, the stress is proportional to strain according to Hooke's law, followed by yielding at E and subsequent strain (or work) hardening up to F. In the latter process, the flow stress required to maintain plastic flow increases with increasing strain. Figure 4.6(c) is typical of many bcc polycrystalline metals that do not yield uniformly. The curve can be divided into four regions: (1) OE, elastic and pre-yield microplastic deformation; (2) EC, yield drop; (3) CD, yield propagation, and (4) DF, uniform hardening. The deformation between E and D is not homogeneous, for plastic flow occurs in only part of the specimen. This band, within which

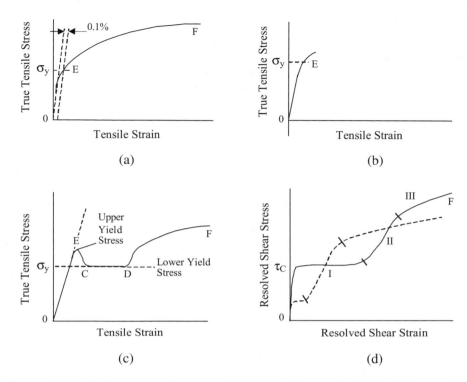

Figure 4.6 *Typical stress-strain curves for (a) ductile material, (b) brittle material, (c) body-centered cubic polycrystalline metals, and (d) single crystals (Hull and Bacon).*

dislocations have multiplied rapidly, extends to occupy the entire length at D. In tests on single crystals, it is usual to resolve the stress and strain onto the plane and direction for which the slip first occurs. The resulting stress-strain curve often has the form shown in Figure 4.6(d). Above the critical resolved shear stress τ_c, the curve has three parts: Stages I, II, and III. The length of these stages depends on the crystal orientation. Polycrystalline specimens of the same metal (Figure 4.6(a)) do not show Stage I but deform in a manner equivalent to Stages II and III.

For most materials, the change from elastic to plastic behavior is not abrupt, and the yield stress σ_y is not unique. This is because some nonlinear microplasticity occurs in the pre-yield region OE due to limited dislocation motion. Then, as shown in Figure 4.6(a), yielding is defined to occur when the plastic strain reaches a prescribed value, say, 0.2%. The corresponding proof stress is taken as the yield stress. Note that the form of all four curves in Figure 4.6 is dependent on test variables such as temperature and applied strain rate. Material parameters such as crystal structure, alloy composition, dislocation arrangement, and grain size also affect the yield and flow stresses. Therefore, it is possible to modify materials to improve their performance.

Although various strengthening mechanisms are discussed in the following sections, it is important to realize that more than one strengthening mechanism often must be used in engineering materials. Because dislocations are responsible for crystalline plastic flow, any factor that reduces dislocation mobility increases strength. A number of microstructural features can be used to decrease this mobility. Collectively, these are referred to as strengthening mechanisms. The following sections discuss the most common and important of these: solid solution hardening, particle hardening, grain boundary hardening, work hardening, and phase transformation hardening.

4.2.1.4.2 Solution Strengthening

Impurity (solute) atoms sometimes strengthen metals substantially. The impurity atoms have a different size than the host atom and a different atomic "stiffness" (i.e., atomic bonding is not the same in the solute and solvent). Both effects alter the crystal lattice in the vicinity of a solute atom. As a result, a moving dislocation is either attracted to or repelled by the impurity. Either situation results in a strength increase. Substitutional atoms (Figure 4.7) take the place of matrix atoms. Because of the mismatch in atomic size between the substitutional and matrix atoms, the lattice may become locally strained. This lattice strain may impede dislocation motion and conventionally is considered to be the source of solid solution strengthening in metals. In general, the strengthening increment varies proportionally with the mismatch in atomic size and properties (specifically modulus) between the solute and solvent atoms.

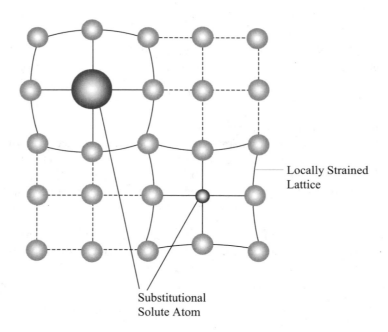

Locally Strained
Lattice

Substitutional
Solute Atom

Figure 4.7 *Substitutional strengthening.*

Interstitial atoms also can be present within the metal (Figure 4.8). In this case, the atom is much smaller than the matrix atoms and is located in the gaps (or interstices) in the crystal lattice.

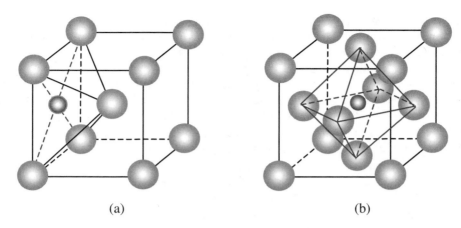

Figure 4.8 *Interstitial strengthening: (a) in a bcc unit cell, and (b) in an fcc unit cell.*

Most often, interstitial atoms can diffuse to the dislocation core due to the more open structure and the local tensile stresses in this region of the crystal lattice. The presence of the interstitial can inhibit dislocation motion, leading to dislocation "locking." This locking necessitates larger applied stresses to produce dislocation motion and further plastic deformation. In the classic example of carbon in iron, such a mechanism can result in discontinuous yielding, and a sharp upper yield point typically is observed, followed by yielding at a constant stress.

4.2.1.4.3 Aging Precipitation Strengthening

There is usually a maximum concentration to which the atoms of element B can be dissolved in solution in a crystal of element A. For concentrations above this solubility limit, the excess B atoms tend to precipitate within particles of either element B or a compound of A and B. The solubility limit varies with temperature, however. Therefore, it is possible to have an alloy composition that is less than the limit at one temperature but exceeds it at another. An example is the iron-carbon system. The α-phase is a solid solution of carbon atoms dissolved interstitially in iron. At 723°C, the solubility limit is approximately 0.025% wt; however, at room temperature, it is less than 0.008% wt. Thus, the alloy containing 0.025% wt carbon is a homogeneous solid solution at the solution treatment temperature of 723°C. On slowly cooling the alloy, the second phase starts to precipitate out, and at 20°C, the alloy consists of large Fe_3C precipitates in equilibrium with the iron-rich matrix. However, if the solution-treated alloy is quenched to room temperature, insufficient time is allowed for precipitation, and a super-saturated solid solution is obtained.

On subsequent annealing, there is sufficient thermal energy available for only precipitate nucleation to occur. Initially, a fine dispersion of numerous small precipitates is formed; however, with increasing aging time at the elevated temperature, the dispersion becomes more coarse. The number of precipitates decreases, and their size and spacing increase. Eventually, after long times, the equilibrium structure will be reestablished.

In many alloys, metastable phases occur during the transition of the precipitates to equilibrium. The structure of the interface between these zones and the matrix governs the kinetics of their formation and growth. The interface may be coherent, semi-coherent, or incoherent. The first zones to form are small and coherent. As the zones grow by bulk diffusion, they thicken, and the large elastic energy associated with the coherency strains is reduced by the transition to semi-coherent zones with dislocations in the interface. These precipitates have crystal structures that may be coherent with the matrix in some directions but not all. Eventually, after further growth, coherency is lost completely, and precipitates of the equilibrium phase are produced. Therefore, aging results in a transition from (a) a solid solution of misfitting atoms to (b) a fine dispersion of coherent precipitates surrounded by elastic strain to (c) a coarse dispersion of particles with incoherent interfaces and negligible lattice strain. These structures offer differing resistance to dislocation motion; thus, the yield strength of the alloy changes with time, as illustrated in Figure 4.9. The solid solution is stronger than the pure metal, but by aging, it is possible to further increase the strength. A peak strength is encountered corresponding to a critical dispersion of coherent or semi-coherent precipitates. Beyond this, the strength falls as overaging occurs. Selection of the optimum heat-treatment conditions therefore is crucial to maximizing the strength of alloys.

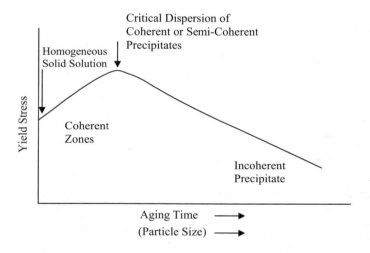

Figure 4.9 *Variation of yield stress with aging time for a typical alloy of an age hardening.*

Solution treating and aging precipitation strengthening commonly are used in austenitic stainless steel and nickel-based superalloys in exhaust valve applications. Austenitic valve alloys are most often solution treated and aged to hardness readings in the range of 25–40 HRc. This is a compromise among good strength, adequate ductility, impact performance, and wear resistance along the stem. Typical austenitic valve alloys such as 21-4N, 21-2N, or 23-8N are solution treated at approximately 1050–1200°C for 30–90 minutes until carbides and nitrides are dissolved in solution. Then they are quenched into cold water to acquire a hardness of 22–35 HRc.

Subsequently, age hardening for these alloys is performed at temperatures above 700–850°C for 10–16 hours for improved hardness or yield strength, reduced residual stress, and dimensional stability.

For high-stress and high-temperature applications, a number of nickel-based superalloys commonly are used. Iron-based austenitic valve steels are hardened by carbonitride precipitation. Nickel-based superalloys are hardened by precipitation of aluminum, nickel, niobium, tantalum, and titanium in the form of intermetallic compounds called gamma-prime phase or γ', consisting of $Ni_3(Al,Ti)$ precipitates. The aging characteristics of nickel-based superalloys are similar to those of austenitic steels. Typical nickel-based valve alloys such as Inconel 751, Nimonic 80A, Nimonic 90, and Pyromet 31 are solution treated at approximately 1040–1150°C for 15–90 minutes until intermetallic compounds are dissolved in solution. Then they are air cooled to acquire a hardness of 22–35 HRc. Subsequently, age hardening for these alloys is performed at temperatures above 700–850°C for 1–4 hours for improved hardness, reduced residual stress, and dimensional stability. A second aging sometimes is performed to further improve the hardness, and a third aging occasionally is applied. However, too high of aging temperature or too long of aging time could lead to overaging, which would result in a coarse γ' phase that will deteriorate strength.

4.2.1.4.4 Work Hardening Strengthening

The glide resistance of dislocations increases when they interact and change their distribution and density. Work hardening results from the interaction of dislocations, which often can be exploited to advantage by raising the strength of the solids shaped by plastic deformation. Typically, the work hardening of a metal is calculated by assuming a parabolic fit to the true stress versus the strain data, as suggested by (Dieter)

$$\sigma = K\varepsilon^n \tag{4.5}$$

where K is the strength coefficient, and n is the strain-hardening exponent. The true stress and true strain measured can be used to determine the strain-hardening exponent (n value). This exponent is simply the slope calculated after plotting the logarithm of true stress versus the logarithm of true strain. The value of the strain-hardening exponent becomes important when predicting the response of metals to straining during primary metalworking as well as forming operations for final components.

4.2.1.4.5 Grain Boundary Strengthening

In polycrystalline materials, the individual crystals have different orientations, and the applied resolved shear stress for slip varies from grain to grain. A few grains yield first, followed progressively by others. The grain boundaries, being regions of considerable atomic misfit, act as strong barriers to dislocation motion, so that unless the average grain size is large, the Stage I easy glide exhibited by single crystals in Figure 4.6 does not occur in polycrystals. Therefore, the stress-strain curve is not simply a single-crystal curve averaged over random orientations. Furthermore, the internal stresses around piled-up groups of dislocations at the boundaries of grains that have yielded may cause sources in neighboring grains to operate. Thus, the macro-scopic yield stress at which all grains yield depends on grain size. Finally, a grain in a polycrystal is not free to deform plastically as though it were a single crystal, for it must remain in contact with, and accommodate the shape changes of, its neighbors.

As the grain boundaries provide an obstacle to dislocation motion, the grain boundary area increases as the grain size becomes smaller. Therefore, as the grain size decreases, the strength of the metal typically increases. From experimental measurement of the yield stress of poly-crystalline aggregates in which grain size d is the only material variable, it has been found that the Hall-Petch relationship is satisfied (Hull and Bacon)

$$\sigma_y = \sigma_o + k_y d^{-n} \qquad (4.6)$$

where

    n   = approximately 0.5
    k_y = a material constant
    σ_0 = a constant stress of uncertain origin

The flow stress beyond yield follows a similar form. Hence, at cold-working temperatures, where grain boundaries do not contribute to creep, high yield and flow stresses are favored by small grain size. One rationalization of Eq. 4.6 is that a pile-up at a grain boundary in one grain can generate sufficiently large stresses to operate sources in an adjacent grain at the yield stress.

4.2.1.4.6 Harden and Temper Strengthening

To understand this strengthening mechanism, a basic iron-carbon phase transformation is depicted. Carbon does not appreciably dissolve in the low-temperature bcc form of iron. However, carbon in excess of the equilibrium solubility can be "trapped" in it at elevated temperatures.

Phase transformation hardening can be obtained through the process illustrated in Figure 4.10. First, the piece is heated to a high temperature. At high temperature, carbon atoms are dis-tributed in the interstitials of the fcc structure. Afterward, rapid cooling to trap these atoms in the interstitials results in a structure called martensite, which is harder than austenite (fcc) or ferrite (bcc).

Steel martensites are not truly cubic; they are a slightly distorted (tetragonal) form of this structure. However, this alteration is not what causes ferrous martensites to be so strong. One factor is that the carbon content of the martensite is the same as that of its parent austenite. This produces substantial solid-solution hardening. The large effect is due to the carbon atoms residing in interstitial sites (this causes the tetragonal distortion), and these atoms interfere markedly with dislocation motion. In addition, a slight volume increase accompanies martensite formation. The accompanying deformation generates a high dislocation density in as-quenched martensite, similar to what is found in cold-worked metals.

Although yield strengths of quenched martensites are high, these materials cannot be used effectively because they are brittle and must be tempered to be useful. Consider steel during

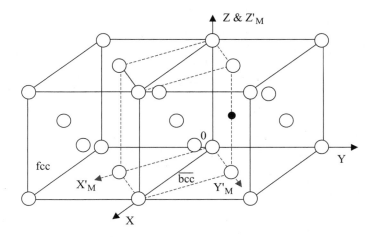

Figure 4.10 *Transformation hardening, from austenite (fcc) to ferrite (bcc) and then to martensite (bct); the circles are Fe, and the black dot is a C atom.*

tempering, for example. Carbide particles precipitate, and the dislocation density also is reduced. Tempering results in a strength reduction but reasonable fracture toughness.

Larger-diameter sections experience a moderate cooling rate in their interiors, regardless of how rapidly their surfaces might be cooled, and the section interior thus is likely to have a ferritic-pearlitic structure. In fact, with plain carbon steels (i.e., those containing only carbon and manganese as alloying additions), the centers of even fairly thin pieces are ferritic-pearlitic. Alloying elements are added to steel to improve their hardenability (i.e., the propensity for forming martensite during cooling). Elements such as chromium, molybdenum, and nickel reduce the rate at which the diffusional transformations from γ to α and/or pearlite take place, but the added elements have minimal effect on the tendency to form martensite. Thus, these elements permit martensite to be formed in sections having respectable diameters, and such alloy steels are the ones used when strength is required throughout large sections.

4.2.1.5 Effect of Temperature and Strain Rate on Deformation

At low temperatures, the thermal activation of dislocations is minimal; therefore, a large applied stress is required for deformation. At high temperatures, thermal activation will assist in dislocation motion "around" the thermal barriers. The applied stress necessary for plastic flow is lowered, which reduces the measured strength. Above a critical temperature, thermal activation provides a substantial portion of the driving force for dislocation motion, such that the strength of the material will be primarily determined by athermal barriers.

The bcc metals show a transition in fracture mode from ductile (microvoid coalescence or shear) to brittle (e.g., cleavage) with decreasing temperature. Figure 4.11 illustrates this transition. The brittle fracture stress (the cleavage stress) varies weakly with temperature and may be

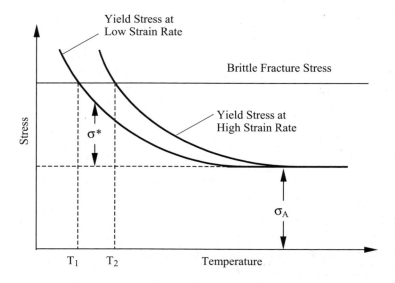

Figure 4.11 *Schematic illustration of the ductile-to-brittle transition in bcc metals (Knott).*

considered to be approximately independent of temperature. However, the yield strength will increase with decreasing temperatures. The temperature where the two curves intersect (T_1 in Figure 4.11) is considered to be the ductile-to-brittle transition temperature for the metal. Above this temperature, the metal will yield prior to fracture; below this temperature, cleavage occurs without macroscopic yielding.

In addition to temperature, the rate of loading (i.e., strain rate) during testing will greatly affect the measured mechanical properties of bcc metals. In general, an increase in strain rate is analogous to a decrease in temperature. The combined effect of strain rate ($\dot{\varepsilon}$) and temperature (T) can be seen as (Hull and Bacon)

$$\dot{\varepsilon} = \dot{\varepsilon}_0 = \exp\left[\frac{-\Delta G^*(\tau^*)}{kT}\right] \tag{4.7}$$

where ΔG^* is the Gibbs free energy associated with the shear stress (τ^*) required to overcome short-range obstacles, and $\dot{\varepsilon}_0$ is the product of the mobile dislocation density, the vibration frequency for the dislocation segment, and the Burgers vector for the dislocation, which is the distance that the dislocation may "jump." This relationship illustrates that sufficiently high temperatures or low strain rates increase the probability for exiting dislocation motion through a thermal activation event in the presence of an applied load. On the other hand, low temperatures and high strain rates can lead to significant strengthening due to smaller contributions by thermal activation.

For the materials exhibiting a ductile-to-brittle transition, increasing the strain rate can have an additional effect. The yield strength increases at a higher strain rate, which shifts the temperature dependence of the yield strength and results in an intersection with the brittle fracture stress at a higher temperature (T_2 in Figure 4.11). The result is that the measured ductile-to-brittle transition temperature will be greater at a high strain rate (T_2) than at a low strain rate (T_1).

4.2.1.6 Notches and Cracks on Fracture

The deformation processes governing fracture in metals are affected by both the stresses and strains experienced in the specimen. In a simple tension test, the stresses are designed to be uniform throughout the cross section of the sample. When stress is applied to a component with a notch, crack, or other stress concentration, regions in the vicinity of these features will always experience much higher stresses compared to unaffected regions, and the strains produced can differ from what would be predicted by the uniform stresses. The stress fields created around stress concentrations are controlled by three factors: (1) the extent of deformation prior to failure, (2) the mode of loading (i.e., the relative orientation of the applied load with respect to the plane of the crack), and (3) the constraints, if any, on the cracked body. As a result, the mechanical properties measured when testing specimens with notches or cracks will be much different than those observed in uniaxial tension tests without notches. In the case of notched tensile specimens, the measured tensile yield strength often will be greater than that observed in a uniaxial tension test. However, the ductility and load-carrying capacity will be decreased. As the sample is loaded, the notched region will yield first due to the elevated local stresses and strains associated with the notches. The maximum stress ahead of the notch will be a function of the geometry of the notch and the applied loading. Furthermore, the stresses are no longer purely uniaxial (such as are developed in a tensile test) but now become triaxial (i.e., tensile stresses in the three primary directions of space). If ductile fracture via microvoid coalescence is reconsidered, the elevated stress and strain fields may accelerate the nucleation of secondary voids. The void growth rate also will increase proportionally to the level of the triaxial stresses, resulting in reduced ductility for notched samples compared to smooth, uniaxial tensile specimens.

To understand the effects of cracks in ductile metals, the interactions between microstructural features and the elevated stress fields around the crack tip must be considered. Ahead of a sharp crack, a finite volume of material is subjected to deformations at high stress values. To a first-order approximation, this volume of material, or the "plastic zone" in plane strain, can be represented as the radius of a circle described as (Dieter)

$$r_P = \frac{1}{6\pi} \left(\frac{K_I}{\sigma_{ys}} \right)^2 \tag{4.8}$$

where

r_P = the distance from the crack to the elastic-plastic boundary
K_I = the stress intensity calculated from the geometry and loading condition
σ_{ys} = the uniaxial yield strength of the material

In general, the stresses are highest in a plastically deforming material ahead of the crack tip. In contrast, the plastic strains are highest at the notch tip and decrease after a critical distance, which is approximately equivalent to the crack-opening displacement (i.e., the relative displacement of the "mouth" of the crack). The extent of the strained region often becomes comparable to microstructural features (e.g., grain size, interparticle particle spacing) and can initiate failure. When large grains are required for fracture, the crack-opening displacement must reach a critical size to envelop the microstructural features responsible for void nucleation. Depending on the intrinsic fracture resistance of the material, void growth and failure will occur when this zone becomes 1–2.7 times the microstructural feature responsible for fracture (e.g., the grain size or the mean spacing of second-phase particles).

An example of this type of fracture process can be seen in the case of metal matrix composites (i.e., a ductile metal matrix with brittle reinforcement particles). When a crack in the ductile matrix is loaded, the large stresses ahead of the notch promote void nucleation by particle fracture or interface decohesion. This void nucleation limits the straining capacity of the metal in the vicinity of the crack tip. The high strain field ahead of the tip then allows for continued growth of the nucleated voids to the point of instability, as the blunted crack links with the microcrack. This process of microcracking, crack-tip blunting, and failure of the matrix (void formation) among the particles continues as the crack propagates. This mechanism gives cracks an easy path for failure and clearly shows that the presence of a stress raiser exacerbates the processes of fracture compared to the case of uniaxial tension (Rice and Johnson).

In this example, crack propagation from a notch or crack tip is related to the spacing of micro-structural features. As a result, a metal with a reduced volume fracture of particles (and the assumed increased interparticle spacing) can exhibit a greater resistance to fracture than a metal with a larger amount of particles.

4.2.2 Mechanical Tensile Testing

4.2.2.1 Engineering Stress-Strain Curves

The shape and magnitude of the stress-strain curve of a metal depend on its composition, heat treatment, prior history of plastic deformation, and the strain rate, temperature, and state of stress imposed during testing. The parameters that are used to describe the stress-strain curve of a metal are the tensile strength, yield strength or yield point, percent elongation, and reduction in area. The first two are strength parameters; the latter two indicate ductility.

In the elastic region, stress is linearly proportional to strain. When the stress exceeds a value corresponding to the yield strength, the specimen undergoes gross plastic deformation. If the load is subsequently reduced to zero, the specimen will remain permanently deformed. The stress required to produce continued plastic deformation increases with increasing plastic strain; that is, the metal strain hardens. The volume of the specimen (i.e., the area times the length) remains constant during plastic deformation. As the specimen elongates, its cross-sectional area decreases uniformly along the gage length.

In the conventional engineering tension test, an engineering stress-strain curve is constructed from the load-elongation measurements made on the test specimen (Figure 4.12). The

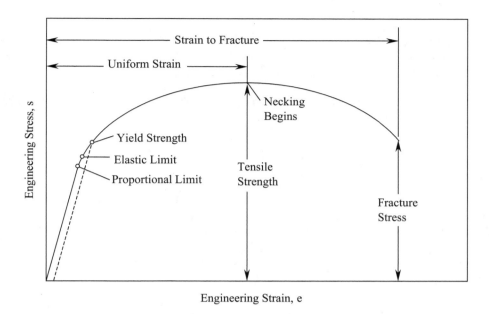

Figure 4.12 *Engineering stress-strain curve (Dieter).*

engineering stress (s) used in this stress-strain curve is the average longitudinal stress in the tensile specimen. It is obtained by dividing the load (P) by the original area of the cross section of the specimen (A_0)

$$s = \frac{P}{A_0} \tag{4.9}$$

The strain, e, used for the engineering stress-strain curve is the average linear strain, which is obtained by dividing the elongation of the gage length of the specimen (δ) by its original length (L_0),

$$e = \frac{\delta}{L_0} = \frac{\Delta L}{L_0} = \frac{L - L_0}{L_0} \tag{4.10}$$

Because the stress and the strain both are obtained by dividing the load and elongation by constant factors, the load-elongation curve has the same shape as the engineering stress-strain curve. The two curves frequently are used interchangeably.

Initially, the strain hardening more than compensates for this decrease in area, and the engineering stress (proportional to load P) continues to rise with increasing strain. Eventually, a point is reached where the decrease in specimen cross-sectional area is greater than the increase in deformation load arising from strain hardening. This condition will be reached first at some point in the specimen that is slightly weaker than the rest. All further plastic deformation is concentrated in this region, and the specimen begins to neck or thin down locally. Because the cross-sectional area now is decreasing more rapidly than the deformation load is being increased by strain hardening, the actual load required to deform the specimen falls off, and the engineering stress defined in Eq. 4.9 continues to decrease until fracture occurs.

The tensile strength, or ultimate tensile strength (s_U), is the maximum load divided by the original cross-sectional area of the specimen and is the value most frequently quoted from the test,

$$s_U = \frac{P_{max}}{A_0} \tag{4.11}$$

However, it actually is a value of little fundamental significance with regard to the strength of a metal. For ductile metals, the tensile strength should be regarded as a measure of the maximum load that a metal can withstand under the restrictive conditions of uniaxial loading. This value bears little relation to the useful strength of the metal under the more complex conditions of stress that usually are encountered. ASTM A285-03 describes standard test methods for obtaining tensile strengths.

For many years, it was customary to base the strength of members on the tensile strength, suitably reduced by a factor of safety. The current trend is to use the more rational approach of basing the static design of ductile metals on the yield strength. However, due to the long practice of using the tensile strength to describe the strength of materials, it has become a familiar property. As such, it is a useful identification of a material in the same sense that the chemical composition serves to identify a metal or alloy. Furthermore, because the tensile strength is easy to determine and is a reproducible property, it is useful for the purposes of specification and for quality control of a product. Extensive empirical correlations between tensile strength and properties such as hardness and fatigue strength often are useful. For brittle materials, the tensile strength is a valid design criterion.

4.2.2.1.1 Measurement of Yielding

The stress at which plastic deformation or yielding starts depends on the sensitivity of the strain measurements. With most materials, there is a gradual transition from elastic to plastic behavior, and the point at which plastic deformation begins is difficult to define with precision. In tests of materials under uniaxial loading, three criteria for the initiation of yielding have been used: (1) the elastic limit, (2) the proportional limit, and (3) the yield strength.

The elastic limit is the greatest stress the material can withstand without any measurable permanent strain remaining after the complete release of load. With the sensitivity of strain typically used in engineering studies, the elastic limit is greater than the proportional limit. Proportional limit is the highest stress at which the stress is directly proportional to the strain. Yield strength is the stress required to produce a small specified amount of plastic deformation. The usual definition of this property is the offset yield strength determined by the stress corresponding to the intersection of the stress-strain curve offset by a specified strain, typically 0.2 or 0.1% in the United States. ASTM E238 describes standard test methods for obtaining yield strengths of materials.

4.2.2.1.2 Measurement of Ductility

The conventional measures of ductility that are obtained from the tension test are the engineering strain at fracture (e_f) (usually called the elongation) and the reduction in the area at fracture (q). Elongation and reduction in area usually are expressed as percentages. Both of these properties are obtained after fracture by putting the specimen back together and taking measurements of the final length, L_f, and the final specimen cross section, A_f:

$$e_f = \frac{L_f - L_0}{L_0} \qquad (4.12)$$

$$q = \frac{A_0 - A_f}{A_0} \qquad (4.13)$$

Because an appreciable fraction of the plastic deformation will be concentrated in the necked region, the value of e_f will depend on the gage length (L_0) over which the measurement was taken. The smaller the gage length, the greater the contribution to the overall elongation from the necked region and the higher the value of e_f. Reduction in area does not suffer from this difficulty. ASTM E290 describes standard test methods for ductility of materials.

4.2.2.1.3 Modulus of Elasticity

The slope of the initial linear portion of the stress-strain curve is the modulus of elasticity, or Young's modulus. The modulus of elasticity (E) is a measure of the stiffness of the material. The greater the modulus, the smaller the elastic strain resulting from the application of a given stress. The modulus of elasticity is determined by the binding forces between atoms. Because these forces cannot be changed without changing the basic nature of the material, the modulus of elasticity is one of the most structure-insensitive of the mechanical properties. Generally, it is affected only slightly by alloying additions, heat treatment, or cold working. However, increasing the temperature decreases the modulus of elasticity. Figure 4.13 gives typical values of the modulus of elasticity for valvetrain component materials at different temperatures. ASTM E111 describes standard test methods for Young's modulus.

4.2.2.2 True Stress-True Strain Curve

The engineering stress-strain curve does not give a true indication of the deformation characteristics of a metal because it is based entirely on the original dimensions of the specimen, and these dimensions change continuously during the test. Also, ductile metal that is pulled in tension becomes unstable and necks downward during the test. Because the cross-sectional area of the specimen is decreasing rapidly at this stage in the test, the load required to continue deformation falls off. The average stress based on the original area likewise decreases, and this produces the fall-off in the engineering stress-strain curve beyond the point of maximum load. Actually, the metal continues to strain harden to fracture, so that the stress required to produce further deformation also should increase. If the true stress is used, based on the actual cross-sectional area of the specimen, the stress-strain curve increases continuously to fracture. If the strain

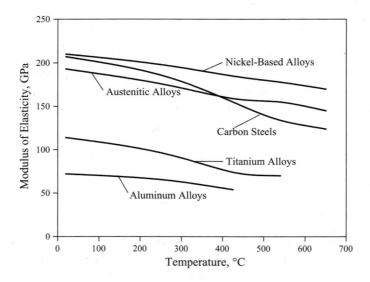

Figure 4.13 *Typical values of modulus of elasticity as a function of temperature.*

measurement also is based on instantaneous measurement, the curve that is obtained is known as the true stress-true strain curve. This also is known as a flow curve because it represents the basic plastic-flow characteristics of the material. Figure 4.14 compares the true stress-true strain curve with its corresponding engineering stress-strain curve.

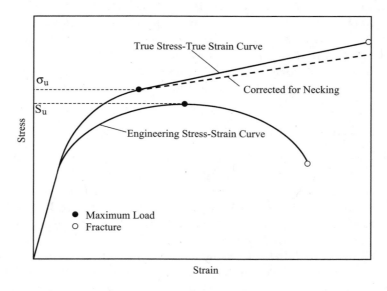

Figure 4.14 *Comparison of the engineering stress-strain curve and the corresponding true stress-true strain curve (Dieter).*

Any point on the flow curve can be considered the yield stress for a metal strained in tension by the amount shown on the curve. Thus, if the load is removed at this point and then is reapplied, the material will behave elastically throughout the entire range of reloading. The true stress (σ) is expressed in terms of engineering stress (s) and engineering strain (e) by

$$\sigma = \frac{P}{A_0}(e+1) = s(e+1) \tag{4.14}$$

The derivation of Eq. 4.14 assumes both constancy of volume and a homogeneous distribution of strain along the gage length of the tension specimen. Thus, Eq. 4.14 should be used only until the onset of necking. Beyond the maximum load, the true stress should be determined from actual measurements of load and cross-sectional area,

$$\sigma = \frac{P}{A} \tag{4.15}$$

The true strain, ε, may be determined from the engineering or conventional strain (e) by

$$\varepsilon = \ln \frac{L}{L_0} = \ln(e+1) \tag{4.16}$$

This equation is applicable only to the onset of necking for the reasons previously discussed. Beyond maximum load, the true strain should be based on actual area or diameter (D) measurements

$$\varepsilon = \ln \frac{A_0}{A} = \ln \frac{\left(\frac{\pi}{4}\right)D_0^2}{\left(\frac{\pi}{4}\right)D^2} = 2\ln \frac{D_0}{D} \tag{4.17}$$

However, beyond maximum load, the high localized strains in the necked region that are used in Eq. 4.17 greatly exceed the engineering strain calculated from Eq. 4.12. Frequently, the flow curve is linear from maximum load to fracture; in other cases, its slope continuously decreases to fracture. The formation of a necked region or mild notch introduces triaxial stresses that make it difficult to accurately determine the longitudinal tensile stress from the onset of necking until fracture occurs.

The true stress at maximum load corresponds to the true tensile strength. For most materials, necking begins at maximum load at a value of strain where the true stress equals the slope of the flow curve. Let σ_u and ε_u denote the true stress and true strain at maximum load when the cross-sectional area of the specimen is A_u. The ultimate tensile strength can be defined as

$$s_u = \frac{P_{max}}{A_0} \tag{4.18}$$

$$\sigma_u = \frac{P_{max}}{A_u} \tag{4.19}$$

Eliminating P_{max} yields

$$\sigma_u = s_u \times \frac{A_0}{A_u} \tag{4.20}$$

$$\sigma_u = s_u \times e^{\varepsilon_u} \tag{4.21}$$

4.2.3 Hardness Testing

Hardness is a surface property of a material determined by the ability of the material to resist plastic deformation when in contact with an indenter under load. The indenter may be spherical (Brinell test), pyramidal (Vickers and Knoop tests), or conical (Rockwell test). In the Brinell, Vickers, and Knoop tests, the hardness value is the load supported by the unit area of the indentation, expressed in kilogram-force per square millimeter (kgf/mm^2). In the Rockwell tests, the depth of indentation at a prescribed load is determined and converted to a hardness number, which is inversely related to the depth.

Hardness testing is perhaps the simplest and least expensive method of mechanically characterizing a material. Theoretical and empirical investigations have resulted in fairly accurate quantitative relationships between hardness and other mechanical properties of materials, such as ultimate tensile strength, yield strength, strain hardening coefficient, and fatigue strength.

4.2.3.1 Brinell Hardness

In the Brinell hardness test, a hard spherical indenter (2.5, 5, or 10 mm in diameter) is pressed under a constant normal load (usually 500–3000 kgf) onto the smooth surface of a material for a specified time (10–30 seconds). When equilibrium is reached, the load and the indenter are withdrawn, and the diameter of the indentation formed on the surface is measured using a microscope with a built-in millimeter scale. The Brinell hardness is expressed as the ratio of the indenter load W to the area of the concave (i.e., contact) surface of the spherical indentation that is assumed to support the load and is given as the Brinell hardness number (BHN) denoted by HB. Thus,

$$HB = \frac{2W}{\pi D\left(D - \sqrt{D^2 - d^2} \right)} \tag{4.22}$$

where W is the load in kilograms, and d and D are the diameters of the indentation and the indenter, respectively, in millimeters.

The calculation for each test has already been made available in tabular form for various combinations of indentation diameters and loads. Before using the Brinell test, several points must be considered. The size and shape of the work piece must be capable of accommodating the relatively large indentation and heavy test loads. In addition, the maximum range of Brinell hardness values is 16 HB for very soft aluminum to 627 HB for hardened steels (approximately 60 HRc). ASTM E10 describes Brinell hardness test standards.

4.2.3.2 Rockwell Hardness

Rockwell hardness testing differs from Brinell testing in that the Rockwell hardness number is based on the difference of the indenter depth from two load applications. Initially, a minor load is applied, and a zero datum is established. A major load then is applied for a specified period of time, causing an additional penetration depth beyond the zero datum point previously established by the minor load. After the specified dwell time for the major load, the major load is removed while still keeping the minor load applied. The resulting Rockwell number represents the difference in depth from the zero datum position as a result of the application of the major load. The entire procedure requires only 5–10 seconds. Figure 4.15 illustrates the principle of the Rockwell test. The indenter may be either a diamond cone or a hardened ball, depending principally on the characteristics of the material being tested. Although Figure 4.15 illustrates a diamond indenter, the same principle applies for steel ball indenters.

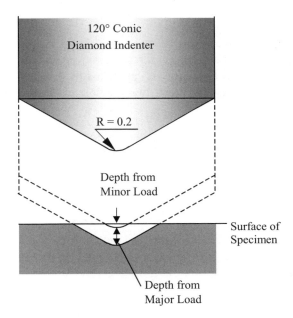

Figure 4.15 *Principle of the Rockwell hardness test (diamond indenter).*

When the surfaces of the specimen are cylindrical or spherical, due to the fact that the resistance to the indenter is reduced around the indent, the hardness measured is lower compared to that of a flat surface. They can be corrected by the following.

For a cylindrical surface:

$$\Delta HRc = \frac{6(100 - HRc')^2}{D} \times 10^{-3} \tag{4.23}$$

For a spherical surface:

$$\Delta HRc = \frac{12(100 - HRc')^2}{D} \times 10^{-3} \tag{4.24}$$

where

    HRc   = hardness measured on the cylindrical or spherical surface
    ΔHRc = additional number to be added to HRc′
    D   = diameter of the sphere or cylinder specimen, in millimeters

The calculation for each test also has been made available in tabular form for various combinations of indentation diameters and loads. The Rockwell hardness test standards are described in ASTM E18 and other standards.

4.2.3.3 Vickers and Knoop Hardnesses

The Vickers test method is similar to the Brinell principle, in that an indenter with a defined shape is pressed into a material, the indenting force is removed, the resulting indentation diagonals are measured, and the hardness number is calculated by dividing the force by the surface area of the indentation. Figure 4.16(a) shows the principle of the Vickers test with a diamond pyramid indenter.

In the Vickers test, the force is applied smoothly, without impact, and is held in contact for 10–15 seconds. After the force is removed, both diagonals are measured, and the average is used to calculate the Vickers hardness (HV) according to

$$HV = \frac{P}{F} = \frac{P}{\dfrac{d^2}{2\sin 68^\circ}} = \frac{1.8544 \times P}{d^2} \tag{4.25}$$

where d is the mean diagonal in micrometers (μm), and P is the applied load in gram-force (gf). Vickers test standards are discussed in ASTM E92.

The Knoop test is based on a rhombohedral-shaped diamond, with the long diagonal approximately seven times as long as the short diagonal. The Knoop indenter is used in the same machine as the Vickers indenter, and the test is conducted in exactly the same manner. However, the

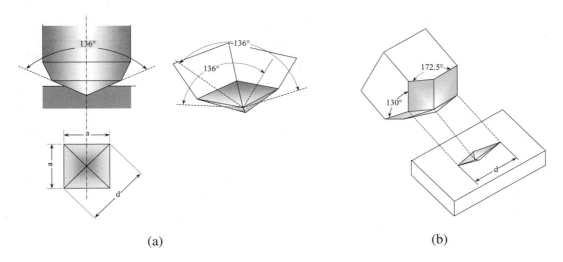

Figure 4.16 *Schematic of (a) a Vickers indenter and (b) a Knoop hardness indenter.*

Knoop hardness (HK) is calculated based on the measurement of only the long diagonal and calculation of the projected area of the indent rather than the surface area of the indent,

$$HK = \frac{P}{c_p d^2} = \frac{1.4229 \times P}{d^2} \tag{4.26}$$

where c_p is the indenter constant, which permits calculation of the projected area of the indent from the long diagonal squared.

The Knoop indenter has a polished rhombohedral shape with an included longitudinal angle of 172.5° and an included transverse angle of 130° (Figure 4.16(b)). The narrowness of the indenter makes it ideal for testing specimens with steep hardness gradients such as case hardening or coatings or phase hardness. In such specimens, it may be impossible to obtain valid Vickers indents because the change in hardness may produce a substantial difference in the length of the two halves of the indent parallel to the hardness gradient. With the Knoop test, the long diagonal is set perpendicular to the hardness gradient, and the short diagonal is in the direction of the hardness gradient. The Knoop hardness test standard is described in ASTM C1326.

4.2.3.4 Other Hardness Testing

In addition to the four major hardness tests previously discussed, a number of test methods were developed for specific applications. These include dynamic or rebound hardness tests, static indentation tests, scratch hardness tests, and ultrasonic microhardness testing.

Dynamic test methods relate hardness to the elastic response of a material; indentation hardness tests determine hardness in terms of plastic behavior. The two most common methods of dynamic hardness testing are the Shore and Leeb testers, which are rebound-type tests. The Shore scleroscope is used frequently for testing large specimens such as forged steel or wrought alloy steel rolls. In this procedure, a diamond-tipped hammer is dropped from a fixed height onto the surface of the material being tested. The height of rebound of the hammer is a measure of the hardness of the metal. The test standard for the Shore scleroscope hardness test is described in ASTM E448.

The Leeb tester is a portable hardness tester that operates on a dynamic rebound principal, similar to the Shore scleroscope. An impact device is propelled into the sample using a spring for the initial energy. The impact device travels a short distance until it contacts the sample. A small indent is formed, and the impact device rebounds away from the test surface according to the hardness and elasticity of the material. An electronic induction coil measures the velocity of the impact device before and after it contacts the sample. The Leeb hardness number (HL) is defined as

$$HL = \frac{\text{Rebound velocity}}{\text{Impact velocity}} \times 1000 \qquad (4.27)$$

The standard test method for Leeb or Equotip hardness testing is described in ASTM A956.

Scratch hardness tests include the Mohs scale, the file hardness test, and the plowing test. The Mohs scale consists of ten minerals arranged in order from 1 (talc), 2 (gypsum), 3 (calcite), 4 (fluorite), 5 (apatite), 6 (orthoclase), 7 (quartz), 8 (topaz), 9 (crundum), and 10 (diamond). Each mineral in the scale will scratch all those below it. The hardness of iron with 0.1% carbon maximum is 3–4 on the Mohs scale, and copper is 2–3 on the Mohs scale. Fully hardened high-carbon tool steel is 7–8 on the Mohs scale.

The file hardness test is useful in estimating the hardness of steels in the high hardness ranges. It provides information on soft spots and decarburization quickly and easily and is readily adaptable to odd shapes and sizes that are difficult to test by other methods. The file hardness test utilizes the calibrated files with hardness, such as HRc 50 and 55, if the test piece cannot be filed by the file with HRc 50 but by HRc 55. This indicates that the hardness of the test piece is HRc 50–55. A number of factors, such as pressure, speed, contact angle, and surface roughness, influence the results of the test. Consequently, its ability to give reproducible hardness values is rather limited.

In ultrasonic microhardness testing, a Vickers diamond is attached to one end of a magnetostrictive metal rod. The diamond-tipped rod is excited to its natural frequency by a piezoelectric converter. The resonant frequency of the rod changes as the free end of the rod is brought into contact with the surface of a solid body. Once the device is calibrated for the known modulus of elasticity of the tested material, the area of contact between the diamond tip and the tested surface can be derived from the measured resonant frequency. The area of contact is inversely proportional to the hardness of the tested material, provided the force pressing the surface is constant. Consequently, the measured frequency value can be converted into the corresponding hardness number. Ultrasonic hardness testing uses a maximum indentation load of approximately 800 gf. Therefore, the indentation depth is relatively small (4–18 μm) and is classified as nondestructive.

4.2.3.5 Hardness of Alloy Microconstituents

Table 4.2 shows the hardness of typical minerals and alloy microconstituents.

TABLE 4.2
HARDNESS OF ALLOY MICROCONSTITUENTS

Microconstitituent	Hardness (Knoop)
Ferrite	200–240
Pearlite	220–450
Austenite	300–600
Martensite	450–800
Quartz (SiO_2)	850
Cementite (Fe_3C)	1000
Chromium Carbides [$(M,Cr)_7C_3$]	1650
Silicon Nitride (Si_3N_4)	1700
Titanium Nitride (TiN)	1800
Tungsten Carbide (WC)	1880
Aluminum Oxide (Al_2O_3)	2100
Titanium Carbide (TiC)	2470
Silicon Carbide (SiC)	2480
Boron Carbide (B_4C)	2750
Vanadium Carbide (VC)	2800
Diamond	7000

4.2.3.6 Relationship of Hardness and Tensile Properties

Because hardness tests are based on different measurement techniques, the relationship between hardness and other mechanical properties is inexact. This is due to characteristics such as force variation, differently shaped indenters, homogeneity of specimens, moduli of elasticity, hardening properties of specimens, and cold-working properties of specimens. Despite these basic limitations, general methods sometimes are used to estimate tensile properties from hardness values when it is impossible to test the material. Table 4.3 shows examples of strength and hardness conversion equations.

One general rule of thumb for quenched and tempered steels is to estimate tensile strength (expressed in units of kilopounds per square inch (ksi), where 1 ksi = 6.9 MPa) as approximately one-half the Brinell hardness number (Bain and Paxton), or HV/2.9 (in units of kgf/mm^2) related to Vickers hardness for carbon and low-alloy steels (Taylor).

TABLE 4.3
EXAMPLES OF STRENGTH AND
HARDNESS CONVERSIONS

Nonhardened Steels	$\sigma_b = 0.362\ HB$	HB > 175
	$\sigma_b = 0.345\ HB$	HB < 175
	$\sigma_b = \dfrac{2640}{130 - HRB}$	HRB < 90
	$\sigma_b = \dfrac{2510}{130 - HRB}$	100 > HRB > 90
Carbon Steels	$\sigma_b = 2.5\ HS$	
	$\sigma_b = \dfrac{5.132 \times 10^5}{(100 - HRc)^2}$	HRc < 10
Grey Cast Irons	$\sigma_b = \dfrac{(HB - 40)}{6}$	
	$\sigma_b = \dfrac{4.886 \times 10^5}{(100 - HRc)^2}$	40 > HRc > 10
Cast Steels	$\sigma_b = (0.3 \sim 0.4)HB$	
	$\sigma_b = \dfrac{8.61 \times 10^3}{100 - HRc}$	HRc > 40

4.2.3.7 Hardness Conversions

Hardness conversions are empirical relationships defined by conversion tables that are limited to specific categories of materials. Hardness conversions are covered in ASTM E140 and are shown here in Table 4.4 for conversion among Rockwell, Brinell, and Vickers hardness for heat-treated carbon and alloy steels, constructional alloy steels, and tool steels in the as-forged, annealed, normalized, and quenched and tempered conditions.

Note that the values in < > brackets in Table 4.4 are beyond normal and are given only for information purposes. Data are for carbon and alloy steels in the annealed, normalized, and quenched-and-tempered conditions; they are less accurate for the cold-worked condition and for austenitic steels.

Other hardness conversion formulas for various materials have been published. Table 4.5 lists some conversion formulas, based on the data shown in Table 4.4.

TABLE 4.4
HARDNESS CONVERSION AMONG ROCKWELL, VICKERS, BRINELL, KNOOP, AND SHORE (ASTM E140)

Vickers	Brinell 3000kg load Standard 10mm Ball	Brinell 3000kg load WC	Rockwell A Scale 60kgf load Diamond	Rockwell B Scale 100kgf load 1/16" dia. Ball	Rockwell C Scale 150kgf load Diamond	Rockwell D Scale 100kgf load Diamond	Rockwell Superficial 15N Scale 15 kgf load Diamond	Rockwell Superficial 30N Scale 30 kgf load Diamond	Rockwell Superficial 45N Scale 45 kgf load Diamond	Knoop >=500gf	Shore Scleroscope
940			85.6		68	76.9	93.2	84.4	75.4	920	97
920			85.3		67.5	76.5	93	84	74.8	908	96
900			85		67	76.1	92.9	83.6	74.2	895	95
880		<767>	84.7		66.4	75.7	92.7	83.1	73.6	882	93
860		<757>	84.4		65.9	75.3	92.5	82.7	73.1	867	92
840		<745>	84.1		65.3	74.8	92.3	82.2	72.2	852	91
820		<733>	83.8		64.7	74.3	92.1	81.7	71.8	837	90
800		<722>	83.4		64	73.8	91.8	81.1	71	822	88
780		<710>	83		63.3	73.3	91.5	80.4	70.2	806	87
760		<698>	82.6		62.5	72.6	91.2	79.7	69.4	788	86
740		<684>	82.2		61.8	72.1	91	79.1	68.6	772	84
720		<670>	81.8		61	71.5	90.7	78.4	67.7	754	83
700		<656>	81.3		60.1	70.8	90.3	77.6	66.7	735	81
690		<647>	81.1		59.7	70.5	90.1	77.2	66.2	725	-
680		<638>	80.8		59.2	70.1	89.8	76.8	65.7	716	80
670		<630>	80.6		58.8	69.8	89.7	76.4	65.3	706	-
660		620	80.3		58.3	69.4	89.5	75.9	64.7	697	79
650		611	80		57.8	69	89.2	75.5	64.1	687	78
640		601	79.8		57.3	68.7	89	75.1	63.5	677	77
630		591	79.5		56.8	68.3	88.8	74.6	63	667	76
620		582	79.2		56.3	67.9	88.5	74.2	62.4	657	75
610		573	78.9		55.7	67.5	88.2	73.6	61.7	646	-
600		564	78.6		55.2	67	88	73.2	61.2	636	74
590		554	78.4		54.7	66.7	87.8	72.7	60.5	625	73
580		545	78		54.1	66.2	87.5	72.1	59.9	615	72
570		535	77.8		53.6	65.8	87.2	71.7	59.3	604	-
560		525	77.4		53	65.4	86.9	71.2	58.6	594	71
550	<505>	517	77		52.3	64.8	86.6	70.5	57.8	583	70
540	<469>	507	76.7		51.7	64.4	86.3	70	57	572	69
530	<488>	497	76.4		51.1	63.9	86	69.5	56.2	561	68
520	<480>	488	76.1		50.5	63.5	85.7	69	55.6	550	67
510	<473>	479	75.7		49.8	62.9	85.4	68.3	54.7	539	-
500	<465>	471	75.3		49.1	62.2	85	67.7	53.9	528	66
490	<456>	460	74.9		48.4	61.6	84.7	67.1	53.1	517	65
480	<448>	452	74.5		47.7	61.3	84.3	66.4	52.2	505	64
470	441	442	74.1		46.9	60.7	83.9	65.7	51.3	494	-
460	443	433	73.6		46.1	60.1	83.6	64.9	50.4	482	62
450	425	425	73.3		45.3	59.4	83.2	64.3	49.4	471	-
440	415	415	72.8		44.5	58.8	82.8	63.5	48.4	459	59
430	405	405	72.3		43.6	58.2	82.3	62.7	47.4	447	58
420	397	397	71.8		42.7	57.5	81.8	61.9	46.4	435	57
410	388	388	71.4		41.8	56.8	81.4	61.1	45.3	423	56
400	379	379	70.8		40.8	56	80.8	60.2	44.1	412	55
390	369	369	70.3		39.8	55.2	80.3	59.3	42.9	400	-
380	360	360	69.8	<110>	38.8	54.4	79.8	58.4	41.7	389	52
370	350	350	69.2		37.7	53.6	79.2	57.4	40.4	378	51
360	341	341	68.7	<109>	36.6	52.8	78.6	56.4	39.1	367	50
350	331	331	68.1		35.5	51.9	78	55.4	37.8	356	48
340	322	322	67.6	<108>	34.4	51.1	77.4	54.4	36.5	346	47
330	313	313	67		33.3	50.2	76.8	53.6	35.2	337	46
320	303	303	66.4	<107>	32.2	49.4	76.2	52.3	33.9	328	45
310	294	294	65.8		31	48.4	75.6	51.3	32.5	318	-
300	284	284	65.2	<105.5>	29.8	47.5	74.9	50.2	31.1	309	42
295	280	280	64.8		29.2	47.1	74.6	49.7	30.4	305	-
290	275	275	64.5	<104.5>	28.5	46.5	74.2	49	29.5	300	41
285	270	270	64.2		27.8	46	73.8	48.4	28.7	296	-
280	265	265	63.8	<103.5>	27.1	45.3	73.4	47.8	27.9	291	40
275	261	261	63.5		26.4	44.9	73	47.2	27.1	286	39
270	256	256	63.1	<102>	25.6	44.3	72.6	46.4	26.2	282	38
265	252	252	62.7		24.8	43.7	72.1	45.7	25.2	277	-
260	247	247	62.4	<101>	24	43.1	71.6	45	24.3	272	37
255	243	243	62		23.1	42.2	71.1	44.2	23.2	267	-
250	238	238	61.6	99.5	22.2	41.7	70.6	43.4	22.2	262	36
245	233	233	61.2		21.3	41.1	70.1	42.5	21.1	258	35
240	228	228	60.7	98.1	20.3	40.3	69.6	41.7	19.9	253	34
230	219	219		96.7	<18>					243	33
220	209	209		95	<15.7>					234	32
210	200	200		93.4	<13.4>					226	30
200	190	190		91.5	<11>					216	29
190	181	181		89.5	<8.5>					206	28
180	171	171		87.1	<6>					196	26
170	162	162		85	<3>					185	25
160	152	152		81.7	<0>					175	23
150	143	143		78.7						164	22
140	133	133		75						154	21
130	124	124		71.2						143	20
120	114	114		66.7						133	18
110	105	105		62.3						123	
100	95	95		56.2						112	
95	90	90		52						107	
90	86	86		48						102	
85	81	81		41						97	End

TABLE 4.5
EXAMPLES OF PUBLISHED HARDNESS CONVERSION EQUATIONS

Steels		Cemented Carbides	
$HB = \dfrac{7300}{130 - HRB}$	40–100 HRB	$HRc = 117.4 - \left(\dfrac{2.43 \times 10^6}{HV}\right)^{1/2}$	900–1800 HV
$HB = \dfrac{3710}{130 - HRE}$	30–100 HRE	$HRA = 0.53\left[211 - \left(\dfrac{2.43 - 10^6}{HV}\right)^{1/2}\right]$	900–1800 HV
$HB = \dfrac{1,520,000 - 4500\,HRc}{(100 - HRc)^2}$	<40 HRc	Rockwell from Knoop for steels:	
$HB = \dfrac{25,000 - 10\,(57 - HRc)^2}{100 - HRc}$	40–70 HRc	$HRc = 64.934$ log HK $- 140.38$ (15 gf)	
$HRc = 119 - \left(\dfrac{2.43 \times 10^6}{HV}\right)^{1/2}$	240–1040 HV	$HRc = 67.353$ log HK $- 144.32$ (25 gf)	
		$HRc = 71.983$ log HK $- 154.28$ (50 gf)	
$HRA = 112.3 - \left(\dfrac{6.85 \times 10^5}{HV}\right)^{1/2}$	240–1040 HV	$HRc = 76.572$ log HK $- 163.89$ (100 gf)	
		$HRc = 79.758$ log HK $- 170.92$ (200 gf)	
$HR15N = 117.9 - \left(\dfrac{5.53 \times 10^5}{HV}\right)^{1/2}$	240–1040 HV	$HRc = 82.283$ log HK $- 176.92$ (300 gf)	
		$HRc = 83.580$ log HK $- 179.30$ (500 gf)	
$HR30N = 129.5 - \left(\dfrac{1.88 \times 10^6}{HV}\right)^{1/2}$	240–1040 HV	$HRc = 85.848$ log HK $- 184.55$ (1000 gf)	
$HR45N = 133.5 - \left(\dfrac{3.13 \times 10^6}{HV}\right)^{1/2}$	240–1040 HV	White cast iron:	
		$HB = 0.363\,(HRc)2 - 22.515\,(HRc) + 717.8$	
		$HV = 0.343\,(HRc)2 - 18.132\,(HRc) + 595.3$	
		$HV = 1.136\,(HB)2 - 26.0$	
$HB = 0.951$ HV (steel ball, 200–400 HV)		Austenitic stainless steel:	
$HB = 0.941$ HV (WC ball, 200–700 HV)		$\dfrac{1}{HB} = \dfrac{130 - HRB}{7668.7}$	60–90 HRB, 110–192 HB

4.2.3.8 Hardness Testing at Elevated Temperatures

Elevated temperature or hot hardness testing consists of a standard Rockwell scale tester or Vickers indenter with a heating furnace mounted in the test area in either a vacuum or inert atmosphere chamber. An indenter of diamond, sapphire, or silicon nitride is mounted in tungsten carbide, and any oxidation during the test will shorten the life of the diamond considerably. Hot hardness data of valve materials, especially valve seat hardfacing materials, are useful properties in assessing seat wear resistance at elevated temperatures when wear data are not available.

4.2.4 Fracture Toughness Testing

4.2.4.1 Fracture Toughness and Fracture Mechanics

Fracture may be defined as the mechanical separation of a solid, owing to the application of stress. The toughness of a material is its ability to absorb energy in the plastic range or its resistance to propagation of a crack. Toughness may be considered as the total area under the stress-strain curve. This area is an indication of the amount of work per unit volume that can be done on the material without causing it to rupture. Toughness is a parameter that comprises both strength and ductility.

A crack in a loaded part or specimen generates its own stress field ahead of a sharp crack, which can be characterized by a single parameter called stress intensity (K). K represents a single parameter that includes both the effect of the stress applied to a sample and the effect of a crack of given size in the sample.

Rapid crack propagation is controlled solely by a material constant, called the critical stress-intensity factor (K_C), where crack propagation becomes rapid. The greater the value of K_C, the greater the resistance of the material to crack propagation and brittle fracture. The critical stress intensity factor is determined using relatively simple laboratory specimens. Figure 4.17 defines three modes of loading: Mode I, which is the opening or tensile mode; Mode II, which is the sliding or shear mode; and Mode III, which is the tearing mode. The concepts of fracture mechanics are essentially the same for each mode. However, most actual cracking and fracture cases are Mode I problems.

The nomenclature for K_C is modified to include the loading mode. For example, K_{IC} is the critical stress-intensity factor or fracture toughness under Mode I loading. Most testing to determine fracture toughness is performed in Mode I; therefore, most of the published fracture-toughness values are K_{IC}.

The stress intensity factor, K_I, for a crack tip in any body that is subjected to tensile stresses σ, perpendicular to the plane of the crack (Mode I deformation) is given by the relationship

$$K_I = \sigma\sqrt{\pi a}\, f(g) \qquad (4.28)$$

where a is crack length and f(g) is a function that accounts for crack geometry and structural configuration. Provided the crack length a is small compared to the width of the specimen, then

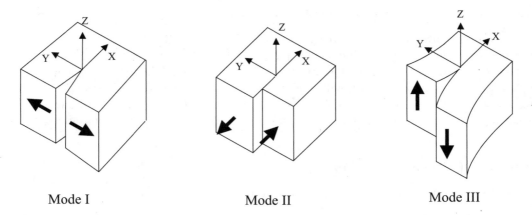

Mode I   Mode II   Mode III

Figure 4.17 Mode of loading. Mode I (opening mode): tension stress in the Y direction, or perpendicular to the crack surfaces. Mode II (edge-sliding mode): shear stress in the X direction, or perpendicular to the crack tip. Mode III (tearing mode): shear stress in the Z direction, or parallel to the crack tip.

$f(g) = 1$. The equation states that the crack propagates or fast fracture will occur when, in a material subjected to a stress σ, a crack reaches some critical size a, or alternatively, when material containing cracks of size a is subjected to some critical stress σ. The key point of Eq. 4.28 is that the critical combination of stress and crack length at which rapid crack propagation commences is a material constant, fracture toughness.

The methods used for toughness testing include linear-elastic and nonlinear loading, slow and rapid loading, crack initiation, and crack arrest.

4.2.4.2 Fracture Toughness Testing (Linear-Elastic Loading)

The linear-elastic methods of fracture toughness testing are used to measure a single-point fracture toughness value. For fracture by a brittle mechanism, this is no problem. Fracture occurs at a distinct point, and the fracture toughness measurement is taken as a value of the fracture parameter at that point. For fracture by a ductile mechanism, the fracture is a process, and the fracture toughness measurement is an R-curve. The R-curve is a crack growth resistance curve, which is defined as a plot of crack extension resistance as a function of slow stable crack extension. As a ductile fracture test, a single point to define the fracture toughness is desired. To accomplish this, a point where the ductile crack extension equals 2% of the original crack length is identified. Figure 4.18 illustrates schematically this criterion with a K-R curve. The ASTM E399 K_{IC} standard describes the construction procedure that is used to obtain a single-point measurement of fracture toughness on the R curve.

Static fracture toughness, K_{IC}, for steels and alloys is so high that it is not a concern in valvetrain component design. Rather, the dynamic impact toughness can be more critical and will be discussed later. Table 4.6 lists typical fracture toughness values for a variety of engineering materials.

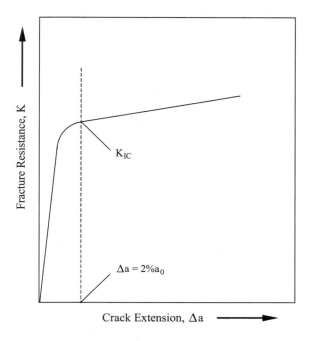

Figure 4.18 *Schematic of K-based crack resistance, R, curve with a definition of K_{IC}.*

TABLE 4.6
FRACTURE TOUGHNESS VALUES FOR VARIOUS
ENGINEERING MATERIALS

Material	K_{IC}, MPa$\sqrt{m}$
Glasses	0.5–1
Most Polymers	1–5
Ceramics (Si_3N_4, SiC, Al_2O_3)	3–7
Intermetallics	1–20
Cast Irons	10–40
Aluminum Alloys	20–50
Titanium Alloys	30–90
Steels	30–200
Superalloys	>100

4.2.4.3 Fracture Toughness Testing (Nonlinear)

The procedure describing a toughness value to be determined based on the arrest of a rapidly growing crack is somewhat different from those for the previously discussed toughness test methods, which determine only initiation toughness values. The preceding linear fracture toughness tests are used to measure fracture toughness for relatively low-toughness materials that fracture under or near the linear-loading portion of the test. For many materials used in structures, it is desirable to have high toughness, a value at least high enough that the structure would not reach fracture toughness before significant yielding occurs. For these materials, it is necessary to use the nonlinear fracture parameters to measure fracture toughness properties. The leading nonlinear fracture parameters, J and δ, are described in ASTM E1820.

4.2.4.4 Dynamic Fracture (Impact) Toughness Testing

Dynamic fracture occurs under a rapidly applied load, such as produced by impact or explosion. In contrast to quasi-static loading, dynamic conditions involve loading rates that are greater than those encountered in conventional tensile tests or fracture mechanics tests. Dynamic fracture includes the case of a stationary crack subjected to a rapidly applied load, as well as the case of a rapidly propagating crack under a quasi-stationary load. In both cases, the material at the crack tip is strained rapidly and, if rate sensitive, may offer less resistance to fracture than at quasi-static strain rates. Values of dynamic fracture toughness usually are lower than those of static toughness (K_{IC}) in comparison.

Some valvetrain components are subjected to high loading rates in service or must survive high loading rates during operating conditions. Therefore, high strain rate fracture testing is of interest, and components must be designed against crack initiation under high loading rates or be designed to arrest a rapidly running crack. Furthermore, because dynamic fracture toughness generally is lower than static toughness, more conservative analysis may require consideration of dynamic toughness.

Measurement and analysis of fracture behavior under high loading rates is more complex than under quasi-static conditions. Also, many different test methods are used in evaluating dynamic fracture resistance in addition to fracture mechanics, such as the Charpy impact test, the Izod impact test, and the drop-weight test (Figure 4.19). The specimen and impact test procedures are described in ASTM E23.

The Charpy and Izod impact tests are both pendulum-type, single-blow impact tests. The principal difference, aside from specimen and notch dimensions, is the configuration of the test setup. The Charpy test involves three-point loading, where the test piece is supported at both ends as a simple beam (Figure 4.19(a)). Because of its simplicity, the Charpy test has been widely used in notch-toughness tests. In contrast, the Izod specimen is set up as a cantilever beam, with the falling pendulum striking the specimen above the notch (Figure 4.19(b)). The drop-weight test is conducted by subjecting a series (generally four to eight) of specimens to a single impact load at a sequence of selected temperatures to determine the maximum temperature at which a specimen breaks (Figure 4.19(c)). The drop-weight test specimen and procedure are described in ASTM E208.

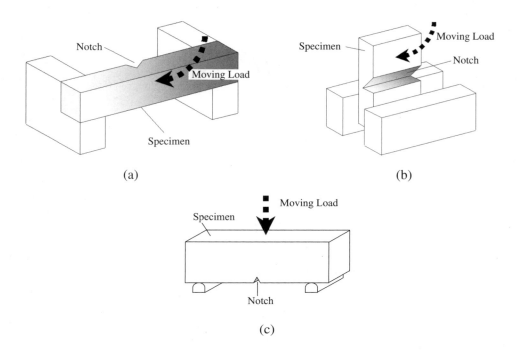

Notch

Moving Load

Specimen

(a)

Specimen

Moving Load

Notch

(b)

Moving Load

Specimen

Notch

(c)

Figure 4.19 *Schematic of the K-based crack resistance, R, curve with a definition of K_{IC}:*
(a) Charpy method, (b) Izod method, and (c) drop-weight method.

Materials undergo a transition from ductile behavior to brittle behavior as the temperature is lowered (Figure 4.20). In the presence of a stress concentrator such as a notch, it takes little loading to initiate a fracture below this transition temperature, and even less to cause such a notch or crack to propagate. These transitions cannot be predicted by tests such as hardness testing, tensile testing, or chemical analysis. The ductile-to-brittle transition temperature could be determined by impact testing using test specimens of uniform configuration and standardized notches.

Table 4.7 shows room-temperature impact toughness data for selected valvetrain component materials.

4.2.4.5 Environmentally Assisted Fracture Testing

Environmentally assisted crack growth is a special form of mechanical degradation that occurs when the combined effect or interaction of the environment and applied or residual stresses causes subcritical crack growth or fracture.

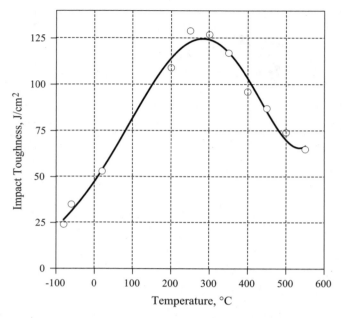

Figure 4.20 *Impact toughness versus temperature for 40 Cr steel.*

TABLE 4.7
EXAMPLES OF IMPACT FRACTURE TOUGHNESS DATA
FOR VALVETRAIN COMPONENTS

Materials	Impact Toughness, α_K (J/cm²)
Sil 1 (1050°C oil quench, 760°C oil cool, ϕ30)	50–70
422SS (1030°C oil quench, 690°C temper for 2 hours)	60
SAE 1547 (830°C oil quench, 550°C temper)	99
SAE 52100 (860°C oil quench, 160°C temper)	33
16 MnCr5 (900°C oil quench, 200°C temper)	72
40 Cr (880°C oil quench, 600°C temper)	53

4.2.4.5.1 Hydrogen Embrittlement Testing

Hydrogen embrittlement is a time-dependent fracture process caused by the absorption and diffusion of atomic hydrogen into a metal, which results in a loss of ductility and tensile strength. Three basic types of hydrogen embrittlement are observed in metals: (1) internal reversible hydrogen embrittlement, (2) hydrogen environment embrittlement, and (3) hydrogen reaction embrittlement. Internal reversible hydrogen embrittlement also has been termed slow-strain-rate embrittlement and delayed failure. Hydrogen that is absorbed from any source is diffusible within the metal lattice. To be fully reversible, embrittlement must occur without the hydrogen undergoing any type of chemical reaction after it has been absorbed within the lattice. The most distinct mechanisms of hydrogen embrittlement in metals include the hydrogen-enhanced decohesion mechanism and the hydrogen-enhanced local plasticity mechanism.

From a prevention standpoint, hydrogen embrittlement failures are reduced or eliminated by controlling the amount of hydrogen introduced during manufacture, processing, and the in-service environment of materials. Test methods and the necessary prevention controls are covered in standards such as ASTM F519.

Figure 4.21 shows the results of a hydrogen embrittlement crack growth rate as a function of applied stress intensity for two different hardnesses and environments for an AISI 4340 steel using a contoured double-cantilever beam test specimen. Higher hardness and a rich hydrogen environment considerably increase the crack growth rate.

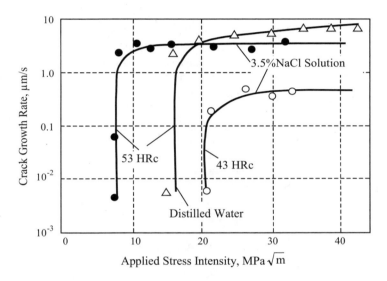

Figure 4.21 *Hydrogen embrittlement crack growth rate as a function*
of applied stress intensity for two different hardnesses and
environments for an AISI 4340 steel (Dull and Raymond).

4.2.4.5.2 Stress-Corrosion Cracking Testing

Stress-corrosion cracking (SCC) is a failure process that occurs because of the simultaneous presence of tensile stress, an environment, and a susceptible material. Failure by SCC frequently is encountered in what may seem to be mild chemical environments at tensile stresses well below the yield strength of the metal. The failures often take the form of fine cracks that penetrate deeply into the metal with little or no evidence of corrosion on the nearby surface. Table 4.8 summarizes typical SCC behavior for various alloy systems. In most cases, the scope of a comprehensive analysis must be designed to fit a given application.

TABLE 4.8
STRESS-CORROSION CRACKING OF SELECTED MATERIALS
(KATZ *ET AL.*)

Material	Environment	Remarks
Carbon steels, ferritic, pearlitic, or tempered martensitic (normally low carbon)	NaOH and nitrate solution, H_2S, ammonia, carbonate, bicarbonate, aqueous solutions, chlorides, seawater	Alloying has significant effects on SCC. Crack propagation rate increases with water vapor. Lowering chloride concentration is beneficial.
High-strength steels (low or high alloy), normally tempered martensite	Water and water vapor, aqueous electrolytes including phosphate ions, NaCl solutions, H_2S, strong acids, various organic compounds	Mutual considerations and compromises regarding alloying elements. Trade-offs in mechanical properties such as yield strength, toughness, and SCC are controlled by alloying.
Ferritic stainless steels	Chloride ions, aqueous solutions	Highly sensitive to alloying elements with synergistic effects.
Austenitic stainless steels	Chloride ions in solutions or steam, NaOH solutions, $NaCl + H_2O_2$ solutions, seawater	Phase-stability aspects. Attention also to carbon concentration and elements affecting sensitization.
Martensitic and precipitation-hardening stainless steels	$NaCl + H_2O_2$, seawater, $NaOH + H_2S$, NaOH solutions, other sulfur compounds, marine environments	Used for applications that require high strength (yield strength values to exceed 1100 MPa).
Aluminum alloys	NaCl solutions, seawater, water vapor, aqueous solutions, halide ions; Cl, Br, and I solutions (synergistic effects)	Combined environment and material group interaction differ regarding severity.
Nickel alloys	Pure steam, NaOH or KOH solutions, fused caustic sulfur compounds, various aqueous halide solutions, aqueous solutions, cupric acid	Sensitive to microstructure and phases such as γ, γ', and carbides.
Copper alloys	Ammonia solutions, amines, water, mercury salt solutions, distilled water	
Titanium alloys	Seawater, organic liquids, N_2O_4; aqueous Cl, B, and I solutions; HCl + methanol solutions; HNO_3; chlorinated hydrocarbons	Microstructure and phase optimization for specific environment. Normally alloy types consider the α (hcp) and the β (bcc phase volume fracture). β at ambient temperatures is stabilized by alloying with Cr, Mo, V, Ta, and W.

If the objective of testing is to predict the service behavior or to screen alloys for service in a specific environment, it often is necessary to obtain SCC information in a relatively short time. This requires acceleration of testing by increasing the severity of the environment or the critical test parameters. The former can be accomplished by increasing the test temperature or the concentration of corrosive species in the test solution and by electrochemical stimulation. Test parameters that can be changed to reduce the testing time include the application of higher stresses, continuous straining, and precracking, which allows bypassing the crack nucleation phase of the SCC process. Stress-corrosion specimens can be smooth and precracked or notched specimens. The loading mode can be constant deflection, constant load, and constant extension or strain rate. The grain flow direction relative to stressing direction, surface condition, and residual stress are important parameters. The nucleation of stress-corrosion cracks depends strongly on initial surface reactions. Several ASTM standards, including ASTM G30, G35, G38, and G41, describe the specimen and procedures to determine the susceptibility of alloys to SCC.

4.2.5 Fatigue Testing

4.2.5.1 Introduction

Fatigue is progressive, localized, and permanent structural damage that occurs when a material is subjected to cyclic or fluctuating strains at nominal stresses that have maximum values less than the static yield strength of the material. The process of fatigue failure consists of three stages:

1. Initial fatigue damage, leading to crack nucleation and crack initiation

2. Progressive cyclic growth of a crack (crack propagation) until the remaining uncracked cross section of a part becomes too weak to sustain the loads imposed

3. Final sudden fracture of the remaining cross section

In general, three simultaneous conditions are required for the occurrence of fatigue damage: (1) cycle stress, (2) tensile stress, and (3) plastic strain. If any one of these three conditions is not present, a fatigue crack will not initiate and propagate. The plastic strain resulting from cyclic stress initiates the crack; the tensile stresses (which may be localized tensile stresses caused by compressive loads) promote crack propagation. In general, the fatigue process consists of a crack initiation phase and a crack propagation phase. However, there is no general agreement regarding when (or at what crack size) the crack initiation process ends, and when the crack growth process begins.

4.2.5.2 Stress-Life Versus Strain-Life

Three basic types of fatigue properties are used in the context of an underlying fatigue design philosophy, as shown in Figure 4.22: (1) stress-life (S-N or high cycle), (2) strain-life (ε-N or low cycle), and (3) fracture mechanics ($\frac{da}{dN}$ – ΔK or cracked). Although fracture mechanics focuses on damage tolerance, and strain-life focuses on safe or finite life, stress-life focuses on safe or infinite life. Standard testing procedures for measuring fatigue properties of stress-life, strain-life, and fatigue crack growth are described in ASTM E466, E606, and E647, respectively.

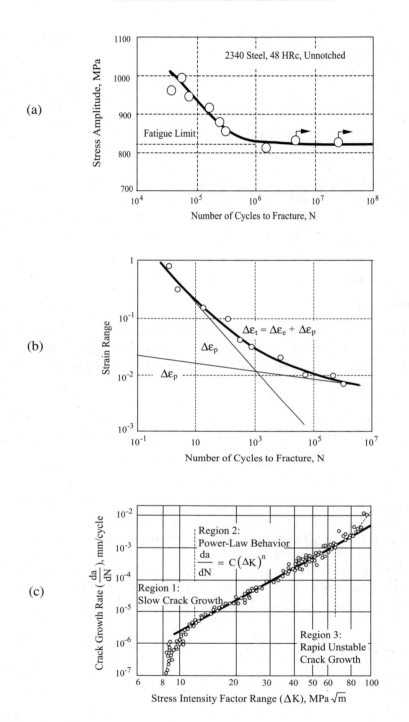

Figure 4.22 *Fatigue data presentation: (a) S-N curve for 2340 steel, (b) ε-N curve, and (c) $\dfrac{da}{dN} - \Delta K$ curve for ASTM A533 B1 steel (Antolovich and Saxena).*

Fracture mechanics methodology applies to the samples with surface irregularities and crack-like imperfections and may eliminate the crack-initiation portion of the fatigue life of the component. Strain life is intended primarily to address the low-cycle fatigue behavior (e.g., approximately 10^2–10^6 cycles), but the stress-life approach is best applied to components that are similar in appearance to the test samples and are approximately the same size. The general property presentation of stress-life is S-N (stress versus log number of cycles to failure). Failure in S-N testing typically is defined by the total separation of the sample. The advantages of this method are simplicity and ease of application, and it can offer some initial perspective on a given situation. It is best applied in or near the elastic range, addressing constant-amplitude loading situations in what has been called the long-life (or infinite-life) regime. Ferrous metals, especially steels, display a fatigue limit or endurance limit at a high number of cycles (typically greater than 10^6) under benign environment conditions, and the stress-life methodology provides a useful and beneficial tool for interpreting the steel infinite fatigue life behavior.

Fatigue test methods are described primarily by the mode of loading: direct (axial) stress, plane bending, rotating bending, alternating torsion, and combined stress. The most frequently used methods are the direct-stress and rotating bending method. Direct-stress fatigue testing subjects a test specimen to a uniform stress or strain through its cross section. Bending testing, such as the rotating beam bending test, can operate up to 10,000 rpm. In all bending-type tests, only the material near the surface is subjected to the maximum stress; therefore, only a small volume of material is under stress. Rotational bending systems effectively apply reversed loading to the outer surface of rods or shafts. By applying a known static force at the end of the shaft, a bending moment can be applied to the test section, the outer surface of which oscillates between tension and compression during each rotation. Figure 4.23 compares S-N curves from rotating beam bending testing and axial tensile pull testing.

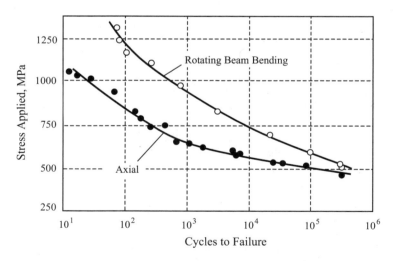

Figure 4.23 *Comparison of axial fatigue testing results to rotating bending fatigue testing results (Socie).*

The external load in fatigue testing usually is applied repeatedly with a chosen frequency while maintaining constant conditions. Figure 4.24 characterizes the load spectrum for the fatigue test.

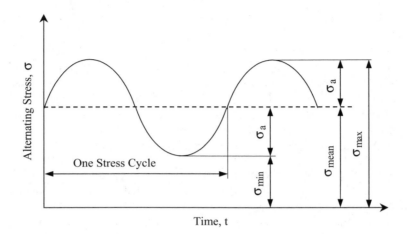

***Figure 4.24** Fatigue test load spectrum.*

The important parameters include mean stress, stress amplitude, and stress ratio. Mean stress is the arithmetic mean of maximal and minimal stresses in the cycle, $\sigma_{mean} = \dfrac{\sigma_{min} + \sigma_{max}}{2}$; stress amplitude, $\sigma_a = \dfrac{\sigma_{max} - \sigma_{min}}{2}$; and stress ratio, the ratio of the maximum value to the minimum value of stress in the cycle $\left(\dfrac{\sigma_{max}}{\sigma_{min}}\right)$. The mean stress affects the permissible alternating stress amplitude for a given life (number of cycles). At zero mean stress, the allowable stress amplitude is the effective fatigue limit for a specified number of cycles. As the mean stress increases, the permissible amplitudes steadily decrease. At a mean stress equal to the ultimate tensile strength of the material, the permissible amplitude is zero.

The horizontal portion of an S-N curve represents the maximum stress that the metal can withstand for an infinitely large number of cycles with 50% probability of failure and is known as the fatigue (endurance) limit. Because the materials do not exhibit a fatigue limit (i.e., the S-N curve continues to drop at a slow rate at high numbers of cycles), fatigue strength rather than fatigue limit is reported. Fatigue strength is the stress to which the metal is subjected for a specified number of cycles.

The presence of elastoplastic deformations causes hysteresis loops to appear in a plot of stress imposed on the specimen versus the resultant strain (Figure 4.25). The area surrounded by the loop is a measure of the energy supplied to the specimen. Hysteresis loops can be recorded in both strain- and load-controlled experiments. The experimental data obtained show that the hysteresis loop usually changes during early stress cycles to stabilize before one-third to one-half of

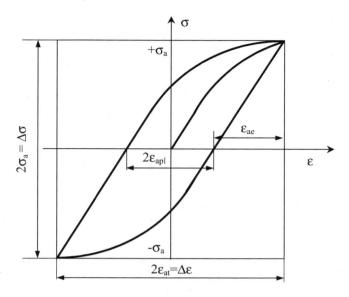

Figure 4.25 _Hysteresis loop:_ σ_a, _amplitude of cyclic stress;_ ε_{ae}, _amplitude of elastic strain;_ ε_{apl}, _amplitude of plastic strain; and_ ε_{at}, _total strain amplitude._

the fatigue life is spent and achieves a saturation state characterized by constant values of stress amplitude or strain. The termination of fatigue life is accompanied by a loss of stability.

4.2.5.3 Fatigue at Elevated Temperatures

At elevated temperatures, the strength of materials is reduced. For common carbon steels, the reduction is moderate below 300–400°C. However, the decrease observed is much more drastic at higher temperatures, and the yield strength and static strength constitute only half the comparable values at a temperature of 627°C than at room temperature. The fatigue strength generally is lower at high temperatures.

Figure 4.26 represents in generalized form the effect of elevated temperatures on smooth bar fatigue strength. The value shown is a percentage of the room-temperature fatigue strength.

In general, the situations encountered in fatigue testing at elevated temperatures can be divided into fatigue-dominated cracking, which is transgranular, and creep-dominated cracking, where a crack links the already existing cavities and minute cracks located at grain boundaries, producing an intergranular macrocrack. The appearance of the fracture surface usually is complicated, and the final crack often is one that started as a transgranular crack but changed into the intergranular mode at higher levels of stress resulting from a reduction in the cross-section load-carrying area and the presence of cavities at grain boundaries. Fatigue-dominated cracking is favored by high

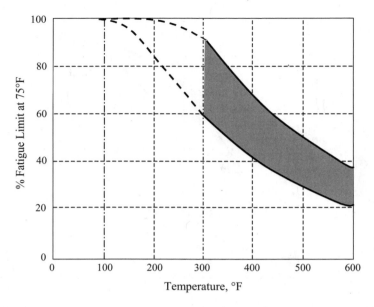

Figure 4.26 *Generalized effect of temperature on fatigue strength at 5 × 10⁸ cycles (smooth, rotating, bending), stress ratio = −1 (SAE Fatigue Design Handbook).*

cyclic amplitudes, frequencies, and relatively low temperatures, whereas intergranular cracking is promoted by the opposite conditions.

Because thermal strains are directly proportional to the coefficient of expansion, α, a low value of α is desirable. A high value of thermal conductivity, k, is beneficial because it reduces the temperature differences in a material that is rapidly heated or cooled. These properties are of importance only when comparing different types of alloys (e.g., ferritic and austenitic steels, which have widely differing values of α and k). Figure 4.27 shows an example of ultimate tensile strength (UTS), elastic modulus (E), and the coefficient of thermal expansion (α) as a function of temperature.

It is desired that the thermal fatigue data be expressed in a cyclic strain range, or its plastic component as a function of number of cycles to failure. However, existing thermal fatigue data of valvetrain components and specifically valve materials in literature are expressed in stress-life form or in S-N curves. Figure 4.28 shows other engineering material fatigue strengths at elevated temperatures on the 10^7 cycle in bending.

4.2.5.4 Thermal Shock/Fatigue Testing

In valvetrain applications, specifically valve applications, the temperatures are not constant. Thus, the temperature cycles at which the tensile strains are imposed during each cycle appear to be of considerable importance to the endurance resulting from repeated cooling and heating

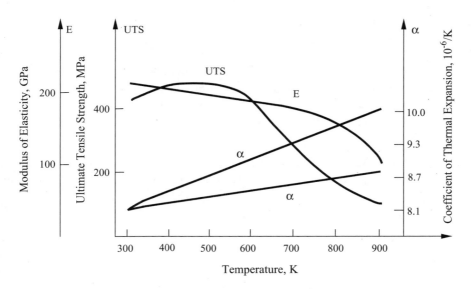

Figure 4.27 *Dependence of mechanical properties of medium-carbon steel on temperature. The upper curve relates to a rapid temperature change; the lower curve relates to a slow change (Weronski and Hejwowski).*

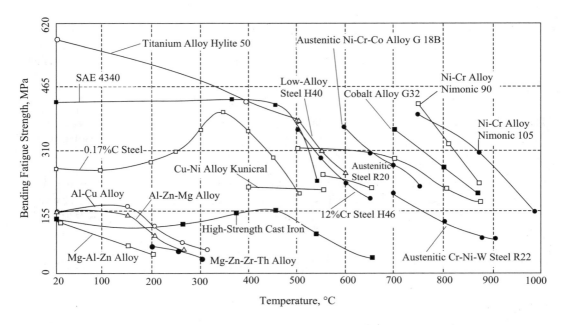

Figure 4.28 *Other engineering material thermal fatigue strength, on the 10^7 cycle in bending (Forrest).*

cycles. Apparently, when insufficient time is available for the mean temperature to attain the edge temperature, a smaller amount of tensile strain will be developed. At high strains, behavior is dominated by ductility at the appropriate temperature; at low strains, the tensile strength is more relevant. Thermal fatigue resistance can be improved if ductility can be increased without a loss of strength. A strain-hardening cyclic stress-strain relationship is desirable so that strains will not become unduly large in regions of strain concentration. Good resistance to oxidation will result in superior strain cycling endurance.

The strength of metals under conditions of low-cycle fatigue usually and most effectively is evaluated by the fatigue curve expressing the relation between the cyclic strain range, or its plastic component, and the number of cycles to failure, which is obtained by strain-controlled low-cycle fatigue tests shown as

$$\varepsilon_p N^\alpha = C \tag{4.29}$$

where α and C are material constants.

However, in the case of high-temperature fatigue, a creep or relaxation phenomenon occurs to some extent, so that the constants in the relationship should be varied according to thermal conditions and cycle rates,

$$\varepsilon_p N^\alpha F(T) = C_1 \tag{4.30}$$

$$F(T) = e^{\left(\frac{-Q}{T_m}\right)} \left\{ 1 + C_2 T_a e^{\left(\frac{Q}{T_m}\right)} \right\} \tag{4.31}$$

where C_1, C_2, and Q are material constants, T_a is the temperature range in degrees Kelvin (°K), and T_m is the mean temperature of the thermal cycle or the isothermal test temperature in degrees Kelvin (°K).

The strong influence of the maximum temperature T_2 is evidenced in Figure 4.29 by the curve obtained by varying ΔT with T_1 constant at 20°C, as compared with the curve obtained by varying ΔT with constant at 920 or 1020°C. For the same ΔT, as long as T_2 is below 920°C, the endurance is increased greatly.

Figure 4.30 shows the effect of temperature range on thermal fatigue strength for H46 stainless steel.

4.2.5.5 Corrosion Fatigue Testing

Similar to stress-corrosion cracking and/or hydrogen embrittlement under monotonic loads, environmentally assisted crack growth occurs under fatigue conditions as well. Corrosion fatigue

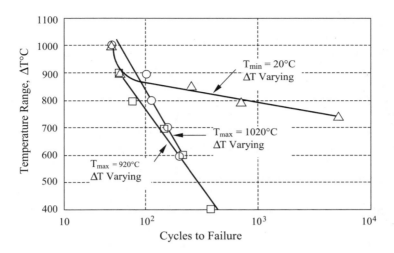

Fig 4.29 *Effect of temperature range on thermal fatigue strength for Nimonic 90 (Glenny and Taylor).*

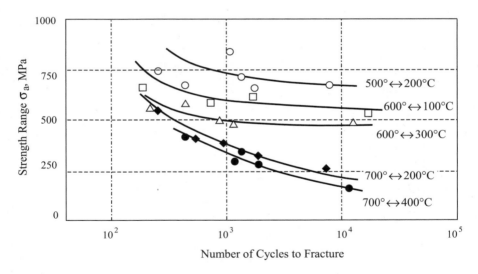

Figure 4.30 *Effect of temperature range on thermal fatigue strength for H46 steel (Udoguchi and Wada).*

is the combined action of repeated or fluctuating stress and a corrosive environment to produce progressive cracking. Usually, environmental effects are deleterious to fatigue life, producing cracks in fewer cycles than would be required in a more inert environment. After fatigue cracks have formed, the corrosive aspect also may accelerate the rate of crack growth.

In corrosion fatigue, the magnitude of cyclic stress and the number of times it is applied are not the only critical loading parameters. Time-dependent environmental effects also are of prime importance. When failure occurs by corrosion fatigue, stress-cycle frequency, stress-wave shape, and stress ratio all affect the cracking processes.

In a corrosive environment, the situation will be different. Disorganized atoms along a gliding plane may require less activation energy to pass into a liquid than more perfectly arrayed atoms elsewhere. Certainly, while the atoms are in motion along a gliding plane, preferential attack may reasonably be expected, even below the fatigue limit. Standard methods of fatigue crack growth are defined in ASTM E647 and generally are applicable to corrosion fatigue crack growth tests. A standard test method for corrosion fatigue testing is described in ASTM F1801. Figure 4.31 shows lead corrosion on the fatigue life of Nimonic 80A at 815°C and indicates that the effect is significant for fine-grained alloys.

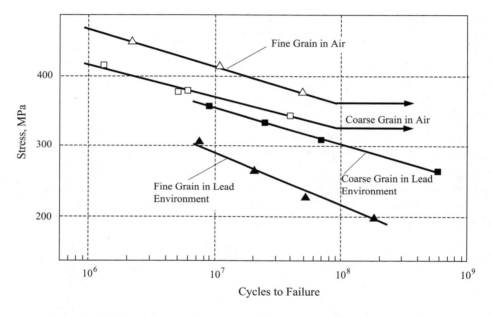

Figure 4.31 *Lead corrosion effect on the fatigue life of Nimonic 80A at 815°C (Caird and Trela).*

Note that although stress-corrosion cracking often is intergranular, corrosion-fatigue cracks usually are transgranular, following gliding planes inclined at such an angle as to provide high resolved shear stress. There are exceptions to both rules.

4.2.5.6 Lifetime Prediction

When the appropriate fatigue properties are not available when needed, it is necessary to estimate them from other available properties or from properties that can be determined quickly by using existing facilities. The fatigue curve prediction models are based on (1) plastic strain (short life), (2) stress (long life), and (3) total strain (intermediate life, plastic plus elastic strain).

For a series of completely reversed tests at different strain ranges, it has been found for ductile metals that a log-log plot of the fatigue life versus the stable plastic strain amplitude gives a band of points that scatter slightly about a straight line, as illustrated in Figure 4.32. Therefore, the equation of fatigue life for the plastic strain model is

$$\frac{\Delta\varepsilon_p}{2} = \varepsilon_f'\left(2N_f\right)^c \tag{4.32}$$

where

$\dfrac{\Delta\varepsilon_p}{2}$ = plastic strain amplitude

ε_f' = fatigue ductility coefficient
$2N_f$ = number of reversals to failure
c = fatigue ductility exponent

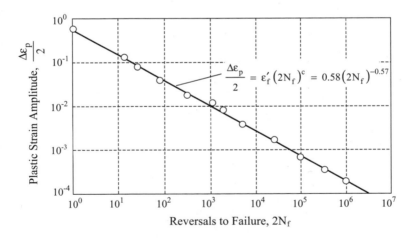

Figure 4.32 *Example of fatigue life prediction for annealed SAE 4340 steel (Smith et al.).*

The equation for the fatigue strength based on a stress model in the long life region is

$$\sigma_a = \sigma_f'(2N_f)^b \tag{4.33}$$

where

σ_a = true fatigue strength
σ'_f = fatigue strength coefficient
b = fatigue strength exponent

Figure 4.33 shows an example of a fatigue life prediction using the stress-life model.

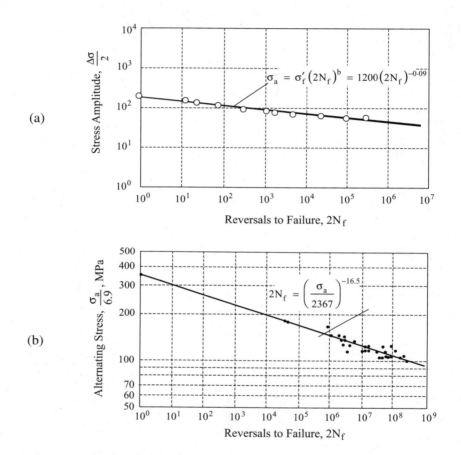

(a)

(b)

Figure 4.33 *Example of a fatigue life prediction using the stress-life model:*
(a) SAE 4340 steel (Smith et al.), and (b) 52100 steel HRc 62 (Sachs et al.).

The fatigue ductility and strength properties that have been defined can be used to express the resistance to total strain, which is dependent on the combined ductility and strength of the metal,

$$\frac{\Delta\varepsilon_p}{2} = \frac{\sigma'_f}{E}(2N_f)^b + \varepsilon'_f(2N_f)^c \qquad (4.34)$$

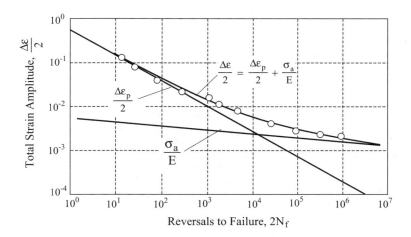

Figure 4.34 *Example of fatigue life prediction using the total strain model, SAE 4340 steel (Smith et al.).*

The total strain amplitude is plotted against life in Figure 4.34.

Engineers often find themselves in a position where they must estimate the S-N curve for a particular type of material or where they wish to compare fatigue data to theoretical values or to past observations by other researchers. Therefore, general information becomes of considerable value if some of the factors that affect fatigue life are taken into consideration. Many researchers have collected fatigue data on steels in an attempt to show the relationship between the static properties and the fatigue limit. Due to the many factors (e.g., surface finish, heat treatment, residual stresses, machining process, and temperature) that affect the fatigue limit, these relationships are not clear. This can be seen in Figure 4.35, where data are shown for small unnotched, laboratory-polished, rotating bending (R = –1) specimens.

Figure 4.36 shows the fatigue limit and tensile strength relationship with various types of finishing processes, including ground, machined, hot rolled, and forged. At varying tensile strengths, the influence of the surface finish becomes more significant than the increase in tensile strength and results in a leveling off or a lowering of the fatigue limit for high-strength steels.

4.2.6 Friction Testing

When two materials are placed in contact, any attempt to cause one of the materials to slide over the other is resisted by a friction force. The force that will cause sliding to start, F_S, is related to the force P acting normal to the contact surface by

$$F_S = \mu_S N \qquad (4.35)$$

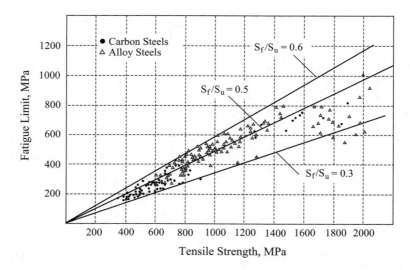

Figure 4.35 *Relation between the rotating bending fatigue limit and tensile strength of wrought steels (Forrest).*

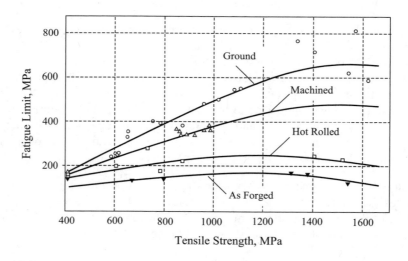

Figure 4.36 *Relationship between the fatigue limit and tensile strength for unnotched specimens in reversed bending (Noll and Erickson).*

where μ_S is the coefficient of static friction. When sliding begins, the limiting frictional force decreases slightly

$$F_K = \mu_K N \qquad (4.36)$$

where μ_K is the coefficient of kinetic friction. The work done in sliding against kinetic friction appears as heat or wear debris, or both.

The friction data of a material characteristically are nonconstant, and the irregularity of data from laboratory test devices is believed to be due to the following: (1) sliding materials are inhomogeneous, and their surfaces are rough at the start, and even more so after some sliding and wearing; (2) all sliding systems, practical machinery, and laboratory devices vibrate and move in an unsteady manner because of their mechanical dynamics; and (3) instrumented sliding systems will show behavior in the data that is affected by the dynamics of the amplifier/recorders (Ludema).

Measurement of the coefficient of friction involves two quantities, namely F, the force required to initiate and/or sustain sliding, and N, the normal force holding two surfaces together. Some of the earliest measurements of the coefficient of friction were done by an arrangement of pulleys and weights, as shown in Figure 4.37(a). Weight is applied until sliding begins, and one obtains the static coefficient of friction with

$$\mu = \frac{F_S}{N}$$

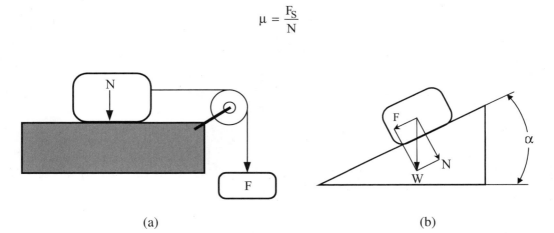

(a) (b)

Figure 4.37 *Simple measurement of friction: (a) dead weight and pulley method, and (b) slippery slope method.*

If the kinetic coefficient of friction is desired, a weight is applied to the string, and the slider is moved manually and released. If sliding ceases, more weight is applied to the string for a new trial until sustained sliding of uniform velocity is observed. The final weight is used to obtain the kinetic coefficient of friction with

$$\mu = \frac{F_K}{N}$$

A second convenient system for measuring friction is the inclined plane shown in Figure 4.37(b). The measurement of the static coefficient of friction simply consists of increasing the angle of tilt of the plane to α when the object begins to slide down the inclined plane. If the kinetic coefficient of friction is required, the plane is tilted, and the slider is advanced manually. When an angle, α, is found at which sustained sliding of uniform velocity occurs, tan α is the operative kinetic coefficient of friction.

As technology developed, it became possible to measure the coefficient of friction to high accuracy under dynamic conditions. Force-measuring devices for this purpose range from the simple spring scale to devices that produce an electrical signal in proportion to an applied force. The deflection of a part with forces applied can be measured by strain gages, capacitance sensors, inductance sensors, piezoelectric materials, optical interference, Moire fringes, light beam deflection, and several other methods. The most widely used—because of simplicity, reliability, and ease of calibration—is the strain gage system. In the same way that many sensing systems are available, many designs of friction-measuring machines also are available. The pin-on-disk geometry is illustrated in Figure 4.38, where the pin is held by a cantilever-shaped force transducer. Although the pin-on-disk geometry rarely is a good simulator of practical devices, it is the most widely used configuration in both academic and industrial laboratories.

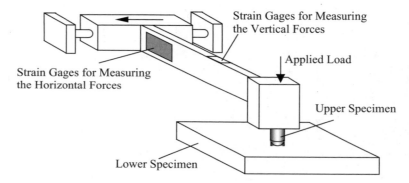

*Figure 4.38 Schematic of a cantilever transducer system
for measuring friction force (Ludema).*

4.2.7 Wear Testing

4.2.7.1 Sliding Contact Wear Testing

Types of surface damage caused by sliding contact include adhesive wear, scuffing or galling, and fretting. These damage mechanisms all are influenced by adhesion of the mating surfaces, but these categories also reflect the nature of the surface damage and the type of sliding contact. Adhesive wear involves the transfer of material from one surface to another. Adhesive wear also typically occurs from the sliding contact of two surfaces, where interfaces in contact are made to slide and the locally adhered regions must separate, leaving transferred material. Breakout of this transferred material will form additional debris. This separation of material results in a wide range of wear rates, depending on the type of contact and the adhesion between the mating surfaces. Galling or scuffing is considered a severe form of adhesive wear that occurs when two surfaces slide against each other at relatively low speeds and high loads. Fretting also is a special case of adhesive wear that occurs from oscillatory motion of relatively small amplitude.

Adhesive wear testing can be carried out with a variety of sliding contact systems. These include four-ball, pin-V-block, pin-on-disk, block-on-ring, and crossed cylinder systems. Table 4.9 shows examples of these test configurations and the corresponding ASTM standards.

TABLE 4.9
SLIDING WEAR TEST CONFIGURATION AND STANDARDS

Test	Schematic	ASTM No.	Title
Four-Ball		D4172	Standard Test Method for Wear Preventive Characteristics of Lubrication Fluid
Pin-V-Block		D2670	Standard Test Method for Measuring Wear Properties of Fluid Lubricants
Pin-on-Disk		G99	Standard Test Method for Wear Testing with a Pin-on-Disk Apparatus
Block-on-Ring		G77	Standard Test Method for Ranking Resistance of Materials to Sliding Wear, Using Block-on-Ring Test
Crossed Cylinder		G83	Standard Test Method for Wear Testing with a Crossed Cylinder Apparatus

A wear test should be run long enough to produce measurable wear. The easiest way to determine measurable wear is to measure weight loss. Dimensional change is a more sensitive method. If a well-defined contact geometry is used, such as ball-on-flat, ball-on-ball, or ring-on-flat, a scar length can be translated to volume loss. Then the normalized wear rate or wear coefficient can be calculated as

$$k = \frac{WH}{LD} \qquad (4.37)$$

$$R = \frac{W}{LD} \qquad (4.38)$$

where

    W = wear volume
    L = normal load or force
    D = distance of sliding
    H = hardness
    k = wear coefficient
    R = normalized wear rate

4.2.7.2 Rolling Wear Testing

Rolling contact damage including wear and fatigue spalling is a major cause of failure in rolling-element bearings, such as in cam and roller followers. It is a particularly insidious form of wear because it sometimes is difficult to detect in its early stages. The precursor flaws may be hidden from view because they grow below the contact surface. By the time cracks grow large enough to emerge at the surface and produce wear particles or delaminations, these particles may become large spalls or flakes, resulting in immediate component loss of function or efficiency. Figure 4.39 shows stress distribution in contact surfaces due to rolling, sliding, and the combined effect of both.

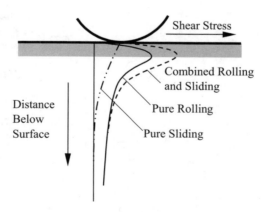

Figure 4.39 *Stress distribution in contact surfaces due to rolling, sliding, and the combined effect of both.*

Rolling contact frequently is accompanied by slip or sliding. The complex motions experienced in many types of rolling contact situations produce at least a small percentage of slip or sliding. Pure rolling is probably the exception rather than the rule in diverse applications or rolling components. The percentage of sliding is defined as

$$\% \text{ Sliding} = 2 \times \frac{|v_1 - v_2|}{|v_1 + v_2|} \tag{4.39}$$

where v_1 and v_2 are the surface velocities of bodies 1 and 2, respectively. At some percentage of sliding, the morphology of the wear changes to one or more characteristics of sliding wear, such as adhesion, abrasion, and scuff.

Pitting and spalling are important contact wear manifestations in rolling-element bearings. The depth of the spall tends to be related to the location of the maximum Hertzian shear stress below the surface. The fatigue spalling life of a bearing, L, usually is defined in terms of the first appearance of a spall and generally is based on the ratio of the equivalent dynamic load, P, to the load capacity, C, obtained from the manufacturer,

$$L = \left(\frac{C}{P}\right)^y \tag{4.40}$$

The exponent y depends on the type of bearing. For ball bearings, $y = 3$; for roller bearings, $y = \frac{10}{3}$.

Figure 4.40 illustrates two test configurations that have been used successfully to address rolling wear. One configuration consists of a pair of driven rollers pressed against one another, and the other configuration is two rollers against a round bar. The typical procedure is to visually

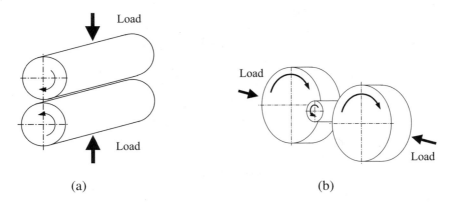

(a)   (b)

Figure 4.40 *Rolling contact testing configurations: (a) simulation of a roller bearing without a needle, and (b) simulation of a roller bearing with a needle.*

monitor the condition of the roller surfaces and determine the number of cycles for a selected level of surface damage to occur.

4.2.7.3 Abrasive Wear Testing

Abrasive wear is caused by hard particles or hard protuberances that are forced against and move along a solid surface, resulting in instantaneous or progressive loss of material. In general, abrasive wear processes typically are divided into two broad regimes: (1) high-stress abrasion, or (2) low-stress abrasion. High-stress abrasion or grinding abrasion occurs when abrasive particles are compressed between two solid surfaces. The high contact pressure produces indentations and scratching of the wearing surfaces and fractures and crushes the abrasive particles. High-stress abrasion often is referred to as three-body abrasion, although two-body, high-stress conditions also can exist. The high-stress abrasion facilitates material removal by cutting. Low-stress abrasion occurs when lightly loaded abrasive particles impinge on and move across the wearing surface, a combination of plastic deformation and tearing as well as fatigue, spalling, or surface fracture if the material is brittle.

Figure 4.41 shows a schematic of a standard abrasive wear test, and ASTM G65 describes the test procedures in detail. The test involves loading a specimen against a rotating rubber-rimmed wheel while a flow of abrasive sand is directed at the contact zone.

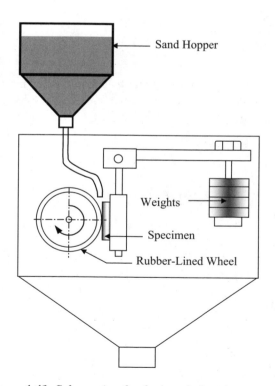

Figure 4.41 *Schematic of a dry sand abrasive wear test.*

4.2.8 Corrosion Testing

Exhaust valve corrosion and burning have constituted a limitation to the durability of many internal combustion engines. The engine exhaust consisting of carbon monoxide (CO), oxygen (O_2), hydrocarbons (HC), oxides of nitrogen (NO_X), and sulfide is found to be very hostile to valves. These components in the exhaust gas passing over the valve fillet at a relatively high velocity (6–9 m/s) will react with metal atoms. The product of the reaction reduces the load-bearing section thickness and thus weakens the valve strength.

4.2.8.1 Oxidation Testing

The valve oxidation process is sensitive to the exhaust gas temperature, composition, velocity, flow pattern, and thermal fluctuations of the environment. The alloys are composed of as many as a dozen elements, each of which affects oxidation behavior. Other factors such as part geometry, service stress, and exposure-induced phase changes within the alloy further complicate the situation. The true concern for valve alloy oxidation is its influence on valve life by simply reducing the load-bearing cross section and introducing sources of stress concentration. The following four distinct oxidation-induced processes can deteriorate material strength: (1) surface scaling, (2) internal oxidation, (3) oxide spalling, and (4) oxide vaporization. When a valve is heated, the rate of oxidation will depend on whether or not the scale formed is protective. If the scale formed is nonprotective, then it will not restrict the access of oxygen to the metal surface. Consequently, the rate of scale thickening will be independent of scale thickness. It will follow the linear relationship (Kubaschewski and Hopkins)

$$y = Kt + C \tag{4.41}$$

where y is scale thickness, t is the time, and K and C are constants.

If the scale is protective, then the diffusion of ions or electrons through the thickening scale will be the rate-determining process, and the rate of scale thickening will follow the parabolic relation

$$y^2 = 2At + C \tag{4.42}$$

where A and C are constants.

Equations 4.41 and 4.42 represent the two most important extremes of oxidation behavior. In reality, most oxidation/corrosion reactions follow Arrhenius' law, as shown in Eq. 4.43 and Figure 4.42. ASTM G54 describes the static oxidation test standard,

$$d = Ae^{-\frac{B}{T}} \tag{4.43}$$

where d is the corrosion/oxidation rate, A and B are constants, T is the temperature, and $B = \dfrac{E}{R}$ where E is the activation energy and R is the gas constant.

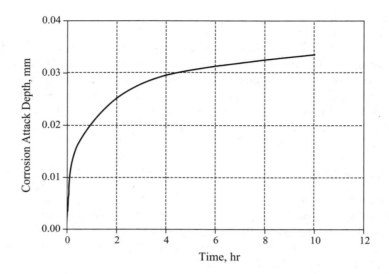

Figure 4.42 *Oxidation attack as a function of time, Nimonic 80A at 1080°C (Betteridge and Heslop).*

4.2.8.2 Hot Corrosion Testing

Exhaust valves at elevated temperatures are exposed to the gas environment containing other reactants in addition to oxygen. Hot corrosion, lead corrosion, and sulfidation are the interchangeable terms used in describing exhaust valve load bearing deterioration under the corrosive deposits on the surface at elevated temperatures. In environments where the gas composition is such that other reactants in addition to oxygen are present, conditions also are conducive to the formation of condensed phases on the surfaces of the alloys. Hot gases may contain sulfur, carbon, and chlorine with metallic constituents of sodium, calcium, magnesium, potassium, or lead. The development of condensed phases on alloy surfaces represents an especially undesirable situation, from a corrosion viewpoint. This type of material degradation is especially severe when the condensed phase is liquid.

During the initiation stage of hot corrosion, alloys are being degraded at rates similar to those that would have prevailed in the absence of the deposit. Numerous factors affect the time at which the hot corrosion process moves from the initiation stage to the propagation stage. These factors are alloy composition, gas composition and velocity, deposit composition and its physical state, temperature, temperature cycles, and the amount of deposit on the surface.

There are two types of hot corrosion. Type I is observed at temperatures of 800–900°C in the presence of Na_2SO_4, $CaSO_4$, $BaSO_4$, or other types of salt deposits on the valve surface. It is characterized by intergranular attack and a denuded zone of base metal. Type II occurs at 600–800°C due to the formation of low-melting eutectics of alkali metal and base alloy metal sulfates. It is characterized by layered corrosion scale, without intergranular attack or a denuded zone. However, as the exposure time increases and the scale is damaged by thermally induced stresses, eventually less protective oxides will be formed, and sooner in the case of alloys with foreign deposits on their surfaces. Another effect of deposits on the corrosion of valve alloys is that the protective oxide scales formed on the alloys may dissolve in the deposits. This can lead to especially undesirable conditions when the deposit is a liquid, as in the case of lead corrosion. In certain cases, conditions can exist where the protective scale dissolves into the liquid deposit at one interface and precipitates at another interface as a nonprotective layer. Such processes have been called fluxing reactions of the protective oxide barrier by the deposit.

4.2.8.2.1 Lead Corrosion

Leaded fuel has been phased out since the early 1970s in the United States and Western countries. However, in some regions of the world today, lead still is added to fuel to increase the octane rating of the fuel, thus increasing the antiknock value of the base fuel. The combustion of this tetraethyl lead in an engine not only is harmful to environment, but also causes the formation of detrimental lead compounds, mainly lead oxide, which results in the so-called "hot corrosion" of valves. To overcome this, ethyl dibromide or ethylene dichloride is introduced with the tetraethyl lead. This additional compound results in the formation of lead bromide, lead chloride, or lead oxide, which is volatile during the combustion of the fuel. This volatile product escapes with the exhaust gases. Unfortunately, the added ethylene compounds do not entirely remove this volatile product. With exhaust gases, the lead compounds produced during combustion, decomposition, and condensation of products can form deposits on valve heads. Owing to the acidic nature of these deposits, this leads to severe cold corrosion, particularly in engines that are allowed to stand for long periods after use.

Corrosive attack by molten lead compounds leads to a breakdown of the protective chromium oxide layer and increases chromium diffusion to form new oxides. The intergranular lead attack in the exhaust valves initially occurs as alloy depletion at the surface and at the grain boundaries. The rate of corrosion penetration then increases as a result of exposure to the lead compounds. This exposure causes severe internal oxidation and the formation of nonprotective scales such as $Pb_3Cr_2O_9$, which in turn provide nucleation sites for subsequent crack propagation.

Nickel-based superalloys such as Nimonic 80A, Inconel 751, and Pyromet 31 are superior in lead corrosion resistance than austenitic alloys such as 23-8N, 21-4N, 21-4N+W+Nb, and 21-2N. The least resistant to lead attack are the ferritic or martensitic alloys such as SAE 1547 and 3140. Sil 1 is ranked better in lead corrosion than SAE 1547 due to higher alloy content such as chromium to form a protective Cr_2O_3 layer. The popularity of large silicon additions in valve steels such as Sil 1 is primarily the result of the improved oxidation resistance. However, it has been shown that silicon concentrations in excess of 0.25% definitely have a deleterious effect on the corrosion resistance of a steel, particularly in the presence of lead oxide. Within each alloy group, the difference may not be significant enough to predict the performance in the engine environment. It is important to emphasize that most of the available superalloys have been developed to achieve resistance to degradation induced by oxygen. Therefore, their compositions are appropriate for selective oxide formation, not necessarily sulfide, carbide, or nitride formation. Hence, such alloys may have extremely poor corrosion resistance when exposed to gases that do not contain oxygen.

Various lead corrosion laboratory tests measure the corrosion resistance of valve alloys. The predominant test that is used depends on lead oxide (PbO) because it has been found to be the major constituent in deposits. The method consists of immersing a test specimen of the valve alloy in molten PbO at 910°C for one hour and then weighing the samples to determine the weight loss. This test may not be realistic because of the high temperatures used, compared to an engine valve temperature of 700–800°C, plus the fact that the samples are continuously exposed to a molten oxide, rather than determining at what critical temperature a molten oxide could be developed. Table 4.10 lists the types of lead compounds formed in engine deposits and the corresponding melting temperatures.

TABLE 4.10
LEAD COMPOUNDS FORMED
IN ENGINE DEPOSITS

Compound	Melting Temperature (°C)
$PbBr_2$	370
$(PbBr_2)_x(PbCl_2)_y$	370/496
$PbCl_2$	496
$PbO \cdot PbBr_2$	497 (decomp.)
$PbO \cdot PbCl_2$	524
$2PbO \cdot PbCl_2$	693
$2PbO \cdot PbBr_2$	709
PbO	888
$4PbO \cdot PbSO_4$	895
$2PbO \cdot PbSO_4$	961
$PbO \cdot PbSO_4$	975
$PbSO_4$	1190

4.2.8.2.2 Sulfidation Corrosion

The corrosive environment of the diesel engine is created by the sulfur in the fuel and the barium, calcium, and sodium in the lubricating oils that combine to produce sulfate salts. Chemical analysis of salt deposits on diesel exhaust valves has revealed a mixture of calcium, barium, and sodium sulfates, as well as some carbon. The carbon, although present only in small amounts and not reacting with the alloy elements, appears to act as a catalyst for the corrosion process. Carbon releases the sulfur into a more active form. Hot corrosion would be negligible without the presence of carbon. The presence of ash, especially alkali, alkaline earth sulfates, or vanadium pentaoxide, will accelerate the corrosion process. The severe sulfidation attack on valve alloys occurs due to contact with chloride-contaminated alkali and alkaline-earth sulfates and when the passive oxide layer becomes nonprotective. The Cr_2O_3 layer either may be destroyed or may be prevented from developing a continuous coating. At this stage, the sulfur will react with free chrome in the alloy, and Cr_3S_4 will form. Next, the chromium sulfide is oxidized, and free sulfur is released to become available for further attack and subsequent deeper penetration into the metal. The chemical reaction mechanism is (Spengler and Viswanathan)

$$2Cr_3S_4 + 9\underline{O} \rightarrow 3Cr_2O_3 + 8\underline{S} \qquad (4.44)$$

$$8\underline{S} + 6Cr \rightarrow 2Cr_3S_4 \qquad (4.45)$$

The result is chromium dilution of the matrix, which increases the susceptibility to corrosive attack. When the level of free chromium is low, the excess sulfur can react with nickel to form a lower melting eutectic phase (Ni-NiS), causing a marked acceleration in corrosion.

ASTM G4 and G111 describe some hot corrosion test standards.

4.3 Valvetrain Component Bench Testing

4.3.1 Component Testing Overview

Valvetrain component testing requires an understanding of service conditions, design, and manufacturing variables and involves a series of processes to validate the product for usage. Computer systems are playing a dominating role in the design of components and in simulating how the components react under different environments. Design engineers use three-dimensional modeling software for various purposes: to design components, to create a three-dimensional image of the components to scale and then to manipulate the image to identify design concerns, to match to models of mating parts to check for interference, and to manufacture the component directly from the model. The design model also can be used to generate a finite element analysis (FEA) model. The FEA model can show how the design reacts under various loading conditions. If areas of the design are suspect, design changes can be made and reevaluated easily. Although FEA models provide an engineer the ability to improve component designs without making a physical part, further test simulations to evaluate durability must be developed to accurately predict the total capability of a valvetrain component. The use of physical tests is required to develop these simulations.

An understanding of the operating environment in which the component must function establishes the basis for testing the component. The environment may include cyclic or static loading, vibration concerns, thermal variations, or many other factors. Duplicating or simulating this environment becomes a challenge at the component level. Elaborate test systems can be produced to incorporate multiple environments, but in most instances, the basic component design functionality is all that must be evaluated. Developing tests to perform the evaluations typically involves developing a set of fixtures to hold the component and to impart the loading into the part on some type of test stand. The design of fixtures is critical to the repeatability of the overall testing. For a test to be developed, the loading characteristics that the component experiences in its application must be understood.

After the test environment is understood, fixtures developed, and the components manufactured, a test can be performed on the component. The testing is performed either to correlate the output of the FEA model or to validate the component design and manufacturing. Testing that is performed to compare the results from math-based FEA models to real-life test results allows engineers the ability to further develop the capability of the models to predict design concerns. This type of testing typically requires fewer samples and can provide long-term cost benefits to an organization. This iterative process of design analysis and testing ultimately leads to product designs that may require little or no component testing—only testing as part of an assembly to validate the system. Bench simulation testing can significantly improve material selection, save design cost, and shorten the time to production.

Testing performed to validate valvetrain component design and manufacturing requires knowledge of the duty cycle in which the valvetrain component must operate. To develop component-level validation tests, test procedures and methods must be correlated to this duty cycle, and the test stand must be able to duplicate these load inputs. Validation tests require multiple samples to be subjected to duty-cycle loading. Results of these tests then are evaluated, using statistical methods to determine whether they meet the product design requirements. This method of testing can be costly if several redesigns must be done.

4.3.2 Cam and Follower Wear Testing

In many engines, the contact stresses at the interface of the cams and cam followers have reached a critical level beyond which significant wear or a breakdown of the surfaces is likely. This has led to the situation in which the further development of engines of this type is becoming restricted by the limitations of the cam and cam follower. Three fields in which progress might be made toward the relief of these limitations are (1) design, (2) materials, and (3) lubricants. The cam and follower wear rig test is intended to evaluate the cam and follower design, the compatibility of the cam and follower material, and the effectiveness of lubricants and additives. Figure 4.43 schematically shows examples of cam and follower testers. The test rig typically comprises a single lobe that is cam driven by a motor and mounted on roller bearings. The load on the cam follower may be varied by alerting the spring setting.

The preceding type of rig testing can quickly validate the contact stresses calculated based on the Hertzian equation, can rank the material compatibility of the cam and follower, and can evaluate the effectiveness of lubricants and additives. The results can help the designer modify the design, screen the material, and change the lubricant chemistry and viscosity to suit the specific engine requirements without committing significant resources and time in typical engine testing.

4.3.3 Valve Seat and Insert Wear Testing

Valve seat and insert wear has been and continues to present a serious challenge to engine and valvetrain designers. Although new valve materials and processing techniques are being developed constantly, these advances sometimes have been outpaced by the demands for increased engine performance that results in high combustion pressure and temperature, for using alternative fuels such as compressed natural gas (CNG), and extended durability. Excessive valve seat wear

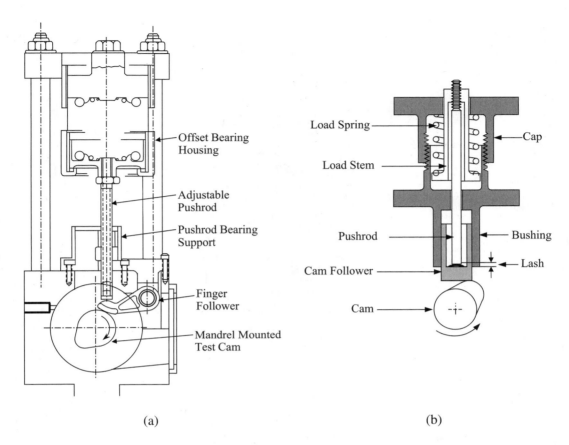

(a)   (b)

Figure 4.43 *Schematic diagram of cam follower wear test machines:*
(a) cam and finger follower wear test rig (Love and Wykes), and
(b) cam and tappet follower wear test rig (Etchells et al.*).*

may result in diminished lash and inadequate seal, and consequentially lost power and valve guttering. Contact stress at the valve seat interface may not be as high as at the cam/follower interface. Furthermore, the harsh environment, including nonlubrication and high temperatures, is to blame for causing greater valve seat wear problems. Most valve seat wear simulators are designed to simulate valve and seat insert wear produced in internal combustion engines in a time and cost-effective manner (Malatesta *et al.*, Lewis *et al.*, Onoda *et al.*, and Wang *et al.*). Figure 4.44 illustrates an example of this type of simulator. Production valves, inserts, and guides are used in the simulation testing. The hydraulic actuator located at the bottom and the return spring located at the top are used to actuate the valve and to apply the load to the valve head to simulate engine combustion pressure. A gas burner is used to heat the valve and the insert; therefore, the simulator does not simulate chemical wear that is seen on field-run valves. This type of simulation test is a viable tool for screening the compatibility of the valve seat and insert material before an engine durability validation test.

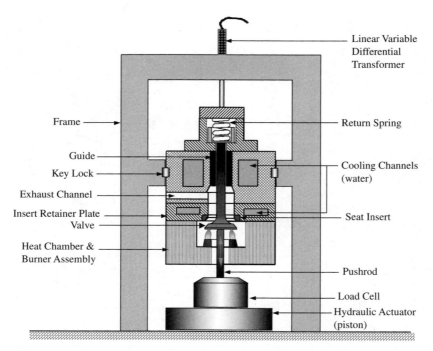

Linear Variable Differential Transformer

Frame

Guide

Key Lock

Exhaust Channel

Insert Retainer Plate

Valve

Heat Chamber & Burner Assembly

Return Spring

Cooling Channels (water)

Seat Insert

Pushrod

Load Cell

Hydraulic Actuator (piston)

Figure 4.44 *Schematic of a valve seat and insert wear test rig (Wang et al.).*

4.3.4 Valve Stem and Guide Wear Testing

Valve stem and guide scuffing have been sporadic but serious problems and continue to challenge engine design engineers. In the past, thanks to powder metallurgy (PM) guide material with solid lubricant and oil impregnation, plus lubricant being fed downward to the stem/guide interface as the valve reciprocates in the guide, guide wear is hardly a problem. However, the drive for reduced oil consumption and exhaust emissions has led to a reduction in the amount of lubricant present in the stem and guide interfaces. These changes have led to a decrease of oil metering downward to the stem/guide interface, and thus an increase in wear of the valve stem and guide, because the correct amount of lubricant is difficult to control, even with a lip seal of metering capability. Most valve stem and guide wear simulators are designed to simulate valve and guide wear produced in internal combustion engines in a time and cost-effective manner (Funabashi *et al.*, Newton *et al.*). An example of this type of simulator, illustrated in Figure 4.45, can be used to simulate the effect of valve stem coatings, surface finishes, guide materials, and side load on guide wear and seizure characteristics. Production valve stems and guides are used in the simulation testing. A side load is applied to the valve stem through a roller bearing. The valve stem and guide are heated electrically to simulate the temperatures required.

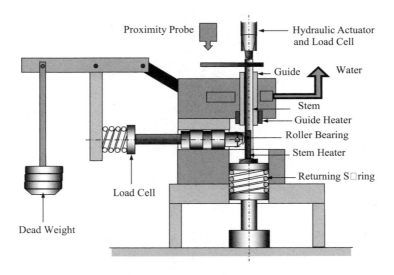

Figure 4.45 *Schematic of a valve stem and guide wear test rig (Rodrigues).*

4.3.5 Flow Testing

Volumetric efficiency at various engine speeds significantly affects engine performance. Many methods (Aoi *et al.*, Heisler), including the one shown in Figure 4.46, have been used in measuring the discharge coefficient and the airflow rate. The rig should be able to evaluate the effect of port shape, valve lift, seat angle, and valve head size and shape on the flow. The isentropic area is the calculated flow area of the equivalent isentropic nozzle, corresponding to the measured mass flow and pressure ratio if the flow is isentropic.

During the initial stages of valve lift, the isentropic and geometric areas are similar, resulting in high discharge coefficients; at larger lifts, the isentropic area falls well below the geometric area to give lower discharge coefficients (Figure 4.47). On a purely geometric basis, the shallow seat angle is superior to the large angles; however, due to the sharper change in flow direction with the 30° seat, it generally has an inferior discharge coefficient. The effect of this is that the 30° seat gives the larger isentropic area at low valve lifts but is inferior to the 45° seat at higher lifts. The 60° seat clearly is inferior to the other two at all practical valve lifts. Thus, the choice of seat angle, from the viewpoint of maximum airflow, is largely one of compromise, although it seems likely that the beneficial effects of the 30° seat during the opening period outweigh its inferior characteristics during the dwell period. SAE J244 describes the flow test standard.

4.3.6 Hydraulic Lifter Leak-Down Testing

In a hydraulic lifter, a very small displacement of the lifter plunger during the valve event, followed by subsequent recovery, is the mechanism by which automatic lash adjustment takes place. The amount of displacement is controlled by the clearance between the plunger and its mating

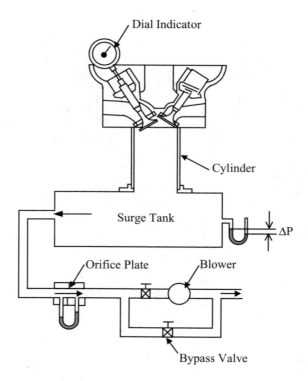

Figure 4.46 *Schematic of a flow test (Aoi et al.).*

surface. Because of the nature of the clearance fit and the accuracy required, normal gaging techniques are not a practical method of specifying this parameter. Instead, the "leak-down test" has been developed over the years to evaluate the time required to displace a predetermined distance. Leak-down time is proportional to land length, inversely proportional to bore diameter, and inversely proportional to pressure in a high-pressure chamber. The leak-down time can have a profound effect on the amount of lash for which the hydraulic lifter can compensate at various engine speeds.

Figure 4.48 shows the principle of a typical leak-down test apparatus to evaluate the leak-down time for hydraulic lifters. It should be emphasized that this test is not a functional test because it does not simulate the engine situation where cyclic impulse loading is applied by the camshaft. The leak-down test is simply a clearance evaluation tool. The variance of the measurement of fluid flow through an annular clearance remains a "statistical" quantity due to geometric tolerance and surface finish variations at the interface of the plunger and lifter body.

Primarily due to the inverse cubic relationship between clearance and leak-down readings, the variances associated with fast readings (high clearance) are much less than those associated with slow readings (low clearance). Similarly, changes in leak-down over time are much less with slow leak-down units. If leak-down time is too short, it may jeopardize the applications; however, if the leak-down time is too long, it could impose challenges to manufacturing. Leak-down

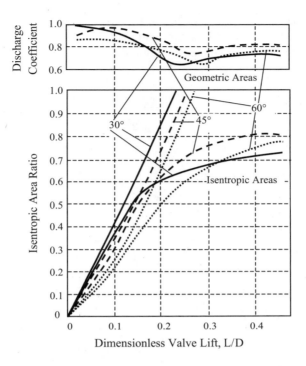

Figure 4.47 *Comparison of flow and discharge characteristics with 30°, 45°, and 60° seat angles (Pope).*

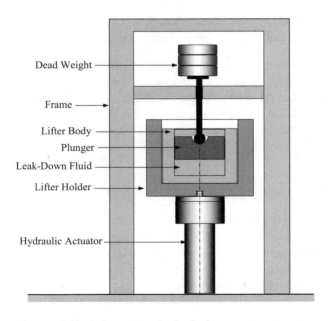

Figure 4.48 *Schematic of a leak-down test apparatus.*

time is a function of load and oil viscosity, in addition to clearance. Figure 4.49 shows a typical acceptable leak-down rate versus load for a hydraulic lifter using SAE 15W-40 oil at 16°C. The results in a specified leak-down range can be tailored for each particular application.

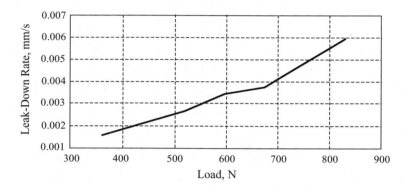

Figure 4.49 *Typical acceptable leak-down rate versus load for a hydraulic lifter at 16°C with SAE 15W-40 oil (Zhao et al.).*

4.3.7 Thermal Shock Testing

One valve failure mode is thermal fatigue or radial cracking. The nuclei for this type of failure originate at the valve seat or head O.D. and appear as one or more radial cracks. The radial cracks eventually can lead to catastrophic failure of chordal fracture or guttering. Thermal fatigue is the most frequent cause of valve breakage in diesel engines.

Figure 4.50 shows a schematic of a valve thermal fatigue test apparatus. The valve is heated alternately by an induction coil and then cooled when it seats on a water-cooled seat. The temperature gradient created by rapid heating and cooling can produce considerable hoop stress and radial cracks on the valve seat. Variations in the time cycle for heating and cooling are desired to produce thermal fatigue in a reasonable amount of time. The relative cracking tendencies of various engine valve materials can be ranked using the rig shown in Figure 4.50 or similar types of rigs.

4.3.8 Stem Seal Oil Metering Testing

The stem seal plays a vital role in engine performance and emissions control. Excessive oil leads to pollution from emissions, and inadequate lubricant metered downward at the stem and guide interface results in stem and guide wear and, more seriously, stem seizure and engine catastrophic failure. Figure 4.51 shows a schematic of a seal rig test apparatus used for screening and validating seal designs and materials.

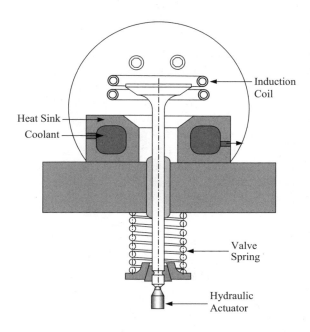

Figure 4.50 *Schematic of a thermal shock test rig (Newton* et al.*).*

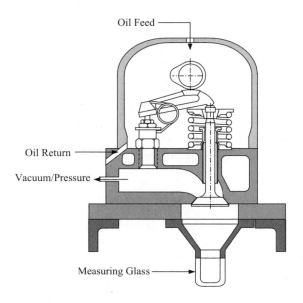

Figure 4.51 *Measurement of oil metering from a stem seal (Netzer and Maus).*

4.3.9 Strain Gage Testing

Finite element analysis (FEA) has been effective in determining the stress distribution of valvetrain components with sufficient input of material information and boundary conditions. However, the stresses at critical locations of the valvetrain components must be validated, and strain gage instrumentation has been widely used. When the Young's modulus of the material is known, stress can be calculated based on the measured strain using Hooke's law. The valve, rocker arm, and pushrod are the valvetrain components that often are strain gaged due to concerns about stress and strain in these components. The stem fillet interface is the location of valve failures that occurred in the field. The maximum stress location on the valve fillet can be determined by strain-gaging the fillet and loading the valve head against its insert seat in a room-temperature fixture. The load simulates the force generated on the valve head by combustion pressure.

4.4 Nondestructive Testing

4.4.1 Overview of Nondestructive Testing

Flaws and cracks in valvetrain components can lead to catastrophic failure of an engine. Therefore, the detection of defects in valvetrain components is an essential part of quality control for the safe and successful use of valvetrain components in practical situations. Nondestructive testing (NDT) also is known as nondestructive evaluation (NDE), nondestructive inspection, or nondestructive characterization. Established NDT methods include x-ray (radiography in general) inspection, ultrasonic inspection, magnetic particle inspection, liquid penetrant inspection, thermography, electrical and magnetic methods, and visual-optical testing. A survey has been conducted by the Institute of Metallurgists of the many different NDT techniques used in the engineering industry. The study found that liquid penetrant and magnetic particle testing accounts for approximately half of all NDT testing. Ultrasonic and x-ray methods account for another third, eddy current testing accounted for approximately 10%, and all other methods accounted for only approximately 2%.

Nondestructive testing examinations are concerned with detecting cracks, tears, imperfect welds, junctions, inclusions, and surface contamination effects in valvetrain components, without altering the piece in any way. Table 4.11 is a simplified breakdown of the complexity and relative requirements of the five most frequently used NDT techniques.

Table 4.12 is a comparison of common NDT methods as judged by the Office of Nondestructive Evaluation, National Institute of Standards and Technology (NIST), in the United States.

Many NDT methods are highly sophisticated, but other techniques are relatively simple. One such method is visual examination with a hand lens. The simple methods are stressed because it is important not to overlook the obvious in examining valvetrain components.

4.4.2 X-Ray Inspection

Figure 4.52 shows a schematic of an x-ray tube used for x-ray NDT inspection. The vacuum is of the order of 10^{-2} Pa. A small voltage is used to pass a current of several amperes through

TABLE 4.11
RELATIVE USES AND MERITS OF VARIOUS NDT METHODS
(THE INSTITUTE OF METALS)

	Test Method				
	Ultrasonics	X-Ray	Eddy Current	Magnetic Particle	Liquid Penetrant
Capital Cost	Medium to high	High	Low to medium	Medium	Low
Consumable Cost	Very low	High	Low	Medium	Medium
Time of Results	Intermediate	Delayed	Intermediate	Short delay	Short delay
Effect of Geometry	Important	Important	Important	Not too important	Not too important
Access Problems	Important	Important	Important	Important	Important
Type of Defect	Internal	Most	External	External	Surface breaking
Sensitivity	High	Medium	High	Low	Low
Formal Record	Expensive	Standard	Expensive	Unusual	Unusual
Operator Skill	High	High	Medium	Low	Low
Operator Training	Important	Important	Important	Important	Important
Training Needs	High	High	Medium	Low	Low
Portability	High	Low	High to medium	High to medium	High
Dependency of Material	Very	Quite	Very	Magnetic only	Little
Ability to Automate	Good	Fair	Good	Fair	Fair
Capabilities	Thickness gaging; some composition testing	Thickness gaging	Thickness gaging; grade sorting	Defects only	Defects only

TABLE 4.12
COMPARISON OF SOME NDT METHODS
(NATIONAL INSTITUTE OF STANDARDS AND TECHNOLOGY)

Method	Characteristics Detected	Advantages	Limitations	Example of Use
Ultrasonic	Changes in acoustic impedance caused by cracks, non-bounds, inclusions, or interfaces	Can penetrate thick materials; excellent for crack detection; can be automated	Normally requires coupling to material either by contact to the surface or immersion in a fluid such as water; surface must be smooth	Adhesive assemblies for bond integrity; laminations; hydrogen cracking
Radiography	Changes in density from voids, inclusions, material variations; placement of internal parts	Can be used to inspect a wide range of materials and thicknesses; versatile; film provides record of inspection	Radiation safety requires precautions; expensive; detection of cracks can be difficult unless perpendicular to x-ray film	Pipeline welds for penetration, inclusions, voids; internal defects in castings
Visual-Optical	Surface characteristics such as finish, scratches, cracks, or color; strain in transparent materials; corrosion	Often convenient; can be automated	Can be applied only to surfaces, through surface openings, or to transparent material	Paper, wood, or metal for surface finish and uniformity
Eddy Current	Changes in electrical conductivity caused by material variations, cracks, voids, or inclusions	Readily automated; moderate cost	Limited to electrically conducting materials; limited penetration depth	Heat exchanger tubes for wall thinning and cracks
Liquid Penetrant	Surface openings due to cracks, porosity, seams, or folds	Inexpensive; easy to use; readily portable; sensitive to small flaws	Flaw must be open to surface; not useful on porous materials or rough surfaces	Turbine blades for surface cracks or porosity; grinding cracks
Magnetic Particles	Leakage magnetic flux caused by surface or near-surface cracks, voids, inclusions, material, or geometry changes	Inexpensive or moderate cost; sensitive both to surface and near-surface flaws	Limited to ferromagnetic material; surface preparation and post-inspection demagnetization may be required	Railroad wheels for cracks; large castings

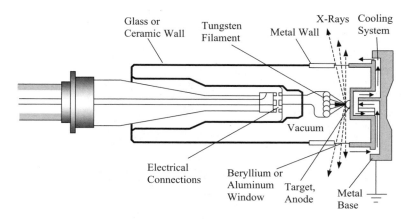

Figure 4.52 *Schematic of an x-ray tube.*

the tungsten filament, which is at negative potential and emits electrons. These electrons are directed by the negatively charged focusing system toward the anode target (anticathode), which usually is at ground potential. The window is made of beryllium or aluminum, with both metals having low x-ray absorption properties. The target must be cooled carefully because nearly 99% of the energy is converted to heat. The target is made of a metal such as tungsten, which has a high melting point, good thermal conductivity, and a high atomic number.

X-rays are generated when an electron beam impinges on a solid target. As the target is cooled, a high percentage of the electrical energy is converted to heat, particularly when generating low-energy x-rays. Electrons from a heated filament are accelerated from a high voltage (20 kV to 20 MV) to the anode target. A window of a low-absorbing foil such as beryllium or aluminum is provided in the vacuum tube to allow the x-rays to exit from the tube. The anode should be of a high-melting-point metal; a typical anode material is tungsten.

As the incident electron loses its energy in the target, x-rays of a broad band of energies are emitted, known as the continuous, white, or *bremsstrahlung* radiation. In radiography, the broad-band spectrum of x-rays is employed. Monochromatic x-radiation, characteristic of the target atoms, also is emitted when the excitation voltage is sufficient to ionize the atoms.

X-rays travel in straight lines and cannot be focused under normal conditions. This determines the principles of x-radiography, which uses a spot source of x-rays or as nearly a point as possible. The size of the source, the focal spot, is important in defining the image, and focal spot sizes are in the range of 1.5–5.0 mm in conventional x-ray tubes. This x-ray source provides a diverging broad beam through the specimen onto a radiographic film, ionization counter, photon converter or counter, or a fluorescent screen. Figure 4.53 illustrates this, and where voids may be present with different orientations, the thickness of the object may vary, as well as its composition and density. The x-ray shadow image should be a faithful reflection of the inhomogeneities of the object. The density levels in the x-ray film are represented by the intensity profile cross section. It is necessary to have access to opposite sides of the specimen. A near-point source of x-rays,

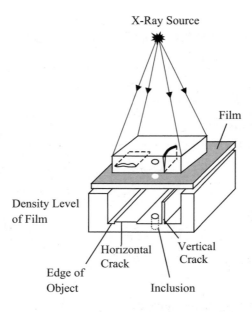

X-Ray Source

Film

Density Level
of Film

Horizontal
Crack

Vertical
Crack

Edge of
Object

Inclusion

Figure 4.53 *Schematic of an x-radiographic system (Cartz).*

from the (tungsten) target of the x-ray tube, diverges through the specimen, projecting a shadow image onto the film. A dense inclusion, a void, and oriented cracks are illustrated, with their resulting effects on the intensity profile of the radiograph. The horizontal crack must provide several change of percent in attenuation to readily be visible on the film.

The attenuation of an x-ray beam by a solid is partly by absorption and partly by scattering; therefore, great care must be taken in selecting x-rays of the appropriate energy to obtain a reasonable and optimum contrast in the image. The excitation voltages typically used in radiography are from approximately 20 kV to 25 MV. High-energy x-rays are required to penetrate great thicknesses of objects, whereas low-energy x-rays provide a different range of contrast in the image. The x-ray tube targets used for radiography are almost always tungsten. X-radiography or radiographic testing standards are described in ASTM E94 and CT-6-6.

4.4.3 Ultrasonic Inspection

Sound travels by the vibrations of the atoms and molecules present, traveling with a velocity that depends on the mechanical properties of the medium. Imperfections and inclusions in solids cause sound waves to be scattered, resulting in echoes, reverberations, and a general damping of the sound waves. When a disturbance occurs at one end of a solid, it travels through the solid in a finite time as a sound wave by the vibrations of the molecules, atoms, or particles present. These vibrations lead to a propagated wave traveling through the medium, with wavelength λ ranging from as long as 10,000 m to a very short λ of 10^{-5} m. Wavelength λ and frequency υ are related to the velocity of the waves by $V = \lambda\upsilon$.

Nondestructive testing via ultrasonic testing is performed using ultrasonic waves of high frequency above the audible range, that is, greater than 20 kHz. Sound waves at frequencies that are greater than 20 kHz also are known as ultrasound or ultrasonics. The normal audible range is approximately 20–20,000 Hz. The velocity (V) of the sound waves depends on the medium and varies from approximately 300–6000 m/s. The range of frequencies used in ultrasonic testing is from less than 0.1 to greater than 15 MHz, and typical values of wavelengths in ultrasonic testing are 1–10 mm. Sounds travel at different velocities through different media; sound velocities vary little with frequency in most metals. The wavelengths, frequencies, and wave velocities used in NDT generally are $\lambda = 1$–10 mm, $\upsilon = 0.1$–15 MHz, and V = 1–10 km/s.

Ultrasonic waves are transmitted through solids over distances of several meters in fine-grained steels, but only approximately 10 cm in some cast irons. Discontinuities or defects cause scattering and reflection of the waves, and the detection of the reflected or transmitted waves permits defects to be located. Figure 4.54 shows various ultrasonic wave NDT arrangements. Figure 4.54(a) shows only one transducer, using pulse echo (PE) techniques (pulse transit time). In Figure 4.54(b), the transmission method, pulsed through-transmission testing, requires two transducers with access to both sides of the specimen. Figure 4.54(c) shows the reflection method, which requires two transducers (pitch-catch) but access to only one side of the specimen. In Figure 4.54(d), an alternative transmission-reflection crack tip diffraction technique compares the time taken by the surface wave to the time taken by waves diffracted from cracks within the specimen. Transmission methods require access to both sides of the specimen, whereas back reflection methods can manage when access is restricted to only one side of the specimen. The coupling transducer-specimen surface must be good. This transmission method is used when small defects are present. The pulsed through-transmission signal is reduced in intensity if small defects are present. Reflections from the far side of the specimen, or from flaws within the specimen, can be picked up by a second transducer. This method of angular transmission permits the selection of the shear mode only to traverse and be reflected by the specimen, as shown in Figures 4.54(c) and 4.54(d). A continuous wave (CW) or a pulse echo (PE) method can be employed. The distance between the transmitter and detector transducers is kept constant with good surface coupling. When a pulse technique is used, an oscilloscope is required to detect the time and amplitude of the transmitted pulse. The angle-beam pulse reflections method (Figure 4.54(d)), is known as crack-tip diffraction. In this method, the time for the surface wave traveling to the receiver is compared to other signals scattered from the far surface or diffracted from a crack as illustrated. The size of the crack can be derived if the scattered (diffracted) beams for the two ends of the crack can be time resolved in the detector system.

The pulse echo method is used extensively. A transducer touches the surface of the specimen through a coupling liquid, sending a pulse of ultrasonic waves traveling through the medium. Alternatively, the specimen may be immersed in water, with the probe immersed as well. The pulse is reflected by a discontinuity or from the rear surface and is detected by the same transducer. The time required for the pulse to travel out and back can be displayed on an oscilloscope, and it is common to use a regular train of ultrasonic pulses so that the oscilloscope signal is more easily observed. The initial pulse may last 1 μs (1–10 cycles of vibration), and the pulses repeat every millisecond. Table 4.13 lists the velocities, V, and wavelengths, λ, of ultrasonic waves used in NDT for several materials. The waves are nondispersive, that is, the velocities are virtually independent of frequency, υ, for the range 0.1–15 MHz used in NDT for water, oil,

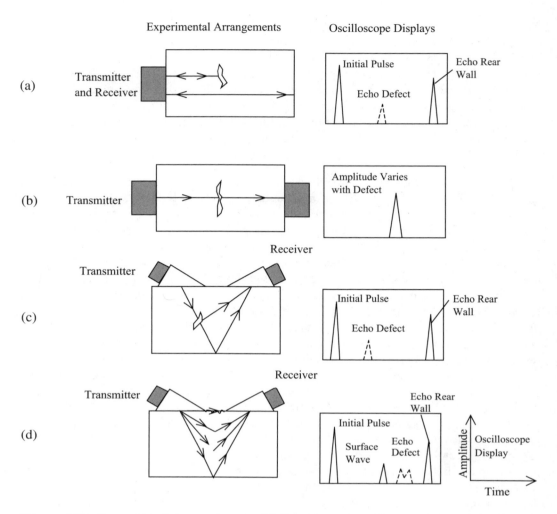

Figure 4.54 *Examples of ultrasonic wave NDT arrangements (Cartz): (a) one transducer, pulse echo (PE) techniques; (b) pulsed-through transmission testing, two transducers with access to both sides of the specimen; (c) reflection method, two transducers and access to one side of the specimen; and (d) alternative transmission-reflection crack tip diffraction technique.*

and most metals. Defect lengths as small as $\frac{\lambda}{4}$ can be detected by ultrasonic testing, although this depends on the orientation of the crack to the beam. Lengths that can be measured depend on the ultrasonic beam diameter.

Standards for ultrasonic testing include ASTM E114 and CT-6-4.

TABLE 4.13
WAVELENGTHS OF ULTRASONIC WAVES
USED IN NDT (CARTZ)

Material	Longitudinal Wave Velocities (km/s)	Wavelength (mm) (1 MHz)
Air	0.33	0.33
Water (20°)	1.49	1.49
Oil (Transformer)	1.38	1.38
Aluminum	6.35	6.35
Copper	4.66	4.66
Magnesium	5.79	5.79
Steel (Mild)	5.85	5.85
Polyethylene	1.95–2.40	1.95–2.40
Lucite/Plexiglass	2.67	2.67

4.4.4 Liquid Penetrant Inspection

Liquid penetrant inspection (LPI) is a simple but effective method of examining surface areas for cracks, defects, or discontinuities and is used extensively. Several delicate stages of preparation involve precleaning of the inspection surface, application of the penetrant, observation of a dwell time to allow the penetrant to seep into flaws, removal of excess penetrant, and application of a developer, again with a dwell time to allow the penetrant to seep out of any surface flaws to form visible indications. Figure 4.55 shows the actions of the penetrant and the developer. When the penetrant liquid is applied, the liquid is drawn downward into the open crack by capillary action. Excess surface penetrant is removed. This can be conducted by wiping the area with a cloth. Developer powder on the surface draws out the penetrant liquid that seeps into and discolors the powder. A colored dye or a fluorescence compound usually is added to the penetrant liquid. The liquid should not seep too far from the crack.

The finest cracks that can be detected using liquid penetrant inspection have been estimated to be approximately 5 μm wide by 10 μm deep. The sensitivity of crack detection depends on the wetting of the solid by the penetrant, the geometry of the crack, and the cleanliness of the crack. Table 4.14 shows two examples of commercial liquid penetrant systems.

Standards for liquid penetrant inspection include ASTM E165 and CT-6-2.

4.4.5 Magnetic Particle Inspection

Iron filings have long been used to display the magnetic field of a bar magnet, and Figure 4.56 shows typical patterns of filings around bar magnets. A magnetic line of force is an imaginary

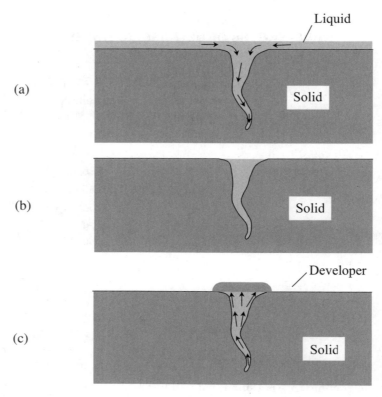

(a)

(b)

(c)

Figure 4.55 Actions of penetrant and developers: (a) liquid applied,
(b) liquid removed, and (c) developer applied.

TABLE 4.14
PROPERTIES AND METHODS OF USE OF A
PENETRANT/DEVELOPER SYSTEM (BETZ, 1963)

Trade Name	Spotcheck	Zyglo
Manufacturer	Magnaflux	Magnaflux
Penetrant Type	Liquid red dye	Fluorescent liquid
Applied by	Dip, burrs, or spray can	Dip, brush, or spray can
Temperature Range	Above 50°F	Above 50°F
Penetration Time	3–20 minutes	3–20 minutes
Removed by	Solvent-based cleaner (some by water)	Water or solvent based
Developer	Volatile liquid suspension of white powder	Powder or liquid
	White light	
Indication Visibly by	None	Near-ultraviolet light
Auxiliary Apparatus		Sources of water and electric power

line whose direction at each point gives the direction of the magnetic field at that point. The magnetic lines of force of a bar magnet can be plotted using a small compass or can be displayed by using a fine powder of iron filings. Iron filing plots can be prepared by spreading wax paper over a magnet and sprinkling iron filings over the paper. The filings become magnetized and link to form chains along the magnetic lines of force, particularly if the paper is vibrated gently to enable the filings to undergo small displacements. By mildly heating the paper to soften the wax and then cooling it, a permanent record can be obtained. If a bar magnet is imperfect, cracked, or broken into two parts, each part will act as a separate magnet with north-south poles so that opposite the two broken ends, or across a crack, the lines of force will be disturbed. Figure 4.56(a) shows filings around an individual bar magnet. The north pole is by convention the end of the magnet that lines up in the direction of the North Pole of the Earth. Figures 4.56(b) and 4.56(c) show two bar magnets arranged either with north-south poles adjacent or with north poles adjacent, respectively. Figure 4.56(d) shows that a crack in a magnet or magnetized part will give rise to north-south poles on two sides of the crack.

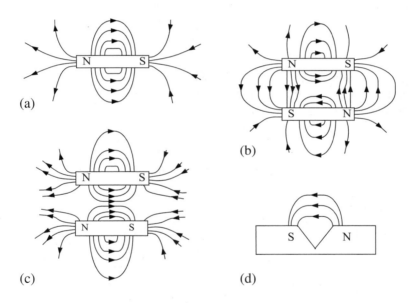

Figure 4.56 *Magnetic lines of force (Cartz): (a) filings around an individual bar magnet; (b) two bar magnets arranged with north-south poles adjacent; (c) two bar magnets with north poles adjacent and south poles adjacent; and (d) crack in a magnet, giving rise to north-south poles on two sides of the crack.*

The magnetic lines of force are highly distorted across the crack; this is known as the magnetic leakage field. When the direction of magnetization is perpendicular to the flaw, as in Case A of Figure 4.57, the magnetic lines of force are distorted the most so that the probability of detection of the flaw is greatest. Flaws such as Case B, parallel to the magnetic field direction, do not result in any magnetic leakage field and therefore are not detectable with the magnetic particle (MP) inspection method.

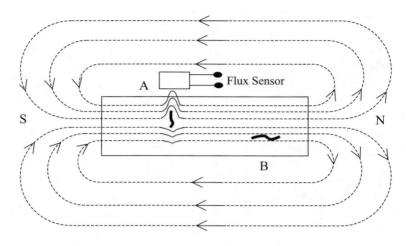

*Figure 4.57 Magnetic leakage field due to defects perpendicular
to the magnetic lines of force (Cartz).*

One method of magnetizing the work piece is by the use of a yoke, which is a U-shaped electro-magnet or permanent magnet (Figure 4.58). The magnetic lines of force are shown across the surface perpendicular to a weld direction and below the surface perpendicular to a surface crack. This method has no electrical contact, is highly portable, and can locate discontinuities in any direction.

Standards for magnetic particle inspection are shown in ASTM E125, E709, CT-6-3, and Betz (1967).

4.4.6 Eddy Current Inspection

A varying electric current flowing in a coil gives rise to a varying magnetic field. A nearby conductor resists the effect of the varying magnetic field, and this manifests itself by an eddy current flowing in a closed loop in the surface layer of the conductor to oppose the change, causing a back electromotive force (emf) in the coil (Figures 4.59(a) and 4.59(b)). Cracks and other surface conditions modify the eddy currents generated in the conductor so that the back emf is altered (Figure 4.59(c)). Eddy currents also are known as Foucault currents or induced currents and can exist only in conducting materials.

An eddy current inspection system consists of the following:

1. A source of varying magnetic field, such as a coil carrying an alternating current of frequencies ranging from less than 1 kHz to greater than 10 MHz. A pulsed source also may be used.

2. A sensor to detect minute changes in the magnetic field (~0.01%), such as an inspection coil or Hall gaussmeter.

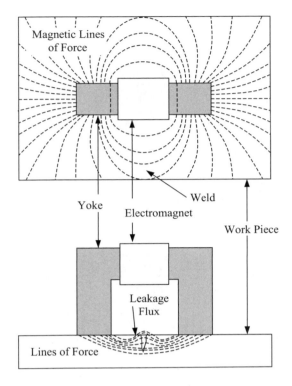

Figure 4.58 *A yoke is a U-shaped electromagnet or a permanent magnet used to induce a magnetic field in the work piece (Cartz).*

3. Electronic circuitry to aid the interpretation of the magnetic field change.

The inspection system can be manual or fully automated. For a flat specimen, a pancake-type inspection coil is used (Figure 4.59(a)). Tubes and cylinders are examined using solenoid coils, as shown in Figures 4.59(b) and 4.59(c). The solenoid can be located on either the outside or inside of the tube.

An ac source is applied to an inspection excitation coil so that the magnetic lines of force penetrate into the specimen-conducting surface, providing good magnetic coupling. Eddy currents circulate in the specimen surface and are modified by the presence of discontinuities (e.g., surface cracks). Eddy currents flow parallel to the plane of the windings of the coil. The detector instrument will note changes in the time variation of the voltage and current in the inspection coil (Figure 4.59(d)).

Alternating currents and eddy currents travel along the surface of conductors and penetrate very little into the specimen. Figure 4.60 illustrates the decrease with depth of the electric current, where the drop-off with depth is exponential. The standard depth of penetration δ is where the electric current has decreased by the inverse of the exponential factor.

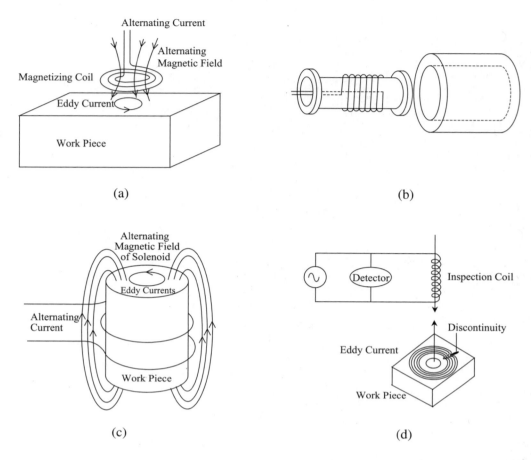

Figure 4.59 *Schematic of eddy current testing of a specimen and coil configurations: (a) a flat surface normally is examined by a flat pancake-type coil; (b) the interior of a tube can be examined by an inside, inserted, or bobbin coil; (c) instantaneous view of an alternating current in the testing coil, producing a magnetic field; and (d) the magnetic field from the inspection coil is opposed by an induced magnetic field from the eddy current. A surface crack modifies the eddy current and hence the induced magnetic field.*

The ac frequencies used in eddy current testing range from a few kilohertz (kHz) to more than 5 MHz. At lower frequencies, the penetration δ is relatively high, but the sensitivity to the detection of discontinuities is relatively low; the reverse is true at high frequencies. In the case of ferromagnetic materials where δ is reduced, the use of low frequencies is inevitable. Figure 4.61 plots values of δ versus frequency.

Eddy current testing standards are described in ASTM E309 and CT-6-5.

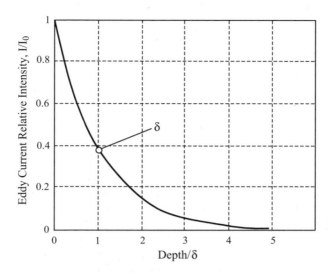

Figure 4.60 *Depth of penetration (CT-6-5).*

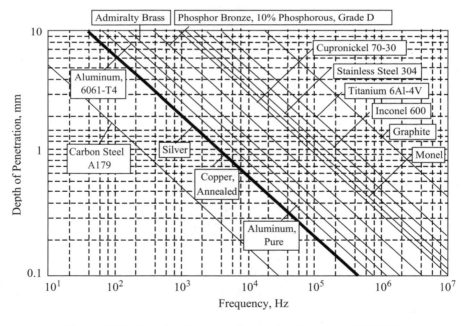

Figure 4.61 *Depth of penetration versus frequency (CT-6-5).*

4.5 Engine Testing

4.5.1 Valvetrain System Testing Overview

The motion of a valvetrain system significantly affects engine performance, durability, and noise levels. The design of a valvetrain system with the parameters such as camshaft maximum acceleration and positive acceleration pulse typically should result in a system that approaches acceptability. However, the engine can produce operating conditions that cannot be foreseen in the design stage. This is particularly true as the engine speed and performance requirements increase. As the levels of camshaft acceleration, maximum lift, and so forth are increased to maximize engine performance, the danger of introducing unacceptable vibration into the valve-train system is increased. Advanced valvetrain dynamic analysis software has made the prediction of valvetrain motion over the engine speed range, total noise estimate, deflection, natural frequency, and component load (strain or stress) more accurate. It also reduces the number of dynamometer tests and avoids the expense of costly engine dynamometer failures or vehicle failures. However, validation of the valvetrain design, performance, durability, and reliability as a system in a fired engine or motorized fixture testing is still required.

4.5.2 Valvetrain Dynamic Testing

One objective of valvetrain dynamic testing and analysis is to measure the valvetrain displacement, velocity, or acceleration to determine how far they deviate from the design profiles as the cam rotates in a running engine. Engine valvetrain dynamics typically are evaluated by using a strain gage in the rocker arm or other valvetrain components or by using noncontacting proximity probes to measure displacement, as shown in Figure 4.62. The velocity and acceleration

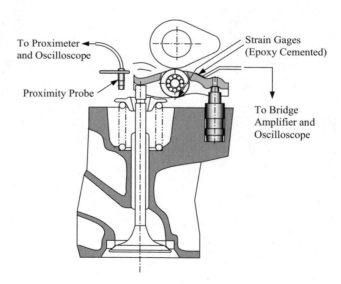

Figure 4.62 *Valvetrain instrumentation strain gage and proximeter to measure valvetrain displacement, deflection, and load.*

then must be differentiated from the displacement. Sometimes, velocity and/or acceleration at various points in the valvetrain can be measured directly, and then acceleration can be obtained from the differentiation of velocity or displacement from the integration of velocity, respectively. A piezoelectric accelerometer can be attached to the rocker arm to obtain acceleration measurement in a running engine.

4.5.2.1 Valvetrain Deflection Testing

Valvetrain deflection affects the valvetrain dynamics and consequently engine performance. Figure 4.63 shows examples of camshaft deflection measured between the end of the intake camshaft and the last intake cam of a double overhead cam (DOHC) engine, the location where the maximum camshaft deflection is expected because no bearing restricts the camshaft movement in the radial direction. Figure 4.63(a) shows the camshaft deflection close to idle speed. Here, at the beginning of the event, the reaction force caused by the valvetrain deflects the camshaft, and the deflection increases during the event. The maximum value appears in the region on top of the cam, corresponding to the point of the maximum spring forces. At 6000 rpm (Figure 4.63(b)), the dynamic effects play a dominant role. The first major peak corresponding to the first acceleration peak at the beginning of lift initiates a damped oscillation of the camshaft. At 90°, the oscillation almost dies out. Then, the closing acceleration peak at the end of the event excites a new oscillation.

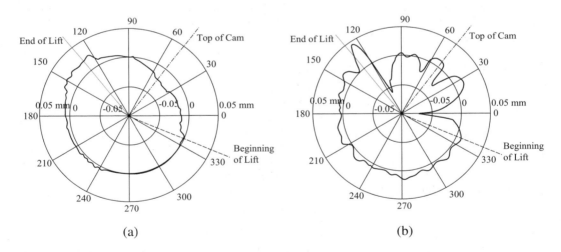

Figure 4.63 *Measured camshaft deflection (Schamel et al.): (a) at 1000-rpm engine speed, and (b) at 6000-rpm engine speed.*

Figure 4.64 shows an example of the deflection of valvetrain components on a typical Type V (pushrod) engine. Note that most of the deflection occurs in the pushrod and rocker arm, indicating that the stiffness of these components is low.

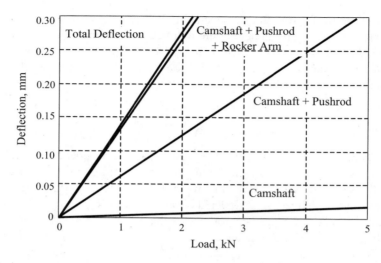

Figure 4.64 *Valvetrain deflection on a typical Type V engine.*

Any large deflection in the valvetrain occurring in the opening phase will affect the valve motion throughout the event length. Because applying acceleration loads at this point causes the valvetrain to vibrate, the higher deflection causes a larger departure of the valve motion from the theoretical over the cam contour. This vibration occurs all around the valve lift and at closing results in a large deflection. This large deflection occurs because of the previously imposed vibration plus the closing acceleration load. At certain resonant speeds, this deflection can cause a severe valve bounce condition. This large bounce causes high valve seating loads that result in excessive noise and valve breakage.

4.5.2.2 Dynamic Stress Analysis

In addition to deflection, displacement, velocity, and acceleration of valvetrains, the force measurement using a load cell or strain gage is another important parameter in valvetrain dynamic analysis. Valvetrain "dynamic performance" typically is defined as the level of strain generated in the critical locations of the valvetrain components because these points are the main locations of failures.

Strain gages usually are placed in critical locations such as in the valve fillet area for testing. Strain gage testing of valve springs during actual engine operation has shown that the dynamic stress may be amplified over the engine revolutions-per-minute range at certain resonant conditions. Consequently, in the design of valve springs, the dynamic stress must be measured to ensure that it is within acceptable limits. The high-temperature strain gages for exhaust valves are operated within a standard potentiometric circuit. A low-pass filter is used to eliminate the large zero shifts, which are due to the dc voltage created by the thermocouple and thermistor effects between the nichrome gage wire and the chromel-alumel gage leads (Worthen and Rauen). As a consequence of this filtering action, only dynamic strains are registered (i.e., above a frequency of 5 Hz). Any static strains, such as valve spring preload and distortion of the valve due to thermal expansion, will not be registered. The strains seen by a closed valve during combustion consist of thermal effects and pressure loading. The thermal loading consists of a steady-state portion induced by the temperature gradients on the valve. Firing of the engine produces a thermal cyclical load as the skin, generally estimated at less than 0.5 mm, is heated and then cooled. The cycle effects may explain some of the large once-per-event strains that are seen.

The strain field on the fillet is biaxial, so the principal stresses are defined by well-known equations as

$$\sigma_r = \frac{E(\varepsilon_r + \nu\varepsilon_h)}{1 - \nu^2} \tag{4.46}$$

$$\sigma_h = \frac{E(\varepsilon_h + \nu\varepsilon_r)}{1 - \nu^2} \tag{4.47}$$

where

σ_r = near radial stress
σ_h = hoop stress
E = Young's modulus
ε_r = near radial strain
ε_h = hoop strain
ν = Poisson's ratio

An example of the firing engine hoop and radial strain gage data at the maximum stress location with the engine operating at maximum speed and load is plotted versus the cam angle in Figure 4.65. The data also clearly show that the strains from the pressure spike are of approximately 80 crank degrees duration, which corresponds to the combustion portion of this engine cycle.

Inspection of the data in Figure 4.65 shows relatively large combustion strains, whereas the closing and opening loads show as being much lower but identifiable strain spikes on the valve fillet. The hoop portion of the combustion stress appears as a fairly wide spike of 200 microstrains on top of an oscillating signal with an amplitude of 50 microstrains. The radial strain is a well-defined spike of 250 microstrains at peak combustion pressure, coinciding with engine firing. The spikes are the combustion pressure strain on the valve and are independent of valve closing.

4.5.2.3 Valvetrain Noise and Vibration

In addition to valve motion, noise generation is another noise, vibration, and harshness (NVH) criterion of engine performance. Valvetrain noise can produce a large part of the total mechanical noise of an engine. Valvetrain noise typically is generated at the point of valve opening and valve closing. The noise that occurs at valve opening is caused by cam impact loads applied to the valvetrain. The noise that occurs at valve closing results from the valve impacting the seat insert. Proper design of the camshaft ramps and control of valvetrain vibration will prevent

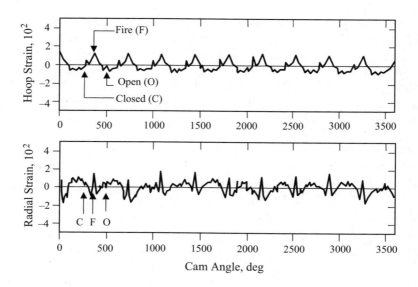

Figure 4.65 *Hoop and radial strain or stress*
at the valve fillet (Worthen and Rauen).

unacceptable valvetrain noise. Figure 4.66 is an example of oscilloscope signals showing valve lift and valvetrain noise. The natural frequency of the valvetrain system can be obtained by observation of the rocker arm stress waves.

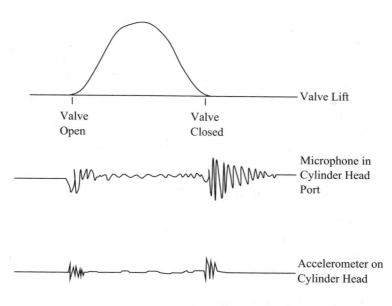

Figure 4.66 *Example of noise signal.*

4.5.2.4 Valvetrain Jump and Bounce

The limitation or improvement in engine performance lies in the ability of higher valve lift and rotation limit. However, components such as the hydraulic lifter and roller follower tend to raise inertia and reduce valvetrain stiffness, thus lowering the revolutional limit. Higher valve lift and cam revolutions per minute can result in the valvetrain jumping and bouncing, which can lead to valve breakage. The primary causes of jump and bounce are valvetrain stiffness and close-side characteristics of the valve lift curve. Because of flexibility in the valvetrain parts, a deviation will exist between the actual and desired valve lift curve. If the valve seats before the closing ramp is reached, high impact velocity and valve bounce will occur.

The valve lift curve deviates from the cam lobe profile at high engine speeds. This phenomenon is called "jump." This abnormal behavior tends to occur when the inertia force working on the valve exceeds the valve spring force with null load between the elements. Figure 4.67 shows the valve lift and the corresponding rocker arm force and cam force at 6200 rpm. The data shown

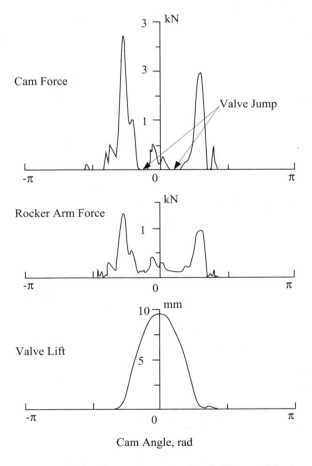

Figure 4.67 *Typical valvetrain motion at 6200 rpm (Kurisu et al.).*

also provide information on the magnitude of delayed valve opening, the maximum level of deflection in the valvetrain, and the magnitude of valve bounce (or valve seating impact) resulting from the valvetrain deflection.

When the speed exceeds a certain limit for a specified valvetrain, a clearance develops between the cam and follower. In the low-speed range, the load increases linearly; however, it rises sharply when it exceeds a certain speed limit, indicating that jump occurs (Figure 4.68).

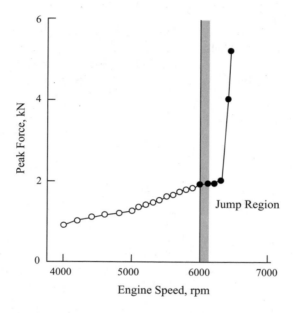

Figure 4.68 *Peak load on the cam (Kurisu et al.).*

The valve jump lasts longer with increased engine speed, and the location of valve return to follow the cam profile reaches the closed-side positive acceleration region. As demonstrated in Figure 4.69, the location of the closed-side cam peak force appears in the closed-side positive acceleration region. When the jump occurs, the valve results in clashing against the cam with high impulse force, and breakage of follower components is likely to occur.

Valve bounce is the term that describes the motion as the valve touches and rebounds from the valve seat. Figure 4.70 shows a typical valve motion and valve seat force in quasi- and real bouncing situations. In a quasi-bouncing situation, the dynamic valve lift at the rebound position is not seen to have reached zero with no impulse force on the valve seat present. It is considered that the valve rebounds by dint of the closed-side force produced by deflection energy of the follower without touching the valve seat. However, in a situation with valve bouncing, it is seen that the impulse forces were generated at both the bounce and return positions. It is assessed

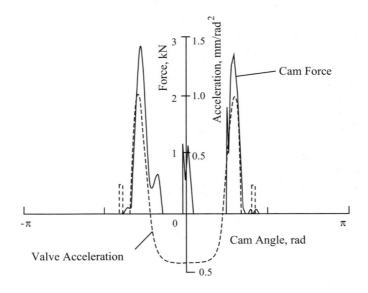

Figure 4.69 *Comparison of valve acceleration and cam force profile (Kurisu et al.).*

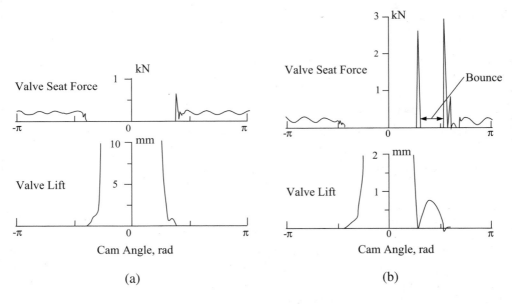

Figure 4.70 *Valve motion and seat force: (a) in quasi-bouncing, and (b) in bouncing (Kurisu et al.).*

that the generation of the force at the bounce position is due to an inability to absorb dynamic energy by too large a deflection during the deceleration period, thereby causing the valve to impact against the valve seat at high velocity.

A properly designed valvetrain system has a camshaft with a closing ramp that allows valve seating at a low velocity. However, it is obvious that any significant deflection in the system results in valve seating on the lift curve well above the normal closing ramp. At this point, the valve is moving at a much greater velocity, and impact with the seat results in severe impact loads.

Most problems related to high valve bounce occur because of excessive valvetrain deflection, but steps can be taken to increase the valvetrain stiffness (frequency), reduce the component weights, or reduce the acceleration forces generating the deflection by modifying the cam closing ramp. A reduction in the acceleration forces can be accomplished at any revolutions per minute by reducing the opening and the closing accelerations, and by increasing the open positive acceleration pulse. Figure 4.71 shows the effect of cam speed and valve lash on the impulse forces.

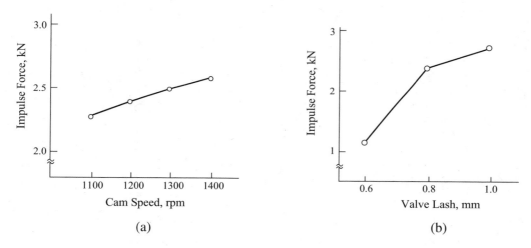

Figure 4.71 *Factors affecting valve seating or impulse force: (a) cam revolutions-per-minute effect, and (b) valve lash effect at 1200-rpm cam speed (Akiba and Kakiuchi).*

Poor valvetrain dynamics could result in valvetrain component failures, including valve seat recession, valve seat pound-in, and valve head breakage. When a valve head breaks off, it falls into the combustion chamber and severely damages or destroys the engine (i.e., damaging the cylinder head, cylinder block, and pistons). Impulse force or peak seat force affects valve seat wear significantly (Figure 4.72).

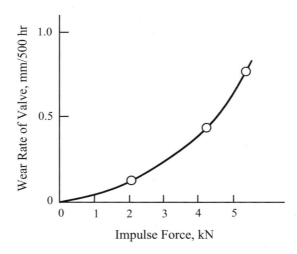

Figure 4.72 *Effect of valve seating or impulse force on valve seat wear (Akiba and Kakiuchi).*

4.5.2.5 Valve Seat Velocity

An additional measure of valvetrain performance, which is the valvetrain natural frequency (VTNF), influences valve closing dynamics. Because the VTNF may vary from cylinder to cylinder, strain-gaged rocker arms often are used to detect the VTNF.

The strain data of the rocker arms then are examined with a Fourier analyzer for their spectral content. A revolutions-per-minute spectral map reveals the VTNF of each cylinder (Worthen and Rauen).

4.5.3 Lash Measurement

The amount of lash can affect valve sealing, impulse force, valvetrain noise, and overall engine performance. Figure 4.73 shows one method of making direct measurements. The motion is indicated by two strain gages fitted on the top and bottom faces of the cantilever spring to provide a half bridge with both arms active.

Figure 4.74 shows plots for the intake and exhaust clearance at three loads over the speed range of a moderately high-speed diesel engine. Note that the exhaust valve lash decreases slightly with load and speed, whereas the intake valve lash increases. Presumably, the intake valve head ran hotter than the stem, so that the head grew in diameter more than the valve stem length extended. The result was that the whole valve moved down the minor seat, with a consequent increase in lash.

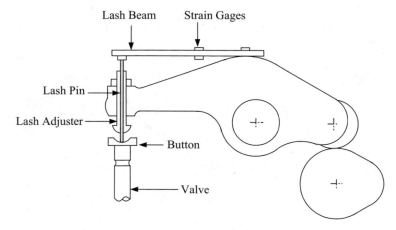

Figure 4.73 *Schematic of a lash measurement technique (Beard and Hempson).*

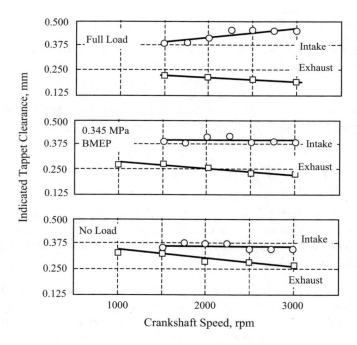

Figure 4.74 *Plots of valve lash determined for a diesel engine (Beard and Hempson).*

4.5.4 Valve Temperature Measurement

Typically, valve temperature increases as the engine speed and torque increase. Figure 4.75 shows variations of valve temperatures with torque.

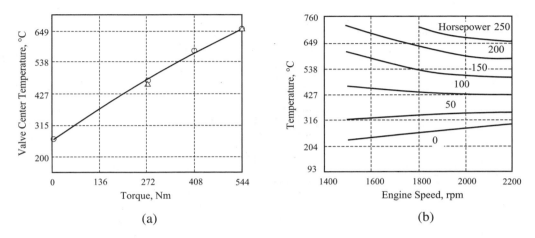

Figure 4.75 *Exhaust valve temperature: (a) effect of torque (Worthen and Rauen), and (b) effect of engine speed (Johnson and Galen).*

The hottest temperature of a valve that is run in an engine can be determined by using temperature-sensitive valves; this is known as the temperature check method. The temperature-sensitive valves, typically made of martensitic steels such Sil 1, are hardened before the test and are tempered as the engine is run for one hour at the speed and load point to be evaluated. Comparisons with test samples that have been placed in the furnace for one hour at different temperatures to allow the valve temperature profiles to be determined and temperature levels to be identified. The engine, with a full set of temperature-sensitive valves, then is run for one hour at maximum power to define the hottest cylinder. Figure 4.76 shows typical results of this test. The temperature-check technique is used to evaluate the hottest temperature that occurred during operation.

Valve and insert temperatures can be measured directly using the thermocoupling method, which allows their temperatures to be evaluated over a full range of engine operating conditions. For valves, stainless steel wires for each thermocouple are run down the valve stem O.D. in a specially machined groove until they are below the guide region. Then, the thermocouples are welded in the fillet, near the seat, or at any desired locations. For the valve combustion face, a hole is drilled through the valve head in the fillet side, and a thermocouple is welded at the valve head approximately 1.5 mm from the combustion side surface. Thermocouple wires for inserts are covered with stainless steel sheet in the port area to protect the thermocouples from being destroyed. Outside the exhaust port and above the spring retainer, the stainless steel wires are joined to more flexible fiberglass-insulated wires. The wires then are led to and fixed at a steel

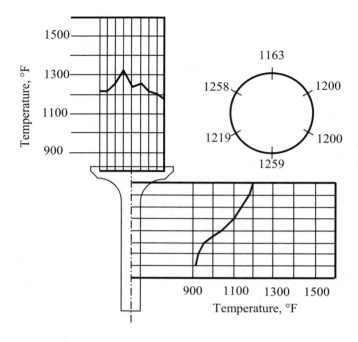

Figure 4.76 *Temperature distribution for the temperature check method (Worthen and Rauen).*

plate to minimize wire motion and vibration in order to obtain adequate durability (Worthen and Rauen).

A thermocouple readout of temperature versus time can be recorded. The engine must be operated at each condition until steady-state temperatures are reached. Depending on the previous test point, the time to achieve stable operation can vary. Because wire durability is short, it is desirable to move to the next point as soon as stable operation is achieved. The advantage offered by the thermocouple method is the ability to map the temperature envelopes of an engine operation.

4.5.5 Valvetrain Friction Measurement

Valvetrain friction fixture testing can help engine designers to design valvetrain systems achieving higher engine fuel efficiency together with an acceptable emissions level by reducing valvetrain friction. A fixture can use either a rotary torque transducer directly coupled to the camshaft (Armstrong and Buuck), a torque transducer and slip-rings to read the torque acting on the camshafts directly (Ball *et al.*), or an instrumented sprocket (Baniasad and Emes) to measure the valvetrain friction of various cylinder heads. The effect of lubricant, engine speed, valvetrain type, roller follower, spring rate, preset load, and so forth on friction loss in the valvetrain can be evaluated using this type of fixture.

4.6 References

Akiba, K. and Kakiuchi, T., "A Dynamic Study of Engine Valving Mechanism: Determination of the Impulse Force Acting on the Valve," SAE Paper No. 880389, Society of Automotive Engineers, Warrendale, PA, 1988.

Antolovich, S. and Saxena, A., *Fatigue Failures, ASM Handbook, Vol. 11, Failure Analysis and Prevention,* ASM International, Materials Park, OH, 2000.

Aoi, K., Nomura, K., and Matsuzaka, H., "Optimization of Multi-Valve, Four Cycle Engine Design—The Benefit of Five-Valve Technology," SAE Paper No. 860032, Society of Automotive Engineers, Warrendale, PA, 1986.

Armstrong, W. and Buuck, B., "Valve Gear Energy Consumption: Effect of Design and Operation Parameters," SAE Paper No. 810787, Society of Automotive Engineers, Warrendale, PA, 1981.

ASTM A285-03, "Standard Specification for Pressure Vessel Plates, Carbon Steel, Low and Intermediate, Tensile Strength," ASTM International, West Conshohocken, PA, 2003.

ASTM A956, "Standard Test Method for Leeb Hardness Testing of Steel Products," ASTM International, West Conshohocken, PA, 2002.

ASTM C1326, "Standard Test Method for Knoop Indentation Hardness of Advanced Ceramics," ASTM International, West Conshohocken, PA, 1999.

ASTM D2670, "Standard Test Method for Measuring Wear Properties of Fluid Lubricants," ASTM International, West Conshohocken, PA, 1999.

ASTM D4172, "Standard Test Method for Wear Preventive Characteristics of Lubrication Fluid," ASTM International, West Conshohocken, PA, 1999.

ASTM E10, "Standard Test Method for Brinell Hardness of Metallic Materials," ASTM International, West Conshohocken, PA, 2001.

ASTM E18, "Standard Test Method for Rockwell Hardness and Rockwell Superficial Hardness of Metallic Materials," ASTM International, West Conshohocken, PA, 2003.

ASTM E23, "Standard Test Method for Notched Bar Impact Testing for Metallic Materials," ASTM International, West Conshohocken, PA, 2002.

ASTM E92, "Standard Test Method for Vickers Hardness of Metallic Materials," ASTM International, West Conshohocken, PA, 2003.

ASTM E94, "Radiographic Testing," ASTM International, West Conshohocken, PA, 2000.

ASTM E111, "Standard Test Method for Young's Modulus, Tangent Modulus, and Chord Modulus," ASTM International, West Conshohocken, PA, 1997.

ASTM E114, "Ultrasonic Pulse-Echo Straight-Beam Examination by the Contact Method," ASTM International, West Conshohocken, PA, 2001.

ASTM E125, "Magnetic Particle Indications of Ferrous Castings," ASTM International, West Conshohocken, PA, 2003.

ASTM E140, "Standard Hardness Conversion Tables for Metals," ASTM International, West Conshohocken, PA, 2002.

ASTM E165, "Liquid Penetrant Examination," ASTM International, West Conshohocken, PA, 2002.

ASTM E208, "Standard Test Method for Conducting Drop-Weight Testing to Determine Nil-Ductility Transition Temperature of Ferritic Steels," ASTM International, West Conshohocken, PA, 2000.

ASTM E238, "Standard Test Method for Pin-Type Bearing Test of Metallic Materials," ASTM International, West Conshohocken, PA, 2002.

ASTM E290, "Standard Test Method of Material for Ductility," ASTM International, West Conshohocken, PA, 1997.

ASTM E309, "Eddy Current Examination of Steel Tubular Products Using Magnetic Saturation," ASTM International, West Conshohocken, PA, 2001.

ASTM E399, "Standard Test Method for Plane-Strain Fracture Toughness of Metallic Materials," ASTM International, West Conshohocken, PA, 1997.

ASTM E448, "Standard Practice for Scleroscope Hardness Testing for Metallic Materials," ASTM International, West Conshohocken, PA, 2002.

ASTM E466, "Conducting Force Controlled Constant Amplitude Axial Fatigue Tests of Metallic Materials," ASTM International, West Conshohocken, PA, 2002.

ASTM E606, "Strain Controlled Fatigue Testing," ASTM International, West Conshohocken, PA, 1998.

ASTM E647, "Standard Practice for Measurement of Fatigue Crack Growth Rates," ASTM International, West Conshohocken, PA, 2000.

ASTM E709, "Standard Guide for Magnetic Particle Examination," ASTM International, West Conshohocken, PA, 2001.

ASTM E1820, "Standard Test Method for Measuring Fracture Toughness," ASTM International, West Conshohocken, PA, 2001.

ASTM F519, "Standard Test Method for Mechanical Hydrogen Embrittlement Evaluation of Plating Process and Service Environments," ASTM International, West Conshohocken, PA, 1997.

ASTM F1801, "Standard Practice for Corrosion Fatigue Testing of Metallic Implant Materials," ASTM International, West Conshohocken, PA, 1997.

ASTM G4, "Standard Guide for Conducting Corrosion Coupon Test in Field Application," ASTM International, West Conshohocken, PA, 2001.

ASTM G30, "Standard Practice for Making and Using U-Bend Stress Corrosion Test Specimens," ASTM International, West Conshohocken, PA, 1997.

ASTM G35, "Standard Practice for Determining the Susceptibility of Stainless Steels and Related Nickel-Chromim-Iron Alloys to Stress-Corrosion Cracking in Polythionic Acids," ASTM International, West Conshohocken, PA, 1998.

ASTM G38, "Standard Practice for Making and Using C-Ring Stress Corrosion Test Specimens," ASTM International, West Conshohocken, PA, 2001.

ASTM G41, "Standard Practice for Determining Cracking Susceptibility of Metals Exposed Under Stress to a Hot Salt Environment," ASTM International, West Conshohocken, PA, 2000.

ASTM G54, "Standard Practice for Simple Static Oxidation Testing," ASTM International, West Conshohocken, PA, 1996.

ASTM G65, "Standard Test Method for Measuring Abrasion Using the Dry Sand/Rubber Wheel Apparatus," ASTM International, West Conshohocken, PA, 2000.

ASTM G77, "Standard Test Method for Ranking Resistance of Materials to Sliding Wear, Using Block-on-Ring Test," ASTM International, West Conshohocken, PA, 1998.

ASTM G83, "Standard Test Method for Wear Testing with a Crossed Cylinder Apparatus," ASTM International, West Conshohocken, PA, 1996.

ASTM G99, "Standard Test Method for Wear Testing with a Pin-on-Disk Apparatus," ASTM International, West Conshohocken, PA, 2003.

ASTM G111, "Standard Guide for Corrosion Test in High Temperature and/or High Pressure Environment," ASTM International, West Conshohocken, PA, 1997.

Bain, E. and Paxton, H., *Alloying Elements in Steel,* ASM International, Materials Park, OH, 1966, p. 225.

Ball, W., Jackson, N., Pilley, A., and Porter, B., "The Friction of a 1.6L Automotive Engine—Gasoline and Diesel," SAE Paper No. 860418, Society of Automotive Engineers, Warrendale, PA, 1986.

Baniasad, S. and Emes, M., "Design and Development of Method of Valvetrain Friction Measurement," SAE Paper No. 980572, Society of Automotive Engineers, Warrendale, PA, 1998.

Beard, C. and Hempson, J., "Problems in Valve Gear Design and Instrumentation," SAE Paper No. 596D, Society of Automotive Engineers, Warrendale, PA, 1962.

Betteridge, W. and Heslop, J., editors, *The Nimonic Alloys and Other Nickel-Based High-Temperature Alloys*, Crane, Russak and Company, New York, 1972.

Betz, C., *Principles of Magnetic Particle Testing*, Magnaflux Corporation, Chicago, 1967.

Betz, C., *Principles of Penetrant*, Magnaflux Corporation, Chicago, 1963.

Boyer, H., *Atlas of Fatigue Curves*, ASM International, Materials Park, OH, 1986.

Caird, S. and Trela, D., "High Temperature Corrosion-Fatigue Test Method for Exhaust Valve Alloys," SAE Paper No. 810033, Society of Automotive Engineers, Warrendale, PA, 1981.

Cartz, L., *Nondestructive Testing*, ASM International, Materials Park, OH, 1995.

CT-6-2, "Liquid Penetrant NDT," American Society for Nondestructive Testing, Columbus, OH, 1977.

CT-6-3, "Magnetic Particle Testing," American Society for Nondestructive Testing, Columbus, OH, 1977.

CT-6-4, "Ultrasonic Testing," American Society for Nondestructive Testing, Columbus, OH, 1977.

CT-6-5, "Eddy Current Testing," American Society for Nondestructive Testing, Columbus, OH, 1977.

CT-6-6, "Radiographic Testing," American Society for Nondestructive Testing, Columbus, OH, 1977.

Dieter, G., *Mechanical Metallurgy, 3rd Edition*, McGraw-Hill, New York, 1986.

Dull, D. and Raymond, L., "Stress History Effect on Incubation Time for Stress Corrosion Crack Growth in AISI 4340 Steel," *Metall. Trans.*, Vol. 3, Nov. 1972, p. 2943.

Etchells, E., Thomson, R., Robinson, G., and Malone, G., "The Interrelationship of Design, Lubrication, and Metallurgy in Cam and Tappet Performance," SAE Paper No. 472, Society of Automotive Engineers, Warrendale, PA, 1955.

Forrest, P., *Fatigue of Metals*, Pergamon Press, Oxford, U.K., 1962.

Funabashi, N., Endo, H., and Goto, G., "U.S.–Japan PM Valve Guide History and Technology," *Proceedings of the International Symposium on Valvetrain System Design and Materials*, edited by Bolton, H.A. and Larson, J.M., ASM International, Materials Park, OH, 1997.

Glenny, E. and Taylor, T., *J. Inst. Metals,* Vol. 88, 1960, p. 449.

Heisler, H., *Advanced Engine Technology*, Society of Automotive Engineers, Warrendale, PA, 1995.

Hull, D. and Bacon, D., *Introduction to Dislocations*, Pergamon Press, London, U.K., 1984.

The Institute of Metals, *Ultrasonic Nondestructive Testing, Monograph No. 9*, The Institute of Metals, London, U.K., 1982.

Johnson, V. and Galen, W., "Diesel Exhaust Valves," SAE Paper No. 660034, Society of Automotive Engineers, Warrendale, PA, 1966.

Knott, J., *Fundamentals of Fracture Mechanics*, Butterworths, London, U.K., 1981.

Kubaschewski, O. and Hopkins, B., *Oxidation of Metals and Alloys*, Butterworths, London, U.K., 1962.

Kurisu, T., Hatamura, K., and Omoti, H., "A Study of Jump and Bounce in a Valvetrain," SAE Paper No. 910426, Society of Automotive Engineers, Warrendale, PA, 1991.

Lewis, R., Dwyer-Joyce, R., and Josey, G., "Investigation of Wear Mechanisms Occurring in Passenger Car Diesel Engine Inlet Valves and Seat Inserts," SAE Paper No. 1999-01-1216, Society of Automotive Engineers, Warrendale, PA, 1999.

Love, R. and Wykes, F., "European Practice in Respect of Automotive Cams and Followers," SAE Paper No. 750865, Society of Automotive Engineers, Warrendale, PA, 1975.

Ludema, K., *Friction, Wear, Lubrication—A Textbook in Tribology*, CRC Press, New York, 1996.

Malatesta, M., Barber, G., Larson, J., and Narasimhan, S., "Development of a Laboratory Bench Test to Simulate Seat Wear of Engine Poppet Valves," *Tribology Transactions*, Vol. 36, No. 4, 1993, p. 627.

Myers, H., *Introductory Solid State Physics*, Taylor and Francis, PA, 1990.

Neale, M., *Drives and Seals, A Tribology Handbook*, Society of Automotive Engineers, Warrendale, PA, 1994.

Netzer, J. and Maus, K., "Improvement of Valve Stem Seals to Meet Future Emission Requirements," SAE Paper No. 980581, Society of Automotive Engineers, Warrendale, PA, 1998.

Newton, J., Palmer, J., and Reddy, V., "Factors Affecting Diesel Exhausts Valve Life," SAE Paper No. 190, Society of Automotive Engineers, Warrendale, PA, 1953.

Noll, G. and Erickson, M., "Allowable Stresses for Steel Members of Finite Life," *Proc. Soc. Exptl. Stress Analysis*, Vol. V, No. II, pp. 132–143.

Onoda, M., Kruoishi, N., and Motooka, N., "Sintered Valve Seat Insert for High Performance Engine," SAE Paper No. 880668, Society of Automotive Engineers, Warrendale, PA, 1988.

Petch, N., "The Cleavage Strength of Polycrystals," *J. Iron Steel Inst. Jpn.*, Vol. 174, 1953, p. 25.

Pope, J., "Techniques Used in Achieving a High Specific Airflow for High-Output, Medium-Speed Diesel Engines," *J. of Engineering for Power*, April 1967, p. 265.

Rice, J. and Johnson, M., *Inelastic Behavior of Solids*, edited by Kanninen, M., McGraw-Hill, New York, 1970.

Rodrigues, H., "Sintered Valve Seat Inserts and Guides: Factors Affecting Design, Performance, and Machinability," *Proceedings of the International Symposium on Valvetrain System Design and Materials*, ASM International, Materials Park, OH, 1997.

SAE Fatigue Design Handbook, Society of Automotive Engineers, Warrendale, PA, 1968.

SAE Standard J244, "Measurement of Intake Air and Exhaust Gas Flow of Diesel Engines," Society of Automotive Engineers, Warrendale, PA, 1992.

Schamel, A., Hammacher, J., and Utsch, D., "Modeling and Measurement Techniques for Valve Spring Dynamics in Revving Internal Combustion Engines," SAE Paper No. 930615, Society of Automotive Engineers, Warrendale, PA, 1993.

Smith, R., Hirschberg, M., and Manson, S., "Fatigue Behavior of Materials Under Strain Cycling in Low and Intermediate Life Ranges," NASA TN D-1574, April 1963.

Socie, O., "Fatigue Life Estimation Techniques," Technical Report 145, Electro General Corporation, Department of Mechanical and Industrial Engineering, University of Illinois, Urbana-Champaign, 1977.

Spengler, C. and Viswanathan, R., "Effect of Sequential Sulfidation and Oxidation on the Propagation of Sulfur in an 85 Ni-15 Cr Alloy," *Metallurgical Transactions*, Vol. 3, January 1972, p. 161.

Taylor, W., "The Hardness Test as a Means of Estimating the Tensile Strength of Metals," *J. Royal Aeronaut. Soc.*, Vol. 46, 1942, p. 198.

Udoguchi, T. and Wada, T., "Thermal Effect on Low-Cycle Fatigue Strength of Steels," edited by Littler, D.J., *Thermal Stresses and Thermal Fatigue*, Butterworths, U.K., 1971.

Wang, Y., Narasimhan, S., Larson, J., Larson, J.E., and Barber, G., "The Effect of Operating Conditions on Heavy Duty Engine Valve Seat Wear," *WEAR 210*, 1996, pp. 15–25.

Weronski, A. and Hejwowski, T., *Thermal Fatigue of Metals*, Marcel Dekker, New York, 1991.

Worthen, R. and Rauen, D., "Measurement of Valve Temperatures and Strain in a Firing Engine," SAE Paper No. 860356, Society of Automotive Engineers, Warrendale, PA, 1986.

Zhao, Y., Tong, K., and Lu, J., "Determination of Aeration of Oil in High Pressure Chamber of Hydraulic Lash Adjuster in Valvetrain," SAE Paper No. 1999-01-0646, Society of Automotive Engineers, Warrendale, PA, 1999.

Chapter 5

 Valvetrain Tribology

5.1 Introduction to Tribology

Tribology is the science of friction, wear, and lubrication. It was derived from the Greek language and means "the science of rubbing." Tribology is an interdisciplinary science involving mechanical engineering, chemistry, physics, and metallurgy. A substantial part of the total energy consumption of mankind is expended in overcoming wear and friction losses during sliding.

5.1.1 Valvetrain Tribology Overview

Figure 5.1 summarizes the sources of energy consumed by friction within an engine. It is estimated that approximately 15% of engine energy is lost to internal friction, that is, between interfaces such as the stems and guides; rocker shafts; cams and lifters; cam bearings; timing chains, gears, or belts; and tensioners. Approximately one-third of this estimated 15% energy loss has been consumed by valvetrain friction.

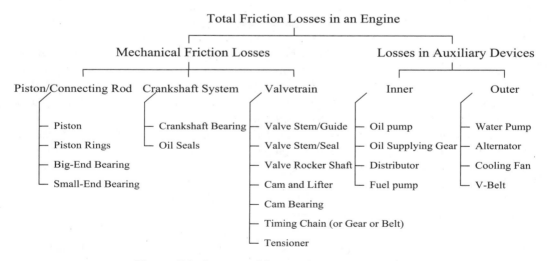

Figure 5.1 *Sources of friction losses in an engine.*

Although the choice of a valvetrain for an engine usually is based on flow, engine height, tooling, and other considerations, friction has been considered as an additional criterion. An increasingly important part of tasks faced by engine designers is to reduce friction and wear, through improved design, the use of more suitable materials, or the application of better lubricants. By blending the information gathered and the friction theory available, it is possible to understand where these differences originate, how their requirements can be met, how a fuel-efficient engine can be built, and how to make valvetrains last longer. Therefore, the goal is to reduce friction in the engine, with consequent improvements in fuel economy and performance. The reduced valve weight requiring lower forces to activate the valve spring is one example of reducing friction forces.

Valvetrain wear has been the subject of study by engine manufacturers and valvetrain component and lubricant suppliers since the early days of the automobile. Valvetrain wear occurs primarily at the interfaces of the valve stem and guide, the valve seat and insert, and the cam and follower. The many factors that influence wear have made it extremely difficult to reach any general conclusions that might be applicable to the variety of conditions existing on the numerous rubbing surfaces of valvetrains in engines.

Valvetrain lubrication has been, is now, and will continue to be one of the key solutions to the friction energy loss and wear problems associated with valvetrains in internal combustion engines.

With proper design and material and lubricant selection, friction and wear of valvetrains in engines can be reduced, and the consequent improvements in fuel economy and performance can be achieved.

5.1.2 Surface Topography and Contact Mechanics

5.1.2.1 Surface Interactions

When two solid materials are placed in contact (Figure 5.2), some regions on their surfaces will be close together, and others will be farther apart because of nonconformed surfaces in contact. It is known that the powerful atom-to-atom forces are of very short range, of the order of magnitude of only a few Angstroms, with this range being approximately the size of the average atom. The regions in contact within the atomic force are referred to as "junctions," and the sum of the areas of all junctions constitutes the real area of contact, A_r (Rabinowicz). The total interfacial area, consisting of both the real area of contact and those regions that appear as if contact might have been made there (but has not), will be denoted as the "apparent" area of contact, A_a. Long-range forces operate at points on the surface separated by distances exceeding 10 Å, but they are negligible in magnitude compared with the short-range forces (Abrikosova and Deryagin). It has been estimated that the diameter of typical junctions is in the range of 1–100 microns (Rabinowicz). The real area of contact generally is much less than the apparent area. The ratio might be as small as 10^{-4} in practical situations (Glaeser). The interaction occurs at these junctions on both the asperity and atomic levels. At the asperity level, the focus is on the type of deformation that occurs at these junctions. The deformation at the junction can be plastic and elastic. How much of each is involved depends on the number of junctions, their size, and the total load, as well as the properties of the materials involved. Although it

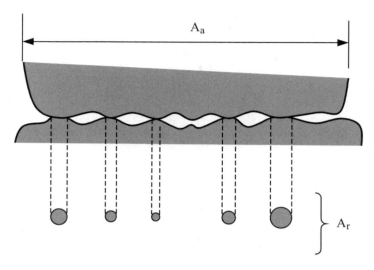

Figure 5.2 *Schematic of the apparent (A_a) and actual (A_r) areas of contact.*

is possible to have only elastic deformation on all junctions, it is not likely. Models based on typical surface profiles indicate that some plastic deformation generally occurs at some of the junctions (Greenwood and Williamson).

5.1.2.2 Real Area of Contact

Because the nature of friction and wear between two surfaces is determined by the real area of contact, it is necessary to derive as much information as possible about the real area. To calculate a minimum value for A_r, a typical junction will resemble the diagram shown in Figure 5.2 if two surfaces are placed in contact. The largest compressive stress that such a region of material can carry without further plastic yielding is known as its penetration hardness H, which will be approximately three times as great as the yield strength σ_y in uniaxial compression. Most surfaces as prepared technically have ridges and valleys, similar to those prevailing in a hardness test but on a smaller scale. When two surfaces are brought into contact and a normal load is applied, plastic deformation will occur so that the initial contact of three points eventually becomes the contact of numerous sizeable areas, and deformation will continue until the total real area of contact reaches a value given by

$$A_r = \frac{L}{H} \tag{5.1}$$

where L is the applied load and H is the hardness. At this stage, deformation will cease. In some circumstances, the real area of contact has been measured directly and has been found to be of the same order of magnitude as the value given by Eq. 5.1 (Kraghelsky and Demkin).

When two surfaces in contact are smooth and there may be no plastic deformation but only elastic deformation, A_r will be greater, perhaps much greater than suggested by Eq. 5.1. Then the area of contact is given by Hertz's equation for elastic deformation for a spherical ball pressed into a flat polished surface,

$$A_r = 2.9 \times \left[Lr \times \left(\frac{1}{E_1} + \frac{1}{E_2} \right) \right]^{2/3} \qquad (5.2)$$

where

L = normal load
r   = sphere radius
E_1 = sphere Young's modulus
E_2 = flat Young's modulus

This assumes that the Poisson's ratio for both surfaces is 0.3 (Tabor).

Shear stress also has a profound effect on the real area of contact. When the shear force first is applied, tangential motion occurs even while the tangential force is quite low (Courtney-Pratt and Eisner). This motion has the effect of increasing the area of contact, which recreates equilibrium under the joint action of the normal and shear forces, so that relative motion of the two surfaces in a normal direction inward ceases. If the shear force is increased continually, the increase in the real area of contact with increased shear force eventually falls short of that required to maintain static equilibrium; hence, sliding motion will occur. For materials that creep, A_r increases with the time of application of the load.

5.1.2.3 Surface Profile

The surface profile may be defined as the departure of the surface shape from some ideal or prescribed form. The surface profile includes surface roughness, waviness, and form. Figure 5.3 shows an example of a two-dimensional surface profile. Roughness is the irregularities inherent in the production process, left by the actual machining agent (e.g., cutting tool, grit, spark). Waviness is that component of texture upon which roughness is superimposed. It may result from factors such as machine or work deflections, vibrations, chatter, various causes of strain in the material, and extraneous influences. Form is the general shape of the surface, neglecting variations due to roughness and waviness.

Figure 5.4 and the following equations further demonstrate typically used roughness parameters and their mathematical definitions.

Average roughness:

$$R_a = \frac{y_1 + y_2 + ... + y_n}{n}$$

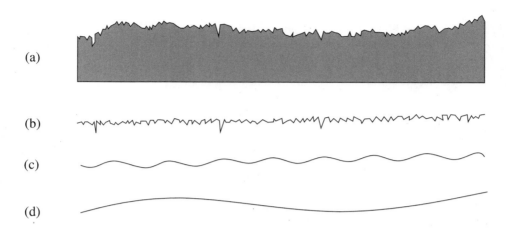

Figure 5.3 *(a) A two-dimensional surface profile, including (b) roughness, (c) waviness, and (d) form (Dagnall).*

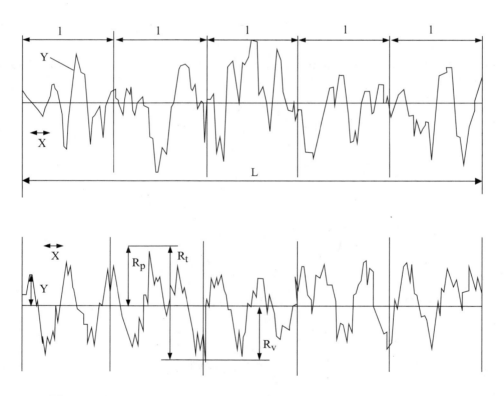

Figure 5.4 *Surface roughness representation, l = equal sampling length, and L = filter cutoff length.*

Root-mean-square roughness:

$$R_q = \left[\frac{y_1^2 + y_2^2 + ... + y_n^2}{n} \right]^{1/2}$$

Roughness total mean:

$$R_{tm} = \frac{R_{t1} + R_{t2} + ... + R_{tn}}{n}$$

Roughness peak mean:

$$R_{pm} = \frac{R_{p1} + R_{p2} + ... + R_{pn}}{n}$$

Roughness valley mean:

$$R_{vm} = \frac{R_{v1} + R_{v2} + ... + R_{vn}}{n}$$

Figure 5.5 shows that the aspect ratio and profile shape vary as the horizontal magnification is reduced or increased.

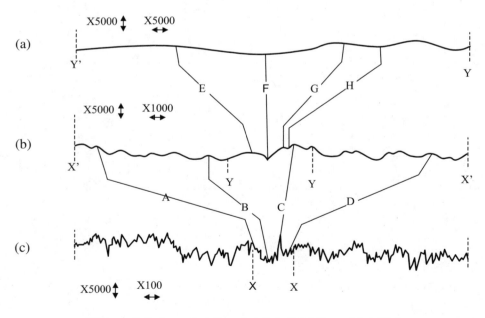

Figure 5.5 *Aspect ratios: (a) 1:1, (b) 5:1, and (c) 50:1.*

5.1.2.4 Micro-Asperity Contact Stress

Contact mechanics are discussed in Section 2.7.2.4 of Chapter 2, and Hertzian contact stress formulas are available in many forms and are listed in Chapter 2. However, because surfaces usually are rough, as shown in Figure 5.3, contact between the surfaces can occur only at limited points. Therefore, the pressure on those points is high. Figure 5.6 shows a schematic of a cam and follower interface, resulting in Hertzian contact pressure and asperity contact pressure based on the apparent and real areas of contact. Micro-asperity stresses can be much higher than macro-Hertzian stresses.

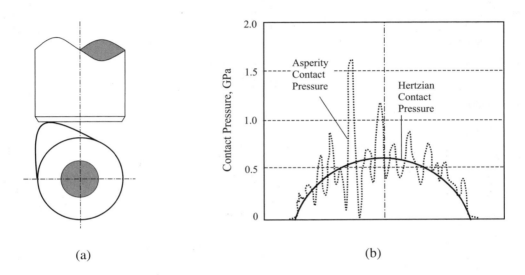

(a)   (b)

Figure 5.6 *Example of macro contact stress versus micro-asperity contact stress: (a) cam-lifter contact, and (b) general nature of the stress in contact situations.*

5.1.3 Friction

5.1.3.1 Friction Laws

Friction can be defined as the resistance to movement between two bodies in contact. The term "coefficient of friction" is defined as the ratio of the tangential force (F) to the normal force (N) between these bodies, where $\mu = \dfrac{F}{N}$. The force required to begin sliding often is greater than the force required to sustain sliding; therefore, the starting friction or static friction often is higher than the kinetic friction.

In the fifteenth century, Leonardo da Vinci conducted extensive studies on friction and established three basic laws that later were confirmed and developed by Amontons and Coulomb in the seventeenth and eighteenth centuries, respectively. These basic laws state the following:

1. Friction force is proportional to the normal force.

2. Friction force is independent of the apparent area of contact.

3. Friction force is independent of the sliding velocity.

These laws generally are valid and provide the quantitative framework within which friction usually is considered by engineers. However, it is important to know how closely these laws apply in actual practice, because in many exceptions, these laws are not obeyed in practice. For example, in some cases, the friction force is not proportional to the load (Whitehead). The explanation of this load-dependent friction coefficient relies on understanding of the nature of surface interaction, the real area of contact, atomic interaction, surface roughness, and surface energy. Figure 5.7 shows how atomic interaction, asperity interlocking, and plowing would affect friction.

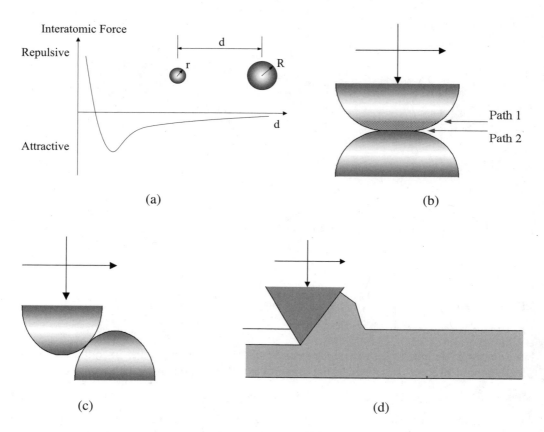

Figure 5.7 *Causes of friction: (a) interatomic force, (b) adhesive forces, (c) interlocking of asperities, and (d) plowing.*

The friction coefficient is independent of sliding speeds over wide ranges of velocity, as stated in the third law. However, the friction coefficient has a positive slope at slow sliding speeds and a negative slope at high sliding speeds (Rabinowicz).

In lubricated systems, when the surfaces slide, lubricant is dragged into the contact region and separates the surfaces. This initially will lower the coefficient of friction; however, at a higher sliding speed, the viscous drag increases, as does the coefficient of friction.

5.1.3.2 Effect of Friction on Contact Stress

The coefficient of friction at the interface can have a significant effect on the distribution and magnitude of contact stresses. Figure 5.8 shows the effect of the coefficient of friction on the tangential stress at a contact surface with a spherical surface sliding on a flat surface. It illustrates that the maximum tensile and compressive stresses increase as the coefficient of friction increases. This phenomenon exists at both the macro contact and asperity contact scales.

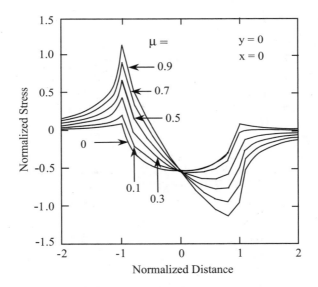

Figure 5.8 *Effect of the coefficient of friction on the tangential stress at a contact surface.*

5.1.3.3 Friction of Materials

A representative surface structure of an unlubricated metal encountered in an industrial environment typically is covered with a whole series of films, as shown in Figure 5.9. Next to the metal interior, the first encountered layer is an oxide layer, produced by the reaction of oxygen from the air with the metal. Next is an absorbed layer derived from the atmosphere, with the main

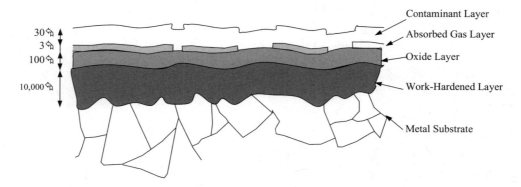

Figure 5.9 *Typical structure of an unlubricated metal surface*
(not to scale) (Rabinowicz).

constituents of this layer generally being molecules of water vapor and oxygen. Outermost, these usually are greasy or oily films.

Metal surfaces of this type generally have initial friction coefficients in the range of 0.1–0.3 when slid together. However, higher values are reached if the surfaces continue to be slid over each other, because under these conditions, the grease film, which is the one with the most drastic influence on friction, eventually will be worn off. Table 5.1 shows typical static and kinetic coefficients of friction for selected materials under various conditions.

TABLE 5.1
COEFFICIENT OF FRICTION DATA FROM
SELECTED PAIRS OF MATERIALS

Material	Condition	μ	Material	Condition	μ
Steel/Steel	Clean	0.58	Tungsten Carbide/WC	Clean	0.17
	Oil	0.1	Tungsten Carbide/Steel	Clean	0.4
Steel/Cast Iron	Clean	0.4		Oil	0.1
	Oil	0.1	Diamond/Diamond	Clean	0.1
Steel/Bronze	Oil	0.15		Oil	0.05
Steel/Tin	Oil	0.2	Graphite/Graphite	Clean/Oil	0.1
Steel/Brass	Oil	0.12	Graphite/Steel	Clean/Oil	0.1
Aluminum/Aluminum	Air	1.9	Glass/Metal	Clean	0.6
Copper/Copper	Air	1.6		Oil	0.2
	H_2 or N_2	4.0	Glass/Glass	Clean	1.0
Iron/Iron	Air	1.2		Oil	0.5
Molybdenum/Molybdenum	Air	0.8		Alcohols	0.1
Nickel/Nickel	Air	3.0	Nylon/Nylon	Clean	0.15
	H_2 or N_2	5.0	Rubber/Solid	Clean	1–4

5.1.3.4 Rolling Friction

Rolling friction is the resistance to motion that takes place when an object is rolled over an abutting surface. The friction force of rolling contact is lower for smooth surfaces than for rough surfaces. The friction force of rolling contact varies inversely with the radius of curvature of the rolling elements. The static friction force generally is much greater than the kinetic, but the kinetic is little dependent on the rolling velocity, although it generally does drop off somewhat as the rolling velocity is increased. The pure rolling friction coefficients generally are in the range of 0.005–0.00001 (Rabinowicz). In practice, however, pure rolling rarely exists. The region of contact is deformed elastically and (in extreme cases) plastically, so that contact is made over an area of some size, with the points within it lying in different planes (Figure 5.10). Thus, it is impossible for pure rolling action to take place except at a very small number of points; rather, at all other points, a combination of rolling combines with a small degree of sliding or slip. To achieve this slipping requires overcoming sliding resistance at the interface, and to accomplish this, a rolling friction force must act. Although the slip velocities generally are small (usually 5% or less of the overall rolling velocity), this small amount of slip produces, in many cases, a major part of the total resistance to rolling. In some rolling contact systems (e.g., ball bearings, gear teeth), the surfaces tend to spin around the region of contact. In others, there is gross slippage. In such cases, the rolling friction coefficient may be quite large, with values greater than 0.001 being common (Rabinowicz).

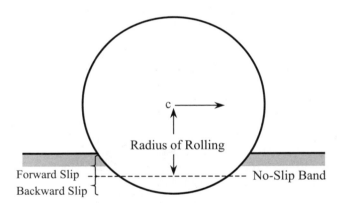

Figure 5.10 *A sphere rolls on a flat surface. Note the regions of forward and backward slippage (Rabinowicz).*

5.1.3.5 Temperature of the Sliding Surface

Surface temperature rises above the bulk temperature due to friction rubbing. It can be substantially higher than the bulk temperature. The friction temperature or flash temperature, T_f, is a function of load, friction, speed, and thermal conductivity of the contacting materials,

$$T_f = f\left(\frac{\mu p V}{k}\right) \tag{5.3}$$

where

$\begin{aligned} T_f &= \text{asperity flash temperature} \\ \mu &= \text{coefficient of friction} \\ p &= \text{Hertzian pressure} \\ k &= \text{thermal conductivity} \\ V &= \text{sliding velocity} \end{aligned}$

The temperature, especially the flash temperature, can significantly affect the tribological characteristics, including friction, wear, and lubricant degradation at surfaces.

5.1.4 Wear

5.1.4.1 Introduction

Wear can be defined from an engineering perspective as damage (including progressive loss or displacement) of material from the operating surface as a result of relative motion at the surface. There are three apparent ways in which wear may be classified (Bayer). One is in terms of the appearance of the wear scar. A second is in terms of the physical mechanism that removes the material or causes the damage. The third is in terms of the conditions surrounding the wear situation. Examples of terms in the first category are pitted, spalled, scratched, polished, crazed, fretted, gouged, and scuffed. Terms such as adhesion, abrasion, delamination, and oxidation are examples of the second type of classification. Phrases commonly used for the third method of classification are lubricated wear, unlubricated wear, metal-to-metal sliding wear, rolling wear, high-stress sliding wear, and high-temperature metallic wear. Although wear generally is described in terms of these three classifications, there is no uniform system in place at the present time. In addition, the same term might be used in the context of more than one classification concept. For example, the term "scuffing" is used in several ways. One may use this term simply to describe the physical appearance. Another may use this term to indicate that the wear mechanism is adhesive wear. A third may use it to indicate wear under sliding conditions. Although relationships exist among these classifications, the classifications are not equivalent nor are the interrelationships necessarily simple, direct, unique, or complete. Because of the complex nature of wear behavior, it has even been argued that it may never be possible or practical to establish complete relationships of this type. Therefore, all three classifications in terms of appearance, mechanism, and environment are useful in providing adequate descriptions of wear situations.

5.1.4.2 Wear Mechanisms

Wear mechanisms can be thought of as typical material failure mechanisms occurring at or near the surface. Because no standards exist, the terminology in this field is unsettled. There is an extremely large number of wear mechanisms, but most tribologists tend to divide basic wear mechanisms into a few major categories. These are adhesion, abrasion, corrosion, surface fatigue, and minor categories.

Adhesive wear is due to localized bonding between the contacting solid surfaces, leading to material transfer between the two surfaces or loss from either surface. Because junctions form as a result of two surfaces being pressed together, the nature of the interatomic forces indicates that bonding occurs at these junctions and that over some portion of the real area of contact, the atoms of the two surfaces must have moved past the point of maximum bonding. This implies that some adhesive forces or bonds must be overcome to separate the two surfaces at these sites. This atomic view of the contact situation at the junctions provides the foundation for the concept of friction forces and adhesive wear mechanisms. The causes of adhesive wear are poor lubrication, high contact stress, incompatible material, wide hardness differences, high surface roughness, and lack of break-in.

The mathematical model for adhesive wear is (Archard)

$$V = k\frac{P}{3H}$$

(5.4)

where

V = wear volume
P = normal load
H = penetration hardness of the softer material
k = wear coefficient

These are listed in Table 5.2 for various material combinations.

The data in Table 5.2 illustrate some of these trends. Clean, unlubricated, and similar metal pairs generally have high values for k. Lubricated conditions giving the lowest values and conditions involving oxides and polymers have intermediate values associated with them. The factor k varies over several orders of magnitude and is influenced by a variety of parameters, such as material pair compatibility, surface energies, and the nature of the asperity contact and load distributions.

TABLE 5.2
TYPICAL WEAR COEFFICIENT (k) VALUES
FOR ADHESIVE WEAR (BAYER)

Combination	k
Self-Mated Metals:	
• Dry	$2 \times 10^{-4} - 0.2$
• Lubricated	$9 \times 10^{-7} - 9 \times 10^{-4}$
Non-Self-Mated Metals:	
• Dry	$6 \times 10^{-4} - 2 \times 10^{-3}$
• Lubricated	$9 \times 10^{-8} - 3 \times 10^{-4}$
Plastics on Metals:	
• Dry	$8 \times 10^{-5} - 3 \times 10^{-7}$
• Lubricated	$1 \times 10^{-6} - 5 \times 10^{-6}$

Abrasive wear causes displacement of material due to hard particles or hard protuberances. The causes of abrasive wear are wide hardness differences, high asperity contact stress, and third particles such as trapped wear debris and oxide particles. Depending on the penetration depth from the hard particles, the abrasive wear processes may differ.

Three general situations for abrasive wear are identified. One situation occurs when hard asperities of one surface are pressed into a softer surface. This abrasive wear situation generally is referred to as two-body abrasive wear. The second situation is one in which hard, loose particles are trapped between the two surfaces, and the forces between the two surfaces are transmitted through these particles. This abrasive wear situation is referred to as three-body abrasion. The third situation occurs when hard particles directly impinge on a surface and is referred to as erosion.

Equation 5.5 generally is used for two- and three-body abrasive wear. This equation was derived by considering the wear produced by a single abrasive grain that can be either an asperity on the mating surface or an individual particle (Bayer),

$$V = \frac{2k\left(\dfrac{\tan\theta}{\pi}\right)L}{H}x \qquad (5.5)$$

where

V	=	wear volume
H	=	indentation hardness of the softer material
$\tan\theta$	=	effective or average sharpness of the abrasive grains
x	=	sliding distance
L	=	total load
k	=	fraction of the total load L supported by those grains

From a practical standpoint, Eq. 5.5 often is written in a simplified form as

$$V = K\frac{L}{H}x \qquad (5.6)$$

where K is a combined factor, taking into account the sharpness, the probability of wear, and the nature of the wear process, as well as additional material properties. Equation 5.7 generally is used for two- and three-body abrasive wear situations. Table 5.3 gives typical values for K for a variety of conditions.

It can be seen that K ranges over several orders of magnitude and that some trends exist. One trend is that two-body abrasive wear situations generally have higher values of K than three-body conditions. A second trend is that the larger the abrasive grain, the larger the value of K. The third trend is the effect of lubrication. Lubrication (i.e., wet versus dry) tends to increase abrasive wear. The presence of a liquid at the interface helps to flush the wear debris from the interface and reduce the shielding effect.

TABLE 5.3
K VALUES FOR ABRASIVE WEAR (BAYER)

Condition	K	
	Dry	Lubricated
Two-Body:		
File	5×10^{-2}	10^{-1}
New Abrasive Paper	10^{-2}	2×10^{-2}
Used Abrasive Paper	10^{-3}	2×10^{-3}
Coarse Polishing	10^{-4}	2×10^{-4}
<100-μm Particles	10^{-2}	
>100-μm Particles	10^{-1}	
Nominal Range, Dry and Lubricated	<1 to >10^{-4}	
Three-Body:		
Coarse Particles	10^{-3}	5×10^{-3}
Fine Particles	10^{-4}	5×10^{-4}
Nominal Range, Dry and Lubricated	<10^{-2} to >10^{-6}	

Contact fatigue is a phenomenon in which cracks form, grow, and eventually link together to form a loose particle that separates from a surface in the form of flakes or chips under cyclic stresses, including mechanical and thermal stresses. This wear mechanism is most evident in rolling and impact wear situations, where it generally is recognized as the principal mechanism. A common feature of fatigue wear is the existence of an incubation period. During this initial period, cracks are formed and propagate to the surface. Some topological changes might be evident during this period, including some evidence of plastic deformation. However, there is no loss of material from the surface or formation of free particles. After this stage, material loss would occur with the formation of free particles. Mathematical prediction of fatigue wear can be challenging.

Corrosive or oxidative wear is a process in which a chemical or electrochemical reaction with the environment predominates due to high temperatures, a reactive and corrosive environment, or the material being susceptible to corrosion and oxidation. Corrosive or oxidative wear does not result directly in the loss or displacement of material. However, when a surface is reacted chemically and forms a layer such as iron oxide, wear then occurs by one or a combination of the previously mentioned three mechanisms, either in this reacted layer or at the interface between the layer and the parent material. The wear rate then is influenced, and possibly controlled, by the growth and formation of that layer. In addition to the chemical corrosion film that forms on metals in reactive environments, another film is derived from the environment. This is the adsorbed film, which will be found on both metallic and nonmetallic surfaces. In air, the main constituent of this film will generally be only of the order of one molecularly thick layer, that is, approximately 3×10^{-8} cm, although thicker water films are found in many instances when the humidity in the environment is high (Bowden and Throssell). Many polymers can take up water vapor below the surface, thus leading to a modification of their bulk properties. Often, there also will be a greasy or oily film that may partially displace the adsorbed layer derived from the atmosphere. This greasy film may be derived from a variety of sources, among them the oil drops found in most industrial environments, the lubricants that were applied while the surface was being prepared, or natural greases from the fingers of the people who handled the solid. The thickness of these grease films may be 3×10^{-7} cm or more. The presence of adsorbed films containing water and other molecules derived from the air serves measurably to reduce the surface interaction of contacting materials. However, the effect of grease films, if present, is even more marked and reduces, often by one or more orders of magnitude, the severity of the surface interaction.

In addition to the preceding four major wear mechanisms, a number of unique wear mechanisms are encountered only in special situations. However, most of those unique wear mechanisms can be explained either by some sequence of events involving the other four, by a combination of the other four, or by a subcategory of the other four. For example, fretting and fretting corrosion were considered as unique wear mechanisms associated with small amplitude sliding, but they also can be described in a two-step sequence. First, wear debris is produced by adhesion, fatigue, or abrasion. The wear debris then acts as an abrasive, accelerating and controlling the wear from that point. By including oxidation, this sequence is used to explain fretting corrosion and fretting.

Another illustration of this is delamination and lamination wear. Each tends to explain the physical wear process in somewhat different manners and emphasizes different aspects. Both involve the idea of crack formation and the eventual formation of a particle or fragment. Both can be viewed as subcategories of fatigue wear because they include the concept of crack formation and growth.

A shear strain or repeated-cycle deformation mechanism involves repeated-cycle plastic deformation of the wearing surface along the sliding direction due to high contact stress, high surface flash temperature, high friction at the interface, and inadequate hot hardness. This mechanism also can be generalized as a fatigue mechanism. Shear strained/deformed wear involves plastic deformation of the wearing surface along the sliding direction due to high contact stress, high surface flash temperature, high friction at the interface, and inadequate hot hardness. It differentiates in a way that can occur even at similar or harder surfaces without any intrusion of hard particles.

There is an element involved with the definitions of adhesive, abrasive, and fatigue wear that frequently is a key in distinguishing between abrasive and fatigue wear. As defined, abrasive and adhesive wear mechanisms are associated with a single action or engagement. When the interaction takes place, the material either deforms, breaks off, or is plucked out at that time. On the other hand, fatigue requires more than one engagement cycle.

Typically, the surfaces of the specimens are examined after, not during, the sliding to determine the types of wear mechanism. A description of the wear process for a given situation generally involves some combination of the primary wear mechanisms, either in a particular sequence or in some parallel combination. It is a fact that wear mechanisms coexist and interact with each other. Each mechanism may play a different role at various stages of wear processes.

A post-examination provides only the evidence on the final stage that wear occurred. These facts make wear prediction or modeling and wear mechanism identification as discussed here simplistic.

The classification of basic wear phenomena into four modes (i.e., adhesion, abrasion, fatigue, and oxidation) is not necessarily complete or undisputed. From certain aspects, it may be inadequate as well. However, it does provide a useful basis for an effective engineering understanding of wear, particularly as it relates to design.

5.1.4.3 Wear Characteristics

Wear is not an intrinsic material behavior, and several factors such as load, roughness, speed, and surface contaminants contribute to the general complex nature of wear behavior. Another factor is the number of basic wear mechanisms involved. Depending on the mechanism and the parameter considered, the general relationship of wear and other parameters often is in some form of nonlinear behavior; sometimes, transitions or maximum/minimum behavior is seen. At the same time, linear relationships or regions of linear behavior often can be found in certain regions.

A frequently encountered behavior is the development of a region of stable wear behavior after some initial wear (i.e., a break-in period) has taken place. In situations where the contact area does not change with wear, this generally is seen as a region of constant wear rate (mm^3/N-m) after an initial period of higher and changing wear rate. From a design standpoint, if suitable break-in is achieved, a stable period of low wear rate may be achieved for a given device and lead to long life.

Another trend is that wear typically decreases as hardness increases (Figure 5.11). However, it may not be universally true. Because wear also is a function of other parameters such as the real

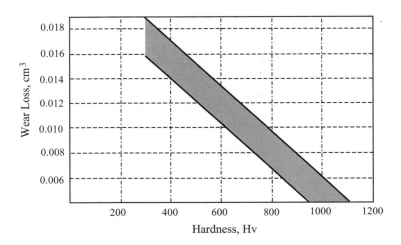

Figure 5.11 *The effect of hardness on wear (Eyre).*

area of contact, flash temperature, and surface roughness, an increase in hardness may reduce the real area of contact, thus increasing asperity contact stresses, and higher wear may result.

The effect of surface finish on wear is that the smoother the surface, the less the wear, because of the increased real area of contact and less severity of asperity colliding with each other as the surface becomes smoother. However, as shown in Figure 5.12, when the surface finish exceeds a certain point, the real area of contact starts to increase, resulting in an increased wear rate.

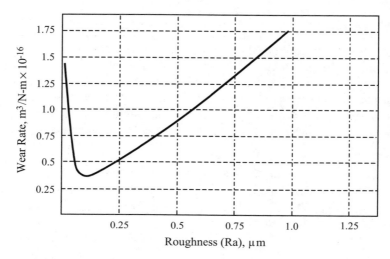

Figure 5.12 *The effect of roughness on wear (Anderson).*

Wear transition is another characteristic encountered and is associated with changes in parameters, such as load, speed, duration, and temperature. Wear tends to be much more abrupt and frequently is manifested by a transition from mild to severe wear. The underlying reasons for the wear transition phenomenon are the changes in relative contributions and interactions of the several wear mechanisms, plus changes in the characteristics of the wearing surfaces. Figure 5.13 shows examples of wear transitions occurring as a function of duration, load, speed, and temperature. It also has been found that changes in friction behavior often are associated with transitions in wear behavior. A drastic friction increase indicates that wear mode starts to change from mild wear to a severe wear region.

5.1.4.4 Wear Measurement

The most common form of wear damage is the loss or displacement of material; thus, volume can be used as a measure of wear—the volume of material removed, or the volume of material displaced or deformed. Mass loss also is used sometimes for wear measurement. However, in engineering applications, the concern generally is with the loss of a dimension, the increase

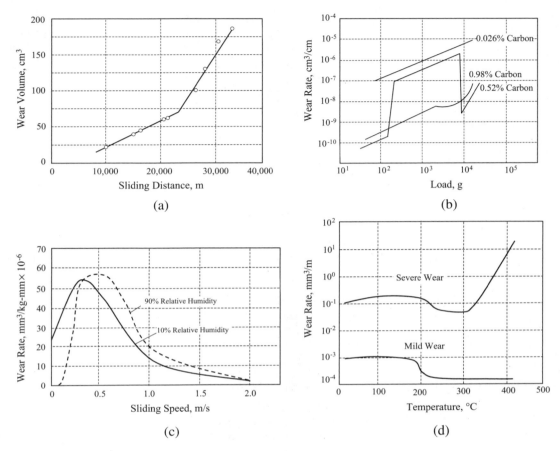

Figure 5.13 *Wear transition phenomena: (a) time dependent (Anderson),*
(b) load dependent (Eyre), (c) speed dependent (Tsuji and Ando), and
(d) temperature dependent (Iwabuchi et al.).

in clearance, or a change in a contour. In an engine valvetrain, the lash in the valvetrain or the clearance between the valve stem and guide is critical.

Wear in each cylinder varies as testing progresses. Many factors contribute to the variation, such as material and manufacturing abnormality of the valve seat and insert, assembly issues, cam lobe profile variations, and cooling variations.

In stem and guide assessment, liner wear and seizure time typically are used to characterize the wear phenomenon.

Other units used in measuring wear are in the area in square millimeters (mm^2) and in the volume in cubic millimeters (mm^3). When wear is in steady state, the wear rate (mm^3/N-m) sometimes is used in an assumption that the wear volume is proportional to the sliding distance and normal load. When different materials are being compared, the wear coefficient is used, with the assumption of the linear relationship of load and distance, and an inverse relationship with hardness,

$$k = \frac{VH}{DL} \qquad (5.7)$$

where

    V = wear volume
    H = hardness
    D = sliding duration
    L = load

5.1.5 Lubrication and Lubricants

5.1.5.1 Introduction

All engines must be lubricated to preserve the integrity of the systems for their designated life-times. The lubricants that are suitable for use in these engines are expected to reduce friction, dissipate heat from internal parts, minimize deposit formation, and prevent corrosion and wear. The extreme temperatures in internal combustion engines make lubrication complex.

The main purpose of lubrication is to reduce friction by forming a film between two moving surfaces. The strength and durability of this film is related to the viscosity of the lubricant and to the speed and load experienced by the moving surfaces. The relationship of the coefficient of friction (μ) and oil film thickness with lubricant viscosity (τ), moving part speed (ν), and load or pressure (p) are shown as the Streibeck curve in Figure 5.14.

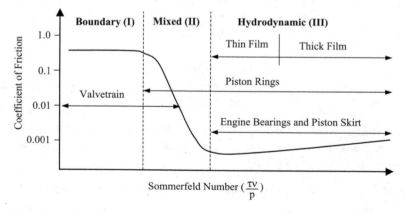

Figure 5.14 *A Streibeck diagram, illustrating various lubrication regions and friction as a function of a composite number $\frac{\tau\nu}{p}$ (Rosenberg).*

The ratio or Sommerfeld number $\left(\dfrac{\tau v}{p} \right)$ is related directly to the oil film thickness but inversely to the coefficient of friction. This implies that high lubricant viscosity, high speed, and low load will allow the formation of thick lubricant film; hence, the moving parts will encounter little or no friction. The initial drop in the coefficient of friction while moving from fluid-film to mixed-film lubrication reflects a decrease in viscous drag because of a decrease in lubricant viscosity. Conversely, low lubricant viscosity, low speed, and high load will create a situation where the film thickness will be inappropriate and the moving parts will encounter high friction. Depending on the lubricating environment, lubrication can be divided into fluid-film (hydrodynamic lubrication), boundary film, and mixed-film types, as shown in Figures 5.14 and 5.15.

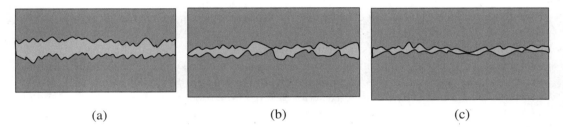

| (a) | (b) | (c) |

Figure 5.15 *Schematics of lubricating film thickness versus lubrication regimes: (a) hydrodynamic lubrication, (b) mixed lubrication, and (c) boundary lubrication.*

Fluid film lubrication, also known as hydrodynamic lubrication (HDL), is most desirable. This type of lubrication depends on the viscosity of the lubricant and is effective only when the load in the contact zone is low. Under these circumstances, the sliding surfaces are separated by a lubricant film that is several times the thickness of the surface roughness. The film develops due to the relative motion of the surfaces and the viscosity of the lubricant. A converging wedge is necessary for HDL and results in the lubricant being pressurized, which allows the film to support a load. Another type of hydrodynamic lubrication, referred to as elasto-hydrodynamic (EHD) lubrication, commonly occurs at high contact pressures and undergoes a large viscosity increase. This results in an extremely rigid lubricant film, which causes elastic deformation of the surfaces in the lubricating zone.

Boundary lubrication simulates the other extreme of the lubricating environment where sliding surfaces are not separated by a lubricant film. High loads and very slow speeds produce extreme pressures that can lead to the lack of effective lubrication and hence promote maximum metal-to-metal contact. If not controlled, the resulting dry metallic friction will cause catastrophic wear and, ultimately, total seizure. Lubricant additives are important in controlling friction and wear in this regime. Reactive chemicals, called anti-wear and extreme pressure agents, provide protection in this kind of lubricating environment. Most valvetrain lubrication is in the boundary lubrication regime.

Mixed-film lubrication falls between the two extremes previously described and contains characteristics of each. There are regions of no metal-to-metal contact, as well as regions of metal-to-metal contact.

5.1.5.2 Lubricant Formation

The inherent ability of mineral oil to act as a lubricant is not sufficient for most modern engine applications. Synthesized chemicals, called additives, are blended in base stocks, such as mineral oils, to help them perform effectively in demanding lubricating environments. A finished lubricant comprises a base fluid and a performance package, and, in the case of multigrade oils, a viscosity modifier. The performance package can constitute as much as 30% of the total composition, depending on the desired performance level and the severity of the end-use requirements.

Base fluids, or lubricant base stocks, are of either mineral or synthetic chemical origin. Mineral oil base stocks are obtained directly from the petroleum fraction, whereas synthetic base stocks are manufactured through transformations of petroleum-based organic chemicals. Partially synthetic base stocks are compatible mixtures of mineral oil and synthetic base stocks. Mineral oil base stocks (lubricating oils) are obtained from the crude oil fraction, which boils at temperatures greater than 340°C through a series of separation processes. These processes include vacuum distillation, solvent extraction, and wax removal. Some base stocks are hydrotreated or hydrocracked to improve color, oxidation resistance, thermal stability, and viscosity. The more highly refined oils have little or no sulfur; therefore, they are more prone to oxidation because of the lack of saturation. The oxidation resistance of oils can be improved greatly through hydrogenation. Most automotive and industrial lubricants, as well as greases, use solvent-extracted base oils. Both solvent-extracted and hydrotreated base oils are used primarily for premium products, such as turbine, hydraulic, and circulating oils.

Mineral oil base stocks typically contain hydrocarbon molecules that contain 20 to 70 or more carbon atoms. These base stocks sometimes are characterized as being paraffinic, naphthenic, or aromatic in type, depending on the nature of the hydrocarbons present. Paraffinic base stocks contain long saturated hydrocarbon chains. Naphthenic base stocks contain five- or six-membered saturated cyclic rings, and the aromatic base stocks contain unsaturated cyclic (aromatic) rings. In reality, none of the base stocks is composed of a single structural type. Instead, they contain mixed structures, with one or another predominating. Paraffinic base stocks, for example, contain all three types of structures, but paraffinic structures are present in the largest amount. The relative distribution of each type of hydrocarbon structure has a significant effect on the viscosity-temperature relationship and the pour point of the base stock.

The viscosity-temperature relationship of an oil is expressed in terms of its viscosity index (VI), which indicates the degree of change in viscosity of an oil with a change in temperature within a given temperature range. A high VI signifies a relatively small change; a low VI indicates a large change. The VI of an oil can be determined by measuring the kinematic viscosity of the oil at 40 and 100°C, and by using ASTM tables. Paraffinic base stocks generally are preferred because of their high viscosity index (HVI). However, because they contain a substantial amount of wax, they have high pour points. Naphthenic base stocks, which have lower viscosity indices, are preferred only in certain applications, because of their lower wax content and lower pour points. Aromatic stocks are the least preferred because of their poor oxidation resistance, tendency to form black sludge at high engine operating temperatures, low VI characteristics, and suspected carcinogenic nature. Table 5.4 shows the relationships between various lubricant properties and structures.

The interest in synthetic lubricants follows from environmental regulations that stress biodegradability, nontoxicity, environmental friendliness, and recyclability. In addition, original equipment manufacturers (OEMs) are requiring lubricants with longer service lives, lower volatilities, and increased energy efficiency. In some cases, such lubricants are required to function in more severe temperature and pressure regimes. Synthetic lubricants are manufactured or "synthesized" in chemical plants by reacting components to make a desired product. In contrast, conventional lubricants, often known as "mineral oils," are obtained from petroleum crude oils by distillation and other refining procedures. Synthetic base stocks are manufactured by chemical reactions such as alkylation, polymerization, and esterification. Commercial synthetic fluids include synthetic hydrocarbons (alkylaromatics and polyolefins), organic esters (dibasic acid esters, polyol esters, and polyesters), and other organic materials (phosphate esters, polyalkylene glycols, and polyphenyl ethers). Polyolefins, dibasic acid esters, polyol esters, and polyalkylene glycols (polyether glycols) are the most widely used synthetic lubricants. Table 5.5 shows their chemical structures and principal applications.

Synthetic base stocks are used in situations where mineral oil base stocks do not provide satisfactory performance because of their inherent limitations. Synthetic base stocks can be devised to possess any number of unique and desirable properties. Commonly sought properties include enhanced thermal stability, high viscosity index, fire resistance, low pour point, low volatility, and low toxicity.

Figure 5.16 compares the useful life of several liquid lubricant classes. It shows that polyphenyl ethers (including C-ethers), polyperfluoroalkyl ethers, and fluoroether triazines are the only synthetic lubricant classes with longer high-temperature service life and higher useful bulk oil operating temperatures than the polyolesters.

Figure 5.17 presents another comparison of liquid lubricant classes, in terms of their potential continuous operating temperature range. The operating temperature range presented in this figure is defined as being bounded on the lower end by the pour point and on the upper end by the maximum useful bulk oil temperature.

Table 5.6 compares some physical properties of hydrofinished HVI stocks and synthetic base stocks. The levels and types of additives used in either synthetic or partially synthetic base stocks (blends of synthetic base stocks and mineral oils) are similar to those used in mineral oils.

5.1.5.3 Additive or Performance Package

A number of chemical compounds (additives) are blended into base stocks to make them suitable for use as engine lubricants. The quality and quantity of additives in the performance package depend on the properties of the base fluid and the intended application. The functions of a lubricant are to limit and control friction, metal-to-metal contact, overheating (from friction and combustion), wear, corrosion, and deposits. The specific properties that the lubricant should

TABLE 5.4
RELATIONSHIP BETWEEN PROPERTIES AND
HYDROCARBON STRUCTURES (SCHILLING)

Hydrocarbon Types	Hydrocarbon Structures	Main Properties
Straight-Chain Paraffin		Viscosity varies little with temperature; good oxidation resistance; high pour point
Branched-Chain Paraffin		Viscosity varies little with temperature (except when divided); good oxidation resistance; may have low pour point
Naphtha Rings with Short Paraffin Side Chain		Good oxidation resistance; low pour point; viscosity varies greatly with temperature; becomes pseudoplastic under cool conditions
Aromatic Rings with Short Paraffin Side Chain		Pour point varies according to structure; good thermal stability; viscosity varies greatly with temperature; easily oxidizes
Naphtha Rings with Long Paraffin Side Chain		Viscosity varies little with temperature; good oxidation resistance; may have low pour point
Aromatic Rings with Long Paraffin Side Chain		Viscosity varies little with temperature; may have good oxidation resistance if cycles are not numerous; may have low pour point

TABLE 5.5
SYNTHETIC BASE STOCKS AND THEIR USES (ASSEFF)

Type	Structural Formula	Principal Applications
Alkylated Aromatics	$R = C_{10}$ to C_{14}	Automotive and industrial
Olefin Oligomer or Synthetic Hydrocarbon Fluid	$CH_3 - CH - CH_2 - CH - CH_2 - CH_2$ $\quad\quad C_8H_{17} \quad\quad C_8H_{17} \quad\quad C_8H_{17}$	Automotive and industrial
Esters of Dibasic Acids	$RO - \underset{O}{\overset{\parallel}{C}} - (CH_2)_n - \underset{O}{\overset{\parallel}{C}} - OR$ $\quad R = C_8$ to C_{13}	Automotive and aircraft
Neopentyl Polyol Esters	$CH_3 - CH_2 - \underset{CH_2 - OR}{\overset{CH_2 - OR}{\overset{\mid}{\underset{\mid}{C}}}} - CH_2OR$ $\quad R = C_5$ to C_{10}	Automotive and aircraft
Polyglycols	$R - O - (CH_2 - \underset{\underset{R}{\mid}}{CH} - O)_n - H \quad R = C_2$ to C_3	Industrial

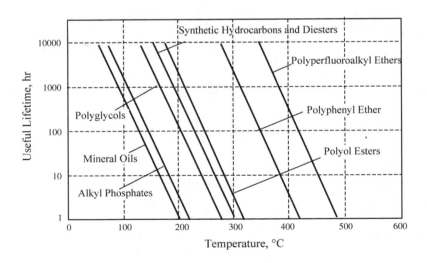

Figure 5.16 Useful life of liquid lubricant classes as a function of temperature (Beerbower).

-461-

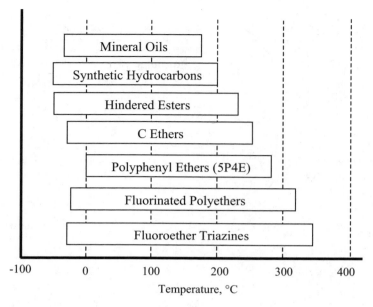

Figure 5.17 *Continuous operating temperature range of liquid lubricant classes (Loomis).*

TABLE 5.6
PHYSICAL PROPERTIES OF HYDROFINISHED (HVI) STOCKS
AND SYNTHETIC BASE STOCKS (ASSEFF)

Type	Kinematic Viscosity (mm²/s)			Viscosity Index	Pour Point (°C)	Flash Point (°C)
	40°C	100°C	−40°C			
Hydrofinished HVI Stocks:						
90 Neutral	17.40	3.68		92	−15	190
100	20.39	4.11		101	−13	192
200	40.74	6.23		99	−20	226
3500	65.59	8.39		97	−18	252
650	117.90	12.43		96	−18	272
150 Bright Stock	438.00	29.46		95	−18	302
Synthetic Base Stocks:						
Alkylated Aromatic	26.84	4.99	9047	119	−18.3	224
Olefin Oligomer	16.77	3.87	2371	126	−43.0	221
Ester of Dibasic Acid						
• Diotyl Sebacate	18.2		3450	176	−15.6	232
Ester of Trimethylol Propane (C₇)	13.94	3.4	2360		−18.3	232
Polyglycol	45.69			150	−21	177

possess to perform these functions include suitable viscosity, slipperiness, high film strength, low corrosivity, low pour point, good cleansing and dispersing ability, nontoxicity, and low flammability. In addition, the lubricant should not foam, that is, it should be capable of getting rid of air (oxygen) to minimize oxidation and to maintain its lubricating characteristics.

5.1.5.3.1 Viscosity Modifiers

Mineral oils respond to temperature changes by becoming less viscous when they are hot and more viscous when they are cold. This change is undesirable because it decreases the stability of the lubricant film at high temperatures and increases viscous drag at low temperatures. The change is greater in aromatic oils and is less in paraffinic oils, hence the benefit of refining to remove aromatics. As a result of the viscosity response to temperature, the choice of oil viscosity is a compromise required to minimize low-temperature starting viscosity and fuel consumption and to maximize high-temperature lubrication. The viscosity of the oil should be such that it can provide proper lubrication, both at low temperatures and at the high temperatures encountered during engine operation. This implies that oil viscosity at low temperatures should not interfere with starting lubrication during the warm-up period, and that the loss of viscosity at high temperatures is minimal. As a general rule, large clearances and high loads demand high viscosity, and high speeds demand low viscosity. Viscosity modifiers or VI improvers are intended to minimize the rate of viscosity change with temperature changes. They are high-molecular-weight polymers such as polyalkyl methacrylates. At low temperatures, the polymer molecules are in a coiled configuration, contributing very little to the oil viscosity. At high temperatures, the configuration is more open and linear and makes a significant contribution to the solution viscosity. This allows oils to span more than one viscosity grade, resulting in so-called multigrade oils such as 10W-30. Mineral oils exhibit Newtonian rheology; the viscosity is independent of the shear stress. However, oil solutions of polymers exhibit complex viscoelastic rheology; the viscosity becomes dependent on the shear stress. When oils containing viscosity modifiers are subjected to moderate shear stress, the viscosity decreases until it approaches the viscosity of the polymer-free oil. When the shear stress is removed, the oil returns to its initial viscosity. One of the results of this shear-dependent viscosity is a decrease in the viscous drag and an improvement in fuel economy compared to single-grade oils, which of necessity are made from high-viscosity base oils. The polymers used as viscosity modifiers, when subjected to high shear stress, can be broken apart irreversibly into smaller segments. A variety of oil-soluble polymers can be used as viscosity modifiers. Their thickening efficiency is a function of polymer structure and average molecular weight. Figure 5.18 shows the viscosity and temperature characteristics of various oils.

5.1.5.3.2 Rust Inhibitors

Rusting is the oxidation of iron surfaces in the presence of water and is worsened markedly by the presence of acids. Water and organic acids are formed during the combustion of gasoline and are blown past the rings into the crankcase, where they become mixed with the engine oil. Together with air, the water and acids cause rusting when appropriate conditions exist, primarily during startup or short-trip driving when the engine is cold and the engine oil retains water. When the engine becomes hot, the water evaporates from the engine oil, and rusting no longer

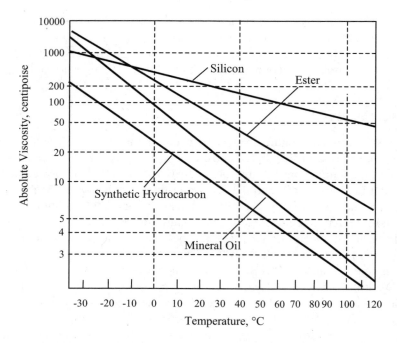

Figure 5.18 *Viscosity and temperature characteristics of various oils (Neale, 1997).*

occurs. The oiliness and film strength of the lubricant are related to its ability to reduce friction and wear under boundary lubrication conditions and to protect against corrosion. The oil should not be corrosive, and it should provide protection against corrosion. The vulnerable parts of an engine valvetrain are the plunger mechanisms of the hydraulic valve lifters, which operate with a very small clearance. Because of this tight fit, a small amount of deposit, either in the form of varnish or rust, will cause the valve lifter to stick and malfunction. Rusting is inhibited universally by using overbased sulfonates, which provide an adsorbed film of sulfonate soap that prevents water and oxygen from reaching the metal surface. In addition, the overbased sulfonates neutralize the acids that catalyze rusting. To be effective, rust inhibitor molecules must adsorb tightly on the iron surface and form a stable film. Thus, rust inhibitors generally are made with the smallest hydrocarbon groups, consistent with having them remain dissolved in the finished oil formulation. Many kinds of polar compounds will inhibit rusting, but the sulfonates are the most widely used because of their cost, capacity for overbasing, and compatibility with other additives. The alkaline earth metals most commonly used in rust inhibitor sulfonates are magnesium and calcium. As little as 0.5% overbased sulfonate can provide adequate rust protection (Watson and McDonnell). Corrosion and rust inhibitors are intended to prevent corrosion and rusting of metal parts in contact with the lubricant. They include ethoxylated phenols and metal sulfonates.

5.1.5.3.3 Anti-Wear Agents

Significant wear occurs in gasoline engines, primarily at the interfaces of the cam and follower. These contacts are under substantial pressure from the valve springs, which close the valves after the cams have opened them. Lubrication in these areas is "boundary" and, in some cases, borders on "extreme pressure." Anti-wear agents are intended to reduce friction and wear and to prevent scoring and seizure. The zinc dialkyl dithiophosphates (ZDDPs) are found to inhibit valvetrain wear and are the principal oil additive for wear inhibition. The ZDDPs are thermally unstable to a degree determined by the structure of the alkyl groups, designated as R, obtained from the hydroxyl raw materials used in their manufacture. The order of thermal stability conferred by these materials is alkylphenols > primary alcohols > secondary alcohols. Generally, higher thermal stability compounds are required in hotter-running engines. These include organic compounds of sulfide, phosphate, and chlorine. However, the phosphorus in ZDDP poisons the emissions system of the vehicle by creating a tenacious coating on the catalyst surface. The trend of phosphorus being permitted in engine oils is declining, as is the use of ZDDP.

5.1.5.3.4 Friction Modifiers

In hydrodynamic lubrication, friction is the result of the viscous drag of the oil film. This drag can be reduced by decreasing the viscosity of the oil; therefore, the lowest practical viscosity base oil will provide the least viscous drag for hydrodynamic lubrication. In the boundary lubrication in which a valvetrain operates, the oil film is not adequate to keep the moving parts separated. The function of friction modifiers is taken over by a film of polar molecules that are strongly adsorbed on the metal or metal oxide surfaces or by reagents that alter the chemical nature of the surface. The drag of boundary lubrication depends on how easily these surfaces slide past one another. Friction modifiers are molybdenum-containing compounds and organic compounds. Examples include sulfurized fats and esters, molybdenum sulfur compounds, colloids of molybdenum disulfide and graphite, polyol monoesters of fatty acids, and amides of fatty acids (Watson and McDonnell).

5.1.5.3.5 Pour Depressants

The pour point is one means used to determine the flow characteristics of lubricants at low temperatures. Lubricants that contain wax will provide little or no lubrication below their pour points, because wax crystals can separate and cause blockage of the oil pump inlet. This causes a starting problem for automobiles in cold weather. Fortunately, the crystallization of wax can be modified to alleviate this problem to an appreciable extent. This is accomplished by adding small amounts of materials that are not waxes but have wax-like segments in their structures. These pseudo-waxes block normal crystallization, but co-crystallize with the forming wax crystals in a way that alters the wax crystal morphology so that they form dense crystals that are much less effective in impeding flow. In addition to being able to co-crystallize with wax, wax modifiers also must be structured to have oil solubility that allows them to come out of solution at the same temperature at which the wax crystals are forming. Wax modifiers frequently are in the form of polymers that have wax-like segments appended to the polymer backbone. Pour depressants include polymethacrylate esters of waxy alcohols and wax alkylated naphthalene (Watson and McDonnell).

5.1.5.3.6 Oxidation Inhibitors

Lubricants should resist oxidation to minimize the formation of harmful products such as acids, lacquer, and sludge. These products will adversely affect engine performance due to oxidative corrosion and thickening of the engine oil. Oxidation problems can be controlled by the use of inhibitors that capture or destroy the active free radicals that are instrumental in the propagation of oxidation reactions. Many kinds of oxidation inhibitors are used in engine oils, including ZDDP, organic sulfides and polysulfides, hindered amine, and phenol.

5.1.5.3.7 Detergents

Detergents are soap-like compounds that have the ability to keep the internal surfaces of engines clean through film formation and by neutralizing the acidic products of combustion and oxidation. Detergents are used to prevent metal attack by acidic by-products of combustion and oxidation and to keep metal surfaces free of deposits. The common detergents are alkaline-earth soaps (most often calcium) of alkylbenzene sulfonic acids, so-called "natural" sulfonic acids made by treating petroleum lubrication stocks with sulfuric acid, alkylphenol sulfides, alkylsalicylic acid, and alkylphosphonic acid (calcium, magnesium, and sodium sulfonates; phenates; salicylates; and carboxylates).

5.1.5.3.8 Dispersants

Dispersants keep insoluble contaminants dispersed in the lubricant. They function somewhat similarly to soap in water to solubilize and disperse the contaminants, and they are particularly effective for low-temperature operation where water accumulates in the crankcase. The ash-less dispersants are designed to have polar chemical heads attached to rather large hydrocarbon groups. The polar heads bind themselves to the sludge particles, and the hydrocarbon groups provide the solubilizing and dispersing action that maintains the sludge in suspension in the oil. Most dispersants currently in use are made from polybutenes of approximately 1000–10,000 molecular weight connected to a polyamine or polyamine derivative. Some also contain alcohol groups. In most cases, the connecting groups are in the form of phenols or succinic acids. These materials generally are manufactured and used as 40–60% concentrates in base oil. They are used in engine oils at 2–8% of diluted product to maintain engine cleanliness (Watson and McDonnell). Because large hydrocarbon polymers are required for viscosity modifiers, it is expedient to incorporate polar chemical groups into these polymers so that they also can perform the dispersant function. The polar groups usually are amines and amine derivatives. In this way, the polymer molecule does double duty, resulting in a net cost savings. When these dispersant-viscosity modifiers are used at the concentration required for viscosity control, they are able to displace half or more of the primary dispersant. Two of the many problems that complicate dispersant technology are (1) the interaction of the amines with the ZDDP to lessen their effectiveness and (2) the attack of the amines on perfluoroethylene oil seals to cause them to stiffen and lose their sealing function.

5.1.5.3.9 Foam Inhibitors

The splashing action of the crankcase and connecting rods provides the necessary mechanical agitation to whip air and other vapors into the oil and subsequently to generate foam. Foam and entrained air can lessen the ability of the oil to provide effective lubrication. Foam inhibitors prevent the lubricant from forming a persistent foam. The most universally used foam inhibitors are the silicone oils, particularly the polymethylsiloxanes, which have extremely low solubility in oil and are effective at remarkably low concentrations. Table 5.7 summarizes the types of lubricant additives, compounds, and chemical structures.

Additives are reactive chemical agents. In the packages and in the blended oil, many physical and chemical interactions occur and affect the finished oil performance, clarity, viscosity, and other more subtle properties. Engine oil additives have a balancing act to perform in reducing engine friction without compromising engine durability, prolonging oil life but at a reasonable cost, improving the effectiveness of oil additives without becoming too aggressive toward seals, and other side effects. One example is the interference effect of the detergent inhibitor package on the functioning of the anti-wear additive (Slater). Table 5.8 shows two examples of additive formulations for top-grade oils: one for use in an SF/CD SAE 30-grade heavy-duty diesel oil, and the other for use in an SF/CC 10W-30 passenger car oil (Watson and McDonnell).

5.2 Engine Lubrication and Lubricants

5.2.1 Engine Lubrication

Figure 5.19 shows some tribological conditions for a typical diesel engine. Most of the materials from which these components are made cannot last long under these stress, speed, and temperature conditions without adequate lubrication.

Engine moving parts, including valvetrain components, are lubricated by one of three systems: a full-pressure system, a splash system, and a combination pressure-splash system. The full-pressure system draws oil from the oil pan by means of an oil pump. The pump then forces oil through drilled and cast passages, called oil galleries, to the crankshaft and camshaft journals. The crankshaft is drilled to permit oil to flow to the connecting rod journals. In some engines, the rods are drilled their full length to allow oil to pass upward to the wrist pin bushings. Bearing throw-off may be helped by spurt holes in the rod to lubricate the cylinder walls and camshaft lobes. Timing gears and chains, lifters, pushrods, and rocker arms also are oiled. In the pressure system, all bearings are oiled either by pumping oil into the bearings or by squirting or dripping it onto them. The lubrication system also provides the pressure to fill the hydraulic lifters.

The most basic splash system supplies oil to moving parts by attaching dippers to the bottom of the connecting rods. These dippers can either dip into shallow trays or into the sump. The spinning dippers splash oil over the inside of the engine. The dippers usually are drilled so that oil is forced into the connecting rod bearings. An oil pump may be used to keep the oil trays full at all times. The basic splash system is used primarily on small one-cylinder engines.

TABLE 5.7
LUBRICANT ADDITIVE TYPES, COMPOUNDS, AND
CHEMICAL STRUCTURES (SCHILLING)

Additive Types	Compounds	Chemical Structure Examples
Detergents (Metallic Dispersants)	Salicylates Sulfonates Phenates Sulfophenates	$R-\bigcirc-\overset{O}{\underset{O}{\overset{\|}{\underset{\|}{S}}}}-O-Ca-OH$
Ashless Dispersants	N-substituted long-chain alkenyl succinimides High-molecular-weight esters and polyesters Amine salts of high-molecular-weight organic acids Mannich base derived from high-molecular-weight alkylated phenols Copolymers of methacrylic or acrylic acid derivatives containing polar groups as amines, amides, imines, imides, hydroxyl, ether, etc. Ethylene-propylene copolymers containing polar groups as above	$CH_2HN[CH_2CH_2NH]_3CH_2CH_2NHCH_2$ (with OH, R substituted phenol groups on each end)
Oxidation, Rust, Corrosion, and Antioxidant Inhibitors	Organic phosphates Metal dithiocarbamates Sulfurized olefins Zinc dithiophosphates Phenolic compounds Aromatic nitrogen compounds Phosphosulfurized terpenes	$R-\bigcirc-\overset{H}{N}-\bigcirc-R$
Viscosity Modifiers	Polymethacrylates Ethylene-propylene copolymers (OCP) Styrene-diene copolymers Styrene-ester copolymers	$-[CH_2CH_2]-[\overset{CH_3}{\underset{}{CHCH_2}}]-$
Foam Inhibitors	Polymethylsiloxane	$-[\overset{R}{\underset{R}{Si-O}}]_n$
Anti-Wear Additives	Organic phosphates Sulfurized olefins Zinc dithiophosphates Alkaline compounds as acid neutralizers	$R-O-\overset{S}{P}\diagdown S \diagup Zn \diagdown S \diagup \overset{S}{P}-O-R$ (with $R-O$ groups)
Pour Point Depressants	Wax alkylated naphthalene Polymethacrylates Cross-linked wax alkylated phenols Vinyl acetate/fumaric-acid-ester copolymers Vinyl acetate/vinyl-ether copolymers Styrene-ester copolymers	$[\bigcirc\bigcirc-R]_n$

TABLE 5.8
EXAMPLES OF ADDITIVE TREATMENT FOR DIESEL
AND GASOLINE ENGINE APPLICATIONS

Additive Type	Diesel Service (%wt.)	Passenger Car Service (%wt.)
Dispersant (Polymer-Amine)	6.0	4.0
Detergent:		
(High Base Phenate)	0.5	–
(Low Base Phenate)	1.5	–
(Low Base Sulfonate)	1.5	0.5
Rust Inhibitor (High Base Sulfonate)	0.5	1.0
Anti-Wear (Zinc Dialkyl Dithiophosphate)	1.5	1.3
Antioxidant (Sulfurized Hydrocarbon)	–	0.5
Viscosity Modifier (Ethylene-Propylene Copolymer)	–	10
Friction Reducer (Sulfurized Fat)	–	0.7

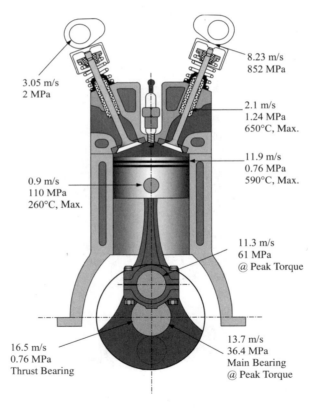

Figure 5.19 Tribological conditions in a diesel engine showing average sliding velocity, contact pressure, and temperature at 4500 rpm. Conditions may vary, depending on the individual engine and application (Committee on Adiabatic Diesel Technology Energy Engineering Board).

The pressure-splash combination system uses an oil pump to supply oil to the camshaft and crankshaft bearings. The movement of the crankshaft splashes oil onto the cylinder walls and other nearby parts. Oil spurt holes in the connecting rods are used to spray oil onto the cylinder walls and piston pins. Oil also is pumped to the cylinder head to lubricate the camshaft, rocker arms, and hydraulic lifters as applicable. Pressurized oil leaving the pushrods sprays oil around the valve cover, lubricating the rocker arms and valve stems. A nozzle sometimes is used to squirt oil onto the timing gears and chain. Some combination systems use a dipper to splash the oil. The combination pressure-splash system best meets the needs of the modern engine and therefore is in almost universal use. Oil is drawn constantly from the bottom of the engine, is pressurized, and then is distributed to the engine parts. The oil then drips to the bottom of the oil pan sump and begins the cycle again.

Most automotive engines use a full-pressure system, in which the oil is pumped from the oil sump to the main bearing and connecting rods, and then up the connecting rods to the piston pin. In Type V valvetrain (overhead valve) engines, a portion of the pumped oil travels through pushrods (in some cases), over rocker arms, and past valve stems, and then meters down the valve guides. In many engines, the cylinder walls and piston pins depend on splash lubrication by the oil that is thrown off the main bearing. Figure 5.20 depicts the lubrication system of an overhead valve engine.

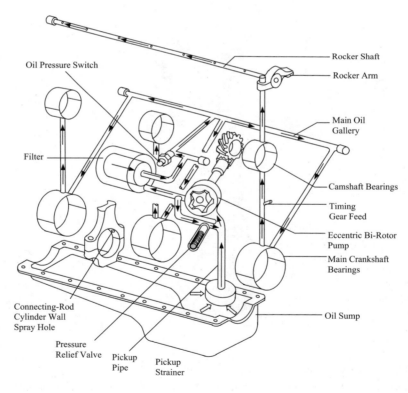

Figure 5.20 *Cutaway showing the oil flow through a pushrod and rocker valve engine (Heisler).*

Oil that is allowed to pass the pressure relief valve is carried to the bearing through the oil galleries. Passages from the main gallery are drilled to the main bearings. The crankshaft and camshaft are drilled to carry oil from the main journals to the rod journals. The valvetrain is lubricated by oil spray and, in the case of overhead valve engines, by oil pumped into the rocker arm shaft, through the rockers to the valves. Oil also can reach the rocker arms via hollow pushrods. The bearing receives oil from a hole drilled through the bearing support. The insert has a hole that aligns with the one in the bearing support. The oil passes through and engages the turning journal and is pulled around between the insert and journal. Some inserts have shallow grooves cut into them to assist in spreading the oil.

Oil is forced constantly through the bearings. Oil leaving a bearing is thrown forcefully outward. This helps to produce the fine oil mist inside the engine that is useful in lubricating hard-to-reach areas. All oil eventually flows back into the sump.

Four principal types of oil pumps have been used on engines: (1) external gear pumps, (2) internal gear pumps, (3) rotary pumps, and (4) vane pumps (Figure 5.21). The gear and the rotary pumps are the most widely used types.

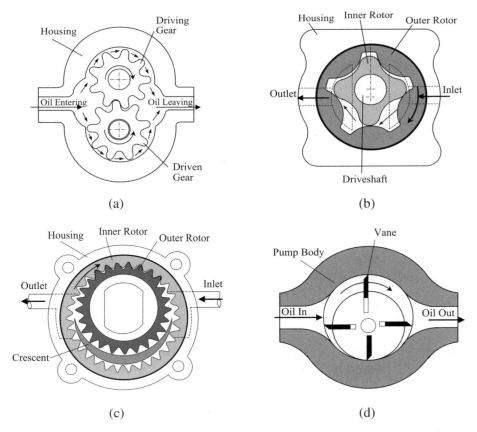

Figure 5.21 *Schematics of oil pumps: (a) external gear pump, (b) rotary oil pump, (c) internal gear crescent pump, and (d) vane-type oil pump (Kovach et al.).*

An oil pump can produce pressures far beyond those necessary for lubrication. The pumps are designed to carry a large flow of oil; however, when the usable pressure limits are reached, a pressure relief valve opens and allows oil to return to the sump (Figure 5.22). The relief valve will allow the pressure reaching the bearing to remain at a predetermined level. By varying spring tension, the pressure can be raised or lowered. The relief valve can be located at the pump or at any spot between the pump and the bearings.

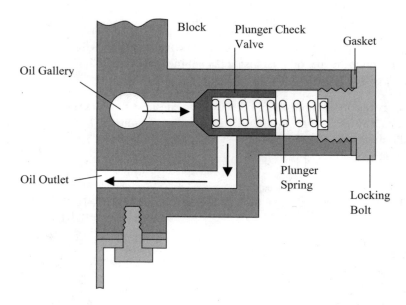

Figure 5.22 *Oil relief mechanism (Stockel* et al.*).*

5.2.2 Engine Lubricants

The physical characteristics and performance of engine oils are defined by the SAE International viscosity grades, the American Petroleum Institute (API) service classifications, U.S. military specifications, and OEM requirements.

Viscosity is one of the most important properties of a lubricant. It can be defined as the resistance of a lubricant to flow or a measure of the lubricant's internal friction or resistance to shear. Viscosity classification systems for engine oils are described by SAE J300 and J1536. There are 11 viscosity grades, ranging from 0W to 25W, and from 20 to 60. Each viscosity grade must meet a number of requirements. Monograde viscosity oils with the letter "W" (winter) must meet the minimum borderline pumping temperature, the maximum low-temperature viscosity (at a prescribed temperature), and the minimum kinematic viscosity requirements at 100°C. Monograde oils without the letter "W" must meet only the high-temperature (100°C) viscosity requirements. Multigrade oils should satisfy the appropriate requirements of both the "W" and the non-"W" grades. The winter requirement of being pumpable at low temperatures and the nonwinter requirement of sufficient viscosity at high temperatures ensure proper lubrication and protection of engine parts during all-season operation. These requirements, along with those of the U.S. military, are summarized in Tables 5.9 and 5.10.

TABLE 5.9
ENGINE OILS AND VISCOSITY
REQUIREMENTS (SAE J300)

Property	0W	5W	10W	15W	20W	25W	20	30	40	50	60
Viscosity, Maximum, Pa•s	3.25	3.5	3.5	3.5	4.5	6.0	...	...	...	...	...
Borderline Pumping Temperature, °C	−35	−30	−25	−20	−15	−10	...	...	...	...	...
Viscosity at 100°C, mm²/s:											
Minimum	3.8	3.8	4.1	5.6	5.6	9.3	5.6	9.3	12.5	16.3	21.9
Maximum	...	...	...	...	...	...	9.3	12.5	16.3	21.9	26.1
Stable Pour Point, Maximum, °C	...	−35	−30	...	...	...	...	...	...	...	...

TABLE 5.10
VISCOSITY REQUIREMENTS FOR ENGINE OILS
(SAE J300, MILITARY GRADE)

Property	10W	30	40	5W-30	10W-30	15W-40
Viscosity at 100°C, mm²/s:						
Minimum	5.6	9.3	12.5	9.3	9.3	12.5
Maximum	7.4	12.5	16.3	12.5	12.5	16.3
Cold Cranking Simulator Viscosity, Pa•s:						
Minimum	3.5	...	...	3.25	3.5	3.5
Maximum	3.5	...	...	3.5	3.5	3.5
Pour Point, Maximum, °C	−30	−18	−15	−35	−30	−23
Flash Point, °C	205	220	225	200	205	215
Stable Pour Point, Maximum, °C	−30	...	...	−35	−30	−23

Engine lubricant viscosities are strongly dependent on temperature and pressure, as shown in Figure 5.23. Viscosity decreases as temperature increases and as pressure decreases.

Figure 5.24 shows the dependency of oil viscosity on shear rate and temperature.

5.2.3 Synthetic Engine Base Oil and Additives

Table 5.11 summarizes some of the properties of typical synthetic lubricants. The assessments refer to formulated products, not necessarily to the base oils.

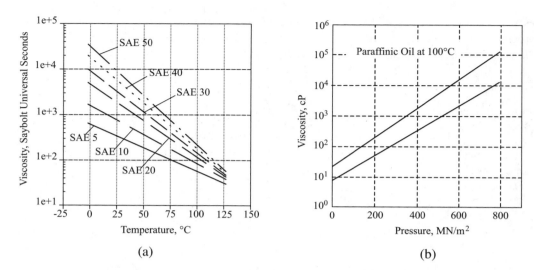

Figure 5.23 *Effects of temperature and pressure on engine oil viscosity: (a) effect of temperature on viscosity (°C), and (b) effect of pressure on viscosity.*

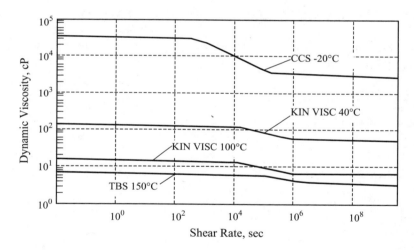

Figure 5.24 *Effect of temperature and shear rate on the viscosity of an SAE 10W-X motor oil.*

The performance advantages of synthetic oils result from their thermal stability, low volatility, and flow properties at low temperatures. Although synthetic engine oils offer certain performance advantages over conventional lubricants, their relatively high price (three to five times more than the cost of conventional mineral oil-based products) has prevented them from capturing a significant share of the domestic engine oil market. In 1991, synthetic lubricants accounted for

TABLE 5.11
RELATIVE PROPERTIES OF SYNTHETIC LUBRICANTS (ASHTON AND STRACK)

	Viscosity Index	High-Temperature Stability	Lubricity	Low-Temperature Properties	Hydrolytic Stability	Fire Resistance	Volatility
Polyalphaolefins (PAOs)	Good	Good	Good	Good	Excellent	Poor	Good
Diesters	Varies	Excellent	Good	Excellent	Fair	Fair	Average
Polyol Esters	Good	Excellent	Good	Good	Good	Poor	Average
Alkylbenzenes	Poor	Fair	Good	Good	Excellent	Poor	Average
Polyalkylene Glycols	Excellent	Good	Good	Good	Good	Poor	Good
Phosphate Esters	Poor	Excellent	Good	Varies	Fair	Excellent	Average
Silicones	Excellent	Excellent	Poor	Excellent	Fair	–	Good
Fluorinated Lubes	Excellent	Excellent	Varies	Fair	Excellent	Excellent	Average

less than 1% of the total motor oil market (Culpon and Mead). Another reason that synthetic engine oils have not achieved greater market share is that engine manufacturers have not permitted longer oil-drain intervals in their warranty maintenance schedules when synthetic motor oils are used. This is because the operating environment of motor oils can vary considerably, depending on engine operating conditions. Crankcase engine oils are constantly contaminated by combustion by-products during engine operation. Some of these by-products are acidic in nature and can cause corrosion, rust, and wear; others contribute to the formation of sludge and varnish deposits. To counteract these harmful effects, engine oils are formulated with additive packages designed to protect the engine from these contaminants. The additive packages are chemically similar for either synthetic or mineral oil-based engine oils and are depleted with use. When the additive packages are depleted, the engine oils are unsuitable for further use, requiring oil and filter changes. Early polyalphaolefin (PAO) based synthetic engine oils tended to harden and shrink elastomer seals, resulting in oil leakage. Conversely, ester-based engine oils—in addition to being more costly than comparable PAO-based products—tended to soften and swell elastomer seals excessively. Most synthetic oils marketed today contain primarily PAO base stocks in combination with lesser amounts of synthetic esters to provide neutral or slightly positive seal swell, thereby preventing oil leaks.

The additive package in an engine oil constitutes as much as 15% of the entire formulation (Sutor and Bryzik). Performance of an engine oil is at least as dependent on the additives as it is on the base stock. Table 5.12 shows oil additives required for engine improvements.

TABLE 5.12
OIL IMPROVEMENTS ANTICIPATED TO MEET
DIESEL ENGINE TRENDS (SAKAMOTO)

Engine Trend	Oxidative Stability	Thermal Stability	Detergency and Dispersancy	Anti-Wear Properties	Multigrade Viscosity
Higher Horsepower	x	x	x		
Retarded Injection Timing		x			
Noise Control by Enclosure	x	x			
Minimized Oil Capacity	x				
Direct Injection Combustion	x	x	x		
Oil Filter Installation	x		x		
Lower Fuel and Oil Consumption					x
Extended Oil Drain Interval	x		x	x	

Fully formulated synthetic oils have better high-temperature properties than those of engine oils and are notably more resistant to thermal and oxidative degradation. This is an important factor in maintaining engine cleanliness because the products of lubricant decomposition and oxidation are precursors to harmful engine sludge and varnish deposits. Oxidative degradation also causes oil thickening. Synthetic oils remain fluid at temperatures of −45°C and lower. This ability makes the fluids attractive in regions where winter temperatures drop below 0°C. At these temperatures, synthetic lubricants permit faster engine starts and provide lubricant flow quickly throughout the engine, thereby maintaining a protective lubricant film over critical valvetrain components. Volatility also is an important physical property of an engine oil because there is a direct correlation between volatility and oil consumption. The low volatility of synthetic motor oils translates to reduced oil consumption, thereby reducing or eliminating the need to add oil between oil changes.

In diesel engines, acidic products of combustion and soot blow by the piston rings into the lubricating oil in the crankcase. Detergent additives chemically neutralize the acidic combustion products and keep the soot suspended in the oil so that it does not settle out. Typical compounds in detergent additives include metal (Mg/Ca) sulfonates or phenates, and dispersant additives contain polyisobutenyl succinimides.

Most additives possess oxidative and thermal stability that are lower than or comparable to those of the mineral oils because many contain long hydrocarbon chains that will not remain unchanged during engine high-temperature operation. For antioxidant, hindered phenol (e.g., 2,4–di-t-butyl-p-cresol or zinc dithiophosphates (ZDP)), zinc dialkyl dithiophosphate (ZDDP) is one of the most effective and most often used anti-wear additives in engine oils. Another anti-wear additive is tricresyl phosphate (TCP). Organic phosphites and sulfurized olefins also are used frequently in anti-wear additives. Likewise, anti-rust additives include metal sulfonate formulations. Although performing outstandingly against wear and oxidation, ZDDP is the prime source for phosphorus poisoning of catalyst systems and is at the top of the list for reduction or elimination.

Cold engine operation in short trips, which permits condensation and accumulation of corrosive products in the oil, leads to rusting of engine components. Figure 5.25 summarizes the effects of additive type and concentration on rusting, based on the reference oil QX as 100 (Pless). The results show that several magnesium sulfonates provided better rust protection than other additives that were evaluated. Rust protection increased greatly and approximately linearly with increasing magnesium content. At lower additive treatment levels, the 286 TBN calcium sulfonate and the calcium phenates also provided good rust protection.

5.2.4 Lubricant-Related Engine Malfunctions

Most problems associated with internal combustion engine lubrication are related to the by-products of combustion, their entry into the crankcase through blowby, and lubricant decomposition. This implies that fuel and lubricant quality and combustion both are important parameters. The major causes of engine malfunction that are due to lubricant quality are deposit formation, oil consumption, lubricant contamination, oil thickening, ring and valve sticking, corrosion, and wear.

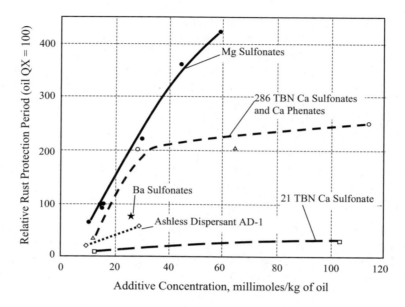

Figure 5.25 *Effects of additive type and concentration on rust protection (Pless).*

The main source of lubricant contamination is blowby gases from the combustion chamber. The blowby is a mixture of nitrogen oxides (NO_x), sulfur compounds (SO_2, SO_3, H_2SO_4), carbon compounds, hydrocarbons (unburned fuel), peroxides, air, water, and carbon dioxide. These species are blown past the piston rings and exhaust valve guides into the crankcase. There, they can interact with each other and with the lubricant to form harmful products. Such products include soot, carbon, lacquer, varnish, and sludge.

Soot is an important particulate contaminant in crankcase lubricants. It can be derived from the inefficient combustion of fuel and the burning of lubricating oil that passes along the pistons into the combustion chamber. Fuel-derived soot, which is more commonly encountered in diesel operation, results from the use of fuel with a wide boiling range, which, under certain operating conditions, does not burn completely. Soot is not pure carbon and contains an appreciable amount of hydrogen, oxygen, and sulfur in combined form. Soot particles basically are hydrocarbon fragments with the hydrogen atoms partly stripped off. These particles are strongly attracted to each other and to the polar compounds in the oil, and they tend to form aggregates. Soot deposits are soft and flaky in texture and commonly are found in the combustion chamber.

Carbon deposits, which are more prevalent in diesel engine operation, are hard and result from the carbonization of the liquid lubricating oil and the fuel on hot surfaces. Carbon deposits have lower carbon content than soot and, in most cases, contain oily material and ash. Carbon deposits are commonly found on the piston top lands and crowns, in the piston ring grooves, and on the valve stems and fillets. They are a source of valve sticking or seizure.

The unburned air/fuel mixture and either oxidized or partially oxidized reactive intermediates in the blowby promote lubricant oxidation. This results in the formation of a variety of oxygenated products which, when exposed to high temperatures, result in lacquer and varnish (Bouman). The term "lacquer" usually describes this type of deposit in diesel engines, whereas the term "varnish" is used to describe such deposits in gasoline engines. Lacquer often is derived from the lubricant and generally is water soluble. On the other hand, varnish is fuel related and is acetone soluble. Lacquer commonly is found on pistons and cylinder walls and in the combustion chamber. Varnish occurs on valve lifters, piston rings, and positive crankcase ventilation (PCV) valves.

The three major causes of sludge formation are (1) oxidation of the lubricant, (2) oxidation and combustion products in the blowby, and (3) the accumulation of combustion water and dirt. Heat can drive off water and thereby cause a change in consistency. Low-temperature sludge, which is more prevalent in gasoline engines, is more watery in appearance and forms at temperatures below 95°C. High-temperature sludge, which is more common in diesel engines, forms above 120°C. Sludge commonly is found in areas of low oil velocity, such as crankcase bottoms and rocker boxes.

The volume of deposits in an engine depends on the fuel quality, engine operating conditions (e.g., speed, load, and temperature), quality of combustion and blowby (its oxygen content, presence of sulfur and nitrogen compounds, and so forth), and the integrity of the seal between the combustion chamber and crankcase. Low-temperature deposits (e.g., soot, varnish, and low-temperature sludge) usually are encountered in gasoline engines subjected to intermittent (stop-and-go) driving, because this type of operation does not allow the engine to achieve the optimum temperature necessary to drive off contaminants. High-temperature deposits (e.g., carbon, lacquer, and high-temperature sludge) typically are found in gasoline engines that experience long continuous operation and in diesel engines. These deposits result from thermal and oxidative degradation of the lubricant and additives. Additional stresses on the lubricant stem from the increased emphasis on cleaner exhaust emissions and efficient combustion of fuel at high temperatures and the use of power accessories such as air conditioning. Nitrogen oxides in the blowby and high temperatures cause oxidative and thermal degradation of the lubricant. The result is the formation of oxygenated products that act as precursors to deposit-forming species. These products are polyfunctional molecules with the ability to polymerize thermally to form products that have higher molecular weights. If the oxygen content of precursors is low, and the polymer product is of low molecular weight and good oil solubility, then only oil thickening is observed. However, if the oxygen content of the precursors is high, and the polymerization results in products of low lubricant solubility, then varnish and resin are formed. When the polymerization occurs on hot metal surfaces, varnish is formed; when it occurs in the bulk lubricant, resin is formed. Because of its low solubility in the lubricant, resin tends to separate out on metal surfaces as lacquer, which can be amber or black in color. At low soot levels, the interaction of resin and soot results in resin-coated soot particles, which separate out on piston surfaces as black lacquer. At high soot levels, extensive interaction between soot and resin forms soot-coated resin particles that have little or no ability to adhere to metal surfaces. Instead, they accumulate as deposits in areas of low oil flow, such as in the grooves behind the piston rings. Sludge results from the interaction of the oxygenates with soot in the presence of oil and water. Deposit levels depend on fuel quality. In the case of sulfur-rich fuels, the oxidation of sulfur results in the formation of sulfur acids, which catalyze the rearrangement of hydroperoxides to carbonyl compounds and their subsequent polymerization to resin. The addition of basic detergents, which are used in diesel engine oils to counter the adverse effects of sulfur, helps to alleviate this problem.

Oil thickening can result from a combination of oxidative degradation of the lubricant and the accumulation of insolubles. The auto-oxidation of the lubricant, accelerated near the oxidation inhibitor depletion stage, can lead to oxygenated products, which can cause a viscosity increase through polymerization. The contaminant-related thickening arises from the suspension of fuel-derived insolubles in the bulk lubricant.

Oil consumption is related mainly to the lubricant that travels past the piston rings and valve guides and burns in the combustion chamber. Burning of the lubricant, along with inefficient fuel combustion, leads to soot and carbon deposits on the inside of the combustion chamber, piston top land, ring grooves, and valve fillets. The extent of oil consumption depends on a number of lubricant and equipment design-related factors. The viscosity, volatility, and seal-swell characteristics of the lubricant play an important role in this regard. A certain minimum amount of oil is necessary for proper lubrication of the cylinder walls and pistons, as well as the valve stems and guides. High oil consumption is indicative of problems in pistons and cylinders, such as increased wear of the cylinder bore, stuck piston rings, or out-of-square grooves. Under these circumstances, the blowby gases have an increased chance of entering the crankcase, which is likely to further complicate the situation. There is some evidence that oil consumption contributes to particulates in diesel emissions. One strategy is to use lower-thickness oil films. This strategy is effective, except that beyond a certain limit, the oil film thickness will be ineffective for proper lubrication, and ring/liner and stem/guide scuffing problems will occur. Lubricant volatility is another important factor responsible for increased oil consumption. Lighter base oils that are used to formulate multigrade diesel oils not only contribute to the formation of a less-effective lubricating film, but also can readily leak past the piston rings and burn.

The major cause of ring sticking is the formation of deposits in the piston grooves. The consequence is a loss of oil seal, which not only favors blowby but also results in poor heat transfer from the piston to the externally cooled cylinder wall. This is quite serious because thermal expansion of the pistons will lead to a loss of compression and, ultimately, to engine seizure.

The deposits also contribute to valve sticking. If the valve is stuck in the open position, the piston could break the valve when the piston reaches top dead center (TDC). Too little lubricant coming down from the guide also can cause valve sticking due to high friction. High friction can lead to overheating of the stem and guide surface, thus causing scuffing and seizure.

Diesel fuel with a high sulfur content can cause corrosive wear at the piston ring/cylinder and valve/insert, especially in large, slow-speed, marine diesel engines. Corrosive wear, which is more commonly associated with combustion and oxidation products, results from the attack of either sulfuric acids or organic acids on ferrous surfaces. This happens when the engine operating temperatures are below the dewpoint of these acids. The organic acids can originate from a thermo-oxidative system (Rizvi).

5.3 Valvetrain Friction Loss or Energy Consumption

5.3.1 Overview

A significant amount of energy is lost in internal combustion engines due to valvetrain friction. Figure 5.26 shows the percentage of friction loss for various components in an engine. The major components contributing to engine friction are the piston ring assembly, crankshaft and oil pump system, valvetrain, and accessory system. Historically, attention has focused on valvetrain durability and dynamics. Today, more efforts are directed toward reducing valvetrain friction to achieve fuel efficiency and emissions reduction.

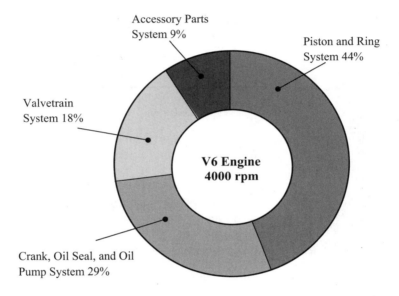

Figure 5.26 *Valvetrain friction loss percentage in engine energy consumption (Nakasa).*

Many factors, including valvetrain types, operating conditions, and the effectiveness of lubrication, can influence friction loss or energy consumption. Optimization of these factors can significantly reduce friction loss, reduce emissions, and improve fuel efficiency.

5.3.2 Speed Effects

Figure 5.27 shows the effect of speed on friction for various engine components. The percentage of friction increases with increasing speed for the crankshaft bearings, connecting rod bearings, and pumps; however, the friction percentage decreases for the piston and ring assembly and the valvetrain assembly.

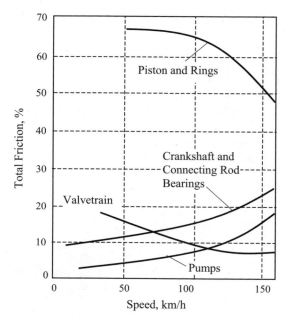

Figure 5.27 *Effect of speed on engine component friction (Cleveland and Bishop).*

Figure 5.28 illustrates the behavior of the components making up the total valvetrain loss, plotted as a percentage of the gross drive torque at various engine speeds. Although the interface between the cam and the follower dominates the total valvetrain friction level at low speeds, other sources of friction tend to gain significance as a percentage at increasing speeds.

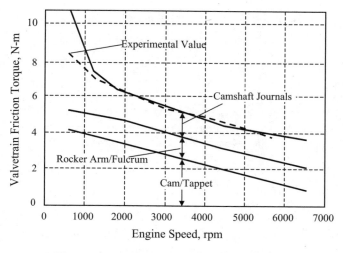

Figure 5.28 *Effect of speed on valvetrain friction (Staron and Willermet).*

The cam/follower contact contributes most to friction loss at low speed. Figure 5.29 shows a comparison of instantaneous camshaft friction torque at 444 rpm and 1469 rpm. Note the maximum friction torque location, which diminishes as speed increases.

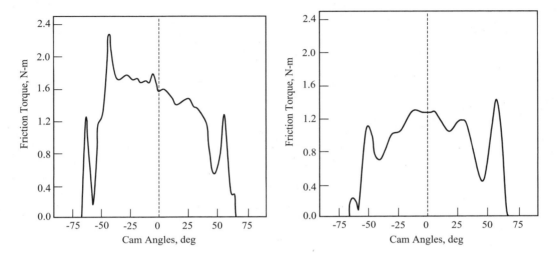

Figure 5.29 *Instantaneous camshaft friction torque at (a) 444-rpm cam speed and (b) 1469-rpm cam speed (Zhu).*

5.3.3 Effects of Valvetrain Type

Various valvetrain designs have inherent advantages or disadvantages over others. The Type I direct-acting valvetrain has lower cam torque than other types of valvetrains. Figure 5.30 shows cam torque as a function of engine speed for mechanical types of valvetrains with flat follower surfaces. The fundamental characteristic exhibited is high torque at low speeds and decreasing torque at high speeds. For all valvetrain types, higher losses are observed at 200–1000 engine revolutions per minute. These high-friction characteristics are theorized to be due to incomplete fluid film formation or boundary lubrication between the cam and the follower, journal, and bearing. The lower surface velocities at the low speeds appear to not form sufficiently thick oil films to fully separate the cam and follower, journal, and bearing, resulting in asperity contact and the higher friction levels. It is known that contact stresses are higher at lower speeds. This fact, coupled with difficulty in complete oil film formation, may lead to the cam/follower wear frequently observed after a low-speed test.

Figure 5.31 shows cam torque as a function of engine speed for hydraulic types of valvetrains with flat-faced followers at various engine speeds, and similar trends remain, that is, friction decreases as the engine speed increases. Type I valvetrains have the lowest friction compared to other types of valvetrains.

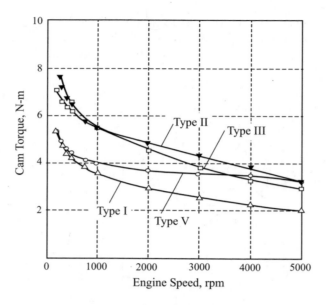

Figure 5.30 *Net drive torque for a mechanical valvetrain varied among the valvetrain types, as well as engine speed (Armstrong and Buuck).*

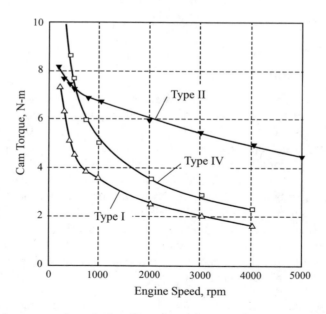

Figure 5.31 *Net drive torque for a hydraulic valvetrain varied among the valvetrain types, as well as engine speed, similar to the mechanical valvetrain. Low operation speed resulted in significantly higher torque requirements (Armstrong and Buuck).*

The direct-acting Type I valvetrain is the simplest valvetrain, with only one heavily loaded area of contact. Because no rocker ratio is involved, these cam lobes tend to be larger and use bigger base circles, resulting in higher surface velocity. This type of valvetrain tends to have fewer moving parts, hence a lower dynamic mass and the lowest spring load. Another contributing factor is the tappet being free to rotate, reducing sliding and increasing entrainment velocity. The combination of high surface velocity, rotating follower, low spring load, and ample oiling seems to account for the direct-acting valvetrain being the lowest in friction.

The Type V pushrod valvetrain is the next lowest in energy consumption. The base circle and over-nose radius of the cam is lower than those of the direct-acting type, but the additional moving components are well designed from a friction viewpoint. The pushrod ends generally are pressure-lubricated ball and sockets, the lifter is free to rotate, and the motion of the rocker on the valve tip is largely rolling.

The Type III (overhead cam [OHC], center-pivot rocker) valvetrain is relatively simple compared to the pushrod type, but the motion between the cam and the rocker has a significant sliding content. Valvetrains with followers that are not contained in a bore are more difficult to lubricate. Extra design effort must be expended in these valvetrain types to ensure an uninterrupted supply of oil on the converging side of the contacts. The higher sliding content is believed to be the major cause of the higher torque requirements. The torque requirements can be reduced considerably if the rolling elements are used in the follower.

The Type IV (OHC, center-pivot rocker with lifter) valvetrain has essentially the same arrangement as the pushrod train, except the lifter contacts the rocker directly. This added interface appears to have additional sliding motion, resulting in higher energy requirements. The lifter is free to rotate and appears well oiled, but the particular design tested had very high spring loads, resulting in high drive requirements.

The Type II (OHC, end-pivot flat rocker) valvetrain tends to have significant sliding content, high spring loads, the highest rocker ratio, and moderate difficulty in lubricating it throughout its valve event. The combination of these factors make the Type II valvetrain appear to be the least efficient. However, if roller rocker arms are used instead of flat arms, the friction torque requirement can be reduced significantly.

5.3.4 Temperature Effects

Friction in all valvetrain types is found to be strongly sensitive to temperature. Within a certain range, low temperatures give low friction (Figure 5.32). This suggests an analogous situation to engine speed, where low-temperature high-viscosity oil permits more complete film formation and hence lower friction. On the other hand, the cam journals lose more energy at low temperatures due to higher viscous shear rates.

5.3.5 Spring Load Effects

Valvetrain load, whether from the valve spring, inertial effects, or oil pressure within the hydraulic lifter, will have a direct effect on valvetrain friction. The spring load acts only during the

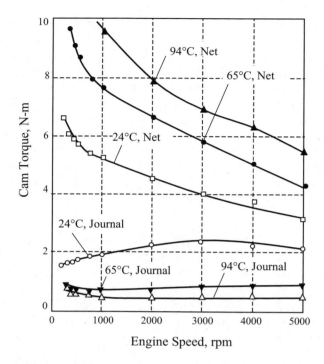

Figure 5.32 *The energy requirements of valvetrain components at various temperatures. The sum would be the gross torque seen at the drive motor. Journals clearly are seen as torque reduced with elevated temperatures, whereas valvetrains require more torque (Armstrong and Buuck).*

lift event and has a proportional effect on the cam drive torque (Figure 5.33). Drive torque can increase when hydraulic components are used. The oil pressure acting over the plunger area and the return spring force supply the motive force in all hydraulic components, but the effect on torque depends on the geometry of the particular valvetrain. If the valvetrain followers are free to rotate, the valvetrain is less sensitive than the valvetrain having followers that slide on the cam. The location of the lash adjuster relative to the cam also affects friction torque.

5.3.6 Rolling Element Effects

One effective method of reducing valvetrain friction is to use rolling-element bearings at critical locations. Such bearings are well suited to carry high loads at low speeds. Roller bearing designs such as roller rocker arms and roller hydraulic lifters have been used widely in both production and racing engines. Rolling-element bearings are included at the cam-lifter interface, the rocker arm pivot, and the interface between the rocker nose and the valve stem. Studies have shown that the valvetrain torque can be reduced by as much as 50% at 1500 rpm using rolling-element bearings; the projected vehicle fuel improvement is 2.9% (Kovach *et al.*). Figure 5.34 shows a

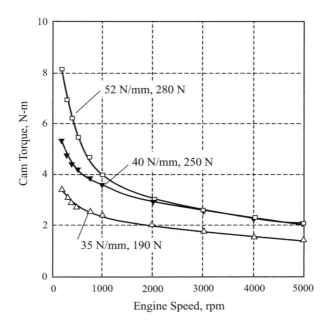

Figure 5.33 *The effect of spring load on cam torque as a function of engine speed (Armstrong and Buuck).*

comparison of flat and roller follower effects on valvetrain friction. Friction also can be reduced drastically with a needle bearing roller in place of a sliding flat follower.

5.3.7 Lubricant Type Effects

Another effective reduction in valvetrain friction is through effective lubrication. Oil viscosity has a relatively small effect on overall friction loss, but a marked reduction in friction can be obtained with oils containing friction modifiers or additives. The results shown in Figure 5.35 suggest that the selection of the proper lubricant may reduce friction levels to those obtainable by component redesign.

Figure 5.36 shows the effect of engine speed and lubricant type on the valvetrain friction and total engine friction. Note that friction torque for valvetrains decreases as engine speed increases; however, total engine friction increases as speed increases. The reduction in fuel economy by reducing friction torque is not the sole benefit: the promotion of a hydrodynamic lubrication regime may significantly reduce valvetrain wear.

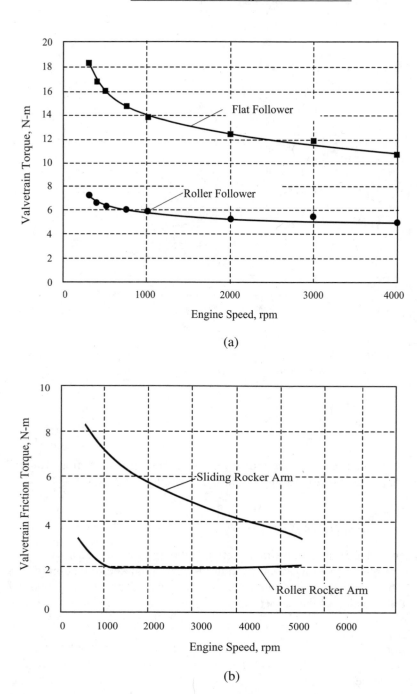

Figure 5.34 *Comparison of valvetrain friction with a roller and flat followers: (a) flat and roller lifter (Type V valvetrain), and (b) flat and roller rocker arm (Type II valvetrain).*

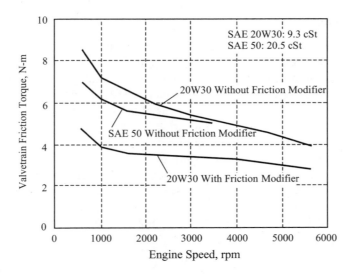

Figure 5.35 *Effects of oil viscosity and additive on friction torque loss (Staron and Willermet).*

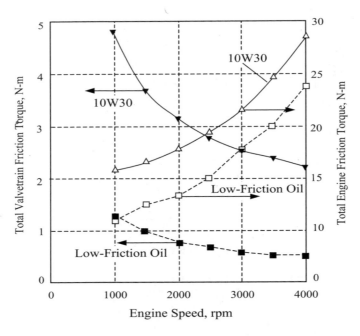

Figure 5.36 *Effect of lubricant on total engine friction and valvetrain friction (Baniasad and Emes).*

5.3.8 Oil Flow Rate Effects

The oil flow rate is another important factor in engine lubrication, especially for the valvetrain system or piston-connecting rod system where the oil flow path is narrow and motions are complex. An excess oil supply should be avoided for obtaining high efficiency of oil pump operation. An increase in oil consumption is another undesirable problem because oil may leak into the combustion area if excessive flows are made in the piston-cylinder area. The friction torque does not show a large difference with various intake oil flow rates at high engine speeds. The friction torque is minimal at 3 l/min (Figure 5.37). However, the torque increases markedly with increasing oil flow rates at 1000 rpm. A small intake oil flow rate less than 2 l/min increases friction torque. However, these conditions may be different according to the size of the engine and the geometry of the lubricating system. To obtain a higher-performance engine, the flow rate of the valvetrain system must be controlled separately from the crankshaft and piston-connecting rod systems. Also, the fraction of oil flow should be controlled to adequately lubricate each component satisfactorily.

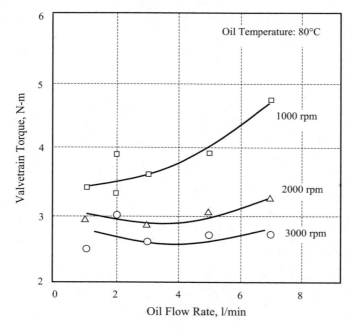

Figure 5.37 *Friction torque of the valvetrain system as function of oil flow rate (Tran et al.).*

5.4 Valvetrain Wear

5.4.1 Introduction

Although minimizing valvetrain friction can lead to a benefit in engine fuel economy, reducing wear of the valvetrain components can increase the durability and reliability of the engine. Several interfaces in the valvetrain system—including the cam and cam follower, stem and guide, and valve seat and insert—are subject to wear and have a significant impact on valvetrain reliability and durability. Wear of valvetrain components can affect valvetrain operation: excessive wear of valvetrain components can lead to excessive or diminished lash, noise, galling, seizure, and catastrophic failure of the valvetrain and engine.

5.4.2 Cam and Follower Interface

The basic forms of cam and follower deterioration are adhesive (mild wear), pitting or contact fatigue (moderate wear), and scuffing (severe wear). Mild adhesive and/or abrasive wear either can be a slow polishing process, lessening with time as a hard smooth surface is produced, or a continuous process at the surface, ultimately resulting in the removal of sufficient material to affect engine performance. In either case, it depends on the Hertzian contact stress and the effectiveness of lubrication between the cam and tappet. The appropriate formulae for the Hertzian stress are listed below (Neale, 1994).

For a flat tappet face on the cam:

$$\sigma_{max} = K \left[\frac{W}{R_c \times b} \right]^{1/2} \tag{5.8}$$

where

σ_{max} = peak Hertzian stress at the point under consideration, N/mm^2
W = load between the cam and tappet, N
b = width of the cam, mm
R_c = cam radius of curvature at the point under consideration, mm
K = cam and tappet material constant, which is 188 for steel on steel, 168 for steel on cast iron, or 153 for cast iron on cast iron

For a spherical-faced tappet:

$$\sigma_{max} = X \times K \left[\frac{1}{R_c} + \frac{2}{R_T} \right]^{2/3} \times W^{1/3} \tag{5.9}$$

where

R_T = tappet radius curvature, mm
X = obtained from Figure 5.38
K = cam and tappet material constant, which is 838 for steel on steel, 722 for steel on cast iron, or 640 for cast iron on cast iron

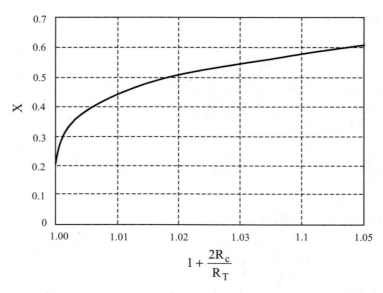

Figure 5.38 *Constant for the determination of contact stresses with a spherical-end tappet (Neale, 1994).*

For curved and roller tappets with a flat transverse face:

$$\sigma_{max} = K\left[\left(\frac{1}{R_c} + \frac{1}{R_T}\right)\frac{W}{b}\right]^{1/2} \qquad (5.10)$$

where

    K = cam and tappet material constant, which is 188 for steel on steel, 168 for steel on cast iron, or 153 for cast iron on cast iron

At the cam and follower interface, contact fatigue and spalling due to high contact stress is a phenomenon in which macroscopic particles separate from a surface in the form of flakes or chips from cyclic stress variations. This is the failure of a surface, manifested initially by the breaking-out of small, roughly triangular portions of the material surface. Failure is due primarily to the high cyclic stresses causing fatigue failure to start at a point below the surface where the highest combined stresses occur or the points of weakness in the material. After initiation, a crack propagates to the surface, and the subsequent failure mechanism may be that the crack then becomes filled with lubricant or one or more of its components, which help to prise out a portion of the material. Heavily loaded surfaces will continue to pit with increasing severity over time. The development of these cracks results in material being removed from the surface to leave a pit of fissure. In an advanced stage, this can result in flaking of the surface, thus affecting engine performance.

Scuffing, scoring, and galling are the most serious of the various forms of failure in which complete cam lobe and follower face surface loss can occur within a short time. Scuffing is caused by the microwelding together of the two contacting surfaces under high pressure, leading to dry friction and high temperatures with oil film breakdown, with the subsequent tearing of part of the welded material. Once scuffing has commenced, it proceeds rapidly and can result in seizure. It is likely to start from high spots due to poor surface finish during early running-in of new parts, particularly when a high degree of relative sliding occurs under poor lubrication conditions. Surface films, formed by the interaction between oil anti-wear additives and the metallic contact surfaces, can reduce the tendency to scuff. Positive rotation of the followers to provide rolling in combination with sliding, and thus a larger contact surface with resulting lower contact stress, also reduces any tendency to scuff.

It is evident from an analysis of service failures that the exact mechanism of surface deterioration is extremely complex. Nevertheless, certain facts are clear. For example, contact stress has a major effect. Thus, cam/follower failures are more pronounced with pushrod engines than with OHC engines because the use of a pushrod system results in higher inertia and requires higher valve spring loads to ensure that contact is maintained between the cam and follower at the maximum engine speed. These high spring loads give rise to high contact stress, particularly at low engine speeds when the spring load is offset only slightly by inertia loading. For this reason, pitting, which is fundamentally a fatigue process, is most likely to occur at low engine speeds. On high-performance engines, an additional form of pitting can occur at high engine speeds. This is due to stresses induced by the deceleration of the follower and gives rise to pitting on the cam flank. In addition to the contact stress and lubrication, another factor that influences the cam and follower durability and the wear life is the cam and follower material and their compatibilities. Chilled cast iron is used mainly where contact stresses are high because it offers good resistance to scuffing. Hardenable cast iron, which provides good resistance to pitting, can be used where contact stresses are lower. Hardenable cast iron is prone to polishing wear and hence is widely used for cams and followers. Overhead cam engines give lower contact stresses because of lower valvetrain inertias and correspondingly lower valve spring loads. This permits the use of steel and nodular iron or malleable iron. Steel normally is used for camshaft applications where contact stresses are not high but vertical gear drives for an oil pump or other auxiliaries are critical.

For chilled cast iron, pitting, spalling, or fatigue is the general failure mode when overstressed. Its surface properties can be improved by various surface treatments. Hardenable cast iron is resistant to pitting but is prone to scuffing in the absence of or with insufficient ZDDP additive in the oil. Carburized steel is resistant to pitting but is sensitive to lubricant and requires ZDDP in the oil to prevent scuffing. Ductile and malleable irons are satisfactory materials but are not good for followers and scuffs in the absence of ZDDP from the oil. Special surface treatments, to assist running-in and to prevent scuffing failures, frequently are applied to cams and/or followers.

Cam lobe surfaces often are treated with phosphate. The value of phosphate-type surface treatments is to inhibit scuffing at first contact and to aid in the breaking-in process. In addition, this process produces a pock-marked type of surface that is roughly similar in cross section to hills and valleys, with the hills having smooth plateaus. The oil carried in the valleys helps to distribute the stresses of the partial or borderline oil film. The process must be controlled rigidly, and the initial surface to be treated must be smooth enough to carry the load; otherwise, early failure will result from scuff and wear. Further improvement can be accomplished when both the cam and followers are phosphate treated under the same conditions. Although phosphating generally is good, improper surface treatment can result in high wear rates and catastrophic failures. Salt bath nitriding also is effective, particularly on chilled iron followers (Tanimoto *et al.*). The roller bearing is the most important factor for wear behavior of cams and followers, and a wear resistance ranking for various combinations of rollers and roller pins can be found in Korte *et al.*

5.4.3 Valve Seat and Seat Insert Interface

Adhesive wear is due to localized bonding between contacting solid surfaces, leading to material transfer between the two surfaces or loss from either surface. Because junctions form as a result of two surfaces being pressed together, the nature of interatomic forces indicates that bonding occurs at these junctions and that over some portion of the real area of contact, the atoms of the two surfaces must have passed beyond the point of maximum bonding. This implies that some adhesive forces or bonds must be overcome to separate the two surfaces at these sites. This atomic view of the contact situation at the junctions provides the foundation for the adhesive wear mechanism. The causes of adhesive wear are poor lubrication, high contact stress, incompatible materials, wide differences in hardness, high surface roughness, and inadequate break-in.

Another type of wear observed at the valve seat is a plastic deformation type of wear, also known as shear strain wear or radial flow. A shear strain or radial flow type of wear involves plastic deformation of the wearing surface along the sliding direction due to high contact stress, high surface flash temperature, high friction at the interface, and inadequate hot hardness.

A limited degree of oxidation on the seating surface generally has been regarded as beneficial in lubricating valve seat interfaces and in preventing direct metal-to-metal contact, thereby limiting friction and wear. However, moderate or severe corrosion usually escalates the wear, which often is observed for certain types of fuel and in certain engines. For example, sulfur and vanadium pentoxide levels must be controlled in fuels to minimize detrimental exhaust valve corrosion in diesel engines. Corrosion by these agents involves removal of the tenacious protective oxides (e.g., Cr_2O_3), followed by catastrophic oxidation of the base alloy. Continued mechanical contact aids oxide removal and tends to aggravate wear. Valve seat and insert material compatibility, hardfacing, and nitriding are the options providing the wear resistance needed in combatting adhesive, shear strain, and corrosive types of wear at the valve seat.

5.4.4 Valve Stem and Guide Interface

Abrasive wear causes displacement of material due to hard particles or hard protuberances. The causes of abrasive wear are wide differences in hardness, high asperity contact stress, and particles such as trapped wear debris and oxide particles. Depending on the penetration depth from hard particles, the abrasive wear process may differ. A brittle chrome layer at the valve stem surface can be crushed and embedded on the surface, acting as abrasives and causing abrasive wear on the valve guide.

5.4.5 Valve Tip Wear

Wear at the valve tip has the opposite consequences as wear at the valve seat. Valve seat wear results in diminished valvetrain lash, whereas valve tip wear increases valvetrain lash. Therefore, excessive tip wear can increase noise and the valve seating velocity. Other drawbacks are increased friction in the valvetrain when the tip is pitted and increased side load on the valve due to higher friction at the valve tip. A hardened tip and a welded wafer with wear resistance material are typical solutions to the tip wear or pitting problem.

Another valvetrain interface having wear concerns is that of the stem and seal. Wear at the seal lip can change the seal metering capability and can cause excessive oil flow to the combustion chamber and exhaust manifold, thereby increasing emissions.

5.5 Valvetrain Lubrication

Although the poppet valvetrain has become the favored method of introducing the combustible charge and exhausting the used gases, primarily because of tribological problems with the alternatives such as rotary and sleeve valves, it too has tribological difficulties. The introduction of the overhead camshaft exacerbates design difficulties because the lubrication, or tribological performance, of such an arrangement is inherently deficient. The operating problems with cam and follower lubrication and the engineering science background to this have been studied in detail and can be found in literature (Taylor, 1991 and 1994).

5.5.1 Film Thickness

The lubrication mechanism in nonconforming contacts, such as in cams and followers, is elastohydrodynamic lubrication (EHL). This mechanism can generate oil films having thicknesses up to the order of 1 μm. Figure 5.39 shows typical film thicknesses at the cam/follower interface at various cam angles. Minimum film thicknesses at the shoulders and noses are the prime concern for potential high stress and resultant wear.

There is a long formula for accurately calculating film thickness, but a simple formula is shown in Eq. 5.11, which gives sufficient accuracy for assessing the lubrication quality of cams and followers (Neale, 1994). This simple formula, which applies only to iron or steel components with mineral oil lubrication, is

$$h = 5 \times 10^{-6} \times \left(\eta u R_r\right)^{0.5} \tag{5.11}$$

where

h = EHL film thickness, mm
η = lubricant viscosity at working temperature (Poise)
u = entrainment velocity, mm/s
R_r = relative radius of curvature, mm

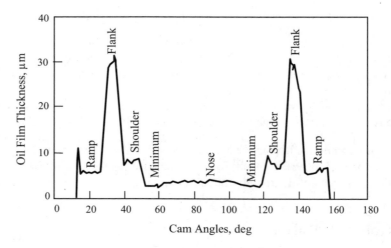

Figure 5.39 *Oil film thicknesses at the cam and follower interface (Zhu).*

For flat tappets, $R_r = R_c$; for curved tappets, $R_r = \left(\dfrac{1}{R_c} + \dfrac{1}{R_T}\right)^{-1}$, where R_c is the cam radius of curvature at the point under consideration, and R_T is the tappet radius curvature. For spherical or barrelled roller tappets, assume $R_T = R_{T1}$, which is the tappet radius of curvature in the plane of the cam.

The entrainment velocity u can vary greatly through the cam cycle, reversing in sign, and in some cases remaining close to zero for part of the cycle. This latter condition leads to very thin or zero-thickness films. For roller followers, u can be taken as being approximately the surface speed of the cam. Calculation of the EHL film is required only at the cam nose and on the base circle. Roller followers usually have good lubrication conditions at the expense of high contact stress for a given size. For plain tappets, the entrainment velocity u at any instant is the mean of the velocity of the cam surface relative to the contact point and the velocity of the follower surface relative to the contact point. Therefore, on the base circle where the contact point is stationary, u is half the cam surface speed. At all other parts of the cycle, the contact point is moving. The entrainment velocity u can be calculated as

$$u = \omega\left[\frac{R_b}{2} + \frac{y}{2} - R_c\right] \tag{5.12}$$

where

$\quad \omega \;=$ cam speed, rad/s
$\quad R_b \;=$ base circle radius, mm
$\quad y \;\;=$ cam lift, mm
$\quad R_c \;=$ cam radius at the point of contact, mm

This applies for flat tappets. For curved tappets with a radius much larger than the cam radius, Eq. 5.12 can be used as a reasonable approximation.

For cams with curved sliding contact followers, Eq. 5.13 for u is complex. However, to check the value of u at only the maximum lift position, use the approximate formula of

$$u = \omega \left[\frac{R_b + y}{2} + \frac{3280 y'' \, R_F}{R_F + R_b + y} \right] \tag{5.13}$$

where

y = maximum cam lift, mm
y'' = maximum cam acceleration at nose, mm/deg^2, which is a negative value
R_F = follower radius, mm

After a value for the film thickness has been calculated, the mode of lubrication can be determined by comparing it to the effective surface roughness of the components. The effective surface roughness generally is taken as the combined surface roughness R_{qt}, defined as

$$R_{qt} = \left(R_{q1}^2 + R_{q2}^2 \right)^{0.5} \tag{5.14}$$

where R_{q1} and R_{q2} are the maximum surface roughness of the cam and tappet, respectively, typically 1.3 times the R_a roughness values.

If the EHL film thickness is less than approximately 0.5 R_{qt}, then there will be some solid contact, and boundary lubrication conditions will apply. Under these circumstances, surface treatments and surface coatings to promote good running-in will be desirable, and anti-wear additives in the oil may be necessary. Alternatively, it may be appropriate to improve the surface finishes, to change the design to an improved profile giving better EHL films, or to use roller followers that are inherently easier to lubricate. Extremely good surface finishes are desirable for successful operation, because the EHL lubrication film usually is very thin. Typical achievable values are 0.4 μm R_a for the cam and 0.15 μm for the tappets (Neale, 1994).

The crankshaft system and the piston-connecting rod system are in fluid lubrication. Therefore, the lower the oil viscosity, the lower the friction torque. However, it is inverted in the case of the valvetrain system. Friction torque increases with decreasing oil viscosity and engine revolution speed. The reason is that the valvetrain system is in mixed and boundary lubrication. The lower viscosity oil results in even thinner film thickness at the cam shoulders and noses. Therefore, the use of high-viscosity oil is better for the valvetrain system. In addition to oil viscosity, other parameters such as the cam base circle radius, cam speed, cam width, valve spring stiffness, and valvetrain mass all affect film thickness differently. Figure 5.40 summarizes their directional effect on film thickness.

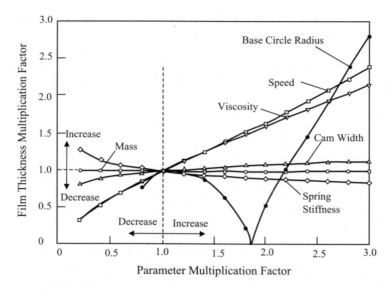

Figure 5.40 *Effect of parameters on film thickness (Zhu).*

5.5.2 Stress Effects

Another important factor that affects valvetrain lubrication is the contact stresses or loads between the cams and followers, which are responsible for many of the difficulties experienced in automotive engines. A design criterion commonly used to avoid wear in valvetrain systems is the maximum Hertzian stress under static conditions on the cam nose (i.e., at the top of the lift). Wear or scuffing problems of the cam and follower systems often are considered exclusively as lubrication difficulties, to be solved by the incorporation of appropriate anti-wear or extreme pressure additives in the lubricating oil. Figure 5.41 shows the effect of tappet curvature on the equilibrium EHD film thickness and on the Hertzian stress, both at the cam nose and both related to a tappet radius of curvature of 25 mm as an arbitrary reference condition. First, consider the critical condition giving minimum stress, where lubrication conditions are at their worst for this critical tappet radius of curvature having zero film thickness. This must be avoided; otherwise, wear would be catastrophic. Then, consider the region in which the curvature is less than the critical value (i.e., the radius of curvature is greater than the critical value). The Hertzian stress rises rapidly as the curvature passes through zero, corresponding to a plane-faced tappet, and it seems inadvisable to use negative curvature (i.e., a concave tappet face). Now consider the right-hand branch of the curve. The film thickness rises to a broad maximum, while the stress rises slowly. For minimum wear, it seems best to work at a curvature of 0.08–0.10 mm^{-1}, that is, a radius of curvature of 10–12.5 mm, where the Hertzian stress still shows only a modest increase over the minimum value and the hydrodynamic effect of tappet wear does not remain too unfavorable.

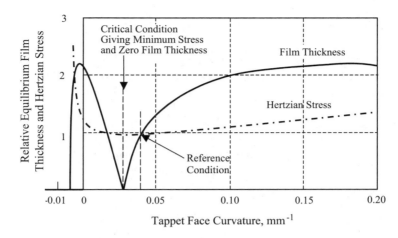

Figure 5.41 _Effect of tappet curvature on relative film thickness and Hertzian stress (Dyson)._

Figure 5.42 shows the cam nose wear as a function of spring loads or contact stresses for various lubricants. The ester-based lubricant offers anti-scuffing behavior positively superior to that of mineral oil, and polyolefins behave worse than mineral oil.

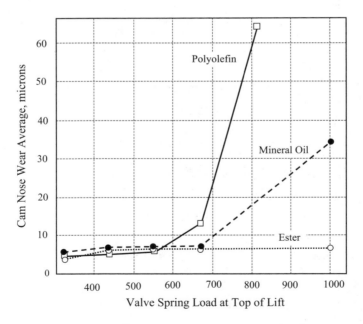

Figure 5.42 _Cam wear versus spring load with various oils (Miorali and Chiarottino)._

5.5.3 Viscosity Effects

The present trend in the formulation of automotive engine lubricants is toward low-viscosity products to minimize power loss due to viscous drag. Consequently, the engines and especially the valvetrains may be exposed to adhesive wear and, in particular, to scuffing or pitting at the cam and follower interfaces. Oil viscosity consideration must be a compromise between viscous drag and scuffing reduction. The effect of mineral oil viscosity on pitting has been studied with oils containing a fixed amount of ZDP corresponding to 0.085% wt zinc (Zn). The results shown in Figure 5.43 clearly confirm the influence of oil viscosity on pitting. By keeping the same concentration of 0.085% wt Zn as alkyl ZDP in the oil, the effect of ester-based lubricants has been checked. The data demonstrate that the ester practically eliminates pitting and behaves as a mineral-based stock of higher viscosity.

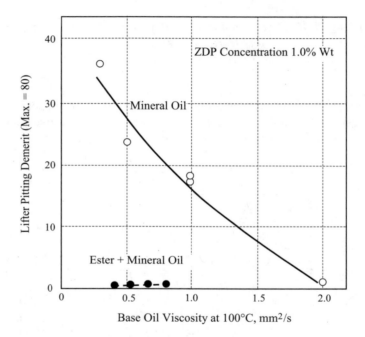

Figure 5.43 *Lifter pitting versus mineral oil viscosity (Miorali and Chiarottino).*

The wear results shown in Figure 5.44 illustrate the best behavior to the extent that no significant wear is observed, even below the critical viscosity of the mineral base oil.

Figure 5.45 summarizes the effect of several parameters, including spring stiffness, cam width, base circle radius, viscosity, and valvetrain mass, on the maximum Hertzian contact stress.

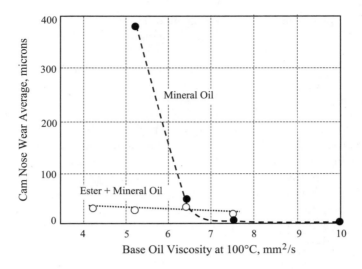

Figure 5.44 *Cam wear versus oil viscosity with mineral and mineral/ester oils (Miorali and Chiarottino).*

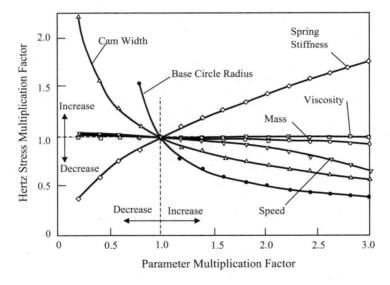

Figure 5.45 *Effect of changes of parameters on maximum Hertzian stress at the cam nose at 1000 rpm (Zhu).*

5.5.4 Anti-Wear Additive Effects

In general, an increase in viscosity decreases the incidence of pitting and scuffing failures because the cam and follower are in the mixed and boundary lubrication regimes, where the role of chemical actions in thin surface films is vital. The additive package or extreme pressure additives, such as ZDDP in particular, are the most common. ZDDP is essential for reducing both scuffing and polishing wear tendencies, but it can promote or retard pitting, depending on the zinc type, concentration, and cam/follower material combination (Jarrett, Abell). Pure paraffinic mineral oils offer the best pitting resistance and are superior to synthetic base stocks and oils containing additives.

Chilled cast iron and hardenable cast iron are preferred materials for both the camshaft and the lifters in OHV engines (Abell). Hardenable cast iron is a material that notably is insensitive to pitting but rather is susceptible to adhesive wear (Figure 5.46). The data show that using a solvent-neutral 450 mineral base stock with ZDP, the cam wear results are extremely low, even at very small dosages of ZDP, while the pitting of the lifter escalates with increasing ZDP concentration. The use of an anti-wear agent cannot be avoided due to the necessity to control cam wear; the base oils evaluation therefore has been centered on the onset of pitting of the lifters.

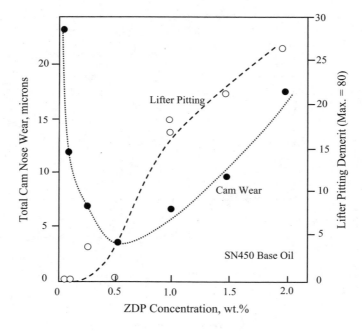

Figure 5.46 *Cam wear and lifter pitting versus ZDP concentration in mineral oil (Miorali and Chiarottino).*

Valvetrain wear is more severe, especially for tappet scuffing under low-temperature and low-speed operating conditions. Some kinds of engine oils containing a lower level of ZDDP are believed to be superior for preventing valvetrain wear to those with a higher level of ZDDP (Torii *et al.*).

Figure 5.47 shows the effect of ZDP concentration and type (aryl versus alkyl) on cam and lifter wear. Generally, the best protection from excessive cam and lifter wear was provided by oils containing either all alkyl ZDP or predominately alkyl ZDP.

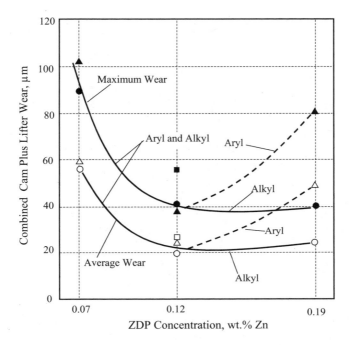

Figure 5.47 *Effect of ZDP concentration and type on wear (Pless and Rodgers).*

Most engine oils contain ZDDP, and increasing the content of ZDDP reduces scuffing and polishing wear tendency. However, too high a treatment with ZDDP can lead to pitting failures. From the standpoint of scuffing, polishing wear, and pitting, alkyl ZDDPs generally are superior to the more thermally stable aryl ZDDPs. Metal detergents tend to adversely affect wear, whereas certain ashless dispersants reduce wear.

5.5.5 Temperature Effects

Figure 5.48 shows the effect of temperature on wear for a base oil and four oils blended with commercial ZDPs with a concentration of 0.1% wt. Wear increases as the temperature increases for all five oils, with and without ZDP additives.

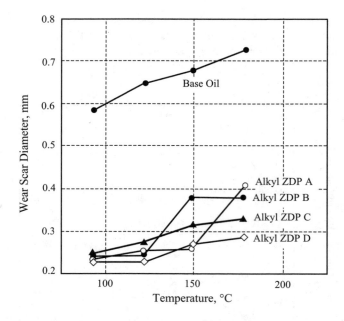

Figure 5.48 *Effect of temperature on ZDP performance (Rounds).*

Previous results show that as viscosity increases, valvetrain wear decreases. Therefore, it is believed that wear that increases when the temperature increases must be attributed to decomposition of the anti-wear additive ZDP and then viscosity reduction.

Engine friction torque becomes smallest when the engine is supplied with an optimum oil flow rate for effective cooling. For an OHC engine, the ratio of oil flow rate typically is 10–20% in the crankshaft system, 30–40% in a piston-connecting rod system, and 50–60% in a valvetrain system. The optimum oil flow rate that minimizes friction torque is 1–2 l/min for crankshaft and piston-connecting rod systems, and 3–4 l/min for a valvetrain system (Tran *et al.*). Again, it is engine and application specific.

5.5.6 Engine Oil Degradation

Engine oil deteriorates in use, through contamination by combustion chamber blowby products, glycol antifreeze, and so forth; by oxidation under relatively high-temperature conditions; or by a combination of these factors. Figure 5.49 shows the cam and lifter wear compared with the contaminant level for both the taxicab test and the Sequence V test on oil drains. Both tests show an increase in cam and lifter wear, with an increase in contaminant level. The total acid number (TAN) is helpful in determining the extent of oil acidic contamination and oil oxidation and under certain conditions may indicate bearing corrosive potential.

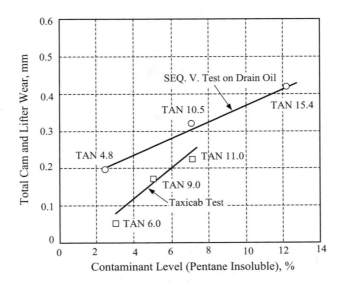

Figure 5.49 *Oil contamination effect on cam and lifter wear (Asseff).*

5.5.7 Soot and Carbon Effects

Figure 5.50 illustrates the effect of diesel soot on wear. As can be seen, wear increases with increasing soot concentration until the wear scar reaches the range of base oils alone; then, at higher soot loadings, no further increase in wear occurred. These results indicate that at soot concentrations greater than 3.5%, the anti-wear performance of ZDP in the engine oil was completely destroyed (Rounds).

This apparently is due to preferential adsorption of the ZDDP decomposition products by the soot, thus preventing the formation of a protective layer on the cams and followers (Rounds, Corso, Nagai, Beshouri).

Figure 5.51 shows the effect of diesel soot and carbon black on the performance of the anti-wear additive ZDP. Diesel soot has a much more detrimental effect on ZDP anti-wear performance than does any of the carbon blacks. However, graphite, which is a crystalline form of carbon, had essentially no effect on ZDP performance.

5.5.8 Lubricant and Material Reactions

ZDDP remains in the oil until asperity contact generates high local temperatures (as high as 900°C or more). At approximately 170°C, the ZDDP (whose principal inorganic constituents are zinc, sulfur, and phosphorus) decomposes. Zinc polyphosphate adsorbs onto the metal surfaces, forming a solid film. This film then melts in the 200–300°C range, forming a fluid "glass" that lubricates the surface and prevents asperity welding. If asperity contact continues to raise the

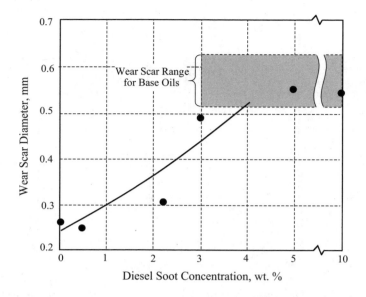

Figure 5.50 *Effect of diesel soot concentration on wear at 93°C with x% ZDP (Rounds).*

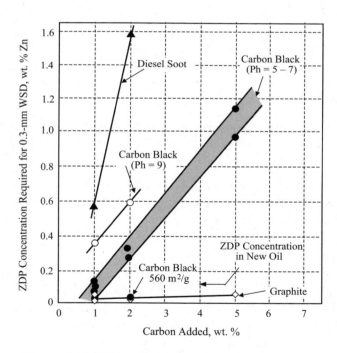

Figure 5.51 *Effect of carbon type and concentration on ZDP anti-wear performance at 93°C (Rounds).*

temperature, a second decomposition product, alkyl sulfides, reacts with the surface iron oxide layer to form iron sulfide. This, in turn, forms a eutectic with the iron oxide, which melts at 900°C and again forms a lubricating fluid surface film (Spedding, Barcrof). Some researchers distinguish between these two mechanisms, calling the first anti-wear and the second extreme pressure.

The current trend is to reduce the amount of ZDDP additive. Of particular concern is the prime source for phosphorus poisoning of catalyst systems.

5.6 Surface Engineering of Valvetrain Components

5.6.1 Overview

Surface engineering is the treatment of the surface and near-surface regions of a material to allow the surface to perform functions that are distinct from those functions demanded from the bulk of the material. These surface-specific functions include protecting the bulk material from hostile environments and providing low-friction contacts with other materials. Although the surface normally cannot be made totally independent from the bulk, the demands on surface and bulk properties often are quite different. For example, in the case of exhaust valves, the bulk of the material must have sufficient hot fatigue strength at the service temperature to provide an acceptably safe service life. On the other hand, the surface of the material at the seat, fillet, stem, and tip must possess sufficient resistance to wear, oxidation, corrosion, and low friction under the conditions of service to achieve that same component life.

In many instances, rather than trying to find one material that has both the bulk and surface properties required to do the job, it often is more economical or absolutely necessary to select a material with the required bulk properties and then modify the surface to achieve the required surface properties. The choice of an appropriate technique may be limited by factors such as chemical or thermal stability, geometric constraints, and cost. The choice of material applied to a surface typically is dictated by the service environment in which the material will be used and the desired physical appearance of the surface. The substrate material being treated usually is chosen for its mechanical properties.

The following sections discuss surface engineering techniques, which arbitrarily are divided into four groups for convenience of discussion: (1) surface modification or phase transformation hardening (no external material introduced except for gas in some instances), (2) vacuum thin coating, (3) thick coating, and (4) other coatings applicable for valvetrain applications.

5.6.2 Selective Surface Hardening

The selective surface hardening typically is achieved by heating and quenching locally, without any chemical modification to the surface. The common methods currently used to harden the surface of steels include induction, flame, laser, and electron beam hardening.

5.6.2.1 Induction Heating and Flame Hardening

Induction heating is an extremely versatile heating method that can perform uniform surface hardening, localized surface hardening, through hardening, and tempering of hardened pieces. Heating is accomplished by placing a valvetrain component in the magnetic field generated by high-frequency alternating current passing through an inductor, usually a water-cooled copper coil. The depth of heating produced by induction is related to the frequency of the alternating current: the higher the frequency, the thinner or shallower the heating. Therefore, deeper case depths or through hardening are produced by using lower frequencies. Flame heating employs direct impingement of a high-temperature flame or high-velocity combustion gases.

Surface hardening of a steel part by either induction or flame heating consists of raising a surface layer above the transformation temperature at which it will be transformed to austenite and then rapidly cooling the part to produce a hard martensitic structure in this region. Induction surface hardening is applied mostly to hardenable grades of steel, including medium-carbon steels such as 1030 and 1040, high-carbon steels such as 1070, and alloy steels such as martensitic valve alloys. Figure 5.52 shows the practical level of minimum surface hardnesses attainable with water quenching for various carbon contents. The curve is applicable for both induction hardening and flame hardening. It also applies for alloy steels, except those containing stable carbide formers such as chromium and vanadium.

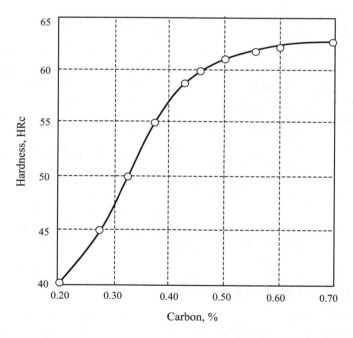

Figure 5.52 *Relationship of carbon content to minimum surface hardness attainable by induction or flame heating and water quenching (Ruglic).*

Another important feature of induction hardening is its ability to impart as-quenched hardness higher than those of conventionally furnace-hardened steels (Figure 5.53). The higher hardness of the induction hardened specimens may be attributable to three sources: (1) residual stresses, (2) smaller amounts of retained austenite, and (3) carbon segregation (Ruglic).

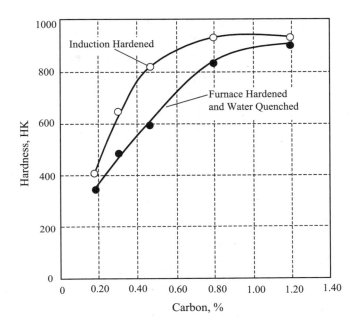

Figure 5.53 *Induction versus furnace hardening in hardness for various carbon contents (Hassell and Ross).*

Tempering of hardened steel structures such as martensite involves the diffusion of carbon atoms. The extent of diffusion increases with both increasing temperatures and time. The major differences between induction and furnace tempering cycles lie in the times and temperatures involved. The results shown in Figure 5.54 illustrate the tempering response for a 1050 steel quench hardened from 855°C and tempered at a variety of temperatures.

5.6.2.2 Laser and Electronic Beam Surface Hardening

When a laser beam impinges on a surface, part of its energy is absorbed as heat at the surface. If the power density of the laser beam is sufficiently high, heat will be generated at the surface at a rate higher than heat conduction to the interior can remove it, and the temperature in the surface layer will increase rapidly. In a short time, a thin surface layer will have reached austenitizing temperatures, whereas the interior of the workpiece is still cool. Self-quenching occurs when the cold interior of the workpiece constitutes a sufficiently large heat sink to quench the hot surface by heat conduction to the interior at a rate high enough to allow martensite to form at

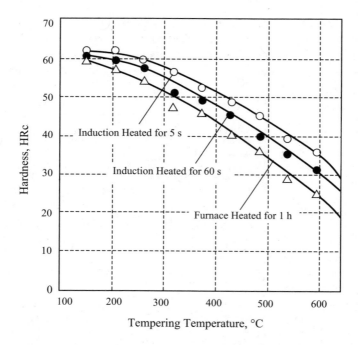

Figure 5.54 _Tempering curves for furnace and induction hardened 1050 steel (Hassell and Ross)._

the surface. By selecting the correct power density and speed of the laser spot, the material will harden to the desired depth. Because of the high heating and cooling rates obtainable, the hardness obtainable by the laser hardening process can, in some instances, be slightly higher than that considered possible with conventional methods. One example of laser surface hardening is a camshaft made of ductile cast iron (Sandven).

For electron beam hardening, the electrons of the beam hit the component surface and penetrate into the metal surface. Because of the intense interaction between the beam electrons and the atoms of the material being bombarded by the beam, the electrons lose their energy rapidly. Most of the energy that is lost by the electrons is transformed into heat at the interaction surface. Typical hardening depths obtained by the electron beam hardening process are in the range of 0.1–1.5 mm (Schiller *et al.*). This is similar to the laser hardening process in many ways, and the hardening occurs through a self-quenching process that is dependent on the thermal conductivity of base materials and that starts after the energy transfer has ceased.

5.6.3 Diffusion Surface Hardening

At high temperatures, elements such as nitrogen or carbon can diffuse readily into the iron surface. Diffusion surface hardening is a heat treatment process that introduces the carbon and/or nitrogen into the surface of valvetrain components at high temperatures, sometimes still in the ferritic region, and afterwards rapidly cooling to trap these atoms in the interstitials or forming compounds with the matrix material to accomplish a structure that is harder than the bulk material. Typical diffusion surface hardening processes include carburizing, nitriding, nitrocarburizing, and carbonitriding. The diffusion processes can be accomplished in gas, liquid, solid pack, or vacuum.

5.6.3.1 Nitriding

The nitriding process introduces nitrogen into the surface of steel while it is in the ferritic condition and does not involve heating into the austenite phase field. Nitriding can be accomplished with a minimum of distortion and with excellent dimensional control. Figure 5.55 shows several possible ways that nitrogen atoms can occupy the matrix, as well as the terminology of the compounds formed.

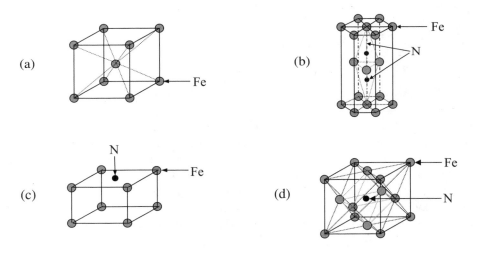

Figure 5.55 *Iron nitriding structures: (a) alpha, α, or bcc structure,*
(b) epsilon iron nitride, ε, Fe$_3$N, (c) zeta iron nitride, ζ, Fe$_2$N,
and (d) gamma prime, γ', Fe$_4$N.

The nitriding processes commonly used include salt bath nitriding, gas nitriding, and plasma (ion) nitriding. Liquid nitriding is performed in a molten, nitrogen-bearing, fused-salt bath containing either cyanides or cyanates. It produces a wear-resistant hard surface and improved fatigue properties. Gas nitriding is a case-hardening process whereby nitrogen is introduced into the surface of an alloy by holding the metal at a suitable temperature in contact with a nitrogenous gas, usually ammonia. The nitriding temperature for all steels is 495–565°C. Because of the absence of a quenching requirement and the comparatively low temperatures employed in this process, nitriding produces less distortion and deformation than either carburizing or conventional hardening. Plasma or ion nitriding is a method of surface hardening that uses glow discharge technology to introduce nascent (elemental) nitrogen to the surface of a metal part for subsequent diffusion into the material. In a vacuum, high-voltage electrical energy is used to form a plasma, through which nitrogen ions are accelerated to impinge on the valve. This ion bombardment heats the valve, cleans the surface, and provides active nitrogen. Ion nitridng provides better control of case chemistry and uniformity and has other advantages, such as lower part distortion, than conventional gas nitriding.

Parts are treated at 350–650°C while in the ferritic condition to produce a compound layer consisting of nitrides. Typical nitrided microstructures of valve alloys with martensite and austenitic microstructures have oxide, compound, and diffusion layers.

Nitriding can be accomplished with minimal distortion and with excellent dimensional control because nitriding does not involve heating into the austenite phase field and a subsequent quench to form martensite. Typically, a nitride layer consists of two zones—a diffusion zone and a compound zone—that have different visual appearances, chemistries, hardnesses, and microstructural characteristics. The diffusion zone of a nitrided case can best be described as the original core microstructure with some solid solution and precipitation strengthening. In iron-based materials, the nitrogen exists as single atoms in solid solution at lattice sites or interstitial positions until the limit of nitrogen solubility (~0.4 wt.% N) in iron is exceeded. This area of solid-solution strengthening is only slightly harder than the core.

The depth of the diffusion zone depends on the nitrogen concentration gradient, the temperature, the amount of time at the given temperature, and the chemistry of the valve material. As the nitrogen concentration increases toward the surface, fine coherent precipitates are formed when the solubility limit of nitrogen is exceeded. The precipitates can exist both in the grain boundaries and within the lattice structure of the grains. These precipitates, in the form of nitrides, distort the lattice and pin crystal dislocations, thereby substantially increasing the hardness of the material. The outer layer is a compound layer of iron and chrome nitrides that provides excellent tribological properties (i.e., low friction and high wear resistance). A diffusion layer of nitrogen below the compound layer provides further support to the compound layer. Figure 5.56 shows the residual stress distribution, various nitride compounds, and hardness profile for nitrided Sil 1 alloy.

The thickness of the nitride compound layer can vary from several microns to several tens of microns, and from a 0.1–1.0-mm-thick layer in which the nitrogen is diffused to enhance wear resistance and fatigue strength. Table 5.13 shows the typical thickness of the compound and diffusion layer for austenitic exhaust valves and martensitic intake valves.

Figure 5.57 shows typical diffusion characteristics of SAE 4140 and 8620 with salt bath nitriding as a function of time and temperature. The figure demonstrates the significant effect of processing temperature on diffusion rates. Although compound layers are detectable at the 400 and 454°C cycles, extended dwell time is required to meet the typically specified minimum depth of 5 µm.

The nitriding process may be limited to iron-based alloys. Nickel-based materials, such as Inconel 751, do not develop a sufficient diffusion and compound layer to improve wear resistance. In most ferrous alloys, the diffusion zone formed by nitriding cannot be seen in a metallograph

Valvetrain Tribology

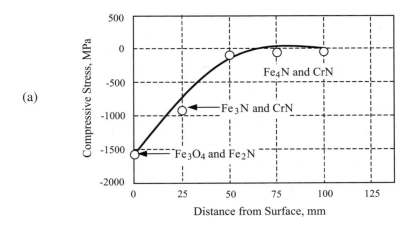

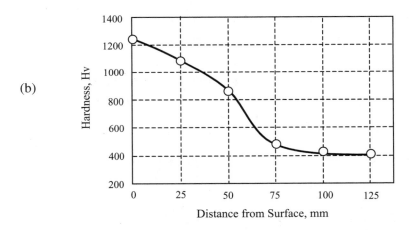

Figure 5.56 *Nitrided Sil 1 valve alloy: (a) residual stress distribution, and (b) hardness profile.*

TABLE 5.13
TYPICAL NITRIDE THICKNESSES
FOR VALVE STEELS

Alloy	Compound Layer (mm)	Diffusion Layer (mm)
Martensitic	0.0127	0.50
Austenitic	0.0102	0.05

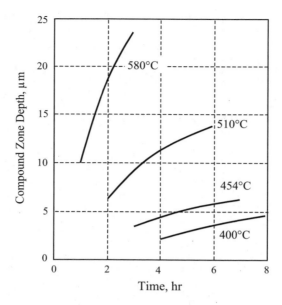

Figure 5.57 *Temperature and time effect on nitride layer for SAE 4140;
SAE 8620 steel has similar diffusion curves (Easterday).*

because the coherent precipitates generally are not large enough to resolve. However, when the chromium level is high enough for extensive nitride formation, it can be seen in the etched cross section.

Nitriding significantly improves fatigue strength. Figure 5.58 shows the S-N curves for SAE 1045 steel in various conditions, including untreated, chrome-plated, and nitrided conditions. Chrome plating reduces fatigue properties due to residual tensile stress and the hydrogen embrittlement effect. However, salt bath nitriding improves the fatigue strength of the material significantly.

5.6.3.2 Carburizing

Carburizing is a case-hardening process in which carbon is dissolved in the surface layers of a low-carbon steel part at a temperature sufficient to render the steel austenitic, followed by quenching and tempering to form a martensitic microstructure and hardened case. The resulting gradient in carbon content below the surface of the part causes a gradient in hardness, producing a strong wear-resistant layer on a material that usually is low-carbon steel, which has higher ductility and is readily fabricated into parts. Figure 5.59 shows the total case depth versus gas carburizing time at four selected temperatures. As with all diffusion hardening, there is greater case depth at higher carburizing temperatures, or it takes longer to achieve the same case depth at lower temperatures.

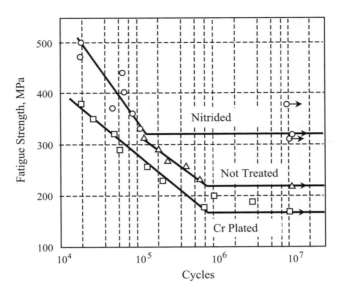

Figure 5.58 _Fatigue strength comparison of SAE 1045 with chrome-plated and salt-bath-nitrided condition, rotating bending with notched test specimen (Mueller, 1985)._

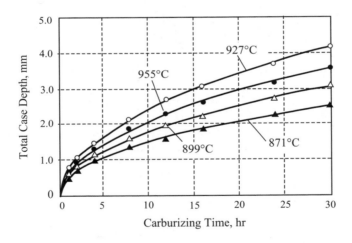

Figure 5.59 _Total case depth versus gas carburizing time at four selected temperatures (Harris)._

Carburizing methods include gas, vacuum, plasma, salt bath, and pack carburizing. In liquid carburizing, carbon diffuses from the bath into the metal, due to the superior heat-transfer characteristics of salt bath solutions. Also, the cycle times for liquid carburizing are shorter than those of gas or vacuum carburizing. A typical case hardness profile of gas carburized SAE 8620 steel can be found in Eerven.

5.6.3.3 Carbonitriding and Nitrocarburizing

Carbonitriding is a modified form of gas carburizing, rather than a form of nitriding. The modification consists of introducing ammonia into the gas carburizing atmosphere to add nitrogen to the carburized case as it is being produced. The nitrogen diffuses into the steel simultaneously with the carbon. Typically, carbonitriding is conducted at a lower temperature and for a shorter time than gas carburizing. Because nitrogen enhances hardenability, carbonitriding makes possible the use of low-carbon steel to achieve a surface hardness equivalent to that of high-alloy carburized steel without the need for drastic quenching, resulting in less distortion and cracking. Figure 5.60 compares the hardness profiles of carburized and carbonitrided SAE 1020 steel.

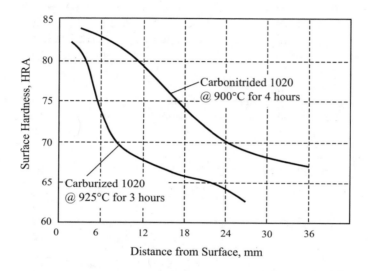

Figure 5.60 *Hardenability comparison curves for SAE 1020 steel carbonitrided at 900°C and carburized at 925°C (Powell et al.).*

Nitrocarburizing also is a thermochemical treatment that is applied to a ferrous object to produce surface enrichment in nitrogen and carbon, which forms a compound layer. Beneath the compound layer is a diffusion zone enriched in nitrogen. It is a modified nitriding process rather than a carburizing process. Depending on the temperature to which it is treated, it can be ferritic nitrocarburizing (typically at 570°C) or austenitic nitrocarburizing (typically above 675°C).

Table 5.14 summarizes the parameters, properties, and characteristics of diffusion hardening treatments, including nitriding, carburizing, carbonitriding, and nitrocarburizing.

TABLE 5.14
CHARACTERISTICS OF DIFFUSION HARDENING TREATMENT

Process	Diffused Element	Process Temp. (°C)	Case Depth (µm)	HRc	Substrate Material	Characteristics
Nitriding:						
Gas	Nitrogen	480–590	125–750		Alloy, nitriding, and stainless steel	Quench not required, low distortion, batch, slow
Liquid		510–565	2.5–750	50–70	Most ferrous metals including compact iron	Quench, batch, fast, cyanide
Ion		340–565	75–750		Alloy, nitriding, and stainless steel	Fast, high equipment cost, batch
Carburizing:						
Pack		815–1090	125–1500			Low equipment cost, difficult to control case depth
Gas	Carbon	815–980	75–1500	50–63	Low carbon	Good control of depth, suitable for continuous operation, gas
Liquid		815–980	50–1500		Steels and alloys	Fast, salt disposal and maintenance
Vacuum		815–1090	75–1500			Excellent control, fast, high equipment cost
Carbonitriding:						
Gas	Carbon and nitrogen	760–870	75–750	50–65	Low carbon	Lower temperature than carburizing, gas control critical
Liquid	Carbon and nitrogen		2.5–125	40–60	Steels and alloys	Batch, salt disposal
Nitrocarburizing:						
Gas	Nitrogen	480–775	50 – 250	60–70	Alloy, nitriding, and stainless steel	Quench not required, low distortion, batch, slow
Ion	Carbon	570	75–750		Alloy, nitriding, and stainless steel	Fast, high equipment cost, batch

5.6.4 Thin Film Coatings

Thin film coatings, including physical vapor deposition (PVD) and chemical vapor deposition (CVD), typically are performed in a vacuum and controlled-atmosphere environment, and the coating thicknesses typically are less than a few micrometers. The properties of a film of a material formed by any PVD or CVD process depend on several factors: the substrate surface condition, details of the deposition process and system geometry, details of film growth on the substrate surface, and post-deposition processing reactions (Mattox).

5.6.4.1 Physical Vapor Deposition

In the PVD coating technique, the coating material is transferred to the vapor phase by physical methods (i.e., evaporation, sputtering) and then condenses as a coating on a substrate at elevated temperatures. The deposition of a solid on a substrate occurs in a vacuum when high-energy vapor particles are accelerated by bias voltage. Figure 5.61 shows a schematic of a PVD coater.

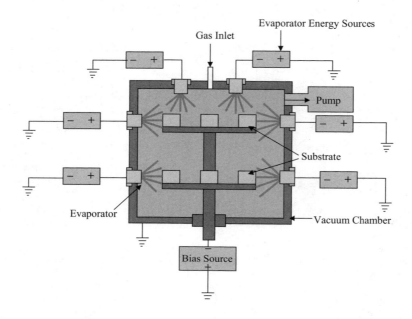

Figure 5.61 *Schematic of a physical vapor deposition (PVD) coater (Mack).*

The advantages of PVD include lower substrate temperatures of 200–550°C; the fact that high-hardness coatings such as TiN, TiAlN, and CrN can be coated; and environmental safety. The disadvantages of the PVD process are that the surface roughness must be less than the coating thickness, and it is restricted to line-of-sight deposition. There are also a size limitation and a high friction coefficient. Typically, PVD is a batching process, but in-line coating is possible. Table 5.15 shows some characteristics of typical PVD coatings for high-speed stainless steel (HSS).

Figure 5.62 shows the hardnesses and coefficients of friction for some PVD coatings. Extreme high hardness can be achieved; meanwhile, high coefficients of friction are observed for typical PVD coatings, except for C-coating.

TABLE 5.15
CHARACTERISTICS OF PVD COATINGS (MACK)

Property	TiN	TiAlN	CrN
Critical Load on HSS, N	70–80	50–60	40–50
Maximum Coating Thickness, μm	10	20	50
Deposit Rates, μm/hr	13	40	16
Surface Roughness	Standard	Higher	Lower
Oxidation Stability, 1 hour in air	550°C	800°C	700°C

(a)

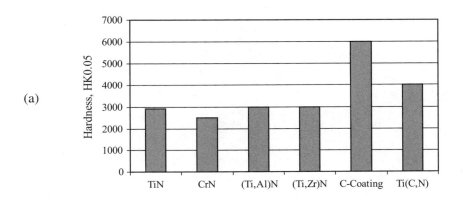

(b)

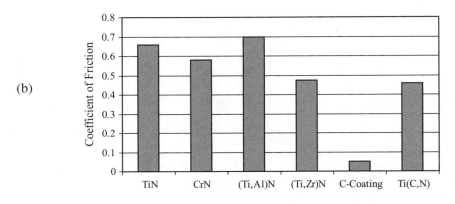

Figure 5.62 *Hardnesses and coefficients of friction of PVD coatings: (a) hardness of PVD coatings on high-speed stainless steel substrate, and (b) coefficients of friction of PVD coatings against a ball of 100 Cr6 (Mack).*

5.6.4.2 Chemical Vapor Deposition

Chemical vapor deposition (CVD) is a coating technique in which hard coatings are produced on a heated surface via a chemical reaction from the vapor or gas phase. The deposition species are atoms or molecules, or a combination of both in nature. In thermal CVD, the reaction is activated by high temperatures (generally above 900°C). Plasma CVD operates at lower temperatures than thermal CVD. The reaction is activated by a plasma at temperatures of 300–700°C. In the plasma CVD process, the stress that is due to thermal expansion mismatch is reduced, and temperature-sensitive substrates can be coated more readily. Typical CVD materials include metals, graphite, ceramics, and diamond-like carbon (DLC). Figure 5.63 shows a schematic of a CVD coater.

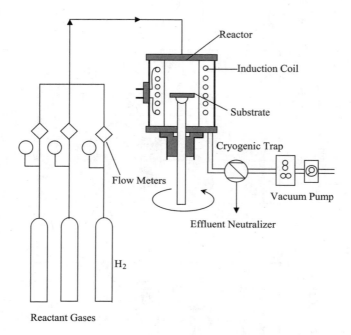

Figure 5.63 *Schematic of a thermal chemical vapor deposition (CVD) coater (Pierson).*

The advantages of CVD include depositing ceramics at temperatures below the melting point, achieving near-theoretical density, controlling grain orientation and size, the possibility of process-ing at atmospheric pressure, and not being restricted to line-of-sight deposition. The disadvantages of CVD include requiring substrate temperatures greater than 900°C. Also, chemical precursors are hazardous or toxic, which necessitates a closed system, and the by-products can be toxic and corrosive. Likewise, CVD has a low efficiency and a high cost. CVD films have residual film stress that can be tensile or compressive and can be very high, and they have a significant effect on the adhesion and properties of coatings. There are several methods of modifying the mechanical stresses developed in films during processing, including limiting the thickness of the stressed film, using a multiple layer with alternating materials, periodically alternating the concurrent bombardment conditions, and periodically adding alloying or reacting materials.

Diamond-like carbon (DLC) represents a new form of carbon coating that is neither diamond nor graphite. Rather, it has an amorphous structure and a wide range of both hardness (6–90 GPa) and coefficient of friction (0.001–0.6). Variation in the coefficient of friction is attributed to the hydrogen content or the ratio of hydrogen and carbon (Figure 5.64). The coefficient of friction for super-hydrogenated (H/C = 10) DLC is as low as 0.003 (Erdemir *et al.*).

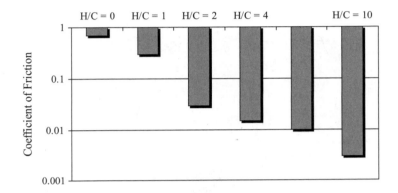

Figure 5.64 *Coefficients of friction for various diamond-like carbons (DLCs) with various ratios of hydrogen to carbon (Erdemir* et al.*).*

Although fundamental differences exist between PVD and CVD, CVD now extensively uses plasma (a physical phenomenon), whereas PVD often is conducted in a chemical environment (reactive evaporation and reactive sputtering). The distinction between the two basic processes becomes blurred.

5.6.5 Thick Film Coatings

Thick coatings in discussion here include thermal spray coatings, plasma transferred arc coatings, and laser cladding.

5.6.5.1 Thermal Spray Coatings

Thermal spraying is a generic term for a group of processes in which metallic and ceramic materials in the form of powder, wire, or rod are fed and heated to near or above their melting points. The resulting molten or nearly molten droplets of material are accelerated in a gas stream and are projected against the surface to be coated. Figure 5.65 schematically illustrates some examples of thermal spray techniques, including flame spray and detonation gun spray. Plasma spray and high-velocity oxy-fuel (HVOF) spray also are used for thermal spray coatings.

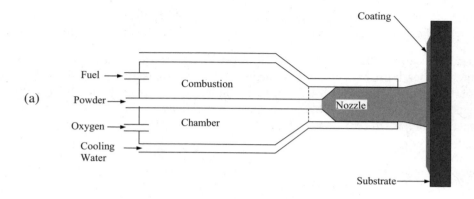

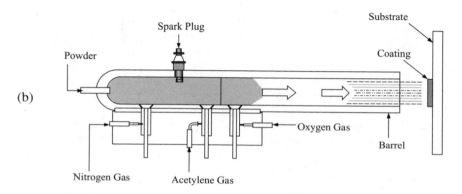

Figure 5.65 *Schematics of thermal spray coatings: (a) flame spray process, and (b) detonation gun spray (Tucker).*

Valve seats sometimes are hardfaced to ensure an adequate margin of safety against mechanical loads, thermal loads, and corrosive environments. In the absence of such protection, failures caused by corrosion, wear, and fatigue can occur. The need for hardfacing also may depend on the designer's choice, durability requirements, and design limitations. Figure 5.66 shows an HVOF device for valve seat hardfacing. Fuel (usually propane, propylene, or hydrogen) is mixed with oxygen and is burned in a chamber. The products of the combustion are allowed to expand through a nozzle, where the gas velocities may become supersonic. Powder is introduced, usually axially, in the nozzle and is heated and accelerated. The powder usually is fully or partially melted and achieves velocities of up to approximately 550 m/s. Because the powder is exposed to the products of combustion, it may be melted in either an oxidizing or reducing environment, and significant oxidation of metallics and carbides is possible. With appropriate equipment, operating parameters, and choice of powder, coatings with high density and with bond strengths frequently exceeding 69 MPa can be achieved. Coating thicknesses usually are

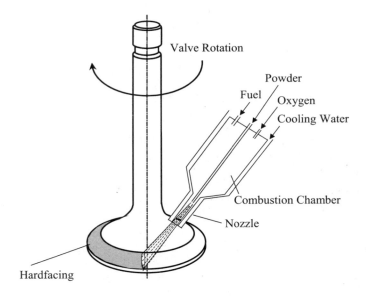

Figure 5.66 *Schematic of the high-velocity oxy-fuel (HVOF) process.*

in the range of 50–500 μm, but substantially thicker coatings occasionally can be used when necessary with some materials. High-velocity oxy-fuel processes can produce coatings of virtually any metallic or most ceramics.

5.6.5.2 Plasma Transferred Arc Hardfacing

Most valve seat hardfacing is accomplished by plasma transferred arc (PTA) hardfacing. Figure 5.67 illustrates a schematic of PTA hardfacing. Plasma transferred arc spraying is a process based on melting material to be deposited in a plasma arc, and then spraying the molten particles on the substrate to produce an adherent coating. The PTA process is one of the most sophisticated and versatile thermal spray methods (such as powder(s), wire, rod, metals, or alloys). Temperatures that can be obtained with commercial plasma equipment have been calculated to be greater than 11,000°C and far above the melting point or even the vaporization point of any known material. Decomposition of materials during spraying is minimized because of the high gas velocities produced by the plasma, resulting in extremely short residence time in the thermal environment.

The plasma process also provides a controlled atmosphere for melting and transport of the coating material, thus minimizing oxidation, and the high gas velocities produce coatings of high density. The plasma gun operates on the principle of raising the energy state of a gas by passing it through an electric arc. The release of energy in returning the gas to its ground state results in exceedingly high temperatures. A gas such as nitrogen or argon enters a direct-current arc between a tungsten cathode and a copper anode that make up the nozzle. Both components are

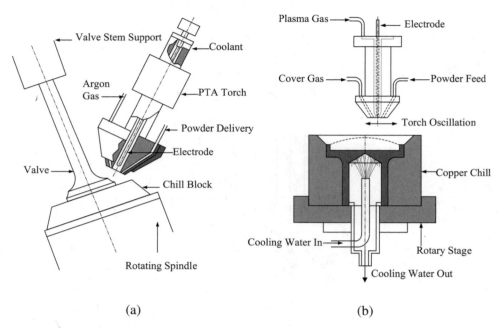

Figure 5.67 *Schematic of the plasma transferred arc (PTA) process: (a) valve seat PTA weld, and (b) lifter PTA puddled weld (Milligan and Narasimhan).*

cooled by a constant flow of water through internal passages. Here, the plasma gas first dissociates (in the case of nitrogen, into two atoms), followed by ionization that releases free electrons. The electrons recombine outside the electric arc, and energy is released as heat and light. In addition, frequent collisions transfer energy from the electrons to the positive ions, accelerating them until the plasma reaches a state of equilibrium. The result is a thermal plasma in which the energy of the electrons has been turned into enthalpy, or heat content. At this point, powdered coating material suspended in a gas is injected into the plasma and subsequently is melted and propelled at high velocity to the valve seat. In practice, a small amount of a secondary gas, such as hydrogen or helium, is mixed with the primary plasma gas to increase operating voltage and thermal energy. The high temperatures and high gas velocities produced by the plasma process result in coatings that are superior in mechanical and metallurgical properties.

5.6.5.3 Laser Cladding

Laser cladding is emerging as a viable alternative to other hardfacing processes. The laser beam is a chemically clean light source that delivers a precisely controlled quantity of energy to localized regions such as the valve seat. The laser beam can be maneuvered easily by optical elements and can be adapted to automation. High power densities and low interaction times result in rapid heating and cooling, which produces a shallow heat-affected zone, low distortion of the valve, and minimal deterioration of the valve head fatigue properties. Rapid solidification effects produce refined and novel microstructures that result in improved properties. Figure 5.68 shows schematics of laser cladding.

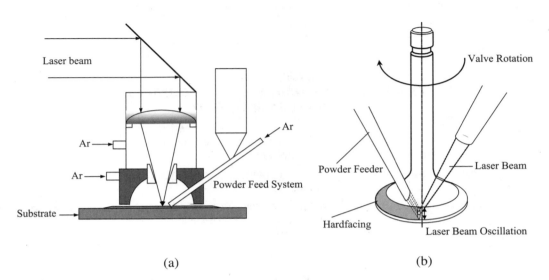

Figure 5.68 *Laser cladding schematics: (a) laser cladding principle, and (b) valve seat laser cladding.*

Laser surface processing frequently utilizes continuous wave lasers instead of a pulsed laser. The most common medium for laser cladding is CO_2 gas and a power output of approximately 5 kW. The shielding gas could be any of the inert gases or a combination of gases, such as He/Ar and H_2/Ar. Clad thickness can vary from 0.05–4 mm. Thermal stresses in the cladding can cause harmful cracking, but this can be eliminated by appropriate preheating practices. Low power densities, large beam diameters, and slow sample translation rates tend to produce crack-free clad layers.

Seat facing alloys are composed of hard precipitates in cobalt-, nickel-, or iron-based austenitic matrices as used in PTA hardfacing in powder form; however, they can be in rod or wire form, too. The hard precipitates generally are chromium, molybdenum, tungsten, or vanadium carbides. Some compositions are intermetallic compounds, such as Laves phase, as the hard precipitates. Chromium is added to the matrix to increase corrosion resistance. Silicon additions provide the fluidity necessary for welding.

Seat facing alloys derive their wear resistance from the volume fraction of precipitates that possess greater hot hardness and strength than the base valve alloys. A significant portion of the adhesive wear resistance in these seat-facing alloys, as well as other high-chromium alloys, is derived from the tenacious chromium oxide layer formed on the wear surface by oxidation during service. When extremely severe conditions of corrosion are encountered, such as those that occur when sulfur-containing fuels are used, cobalt-based hardfacing alloys may be required to assure valve seat durability. Selection of a particular seat-facing alloy usually depends on temperatures, stresses, and corrodents encountered in service. Manufacturing considerations may preclude some facing alloy/valve head alloy combinations. Table 5.16 summarizes the characteristics of current thick-coating technologies.

TABLE 5.16
COMPARISON OF THERMAL SPRAY COATING
AND HARDFACING WELD

Process	Materials	Substrate Temperature (°C)	Particle Velocity (m/S)
Powder Flame Spray	Metallic, ceramic, and fusible	105–160	65–130
Nontransferred Arc Plasma	Metallic, ceramic, plastics, and compounds	95–120	240–560
HVOF	Metallic, cement, and some ceramic	95–150	100–550
D-Gun	Metallic, cement, and ceramic	95–150	730–1000
PTA	Metallic fusible coatings	Fuses base metal	0–500
Laser Cladding	Metallic fusible coatings	Fuses base metal	0–500

5.6.6 Other Surface Treatments

Other surface enhancement methods commonly used in valvetrain applications that are not included in the preceding categories are chrome plating, phosphating, and shot peening.

5.6.6.1 Chrome Plating

The continual sliding motion of the stem along the length of the valve guide can result in wear of the guide and/or stem without adequate lubrication and surface protection. If a high degree of friction exists between the stem and guide, friction torque is lost as useless work. Therefore, the resulting requirement is to provide a minimum friction coefficient between the two surfaces. In the past, allowing oil to run down the valve stem has been effective. However, because of emissions requirements, this passage of oil has been restricted by using various forms of stem seals. Therefore, it is vital to coat stems to reduce the friction coefficient and to provide a hard wear-resistant surface. Chrome plating has been effective in reducing both friction and wear.

Various chrome thicknesses can be used in valve stems, depending on the engine requirements (Table 5.17). Chrome plating possesses high tensile stresses, and this condition can result in a substantial reduction in fatigue strength. As a general rule, flash chrome plating can be deposited in keeper groove areas without detriment. However, a thicker coating, such as the triple flash and full chrome plating, should be avoided in keeper grooves, tips, and the fillet stem blend area. In short, it should be confined to areas within the valve guides.

TABLE 5.17
TYPICAL CHROME PLATING THICKNESSES
FOR VALVE STEM APPLICATIONS

Type	Thickness (μm)
Flash Chrome	0.75–2.0
Triple Flash	2.5–5.0
Full Chrome	6.0–9.0
Heavy Chrome	>18.0

Heavy chrome plating, insufficient cleaning before plating, a rough surface before plating, a chromic acid content that is too high, a low temperature, a low sulfate content, and too high of a current density all could cause nodular deposits to develop on the surface. These hard nodular deposits are detrimental in causing scuffing and abrasion at the guide surface. Thus, they usually are removed by belt polishing.

Chromium plating is produced by electrodeposition from a solution containing chromic acid (CrO_3 hexavalent or Cr_2O_3 trivalent) and a catalytic anion in proper proportion. A plating solution consists of a chromic acid solution and a catalyst, conventionally a mixture of sulfate (SO_4^{2-}) and fluoride solutions. The properties of the electrodeposits are influenced by the ratio of CrO_3 to the catalysts, plating temperature, and current density. Chromic acid to sulfate ratios vary between 75 to 1 and 120 to 1, plating temperatures range from 45–65°C, and current densities range from 10–90 A/dm^2. The metal so produced is extremely hard and corrosion resistant. Hard chromium deposits provide a surface with a low coefficient of friction that resists galling, abrasive and adhesive wear, and corrosion.

The hardness of chromium electrodeposits is a function of the type of chemistry selected and the plating conditions. Typically, bright chromium deposits from conventional plating solutions have hardness values of 850–900 HV, those from mixed-catalyst solutions have values of 900–1000 HV, and those from fluoride-free chemistries have values of 950–1100 HV or higher. A typical coefficient of friction for chrome plating is 0.15 against steel.

During chromium deposition, internal residual stress can stack up to 1000 MPa. Microcracks or, more precisely, grain boundaries on chrome plating surfaces appear. They are considered grain boundaries because they must be etched to reveal and are nonexistent if the deposition rate is high.

Hydrogen embrittlement is a serious issue with chromium-plated valves. The susceptibility of chromium-plated valve stems to hydrogen embrittlement is affected by many factors, including hardening of the steel, surface defects, pickling, cathodic cleaning, and the depth of the plate relative to the thickness and hardness of the part being plated. Surfaces to be chromium plated must be free from stresses included during machining, grinding, or hardening. Stresses from the hardening operation may be increased further during grinding and may result in microcracks. If the hardness of steel is less than 40 HRc, it is unlikely that any damaging effect will occur as a result of residual stress. Materials with hardnesses above 40 HRc should be baked at a temperature of at least 190°C for 4 hours after plating to ameliorate the effects of hydrogen embrittlement. This treatment should be started as soon as possible, preferably within 15 minutes after plating (Newby).

Chemical, electrochemical, or mechanical methods are used to remove hard chromium coatings. Hydrochloric acid at any concentration greater than 10 vol.% and at room temperature or higher removes chromium effectively from valve stems, whether iron-based or nickel-based alloys. In some operations, inhibitors are added to the acid solution to minimize attack on the steel substrate. Most often, chromium is removed electrochemically from iron-based valve stems by the use of any convenient heavy-duty alkaline cleaner at room temperature or higher, at 5–6 V with anodic current. This method is unsatisfactory for nickel-based alloys, which should be stripped chemically in hydrochloric acid. Anodic stripping operations result in the formation of oxide films on the base metal. These films should be removed by one of the conventional deoxidizing processes prior to replating. After stripping the chromium deposits, the valves then are stress-relieved at 190°C for a minimum of 3 hours. The following solutions and operating conditions are recommended for removing chromium deposits from the valve alloys.

Grinding is used occasionally to remove heavy chromium deposits. Because chromium is hard and brittle, a soft grinding wheel is essential. A hard wheel forms a glazed surface, which results in a temperature rise that causes the chromium to crack. Good performance can be obtained with an aluminum oxide resin-bonded wheel of approximately 60 grit. To prevent or minimize glazing, the contact area should be flooded with a coolant. Usually, the coolant is water with a small amount of soluble oil. Because of its hardness, excess chromium cannot be removed as rapidly as when grinding most other materials. The maximum thickness of metal removed should not exceed 5 µm per pass, and this amount should be reduced if there is any evidence of cracking. The optimum grinding speed is approximately 20 m/s. High-strength untempered chromium-plated martensite valves are prone to crack during grinding due to the heat of grinding and hydrogen embrittlement.

5.6.6.2 Phosphate Coating

Phosphate coating is the treatment of bare or galvanized steel with a dilute solution of phosphoric acid and other chemicals in which the surface of the metal, reacting chemically with the phosphoric acid medium, is converted to an integral, mildly protective layer of insoluble crystalline phosphate. The three principal types of phosphate coatings in general use include the following: (1) iron phosphate, (2) zinc phosphate, and (3) heavy phosphate (manganese phosphate). Table 5.18 summarizes the characteristics of these three types of phosphate coating.

TABLE 5.18
CHARACTERISTICS OF PHOSPHATE COATINGS (DAVIS)

Characteristics	Types of Coating		
	Iron Phosphate	Zinc Phosphate	Heavy Phosphate
Coating Weight	0.16–0.80 g/mm^2	1.6–4.0 g/mm^2	7.5–30 g/mm^2
Types	Cleaner/coater Standard Organic phosphate	Standard Nickel modified Low zinc Calcium modified Manganese modified	Manganese phosphate Zinc phosphate Ferrous phosphate
Common Accelerators	Nitrite/nitrate Chlorate Molybdate	Nitrite/nitrate Chlorate Nitrobenzene sulfonic acid	None Chlorate Nitrate Nitroguanidine
Operating Temperature	Room temperature to 70°C		
Free Acid, Points	–2.0 to 2.0	0.5–3.0	3.6–9.0
Total Acid, Points	5–10	10–25	20–40+
Prephosphate Conditioners	None	Titanium phosphate None	Manganese phosphate Titanium phosphate None
Primary Use	Paint base for low-corrosion environments	Paint base for high-corrosion environments	Unpainted applications
Limitations	Low painted corrosion resistance; low unpainted corrosion resistance	Poor unpainted corrosion resistance	Expensive, long processing times
Materials Needed for Tanks	Low-carbon steel	Low-carbon steel, stainless steel, or plastic-lined steel	Stainless steel or low-carbon steel
Application Method	Spray and immersion	Spray and immersion	Immersion only

5.6.6.3 Shot Peening

Shot peening is a cold-working process in which the surface of a part is bombarded with small spherical media called shot. Each piece of shot striking the material acts as a tiny peening hammer, imparting to the surface a small indentation or dimple, thus creating compressive residual stress on the surface. Cracks will not initiate or propagate in a compressively stressed zone. Because nearly all fatigue and stress corrosion failures originate at the surface of a part, compressive stresses induced by shot peening provide considerable increases in part life. Many materials also will increase in surface hardness due to the cold-working effect of shot peening.

The benefits obtained by shot peening are the result of the compressive stress effect and the cold working induced. Compressive stresses are beneficial in increasing resistance to fatigue failures, corrosion fatigue, stress corrosion cracking, hydrogen-assisted cracking, fretting, galling, and erosion caused by cavitation. Figure 5.69 shows the effect of shot peening at different speeds on compressive residual stresses in carburized 16MnCr5 steel. A higher speed of shot peening increases the magnitude of compressive residual stress and the depth of the stressed layer.

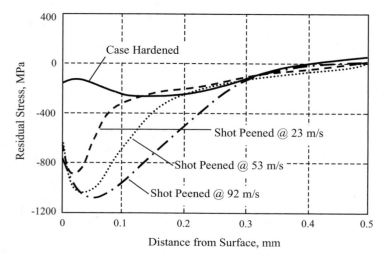

Figure 5.69 *Effect of shot peening at different speeds on compressive residual stresses in carburized 16MnCr5 steel (Scholtes and Macherauch).*

Valves sometimes are shot peened prior to chrome plating to counteract the harmful effects of plating on the fatigue life, which can be residual tensile stresses, hydrogen embrittlement, and brittle notch-sensitive plate and can lead to early failure. When the surface of the base metal is stressed compressively, the cracks cannot propagate into the base metal. Figure 5.70 shows the reduction in fatigue resistance experienced on SAE 4340 steel due to plating and the beneficial effect of shot peening prior to the plating process.

Caution should be exercised if parts are to be heated after shot peening, such as during a post-plating bake, because of the stress-relieving effect on the residual compressive stress of peening. Figure 5.71 shows that the benefits of shot peening can be totally lost when stress relieved at 450°C.

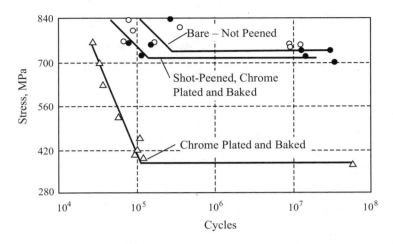

Figure 5.70 *Effect of shot peening on fatigue life of 4340 steel, 52–53 HRc, rotating beam fatigue (Cohen).*

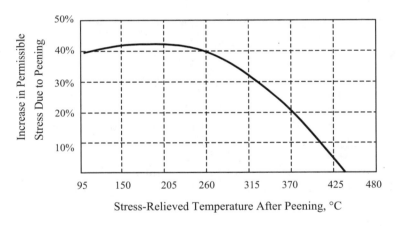

Figure 5.71 *Effect of heat on peened steel springs (Zimmerli).*

5.7 References

Abell, R., "I.C. Engine Cam and Tappet Wear Experience," SAE Paper No. 770019, Society of Automotive Engineers, Warrendale, PA, 1977.

Abrikosova, I. and Deryagin, B., "Direct Measurement of Molecular Attraction Between Solids in Vacuum," *Soviet Physics*, JETP, 4, 2–10, 1957.

Anderson, J., *Tribology International*, Vol. 15, No. 1, 1982, p. 43.

Archard, J., *J. Appl. Phys.*, Vol. 24, 1953, p. 981.

Armstrong, W. and Buuck, B., "Valve Gear Energy Consumption: Effect of Design and Operation Parameters," SAE Paper No. 810787, Society of Automotive Engineers, Warrendale, PA, 1981.

Ashton, W. and Strack, C., "Chlorofluorocarbon Polymers," in *Synthetic Lubricants,* edited by Gundarson, R. and Hart, A., Reinhold, New York, 1962.

Asseff, P., *Lubrication Theory and Practice,* Publication 183-320-59, Lubrizol Corporation, Wickliffe, OH, 1998.

Baniasad, S. and Emes, M., "Design and Development of Method of Valvetrain Friction Measurement," SAE Paper No. 980572, Society of Automotive Engineers, Warrendale, PA, 1998.

Barcrof, T., "The Mechanism of Action of ZDPs as EP Agents," *Wear*, Vol. 77, 1982, pp. 355–384.

Bayer, R., *Mechanical Wear Prediction and Prevention*, Marcel Dekker, 1994.

Beerbower, A., *Predicting Lube Life*, STLE Special Publication SP-15, Park Ridge, IL, 1982.

Beshouri, G., "Lubricant Additive Effects on Cam and Tappet Wear in Medium Speed Diesel Engines," SAE Paper No. 860377, Society of Automotive Engineers, Warrendale, PA, 1986.

Bouman, C., *Properties of Lubricating Oils and Engine Deposits*, McMillan and Co., London, 1950.

Bowden, F. and Throssell, W., "Adsorption of Water Vapor on Solid Surfaces," *Proc. Roy. Soc.*, A 209, London, U.K., 1951.

Cleveland, A. and Bishop, I., "Several Possible Paths to Improved Part-Load Economy of Spark-Ignition Engines," SAE Paper No. 150A, Society of Automotive Engineers, Warrendale, PA, 1960.

Cohen, B., "Effect of Shot Peening Prior to Chromium Plating on the Fatigue Strength of High Strength Steel," Wright Air Development Center, Technical Note 57-178, Wright-Patterson Air Force Base, OH, 1957.

Committee on Adiabatic Diesel Technology Energy Engineering Board, "A Review of the State of Art and Projected Technology of Low Heat Rejection Engines," Commission on Engineering and Technical Systems National Research Council, National Academy Press, Washington, D.C., 1987.

Corso, S., "The Effects of Diesel Soot on Reactivity of Oil Additive and Valvetrain Materials," SAE Paper No. 841369, Society of Automotive Engineers, Warrendale, PA, 1984.

Courtney-Pratt, J. and Eisner, E., "The Effect of a Tangential Force on the Contact of Metallic Bodies," *Proc. Roy. Soc.*, A 238, London, U.K., 1957.

Culpon, D. and Mead, T., "Synthetic Lubricants," *Lubrication*, Texaco Inc., Vol. 78, No. 4, White Plains, NY, 1992.

Dagnall, H., *Exploring Surface Texture*, Rank Taylor Hobson, 2000.

Dyson, A., "Elastohydrodynamic Lubrication and Wear of Cams Bearing Against Cylindrical Tappets," SAE Paper No. 770018, Society of Automotive Engineers, Warrendale, PA, 1977.

Easterday, J., "Low Temperature Ferritic Nitrocarburizing," SAE Paper No. 1999-01-2862, Society of Automotive Engineers, Warrendale, PA, 1999.

Eerven, K., "The Effects of Sulfur and Titanium on Bending Fatigue Performance of Carburized Steels," M.S. Thesis, Colorado School of Mines, Golden, CO, 1990.

Erdemir, A., Eryilmaz, O., Nilufer, I., and Fenske, G., "Effect of Source Gas Chemistry on Tribological Performance of Diamondlike Carbon Films," *Diamond Rel. Mat.*, Vol. 9, 2000, pp. 632–637.

Eyre, T., in *Source Book on Wear Control Technology*, edited by Rigney, D. and Glaeser,W., ASM International, Materials Park, OH, 1978.

Glaeser, W., "Wear Fundamentals Course for Engineering," Intl. Wear of Materials Conf., American Society of Mechanical Engineers, West Caldwell, NJ, 1989.

Greenwood, J. and Williamson, J., "Contact of Nominally Flat Surfaces," *Proc. Roy. Soc.* A 295, London, U.K., 1966.

Harris, E., *Met. Prog.*, Vol. 44, 1943, p. 401.

Hassell, P. and Ross, N., "Induction Heat Treating of Steel," *ASM Handbook*, Vol. 4, ASM International, Materials Park, OH, 1992.

Heisler, H., *Vehicle and Engine Technology, 2nd Edition*, Society of Automotive Engineers, Warrendale, PA, 1999.

Iwabuchi, A., Hori, K., and Kudo, H., *Proc. Intl. Conf. On Wear of Materials*, American Society of Mechanical Engineers, West Caldwell, NJ, 1987, p. 526.

Jarrett, M., "Material Considerations for Automobile Camshafts," SAE Paper No. 710545, Society of Automotive Engineers, Warrendale, PA, 1971.

Korte, V., Glas, T., Lettmann, M., Krepulat, W., and Steinmets, C., "Roller Cam Followers for Heavy-Duty Diesels," *Automotive Engineering International*, Society of Automotive Engineers, Warrendale, PA, 2000.

Kovach, J., Tsakiris, E., and Wong, L., "Engine Friction Reduction for Improved Fuel Economy," SAE Paper No. 820085, Society of Automotive Engineers, Warrendale, PA, 1982.

Kraghelsky, I. and Demkin, N., "Determination of the True Contact Area," *Friction and Wear in Machinery*, CRC Press, New York, 1960.

Loomis, W., "Overview of Liquid Lubricants for Advanced Aircraft," *Solid and Liquid Lubricants for Extreme Environments*, American Society of Lubrication Engineers, Park Ridge, IL, 1982.

Mack, M., *Surface Technology—Wear Protection*, Verlag Moderne Industrie, Landsberg, Germany, 1990.

Mattox, D., "Growth and Growth-Related Properties of Films Formed by Physical Vapor Deposition," *ASM Handbook, Vol. 5, Surface Engineering*, ASM International, Materials Park, OH, 1994.

Milligan, J. and Narasimhan, S., "A Powder Fed Plasma Transferred Arc Process for Hard Facing Internal Combustion Engine Valve Seats," SAE Paper No. 800317, Society of Automotive Engineers, Warrendale, PA, 1980.

Miorali, M. and Chiarottino, A., "Engine Valvetrain Wear: Performance of Synthetic and Mineral Oils," SAE Paper No. 811226, Society of Automotive Engineers, Warrendale, PA, 1981.

Mueller, J., "Review of Salt Bath Nitriding," *Proceedings of Salt Bath Nitriding Seminar*, Kolene Corporation, Detroit, MI, 1985.

Mueller, R., "The Effect of Lubrication on Cam and Tappet Performance," *Motor Tech. Z.*, Vol. 27, No. 58, 1966.

Nagai, I., "Soot and Valvetrain Wear in Passenger Car Diesel Engine," SAE Paper No. 831757, Society of Automotive Engineers, Warrendale, PA, 1983.

Nakasa, M., *Proc. of Int. Tribo. Conf.*, Japan Society of Tribologists, Yokohama, Japan, 1995.

Neale, M., ed., *Drives and Seals, A Tribology Handbook*, Society of Automotive Engineers, Warrendale, PA, 1994.

Neale, M., ed., *Lubrication—A Tribology Handbook*, Society of Automotive Engineers, Warrendale, PA, 1997.

Newby, K., "Industrial (Hard) Chromium Plating," *ASM Handbook Vol. 5, Surface Engineering*, ASM International, Materials Park, OH, 1994.

Pierson, H., "Chemical Vapor Deposition of Non-Semiconductor Materials," *ASM Handbook, Vol. 5, Surface Engineering*, ASM International, Materials Park, 1994.

Pless, L., "The Effects of Some Engine, Fuel, and Oil Additive Factors on Engine Rusting in Short-Trip Service," SAE Paper No. 700457, Society of Automotive Engineers, Warrendale, PA, 1970.

Pless, L. and Rodgers, J., "Cam and Lifter Wear as Affected by Engine Oil ZDP Concentration and Type," SAE Paper No. 770087, Society of Automotive Engineers, Warrendale, PA, 1977.

Powell, G., Bever, M., and Floe, C., "Carbonitriding of Plain Carbon and Boron Steels," *Trans. ASM*, Vol. 46, ASM International, Materials Park, OH, 1954.

Rabinowicz, E., *Friction and Wear of Materials*, John Wiley and Sons, New York, 1965.

Rizvi, S., "Internal Combustion Engine Lubricant," *ASM Handbook Vol. 18, Friction, Lubrication and Wear Technology*, ASM International, Materials Park, OH, 1992.

Rosenberg, R., "General Friction Considerations for Engine Design," SAE Paper No. 821576, Society of Automotive Engineers, Warrendale, PA, 1982.

Rounds, F., "Carbon: Cause of Diesel Engine Wear?," SAE Paper No. 770829, Society of Automotive Engineers, Warrendale, PA, 1977.

Ruglic, T., "Flame Hardening," *ASM Handbook, Vol. 4, Heat Treating*, ASM International, Materials Park, OH, 1991.

SAE Standard J300, "Engine Oil Viscosity Classification," Society of Automotive Engineers, Society of Automotive Engineers, Warrendale, PA, 1989.

SAE Standard J1536, "Two-Stroke-Cycle Engine Oil Miscibility/Fluidity Classification," Society of Automotive Engineers, Warrendale, PA, 1989.

Sakamoto, K., "Trends in Japanese Diesel Engines and Their Lubrication," SAE Paper No. 760720, Society of Automotive Engineers, Warrendale, PA, 1976.

Sandven, O., "Laser Surface Hardening," *ASM Handbook, Vol. 4, Heat Treating*, ASM International, Materials Park, OH, 1991.

Schiller, S., Panzer, S., and Furchheim, B., "Electron Beam Surface Hardening," *ASM Handbook, Vol. 4, Heat Treating*, ASM International, Materials Park, OH, 1991.

Schilling, A., *Motor Oils and Engine Lubrication*, Scientific Publications, Shropshire, Great Britain, 1968.

Scholtes, B. and Macherauch, E., *Residual Stress Determination, in Case-Hardened Steels: Microstructural and Residual Stress Effects*, edited by Diesburg, The Minerals, Metals & Materials Society, Warrendale, PA, 1984.

Slater, B., "European Valvetrain Wear—Some Experience with the Volvo B20 Test," SAE Paper No. 750866, Society of Automotive Engineers, Warrendale, PA, 1975.

Spedding, H., "The Anti-Wear Mechanism of ZDDPs," *Tribology Int.*, Feb. 1982, pp. 9–14.

Staron, J. and Willermet, P., "An Analysis of Valvetrain Friction in Terms of Lubrication Principles," SAE Paper No. 830165, Society of Automotive Engineers, Warrendale, PA, 1983.

Stockel, M.W., Stockel, M.T., and Johanson, C., *Auto Fundamentals*, The Goodheart-Willcox Co., Tinley Park, IL, 1996.

Sutor, P. and Bryzik, W., "Tribological System for High Temperature Diesel Engines," SAE Paper No. 870157, Society of Automotive Engineers, Warrendale, PA, 1987.

Tabor, D., *The Hardness of Metals*, Oxford University Press, Oxford, U.K., 1951.

Tanimoto, I., Kano, M., and Sasaki, M., "Establishment of a Method for Predicting Cam Follower Wear in the Material Development Process," SAE Paper No. 902087, Society of Automotive Engineers, Warrendale, PA, 1990.

Taylor, C., "Fluid Film Lubrication in Automotive Valvetrain," *Proc. IMechE, Journal of Eng. Trib. Part J*, Vol. 208, 1994, pp. 221–234.

Taylor, C., "Valvetrain Lubrication Analysis," *Proc. 17th Leeds-Lyon Symposium on Tribology—Vehicle Tribology*, Elsevier, London, U.K., 1991, pp. 119–131.

Torii, K., Chida, H., Otsubo, K., and Tsusaka, Y., "Anti-Wear Properties of Engine Oils—Effects of Oil Additives on Valvetrain Wear," SAE Paper No. 770635, Society of Automotive Engineers, Warrendale, PA, 1977.

Tran, P., Yamamoto, T., Baba, Y., and Hoshi, M., "An Analysis of Lubricating System of Automobile Gasoline Engine," SAE Paper No. 871659, Society of Automotive Engineers, Warrendale, PA, 1986.

Tsuji, E. and Ando, Y., *Proc. Intl. Conf. on Wear of Materials*, American Society of Mechanical Engineers, New York, 1977.

Tucker, R., "Thermal Spray Coatings," *ASM Handbook, Vol. 5, Surface Engineering*, ASM International, Materials Park, OH, 1994.

Wang, Y., Schaefer, S., Bennett, C., and Barber, G., "Wear Mechanisms of Valve Seat and Insert in Heavy Duty Diesel Engine," SAE Paper No. 952476, Society of Automotive Engineers, Warrendale, PA, 1975.

Watson, R. and McDonnell, T., "Additives—The Right Stuff for Automotive Engine Oils," SAE Paper No. 841208, Society of Automotive Engineers, Warrendale, PA, 1984.

Whitehead, J., "Surface Deformation and Friction of Metals at Light Loads, *Proc. Roy. Soc. of London, Mathematical and Physical Sciences*, A 201, 1950, pp. 109–124.

Zhu, G., *Valvetrain—Design Studies, Wider Aspects and Future Developments, Engine Tribology*, edited by Taylor, C.M., Elsevier Science, London, U.K., 1993.

Zimmerli, F., "Heat Treating, Setting and Shot Peening of Mechanical Springs," *Metal Progress*, Vol. 67, No. 6, 1952, p. 97.

Chapter 6

Valvetrain Failure Analysis

6.1 General Failure Analysis Practice

The objective of failure investigation, including valvetrain failure investigation, and subsequent analysis is to determine the primary cause of a failure. Based on the determination, corrective action should be initiated that will prevent similar failures. Frequently, the importance of contributory causes to the failure must be assessed, and new experimental techniques may have to be developed if necessary. Although the sequence is subject to variation, depending on the nature of a specific failure and component, the principal stages that comprise the investigation and analysis of failure include the following (Ryder *et al.*):

- Collection of background data and selection of samples

- Preliminary visual examination of the failed parts, including selection, identification, and preservation of all specimens, and recordkeeping

- Macroscopic examination and analysis (worn surfaces, fracture surfaces, secondary cracks, and other surface phenomena)

- Metrology measurement, including geometric dimensions and surface finishes

- Mechanical testing, including hardness to see if it meets the specifications

- Microscopic examination and analysis

- Scanning electronic microscopy (SEM) and energy dispersive x-ray (EDX) spectrometry analysis

- Determination of the failure mechanisms

- Simulation test

- Writing a report with conclusions and recommendations

6.1.1 Background Information

The failure investigation initially should be directed toward gaining an understanding with all pertinent details relating to the failure, collecting the available information regarding the manufacturing or processing and service histories of the failed component, and reconstructing as much as possible the sequence of events leading to the failure. The availability of complete service records greatly simplifies the failure analysis process. However, in most cases, only fragmentary service information is available, rather than complete service records. Photographic records of the failed components or structure are usually very important because subsequent investigation most likely will destroy the parts. Also, a failure that appears almost inconsequential in a preliminary investigation later may be found to have serious consequences.

Selection of samples should be suitable for the intended purpose and should represent adequately the characteristics of the failure. It is advisable to look for additional evidence of damage beyond that which is immediately apparent. Often, it is necessary to compare failed components with similar components that did not fail to determine whether the failure was brought about by service conditions or was the result of an error in manufacturing or the material. Generally, it also is necessary to examine other components associated with the failed component, which may help in establishing the severity of the operating conditions. A fractured valve may have been caused by inadequate lubrication at the stem and guide interface, which is attributed to the tight seal. Therefore, in addition to examining the valve, guide, and stem, the seals must be examined closely, too. Also, in failures involving corrosion, stress corrosion, or corrosion fatigue, a sample of the fluid that has been in contact with the components of any deposits that have been formed often will be required for analysis.

In addition to developing a general history of the failed component or structure, it is advisable to determine if any abnormal conditions prevailed or if events occurred in service that may have contributed to the cause of failure. When dealing with valve guttering, it generally is desirable to ascertain the condition of ignition and to determine whether any pre-ignition existed.

The background information concerning the failed parts includes the type of part, part number, engine and valvetrain type, materials used (including heat treatment and coatings), service or testing history, manufacturing history, and any documentation pertinent to the failed part.

6.1.2 Preliminary Examination

The failed part, including all its fragments, should be subjected to a thorough visual examination before any cleaning is undertaken. Deposits and debris found on the part often provide useful evidence in establishing the cause of failure or in determining a sequence of events leading to the failure. The preliminary examination should begin with unaided visual inspection. The unaided eye has exceptional depth of focus and has the ability to examine large areas rapidly and to detect subtle changes of color and texture. Some of these advantages are lost when any optical or electronic microscopes are used. Particular attention should be given to the surfaces of fractures and to the paths of cracks. The significance of any indications of abnormal conditions or abuse in service should be observed and assessed, and a general assessment of the basic design and workmanship of the part should be made.

The preservation of fracture surfaces is vital in preventing important evidence from being destroyed or obscured. Surfaces of fractures may suffer mechanical or chemical damage. Mechanical damage may arise from several sources (e.g., a fractured valve may have dropped into the combustion chamber and may have been struck repeatedly by the piston). The surface of a fracture should not be touched or rubbed with the fingers. Also, no attempt should be made to fit together the sections of a fractured part by placing them in contact. This generally accomplishes nothing and almost always causes damage to the surface of the fracture. The surfaces of fractures should not be cleaned unless absolutely necessary. The evidence of chemical (corrosion) damage to a fracture specimen could be destroyed if the specimen is cleaned improperly.

6.1.3 Macroscopic Examination and Analysis

Following the preliminary visual inspection, the detailed examination of worn surfaces, fracture surfaces, secondary cracks, deposits, pits, corroded regions, dimensional integrity, and other surface phenomena are examined using a stereo binocular microscope or a low-magnification lens (e.g., 10x). Occasionally, it also may be advantageous to use a scanning electron microscope (SEM) at low magnification if the part fits the vacuum chamber. The amount of information that can be obtained from examination of a fracture surface at low magnification is surprisingly extensive. Macroscopic examination usually can determine the direction of crack growth and therefore the origin of failure. For example, brittle fracture determination depends largely on the fracture surface exhibiting chevron marks, and the direction of crack growth is almost always away from the tips of the chevrons. Chevron marks occur because nearly all cracks are stepped at an early stage in their development, and as the crack front expands, the traces of the steps form chevron marks. Chevron marks may result from the nucleation of new cracks ahead of a main crack front. A low-magnification examination of fracture surfaces often reveals regions having a texture that is different from the region of final fracture. Fatigue, stress-corrosion, and hydrogen-embrittlement fractures all may show these differences.

6.1.4 Metrology Measurement

Pertinent metrological measurements, including geometric dimensions and surface finishes, may be made to determine if the part was manufactured or assembled according to the engineering drawings. Changes due to service are documented. Measurements may include any of the following: dimensions, runout, roundness, straightness, surface finish, and surface profiles, including worn surfaces. Sometimes, the mating components must be measured, too. For example, non-uniform insert and guide wear can provide evidence of misalignment of the insert to the guide, which led to a high side load on the stem/guide and caused severe wear, subsequent seizure, and valve breakage.

6.1.5 Microscopic Examination and Analysis

The microscopic examination of failed components typically is conducted using an optical (light) microscope and/or an SEM. Due to the limited resolution and depth of field (i.e., it cannot focus on rough surfaces) of these instruments, the surface must be polished for optical microscopic examination. Metallographic examination of polished and polished-and-etched sections by optical microscopy is a vital part of failure investigation. Because of the surface requirement and the size limitation of the examination tool, it often is necessary to remove from a failed component or section a portion containing a fracture surface. Before cutting or sectioning, the fracture area should be carefully protected. All cutting should be done so that the surfaces of fractures and the areas adjacent to them are not damaged or altered. Cutting must be done at a sufficient distance from the fracture site to avoid altering the microstructure of the metal underlying the surface of the fracture by the heat during sectioning.

Metallographic examination provides the investigator with a good indication of the class of material involved and whether it has the desired structure (e.g., grain morphology and size, phase and volume fraction, heat-affected zone, weldment, cracks, inclusions, case hardening, coatings, and corrosions). If abnormalities are present, they may or may not be associated with undesirable characteristics that predispose the part to early failure. The microstructure also may provide information regarding the method of manufacture of the part under investigation and the heat treatment to which it has been subjected, either intentionally during manufacturing or accidentally during service. Other service effects, such as corrosion, oxidation, and severe work hardening of surfaces, also are revealed. In addition, the characteristics of any cracks that may be present, particularly their mode of propagation, provide information regarding the factors responsible for their initiation and development. In investigations involving overheating in service, the original condition of the material sometimes can be ascertained only from a sample cut from a part far from the heat-affected zone.

In the investigation of fatigue cracks, it is advisable to take a specimen from the region where the fracture originated to ascertain if the initial development was associated with an abnormality, such as a weld defect, a decarburized surface, a zone rich in inclusions, or a zone containing severe porosity.

Sometimes, it is necessary to plate the fracture surface of a specimen with a metal, such as nickel, before mounting and sectioning, so that the fracture edge is supported and can be included in the examination.

6.1.6 Scanning Electron Microscopy and Energy Dispersive X-Ray Spectrometry Analysis

The SEM is one of the most versatile instruments for investigating the failure mechanisms of valvetrain components. Compared to the optical microscope, it expands the resolution range by more than one order of magnitude to approximately 10 nm. Compared to optical microscopy, the depth of focus, ranging from 1 μm at 10,000x to 2 mm at 10x, is larger by more than two orders of magnitude due to the small beam aperture. Fractography is probably the most popular field of SEM. The large depth of focus, the possibility of changing magnification over a wide range, simple nondestructive specimen preparation with direct inspection, and the three-dimensional appearance of SEM fractographs make the SEM a crucial tool in failure analysis.

Most SEMs are equipped with EDX spectrometry detectors, and the combination of SEM and EDX permits high-resolution photography of the specimen and the ability to identify easily the elemental composition of selected areas of the specimen. Scanning electron microscopy with EDX can be used to determine the base material, as well as deposits, corrosion products, and adhered particles. The SEM with EDX becomes an indispensable tool in wear, corrosion, and fracture failure analysis.

6.1.7 Mechanical Testing and Simulation Testing

The purpose of mechanical testing is twofold: (1) to find the root cause of failure, and (2) to determine if the parts as made meet the specifications. Hardness testing is the simplest of the mechanical tests and often is the most versatile tool available. Among its many applications, hardness testing can be used to assist in evaluating heat treatment, to provide an approximation of the tensile strength of steel, to detect work hardening, or to detect softening or hardening caused by overheating and decarburization. Other mechanical tests are useful in confirming that the failed component conforms to specifications or in evaluating the effects of surface conditions on mechanical properties. Where appropriate, tensile, fatigue, and impact tests should be conducted, provided sufficient material for the fabrication of test specimens is available. It may be helpful to test specimens after they have been subjected to particular treatments that simulate the treatment of the failed component in service to determine how this treatment has modified the mechanical properties. For example, nitriding or chromium plating of a valve will change the stress state at the surface, thus altering the tensile and fatigue strength of the material. The determination of residual stress at the surface may be useful in investigating the failure of the components.

Simulated-service testing often is not practical because elaborate equipment is required; even when practical, it is possible that not all of the service conditions are fully known or understood. For example, corrosion failures are difficult to reproduce in a laboratory. Serious errors can arise when attempts are made to reduce the time required for a test by increasing the severity of a test by artificially increasing the severity of one of the factors, such as the corrosive medium or the operating temperature. On the other hand, when its limitations are clearly understood, the simulated testing of the effects of certain selected variables encountered in service may be helpful in planning corrective action that will avoid similar failure or, at least, extend the service life. The evaluation of the efficiency of special additives to lubricants to counteract wear is an example of the successful application of simulated-service testing using a selected number of service variables.

6.1.8 Determination of Failure Mechanisms

Typical valvetrain failures include wear, corrosion, and fracture. Wear and corrosion failures are progressive material deterioration that can lead to catastrophic fracture failure. Although the interpretation of microfractographs requires practice and understanding of fracture mechanisms, only a small number of basic features are clearly recognizable and are indicative of a particular mode or failure: ductile, transgranular, intergranular, and fatigue fractures.

6.1.8.1 Ductile Fracture

Overload fractures of many metals and alloys occur by ductile fracture. Classically, ductile-tensile fracture of a cylindrical specimen involves plastic extension, initially without necking. During this extension, cracking of the included particles, which are present in even the purest metals, or decohesion of the particle/matrix interfaces occurs, creating microvoids. When the ability of the material to work harden is exhausted, necking begins, and triaxial stresses are set up that cause lateral extension of the microvoids, which coalesce to form a central crack. Fracture of the remaining section produces an annulus of slant fracture. Ductile fracture surfaces usually reveal approximately equal-axial dimples, generally with evidence of the particles that originated at the fracture. Fractures that are ductile as seen macroscopically usually are transgranular.

6.1.8.2 Brittle Fracture

Transgranular cleavage of iron and low-carbon steel is the most commonly encountered process of brittle fracture. Transgranular cleavage also can occur in several other body-centered cubic (bcc) metals and their alloys (e.g., tungsten, molybdenum, and chromium) and some hexagonal close-packed (hcp) metals (e.g., zinc, magnesium, and beryllium). The general plane of transgranular brittle fracture is approximately normal to the axis of maximum tensile stress, and a shear lip often is present. The local absence of a shear lip or slant fracture suggests a possible location for initiation of the fracture. The most characteristic feature of the transgranular brittle fracture plateaus is the presence of a pattern of river marks, which consists of cleavage steps and indicates the local direction of crack growth. The rule is that if the tributaries are regarded as flowing into the main stream, then the direction of crack growth is downstream. This is in contrast to macroscopic chevron marks, for which the direction of crack growth, using the river analogy, would be upstream.

6.1.8.3 Brittle Intergranular Fracture

Intergranular brittle fracture usually can be recognized easily, but determining the primary cause of the fracture can be difficult. The first cause of intergranular brittle fracture is the absence of sufficient deformation systems to satisfy the Taylor-von Mises criterion, which states that five independent systems (slip or slip-plus-twining) are necessary for a grain to deform to an arbitrary shape imposed by its neighbors. The second cause is the presence at a grain boundary of a large area of second-phase particles, such as carbides in iron-nickel-chromium alloys. The third cause is the segregation of a specific element or compound to a grain boundary to cause embrittlement. Stress-corrosion cracking (SCC) and hydrogen embrittlement are examples of how brittle fracture follows the intergranular path before final overload fracture occurs.

Stress-corrosion cracking is a mechanical-environmental failure process in which mechanical stress and chemical attack combine in the initiation and propagation of fracture in a metal part. It is produced by the synergistic action of a sustained tensile stress and a specific corrosive environment, causing failure in less time than would the sum of the separate effects of the stress and the corrosive environment. Failure in the form of SCC frequently is caused by exposure to what would seem to be a mild chemical environment while subject to a tensile stress that is well below the yield strength of the metal. Under such conditions, fine cracks can penetrate deeply into the part, although the surface may show only apparently insignificant amounts of corrosion. Therefore, there may be no macroscopic indications of impending failure. Valve alloys such as austenitic stainless steels and nickel alloys of the Inconel type in the presence of very low concentrations of chloride ions, and low-carbon steels usually in the presence of hot concentrated nitrate of caustic alkali solutions, are susceptible to SCC. Stress-corrosion cracks may be intergranular, transgranular, or a combination of both.

Hydrogen embrittles several metals and alloys, but its deleterious effect on steels is most significant, particularly when the strength of the steel exceeds approximately 1240 MPa or the hardness exceeds approximately 45 HRc. A few parts per million (ppm) of hydrogen dissolved in steel can cause hairline cracking and loss of tensile ductility. Hairline cracking usually follows prior austenite grain boundaries and seems to occur when the damaging effect of dissolved hydrogen is superimposed on the stresses that accompany the austenite-to-martensite transformation. Hairline cracking may extend by fatigue and thus initiate catastrophic fracture.

6.1.8.4 Fatigue Fracture

Fatigue fracture results from the application of repeated or cyclic stresses, each of which may be substantially below the nominal yield strength of the material. Many variables influence fatigue behavior, including the magnitude and frequency of application of the fluctuating stress, the presence of a mean stress, temperature, environment, specimen size and shape, the state of stress, the presence of residual stresses, surface finish, microstructure, and the presence of fretting damage. The most noticeable macroscopic features of classic fatigue-fracture surfaces are the progression marks—also known as beach marks, clamshell marks, or tide marks—that indicate successive positions of the advancing crack front. Fatigue-fracture surfaces are smooth textured near the origins and generally show slight roughening as the crack grows. There is little macroscopic ductility associated with fatigue fracture, and there may be some evidence that the crack has followed specific crystal planes during early growth, thus giving a faceted appearance. Unfortunately, many fatigue fractures do not show the classic progression marks. Most fatigue cracks are transcrystalline without marked branching, although intercrystalline fatigue is not particularly uncommon. Corrosion fatigue in most materials also is transcrystalline. Its most striking feature usually is the multiplicity of crack origins, only one of which extends catastrophically. Fatigue initiated by fretting has similar characteristics and generally is diagnosed by the presence of a fretting product filling the multiplicity of cracks and by the presence of a fretting product on the surface of the component. Fretting product appears to be a mixture of finely divided particles of the base metal, its oxides, and its hydrated oxides. Microscopically, surfaces of fatigue fractures are characterized by the presence of striations, each of which is produced by a single cycle of stress. Fatigue striations may propagate in different directions, but they never cross each other.

6.1.9 Root Cause of Failure

If the probable cause of failure is apparent early in the examination, the pattern and extent of subsequent investigation will be directed toward confirmation of the probable cause and the elimination of other possibilities. Otherwise, after gathering together the results of mechanical tests, chemical analyses, fractography, and microscopy, one should be able to form preliminary conclusions and make recommendations. However, in those investigations in which the cause of failure is particularly elusive, a search through published reports of similar instances may be required to suggest possible clues.

Identification of the root cause of failure is the objective of failure analysis. It is a key step toward a solution for avoiding recurrence of similar failures in the future and for prolonging service life. Fishbone diagrams typically are used to assist failure root-cause identification. Figures 6.1 to 6.3 show examples of fishbone diagrams for valve failure, valve/guide wear and seizure failure, and valve seat insert wear, respectively.

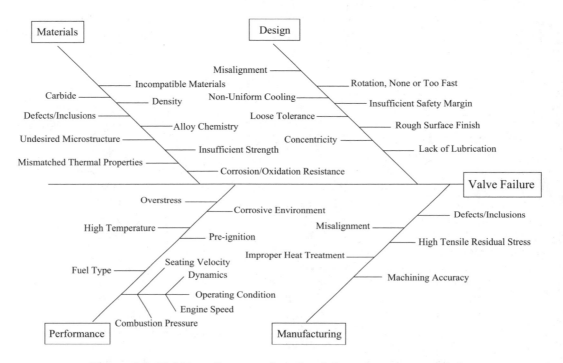

Figure 6.1 *Fishbone diagram of a valve failure root cause analysis.*

6.1.10 Writing the Failure Analysis Report

After all investigations are performed, the evidence revealed by examinations and tests is analyzed and collated, and preliminary conclusions are formulated. The failure analysis report should be written clearly, concisely, and logically. The failure analysis report should include the following sections:

• A description of the failed component with "as-is" macro photographs

• The service or testing conditions at the time of failure

Valvetrain Failure Analysis

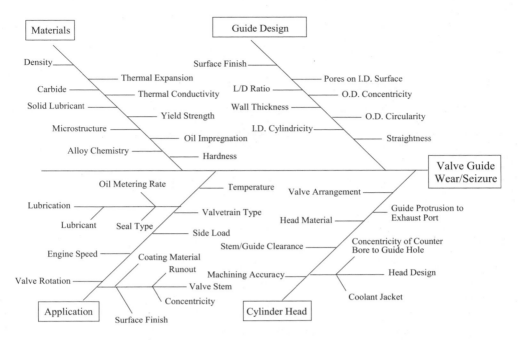

Figure 6.2 *Fishbone diagram of a valve/guide wear and seizure failure root cause analysis (Rodrigues).*

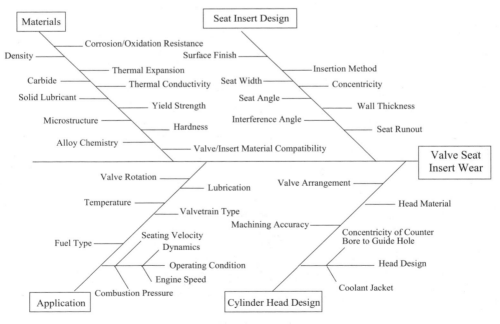

Figure 6.3 *Fishbone diagram of a valve seat insert wear failure root cause analysis (Rodrigues).*

- The service history or testing procedures

- The manufacturing and processing history of the component, including heat treatment

- A mechanical and metallurgical study of the failure

- A summary of mechanisms that caused the failure

- Recommendations for prevention of similar failures or for correction of similar components in service

In subsequent sections of this chapter, various failure examples of valvetrain components are courtesy of Eaton Corporation unless specifically referenced.

6.2 Valve Failures

6.2.1 Overview

Valves failures that occur before the end of the full service life of the valves can be attributed to abusive service, material defects, manufacturing faults, and incorrect design (Figure 6.1). The primary causes for valve failures include excessive temperature, either in magnitude or gradient; high stress, possibly from misalignment or distortion; poor dynamics resulting in high seating velocity; and corrosive environments.

The principal valve failure modes include wear, corrosion, and fracture. Wear may occur at any location that is in contact with another under relative motion. Typical wear locations for valve failure resulting from contact are the valve tip, keeper grooves, stem, and seat.

Corrosion typically results from high temperatures and a corrosive environment. Intake valves often are exposed to operating temperatures of approximately 650°C. Exhaust valves, subjected to higher stresses, often are heated to temperatures higher than 800°C. Corrosion failures occur on surfaces exposed to the combustion process or hot exhaust gases, such as the seat face, stem-fillet region, or top of head. Corrosion also may occur in the stem/guide contact or stem/tip region if corrosive constituents are present in the oil. Valves may corrode as a result of leakage, wear, poor valve dynamics, high temperatures, material selection, or deposit buildup. A special case is stress-corrosion cracking that results from a combination of manufacturing and engine operation processes. Another special case is acidic sulfur corrosion that may form pits in martensitic steels or may assist intergranular oxidation on austenitic stainless steels.

Intake valves are mainly mechanically stressed, whereas exhaust valves also are subjected to thermal and chemical corrosion stresses. Many circumstances and factors can be responsible for a possible breakdown of valves, in the same way as for all highly stressed precision parts. The following are possible causes of valve breakdown fractures: thermal and mechanical overstress, defective manufacture, valve drive disturbances, a fault in the design, improper assembly, incorrect choice of material, material defects, abrupt temperature changes, defect guides, unsuitable fuel, consequential breakdown, and errors in operation. However, burning and breaking are the principal causes of valve failures. These two types of failure also frequently are found in combination, whereas the primary damage cause may be superimposed by a secondary failure cause. For instance, a broken valve stem can be caused by a crack that has been induced by alternating bending stress due to valve distorted seating.

Figure 6.4 shows the valve schematic and nomenclature of valve failure locations. Table 6.1 shows typical valve failure modes at various locations and the possible causes. Typical valve failure modes are described extensively in the literature (Arnold *et al.*, Newton *et al.*, Giles, Hannum, Kelley and Musil, Johnson and Galen, Milbach, and Wang *et al.*).

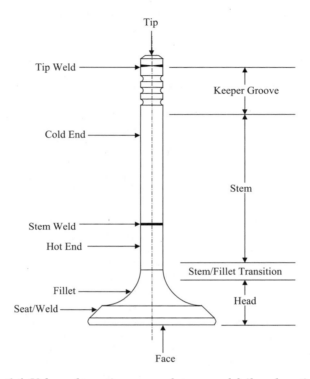

Figure 6.4 *Valve schematic, nomenclature, and failure locations.*

6.2.2 Valve Face Failures

The valve combustion face is subjected to compressive stress during seating and the combustion process. Therefore, it is unlikely to be the location for initiation of mechanical failure. However, both exhaust and intake valve faces are exposed to the high temperatures of the combustion chamber, and oxidation and dealloying are the most frequently observed valve face phenomena. Valve head cupping or "tulipping" occurs due to degradation of the valve head strength at high temperatures. Pre-ignition can raise the valve face temperature so high that it melts the material locally. The rapidly repeated heating and cooling causes thermal fatigue cracks and eventually leads to guttering. Examples of valve combustion face failure due to high temperatures or temperature gradients can be found in Kelley and Musil and in Newton *et al.*

TABLE 6.1
VALVE FAILURE LOCATIONS, FAILURE MODES,
AND POSSIBLE CAUSES

Location	Failure Mode	Possible Cause
Head:		
• Face	Guttering/torching	• Heat affected (pre-ignition)
		• Improper sealing
		• Corrosion assisted
		• Material/manufacturing defects
• Seat	Wear:	
	• Adhesive	• Valve and insert material compatibility
	• Abrasive	• Abusive dynamics and high seating velocity
	• Shear strain	
	• Contact fatigue	• Severe contact stress
	• Corrosive	• Misalignment
		• Thermal distortion
		• Non-uniform heating or cooling
		• Large seat runout
		• Excessive valve rotation
	Guttering/torching	• Heat affected
		• Misalignment and improper sealing
		• Thermal distortion
		• Material selection
	Radial crack	• High hoop tensile stress
		• High temperature gradient
		• Thermal distortion
		• Hardfacing/valve material mismatch
		• Hardfacing/insert material mismatch
		• Positive rotation when seated
• Fillet	Chordal	• High hoop or radial tensile stress
	Radial	• High combustion pressure and/or temperature
	Circumferential	• Material selection
		• Abusive dynamics
		• Material/manufacturing defects
		• Corrosion assisted
		• Assembly distortion
		• Thermal distortion
• Stem/Fillet	Fatigue fracture	• Abusive dynamics
		• Material and heat treatment
		• High stress and temperature
		• Corrosion assisted (acidic sulfur, lead oxide, sulfidation, vanadium pentoxide)
		• Material/manufacturing defects
	Impact fracture	• Abusive dynamics
		• Mis-motion
		• Material selection
		• Material/manufacturing defects
		• Hydrogen embrittlement

TABLE 6.1 *(Continued)*

Location	Failure Mode	Possible Cause
Stem:		
• Weld	Fatigue fracture	• Abusive dynamics • Weld anomalies (including incomplete bond, oxides at weld, internal crack, shallow heat-affected zone—low weld energy) • Roll straightening crack/tensile residual stress • Strain age crack • Stress-corrosion crack
	Impact fracture	• Abusive dynamics • Weld anomalies • Roll straightening crack/tensile residual stress

• Hot End	Wear:
	  • Adhesive (scuffing, galling)
	  • Abrasive (plowing)
	  • Corrosive
		• Lack of lubrication
		• High side load on valve tip
		• Misalignment of insert to guide
		• Thermal distortion
		• Guide material compatibility
		• Inadequate guide length
		• Material/manufacturing defects

	• Fatigue fracture	• Abusive dynamics • Weld anomalies • Roll straightening crack/tensile residual stress • Strain age crack • Stress corrosion crack

• Cold End	Wear:
	  • Adhesive
	  • Abrasive
		• Lack of lubrication
		• Misalignment of insert to guide
		• Thermal distortion
		• Guide material compatibility
		• Inadequate guide length
		• Material/manufacturing defects

	• Fatigue fracture	• Abusive dynamics • Weld anomalies • Roll straightening crack/tensile residual stress • Strain age crack • Stress corrosion crack

TABLE 6.1 *(Continued)*

Location	Failure Mode	Possible Cause
Keeper Groove:		
	Fatigue fracture	• Abusive dynamics
		• Material/manufacturing defects
	Impact fracture	• Abusive dynamics
		• Material/manufacturing defects
		• Mis-motion
Tip:		
• Tip End	Wear:	• Abusive dynamics
	• Adhesive	• High contact stress
	• Abrasive	• Lack of lubrication
	• Contact fatigue	• Inadequate additives
	• Corrosion	• High rocker pad and tip traveling velocity
• Wafer Weld	Fatigue fracture	• Abusive dynamics
		• Weld anomalies
		• High rocker pad and tip traveling velocity
	Impact fracture	• Abusive dynamics
		• Weld anomalies
		• Mis-motion

6.2.3 Valve Seat Failures

6.2.3.1 Valve Seat Wear

Valve seat wear can result in diminished lash, and excessive wear or lash loss can affect valve sealing and combustion pressure loss (Figure 6.5). Typical wear modes at valve seats include adhesion, abrasion, deformation (shear strain or radial flow), corrosion (oxidative), and contact fatigue (Wang *et al.*). Adhesive wear is the most common type of wear for the valve seat. Shear strain wear mode occurs under repeated shearing in the same direction in heavy-duty diesel applications under high combustion pressures. A large seat angle and friction often aggravate the phenomenon. Contact fatigue wear typically is caused by high-asperity contact stress such as hard particles trapped at the interfaces or oxides or deposits. Corrosive and oxidative wear, under a corrosive environment and at high temperatures, occurs after corrosive product forms and flakes under contract stress. In many instances, various wear mechanisms may occur sequentially; in other instances, a combination of wear mechanisms may occur simultaneously, with one mechanism dominating.

6.2.3.2 Valve Seat Guttering and Burning

Valve guttering and burning occur due to the attack of the valve seat material by hot exhaust gases. Guttering and burning are used interchangeably, and it is arbitrarily distinguished that

Figure 6.5 *Severe valve seat recession. Continued operation with this type of wear leads to cupping (Milbach).*

guttering is a severe form of torching or burning. Both result in combustion pressure reduction or loss. The valve is overheating because it is not seating properly due to insufficient valve lash or misalignment of the insert to the guide or insert distortion due to cylinder head non-uniform cooling. The root cause may include excessive seat recession, the larger coefficient of expansion of iron-based valve materials, inadequate valve closed spring load, and severe distortion of the valve head, valve seat, cylinder head, and seat insert. Any one of these problems potentially can cause the valve to be held off the seat, artificially elevating the temperature at the seating surface to the point at which hot corrosion takes place and leak paths form. Valve guttering frequently occurs after a valve has shown signs of cracking from thermal fatigue. Improper selection of materials with inadequate oxidation resistance sometimes bears the blame. Figure 6.6 shows an example of valve guttering failure.

6.2.3.3 Radial Cracks

Radial cracks usually are generated at the valve seat or at the valve seat hardfacing bond lines and propagate radially due to repeated heating and cooling, a thermal fatigue phenomenon. The radial crack can be a leak path or can propagate farther to the valve fillet area. It then follows the maximum stress trajectory and becomes a circumferential crack and then a chordal fracture. Most hardfacing materials contain primary carbides and are too brittle; thus, they are prone to cracking. Because of large temperature gradients and a substantial difference in the coefficients of thermal expansion between the matrix and the hardfacing material, hardfacing materials applied to combat a seat wear problem are to blame under many circumstances. Figure 6.7 shows an example of a valve seat radial crack leading to subsequent chordal fracture of the valve.

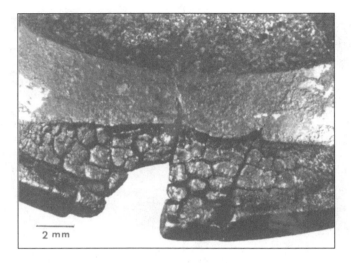

Figure 6.6 *Example of guttered valves, due to poor sealing, misalignment of the insert to the guide, non-uniform cooling/heating, and pre-ignition/misfiring: exhaust valve, radial cracks led to a leakage path (Wang et al.).*

Figure 6.7 *Example of a valve seat radial crack leading to chordal fracture (Wang et al.).*

6.2.4 Valve Fillet Failures

The valve fillet surface is subject to maximum tensile stresses in the valve. High temperatures and a corrosive environment at the exhaust fillet substantially weaken the valve strength. Many valve failures are associated with fillet crack initiation and chordal propagation, and lead to valve fracture. Figure 6.8 shows an example of a valve that has failed at the fillet.

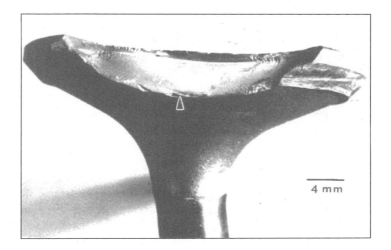

Figure 6.8 *Example of valve fillet fractures due to overstress, at elevated temperatures, and a corrosive environment; the arrow shows the crack initiation site at the fillet (Wang et al.).*

6.2.5 Stem-Fillet Blend Area Failures

The exhaust valve stem-fillet area is subject to high temperatures and exhaust gas impingement. Exhaust valve erosion at the stem-fillet blend area is due to exhaust gas impingement and substantially reduces its load-bearing capability. A large percentage of exhaust valve failures that occur at this location were due to one or a combination of the following factors: abusive dynamics, overstress, and hot corrosion. Examples of stem-fillet blend failures due to high stresses, high temperatures, and corrosive environments can be found in Kelley and Musil and in Milbach.

6.2.6 Stem Failures

One common valve stem failure is scuffing, and severe scuffing can lead to valve seizure in the guide. The seized stem or guide leads to a valve fracture hit by either a piston or cam, and this can cause catastrophic engine failure. Side load and lack of lubrication can aggravate the seizure phenomenon. An internal strain age crack and stem failure can occur when a valve is underaged during heat treatment, is roll straightened, and subsequently is aged further. The straightening operation induces deformation to the surface that may age more fully and rapidly than the core, resulting in surface compression stress, whereas the core goes into tension and crack in an intergranular manner. Strain age cracks can form both in underaged austenitic stainless steels and in nickel-based superalloys.

6.2.7 Keeper Groove Failures

Valve keeper grooves have smaller cross-sectioned areas and thus are subject to higher stress, compared to valve stems. Common failure modes include wear, fatigue fracture, and impact fracture. Abusive dynamics, a loosened key, and inadequate hardness all contribute to valve keeper groove failures. Examples of valve keeper groove failures can be found in literature by Johnson and Galen.

6.2.8 Tip Failures

A typical valve tip failure includes tip wear, pitting, and fracture at the wafer weld and is the result of inadequate lubrication and high contact stress. Tip wear or pitting increases valvetrain lash, noise, and friction, thus adversely affecting fuel economy, engine efficiency, and noise, vibration, and harshness (NVH). Tip wear and pitting typically appear more severe in some valvetrain types than in others. A Type II valvetrain with an end-pivot rocker arm poses higher tip wear due to the greater rocker ratio and sliding distance at the rocker arm pad and valve tip. A direct-acting valve usually does not have problems with tip wear. The fracture in the wafer weld is due to a cold weld and typically exhibits a brittle (intergranular and cleavage) or ductile fracture mode due to overstress. The fracture surface may show original grinding marks on the stem end, as well as some characteristics of cold welding.

6.3 Cam Failures

Typical cam failures include wear, pitting, and scuffing. Cam lobe wear can be either a slow polishing process that lessens with time as a hard smooth surface is produced, or a continuous process that ultimately results in the removal of sufficient material and in a cam profile change that affects the valve lift, velocity, and acceleration profiles, thus affecting engine performance. Scuffing is the most severe form of cam failure caused by localized welding of the surfaces brought into contact (by breakdown of the oil film under high contact pressure). Once scuffing has commenced, it proceeds rapidly. Pitting is a fatigue process caused by repeated contact stresses at the surfaces, resulting in subsurface cracks initiating from the points of highest shear stress. The development of the crack results in material being removed from the surface to leave a pit of fissure. In an advanced stage, it can result in flaking of the surface, thus affecting engine performance. Contact stress has a major effect on cam failure. Cam/follower failures are more pronounced in engines with Type V valvetrains (pushrods) than with other types of valvetrain. Because the pushrod system results in high inertia, it requires high valve spring loads and gives rise to high contact stress, particularly at low engine speeds when the spring load is offset only slightly by the inertia loading and hydrodynamic lubrication can hardly exist. Figure 6.9 shows examples of typical cam failure modes.

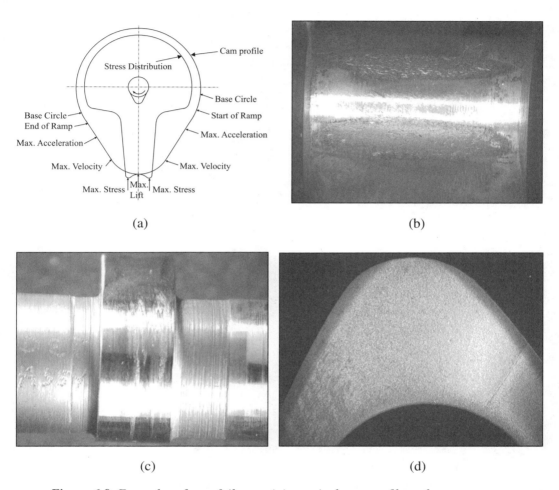

(a)

(b)

(c)

(d)

Figure 6.9 *Examples of cam failures: (a) a typical cam profile and contact stress distribution, (b) moderate wear at maximum lift and maximum stress locations at the cam nose, (c) a cam lobe scuffed from the base circle to the nose from the roller follower, and (d) severe wear at the cam nose and with the case worn through.*

6.4 Lifter Failures

In most cases, failures of cam followers or lifters are the result of overstress and inadequate lubrication. The cam faces of flat-faced cam followers—including the lifter, tappet, and bucket—usually are treated with ferrox coating, phosphating, nitriding, or prelubrication with special lubricants containing an extreme pressure additive for break-in. A typical lifter cam face failure due to inadequate lubrication is scuffing or galling at both the cam lifter face and the cam lobe. This type of failure usually starts in the first few moments of engine operation and will continue to progress as long as the engine runs. Another failure mode due to overstress at the asperity scale is pitting or contact fatigue failure when the lubricant film cannot separate the cam and its follower. One of the more common causes of hydraulic lifter malfunction or failure is dirt or debris. Debris either will cause the plunger to stick in the lifter-barrel and not leak down, or it will unseat the check valve, allowing the lifter to collapse. Figure 6.10 shows examples of lifter/follower failure modes.

(a) (b)

(c) (d)

Figure 6.10 *Examples of lifter/follower failures: (a) a lifter body showing a crack at the strut, (b) a roller exhibiting contact fatigue, (c) a roller bearing surface showing wear, and (d) a section view through the axle, revealing fatigue cracks at the outside diameter (O.D.) surface.*

6.5 Insert Failures

An insert failure typically is wear that leads to diminished lash and poor sealing and, ultimately, to valve guttering and total combustion pressure loss. Insert wear mechanisms are primarily adhesive, sometimes in combination with abrasive, shear strain, and oxidative wear. Contact stress, valve/insert material compatibility, metal-to-metal contact, high temperatures, and a corrosive environment all contribute to insert wear. Figure 6.11 shows some typical insert failure modes.

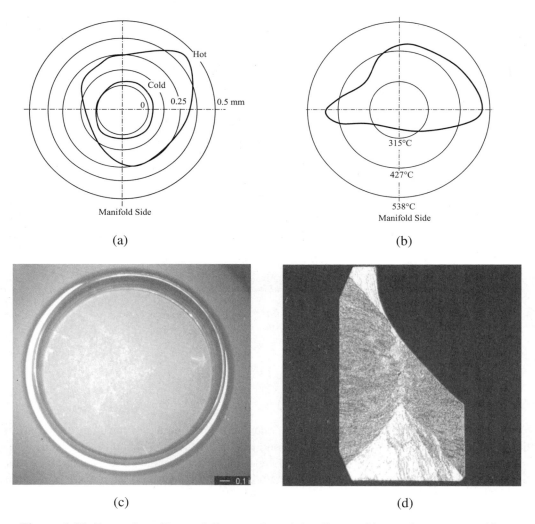

Figure 6.11 *Examples of insert failure modes: (a) a distorted insert due to non-uniform head cooling (Taushek and Newton), (b) a non-uniform temperature distribution around an insert due to a distorted insert (Taushek and Newton), (c) non-uniform wear patterns that indicate misalignment of the insert to the guide, and (d) an insert section showing considerable recess.*

6.6 Guide Failures

Typical guide failures include guide wear and scuffing. Guide wear can be aggravated when insufficient lubricant enters the guide, when debris enters the guide, with misalignment of the insert to the guide, poor concentricity of the guide inside diameter (I.D.) and outside diameter (O.D.) with the insert I.D. and O.D., excessive runout, and side load from the valve tip. Excessive guide wear due to lack of lubrication and high friction can lead to valve scuffing or seizure and catastrophic engine failure. However, too much lubricant entering the guide, especially the exhaust guide, can detrimentally increase emissions when the lubricant is not completely combusted and thus escapes into the exhaust pipe. A delicate balance must be accomplished. Figure 6.12 depicts an example of guide wear failure; discoloration at the hot end indicates a high operating temperature.

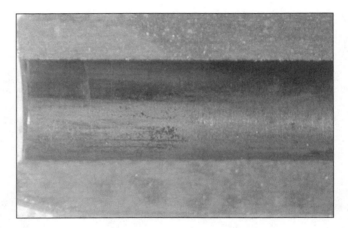

Figure 6.12 *Example of a valve guide failure mode. Oxidation*
at the hot end indicates a high operating temperature.

6.7 Spring Failures

When a spring breaks, other valvetrain components also will break. Typical spring failure is due to fatigue and spring surge at overspeed conditions and excessive dynamics. Surface imperfections such as nicks during installation and handling are factors that raise the external stress, thereby weakening the designed fatigue strength of the spring. Process defects and inclusions in the spring wire are factors that increase the internal stress. Improper heat treatment, insufficient carburized case depth, and corrosion weaken the strength of the spring and shorten the service life of the spring. Another failure is the relaxation of the spring when it operates at high temperatures. When the valve spring suffers set (i.e., the valve spring is permanently deformed) during service, the spring becomes shorter than it is designed. This leads to insufficient spring load, and the spring cannot close the valve properly. Figure 6.13 shows a typical valve spring failure mode.

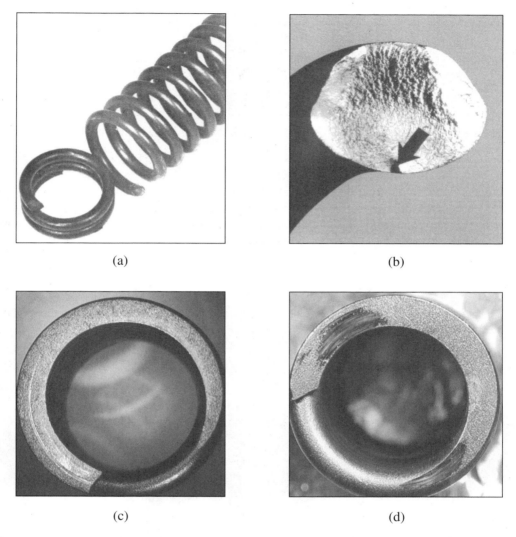

(a)

(b)

(c)

(d)

Figure 6.13 *Examples of spring failure: (a) a broken spring (TRW Handbook),*
(b) a broken spring showing the crack initiation site (TRW Handbook),
(c) the bottom coil of a spring exhibiting moderate wear, and
(d) the bottom coil of a spring, showing end contact and wear.

6.8 Stem Seal Failures

Valve stem seals play a critical role in controlling valve guide lubrication and oil consumption. If the seals do not fit properly, are not installed correctly, or are worn excessively, the guides either may be starved for lubrication or may be flooded with oil. A rough valve stem surface finish is the major contributor of excessive seal wear, and it is recommended that the surface roughness of the stems be smaller than 2 μm (Matsushima *et al.*). High operating temperatures, including friction heat, can cause seal material such as nitrile to harden and become brittle over time. Eventually, this can lead to cracking, loss of oil control, and seal failure. When a valve stem seal loses its ability to control the amount of oil that enters the guide, it can cause a variety of problems. Examples of stem seal failures can be found in literature by Young and Martin and by Matsushima *et al.*

6.9 Rocker Arm Failures

Wear at the socket and valve tip pad is common in rocker arms. This type of failure due to high stress and lubricant starvation usually leads to increased lash, resulting in poor idle quality and increased valvetrain friction, NVH, and dynamics. Fracture is another type of failure that occurs occasionally in rocker arms at the maximum tensile stress location due to poor valvetrain dynamics; it leads to catastrophic engine failure. Examples of rocker arm failures can be found in the literature by Snyder and in the *TRW Handbook*.

6.10 Retainer Failures

The spring retainer wears at the inner diameter in contact with the outer diameter of the keys if the friction force is not large enough to lock the keys but is large enough to cause wear. Loose keys can exacerbate the wear at the interface and can affect the valve lift as designed, thus adversely affecting engine performance. Figure 6.14 shows examples of spring retainer failure modes.

(a)   (b)

Figure 6.14 *Examples of spring retainer failures: (a) a retainer I.D. showing relative longitudinal sliding between the retainer and the key O.D. surface, and (b) a retainer I.D. showing heavy contact and fatigue cracks.*

6.11 Key Failures

Key failure and wear typically represent poor valvetrain dynamics and can lead to catastrophic engine failure. Figure 6.15 shows examples of key failures.

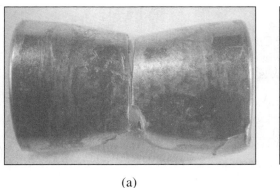

(a)   (b)

Figure 6.15 *Examples of key failures: (a) severe O.D. contact with a spring retainer, and (b) severe contact and damage at the I.D. bottom end.*

6.12 References

Arnold, E., Bara, M., Zang, D., Tunnecliffe, T., and Oltean, J., "Development and Application of a Cycle for Evaluating Factors Contributing to Diesel Engine Valve Guttering," SAE Paper No. 880669, Society of Automotive Engineers, Warrendale, PA, 1988.

Giles, W., "Fundamentals of Valve Design and Material Selection," SAE Paper No. 660471, Society of Automotive Engineers, Warrendale, PA, 1966.

Hannum, A., "Why Valves Fail," SAE Paper No. 249A, Society of Automotive Engineers, Warrendale, PA, 1960.

Johnson, V. and Galen, C., "Diesel Exhaust Valves," SAE Paper No. 660034, Society of Automotive Engineers, Warrendale, PA, 1966.

Kelley, J. and Musil, N., "Valve Failure Fingerprints," *Engine Failure Analyses at Shop Level*, by M. Durella *et al.*, Society of Automotive Engineers, Warrendale, PA, 1962.

Matsushima, A., Hatsuzawa, H., Yamamoto, Y., and Iida, S., "Comments on Valve Stem Seals for Engine Application in Small Size Passenger Cars," SAE Paper No. 850331, Society of Automotive Engineers, Warrendale, PA, 1985.

Milbach, R., *Valve Failures and Their Causes—Practical Experience from Valve Failure Research, 3rd Edition*, TRW, Cleveland, OH, 1984.

Newton, J., Palmer, J., and Reddy, V., "Factors Affecting Diesel Exhaust Valve Life," SAE Paper No. 190, Society of Automotive Engineers, Warrendale, PA, 1953.

Newton, J.A. and Tauschek, M.J., "Valve Seat Distortion," SAE Paper No. 64, Society of Automotive Engineers, Warrendale, PA, 1953.

Rodrigues, H., "Sintered Valve Seat Inserts and Guides: Factors Affecting Design, Performance, and Machinability," *Proceedings of the International Symposium on Valvetrain System Design and Materials*, ASM International, Materials Park, OH, 1997.

Ryder, D., Davies, T., Brough, I., and Hutchings, F., "General Practice in Failure Analysis," *ASM Handbook, Vol. 11, Failure Analysis and Prevention*, ASM International, Materials Park, OH, 1986.

Snyder, J., "Failures of Iron Castings," *ASM Handbook, Vol. 11, Failure Analysis and Prevention*, ASM International, Materials Park, OH, 1986.

TRW Handbook, Teves-Thompson GmbH, Barsinghausen,1979.

Wang, Y., Schaefer, S., Bennett, C., and Barber, G., "Wear Mechanisms of Valve Seat and Insert in Heavy Duty Diesel Engine," SAE Paper No. 952476, Society of Automotive Engineers, Warrendale, PA, 1995.

Young, W. and Martin, J., "Advanced Fluorocarbon Compounds for Valve Stem Seal Applications," SAE Paper No. 920708, Society of Automotive Engineers, Warrendale, PA, 1992.

Index

The letter *n*, *f*, or *t* following a page number denotes a note, figure, or table, respectively.

Abrasive wear, 390, 450, 451*t*, 448–453, 494
Acceleration, valvetrain dynamic testing, 418–420, 424, 425*f*, 426
Acceleration curves, 70, 71*f*, 112–114, 115*f*, 180–190
Acceleration pulse, 105, 112, 114
Actuator
    electrohydraulic, 55–57
    electromagnetic, 53–55, 57
    hydraulic, 36, 44–45
Additives. *See* Lubricant additives
Adhesive wear, 386–387, 448–453, 494, 550
Aftercooler, 23
Aging heat treatment, 157–159
Air pressure, 9
AISI 4340, hydrogen embrittlement, 369*f*
Alfa Romeo
    cam and follower materials, 205*t*
    cam phaser, 36–38
Alloy C, 149*t*, 150*t*, 160*f*, 162*f*, 163*f*, 164*f*, 166*f*, 167*f*
Alloys, 148–174
    chemistries, 241–245*t*
    corrosion resistance, 168–171
    oxidation, 168–170
    wear resistance, 171–174
    see also specific types
Aluminum, 158, 169, 332, 446*t*
Aluminum alloys, 273, 365, 370*t*, 273
Aluminum oxide, 359*t*
Amontons, 443
Archard's equation, 172
Area reduction, 163, 164*f*
Aromatic lubricant base stocks, 458, 459
Arrhenius' law, 391, 392*f*

Asperity contact pressure, 443, 444*f*
ASTM A533 B1 steel, fatigue test results, 372*f*
Atomic forces, 330, 331*f*, 444*f*
Audi
    cam and follower materials, 205*t*
    Type I valvetrain engines, 64
Austenite, 151*t*, 152, 158, 265, 332, 359*t*
Austenitic alloys
    converting hardness results, 362*t*
    nitride thicknesses, 513*t*
    for seal inserts, 240, 250
    strengthening, 152–153, 157, 342–343
    stress-corrosion cracking, 370*t*, 543
    for valves, 133, 148, 149*t*, 150*t*, 152–153, 157–167, 169, 170*f*, 171
Austin Morris, cam and follower materials, 205*t*

Bainite, 151*t*, 152
Barrel lifters, 224–226
Base circle, 176, 200, 201*f*, 219
Base circle radius, 190, 191*t*, 193–194
Base circle runout, 204*f*, 218–219
Bearing journals, 175
Belleville spring, 320, 321*f*
Bench simulation testing, 395–396
Bench testing, 395–402
Bending tests, 373
Beryllium, 332, 542
Bleistahl Produktions GmbH, 241*t*
Block assembly, 3
BMW, 3*n*
    cam and follower materials, 205*t*
    Double VANOS, 39*f*, 48, 49*f*, 59*t*

BMW *(continued)*
    Type I valvetrain engine, 64
Mean intake gas velocity, 143
    Valvetronic, 48, 49*f*, 51, 59*t*
Boron carbide, 359*t*
Bottom dead center (BDC), 4, 5*f*, 28
Bounce, valvetrain dynamic testing, 423–426
Boundary lubrication, 457
Brake mean effective pressure (BMEP), 17
Brake power, 16, 17
Brake thermal efficiency, 18
Brass, 446*t*
Bravais lattices, 332, 334–335*t*
Breathing ability. *See* Volumetric efficiency
Brinell hardness test, 354–355, 359, 360, 361*t*
Brittle fracture, 542–543
Brittle materials, 348, 350, 390
    ductile-to-brittle-transition, 345–347, 367, 368*f*
    stress-strain curve, 338, 339*f*
    valve seat insert materials, 231, 252
Bronze, 446*t*
Bucket lifter, 214, 222–223
Bucket tappet, 44, 46–47
    follower rotation, 75–78
    lifter diameter, 74
    nose radius, 78
Burgers vector, 346
Burn rate, 46

Cadmium, 332
CaF_2, 239, 253, 264
Cam, 69–72
    acceleration, maximum, 72
    lift, 71, 177
    failures, 554–555
    phasing, 34, 35*f*, 36–40, 48
    torque, 483, 484*f*
    velocity, 71, 74–75
    wear
        cam and follower deterioration, 491–494
        and oil, 501*f*, 502*f*
        and spring load, 499
        testing, 396, 397*f*
Cam angle, 27, 69, 71*f*, 72, 73, 105*f*, 108*f*, 110*f*
Cam degree, 33

Cam follower, 43–44, 62, 69–70, 180, 188, 205*t*, 228
    displacement, 69–70
    failures, 555–556
    materials, 203–206
    and pressure angle, 270
    wear testing, 388–390
    see also Finger follower; Flat follower; Roller follower
Cam gear, 33*f*
Cam lifter failures, 555–556
Cam lobe, 26, 46, 175–176
    offset centerline, 75–76
    surface finishes, 201–202
    tapered, 75, 78
    tolerances, 202–203, 204*f*
    wear, 554
Cam lobe design. *See* Cam profile
Cam lobe taper, 191*t*
Cam profile, 27, 46–48, 53, 69, 74–75, 78
    base circle radius, 190, 191*t*, 193–194
    high-speed, 198–200
    lightweight, 200–201
    radius of curvature, 195–198, 199*f*
    ramps, 194–195
    tolerances, 202–203, 204*f*
    valve lift, 176–179
    valve lift curves, 179–190
    and valvetrain dynamics, 220
    and valvetrain dynamics testing, 423–426
    see also Valve timing diagram
Cam profile switching, 40–43, 57*t*
Camless valve actuation, 67*t*, 52–53, 121
    electrohydraulic, 55–57
    electromagnetic, 53–55, 57
Camshaft, 26, 175*f*
    design, 174–203, 217–219
    hollow, 200–201
    hydraulic, compared to mechanical, 217, 219
    hydraulic design, 217–219
    materials, 204–206, 228*t*
    motion, 176
    parts, 175–176
    speed, effect on oil aeration, 218*f*
    weight reduction, 200–201
Camshaft deflection, valvetrain dynamic testing, 419

Index

Camshaft ramps, 148, 217–218
   hydraulic, 218
   profile, 190, 191t, 194–195
   height, 107, 108f, 217, 218t
   velocity, 71, 218t
Carbides, 252–253
Carbon, 148, 151t, 152, 153, 154, 158, 161
Carbon deposits, lubricant, 478, 479, 480, 505, 506f
Carbon diffusion, 510–511
Carbon monoxide emissions, 306, 307f
Carbon scraper, 128
Carbonitrided alloys, 222, 223, 229, 273, 313, 317
Carbonitriding, 516, 517t
Carbonization, and lubricant-related engine malfunctions, 478, 479, 480, 505
Carbonized alloys, 313, 317
Carburized alloys, 204, 207f, 229, 273, 322
Carburizing, 514–515, 517t
Cast alloys, 240, 260, 360t
Cast iron
   in cams, 204, 205t, 206, 207f
   coefficient of friction data, 446t
   and contact stress, 493
   converting hardness results, 362t
   estimating strength from hardness values, 360t
   fracture toughness values, 365
   in valve guides, 240
   in valvetrain lash compensators, 229
Catalytic converter, 138
Caterpillar, Type III valvetrain engine, 65
Cemented carbides, 362t
Cementite, 151t, 359t
Center-pivot rocker arm, 61, 62, 63f, 65–66, 67t, 76t, 266, 268f, 269, 270
Center-pivot rocker arm valvetrain system, 61, 67t
   centerline, 85
   kinematics, 83–87
   pressure angle, 85–87
   rocker ratio, 73, 74f, 83
   roller-follower, 85
   side thrust, 85
   slip-slide, 85
   valve tip contact travel, 83, 84f
   see also Type III, Type IV, Type V valvetrain

Ceramics, 239, 245–246, 251, 255, 365
Cerments, 239
Charge air cooler, 23
Charpy impact test, 366
Check ball, 212, 213, 214f
Check ball spring, 209–211f, 212
Chemical vapor deposition (CVD), 517, 520–521
Chevrolet, Type V valvetrain engine, 66
Chrome, 494
Chrome plating, 136, 229, 514, 515f, 526–528
Chromium, 332, 345, 393, 514, 542
   in valve seat inserts, 252, 254
   in valves, 152, 154, 155, 156, 158, 169, 171
Chromium carbides, 152, 359t
Chromium coatings, removal, 528
Chromium diffusion, 393
Chrysler
   cam and follower materials, 205t
   Type V valvetrain engines, 66
Circular arc curves, 181–182
Citroën, cam and follower materials, 205t
Cladding, 246–249
Cobalt, 154, 332
Cobalt-based alloys
   for seal inserts, 240, 251
   for valves, 149t, 150t, 154–155, 164, 171, 173
Coefficient of friction, 82–83, 443, 445, 446, 456, 457
Coefficient of kinetic friction, 385, 386
Coefficient of static friction, 384, 385, 386
Coefficient of thermal expansion, 129, 160, 249–250, 376, 377f
Coil diameter tolerances, 282–283
Cold forming, 158
Cold start, 138, 213, 216
Combustion chamber, 5f
Component bench testing, 395–402
   cam and follower wear, 396, 397f
   computer simulation, 395
   flow testing, 399, 400f, 401f
   hydraulic lifter leak-down, 399–402
   stem seal oil metering, 402, 403f
   strain gage, 404
   testing fixtures, 395
   thermal shock, 402, 403f
   valve seat and insert wear, 396–397

Component bench testing *(continued)*
    valve stem and guide wear, 398, 399*f*
Compressed natural gas (CNG), 396
Compression ignition engine, 9–10, 31
    compared to spark ignition engine, 9–10,
        122–123
    valve temperature distributions, 123*f*
    valve timing, 147
    see also Diesel engine
Compression ratio, 9, 11–12, 22, 38, 50–51, 58
Compression stroke, 5, 6*f*, 7–8, 9, 16
Computer simulation, component bench test-
    ing, 395
Concave valve head, 126
Constant pulse curves, 185–188
Constant velocity curve, 180
Contact, real area of, 439–440
Contact fatigue, 451, 492–494
Contact mechanics, 438–442
Contact point, 71, 80–81
Contact pressure, 87, 89–93
Contact stress, 87, 94–95, 443, 493
    and cam failures, 554
    and camshaft design, 199*f*, 200, 201–202,
        207*f*
    tappet faces, 228, 229*f*
Continuous combustion gas turbine engine, 2
Coolant, valve, 121, 122*f*, 127
Cooling system, 3*f*, 4
Copper, 158, 239, 247, 251*f*, 261, 265, 332,
    446*t*
Copper alloys, stress-corrosion cracking, 370*t*
Corrosion, 123–124, 128, 494
    and valve failure, 546
    wear mechanisms, 448–453
Corrosion fatigue, 543
Corrosion fatigue testing, 378–380
Corrosion inhibitors, 464, 468*t*, 469*t*, 477
Corrosion resistance, 168–171
Corrosion testing, 391, 392–395
Corrosive wear, 451–452
Coulomb, 443
Covalent bonds, 330
40 Cr, 368*f*, 368*t*
Crack detection. *See* Nondestructive testing
Crack resistance, 365*f*
Cracking, 363–364

Cracks, 492
    effect on test results, 347–348, 367
    failure analysis, 540
    valve seat failure, 551
Crank angle, 33*f*, 34*f*
Crank degree, 33
Crank gear, 33*f*
Crankshaft 3, 4
Crankshaft horsepower, 17
Crazed, wear classification, 448
Critical engine speed, 107–108
    acceleration curve, 112–114, 115*f*
    harmonics, 115–118
    quasi-static deflections, 110–111
    shock response, 111–114
Critical stress-intensity factor, 363
Crosshead rocker arms, 323–324
Crystal structure, 330, 331–332, 334–335*t*
    body-centered cubic (bcc), 332, 333*f*
        and decreasing temperature, 345–346
        and strain rate, 346
        materials, 151*t*, 152, 332
        strengthening, 341*f*, 344
        stress-strain curve, 338–339
        transgranular cleavage, 542
    face-centered cubic (fcc), 151*t*, 152, 332,
        333*f*
        materials, 151*t*, 152, 153, 163, 332
        strengthening, 341*f*, 344
    hexagonal close-packed (hcp), 332, 333*f*
        materials, 332
        transgranular cleavage, 542
    simple cubic, 333*f*
Cubic curves, 185–188
Cu$_2$O, 253, 264
CuS$_2$, 239, 253, 264
Cylinder blowdown, 35–36
Cylinder volume, 9, 11
Cylinders, 4, 5*f*, 6*f*, 10, 11*f*, 14*f*

Daimler-Benz, cam and follower materials,
    205*t*
DaimlerChrysler
    Type II valvetrain engine, 65
    Type V valvetrain engine, 66
Damped vibration, 97, 100
Datsun, cam and follower materials, 205*t*

DCX, VVT system, 59*t*

Debris, 452, 556, 558

Defect detection. *See* Nondestructive testing

Deflection, 386
    and valve bounce, 426
    valvetrain dynamic testing, 418*f*, 419–420, 424, 426

Deflection curve, for linear compression springs, 278

Deflections, 71, 102, 107, 110–111
    and camshaft design, 175, 177, 193, 195, 198, 217, 218, 219, 229*f*
    and contact stress, 228, 229*f*
    equations, 111

Deformation
    effect of strain rate on, 346–347
    effect of temperature on, 345–347
    elastic deformation, 332–333, 338
    fractured metals, 347–348
    plastic deformation, 89, 163, 172, 251, 252, 333, 335, 338, 350, 452
    thermal deformation, 156

Delamination, 448, 452

Detroit Diesel, Type III valvetrain engines, 65

Diamond, 359*t*, 446*t*

Diesel engine
    engine braking mechanism, 58
    air intake, 51
    and anti-wear additive ZDP, 506*f*
    camshaft materials, 204
    compared to gasoline engine, 9–10
        compression ratios, 12, 22
        contact pressure limits, 93
        corrosion, 123, 168
        materials, 150*t*, 228*t*, 240, 246*t*
        ramp rates, 195
        rocker ratio, 76*t*
        valve and port size, 138*t*
        valve overlap, 147
        valve seat inserts, 234–235, 237, 240, 246*t*
        valve timing, 32*t*
    corrosion, 168
    corrosive wear from fuel, 480
    effect of soot on wear, 506*f*
    engine brake, 58, 60*f*
    exhaust valve corrosive deposits, 394–395
    lash measurement, 428*f*
    lubricant additives, 469*t*
    sulfidation corrosion, 394–395
    synthetic oil additives, 476*t*, 477
    tribological conditions, 469*f*
    valve materials, 123, 153, 246*t*
    vertical valves, 139
    see also Compression ignition engine

Diffusion, 510–511

Direct-acting valvetrain system, 27, 61, 67*t*
    kinematics, 69–72, 74–79
    lash, 78–79
    maximum velocity, 74, 76
    see also Type I valvetrain

Dispersed hard phase, 239–240

Displacement
    valvetrain dynamic testing, 418–419, 420

Doming, 139, 140*f*

Drop-weight test, 366, 367*f*

Dry lash, 214–215, 222, 226

Ductile fracture, 542

Ductile metals, 381–383

Ductile-to-brittle transition, 345–347, 367, 368*f*

Ductility measurement, 351

Dynamic deflections, 110–111

Dynamic hardness tests, 357–358

Eatonite, 149*t*, 150*t*, 164, 165*f*, 174*f*

Eatonite 2, 173

Eatonite 3, 149*t*, 150*t*, 165*f*, 173

Eatonite 4, 170

Eatonite 5, 149*t*, 150*t*, 173

Eatonite 6, 149*t*, 150*t*, 155, 160*f*, 161, 164, 165*f*, 170, 173, 174*f*

Eaton's variable displacement rocker arms, 274, 276*f*

Eddy current testing, 404, 405*t*, 406*t*, 414–416, 417*f*

Effective expansion ratio, 35–36

Effective pressure, 88

Efficiency, mechanical, 18

Eight-cylinder engine, 34, 35*f*

Elastic deformation, 332–333, 338

Elastic limit, 350

Elastic modulus, 377*f*

Elasticity theory, 87–89

Elasto-hydrodynamic lubrication, 457, 495, 497

Electrical methodology, nondestructive testing, 404

Electrohydraulic valve actuation, 55–57

Electromagnetic valve actuation, 53–55

Electronic beam surface hardening, 510

Elongation, 161, 163

Emissions, 26, 31, 32*f*, 35, 36, 42, 57*t*, 79, 481, 558

  cold start, 138, 213

  effect of valve timing, 31, 32*f*

  and exhaust valve guides, 299–300

  and lash compensators, 208

  and oil metering rate, 264, 304–306, 307*f*

  and positive valve rotators, 320

  and valve stem seals, 136, 299, 300, 304

  and valve timing, 31, 32*f*, 35, 57–58

Emperial Piston Ring, 242*t*

End-pivot rocker arm valvetrain system, 26*f*, 61, 67*t*

  centerline, 79–80, 82

  contact point, 80–81

  kinematics, 79–83

  lash adjuster side load, 82–83

  pressure angle, 82, 83

  rocker ratio, 73, 74*f*, 75*f*, 81

  slip-slide, 80–81, 85

  valve keeper groove location, 80, 122*f*

  valve tip location, 80

  see also Type II valvetrain

End-pivot rocker arms, 62, 63*f*, 64–65, 67*t*, 76*t*, 266–270, 272*f*, 274, 276*f*

Energy, 1, 8

  interatomic, 330, 331*f*

Energy dispersive x-ray (EDX) spectrometry, for failure analysis, 540–541

Energy loss, 50, 55, 58, 96

Energy recovery, 50, 56–57

Engine, 3*f*, 11, 12*f*

Engine brake, 58, 60–61, 62*f*

Engine clearance volume, 11, 12*f*

Engine displacement, 11, 12*f*

Engine efficiency, 18–23, 48–50

  see also Thermal efficiency; Volumetric efficiency

Engine malfunctions. *See* Lubricant-related engine malfunctions

Engine speed, 9

  critical, 107–108, 110–118

  effect on valve temperature, 429–430

Engine working volume, 11, 12*f*

Engineered Sintered Components, Inc., 241*t*

Entrainment velocity, 496–497

Environment, consideration in component bench testing, 395

Environmentally-assisted fractures

  hydrogen embrittlement testing, 368–369

  stress-corrosion cracking testing, 369

Equations

  abrasive wear, 449

  acceleration pulse, 112

  adhesive wear, 449

  Archard's, 172

  area of contact, real, 439–440

  base circle radius, 193–194

  brake horsepower, 17

  cam velocity, 71

  coefficients of friction, 385, 443

  compression ratio, 12

  contact pressure, Hertzian, 89–90, 92–93, 94–95

  contact travel, 83

  converting hardness values, 362*t*

  critical stress-intensity factor, 363–364

  diesel engine thermal efficiency, 20–21

  elasticity, 87–89

  engine mechanical efficiency, 18

  engine speed, 112, 117

  engine thermal efficiency, 18

  engine working volume, 11

  entrainment velocity, 496–497

  fatigue life prediction, 381–382

  film thickness, lubricant, 495–496

  flow velocity, 143, 146

  Fourier, 116

  friction temperature, 447

  hardness, 354, 356, 357, 358

  Hertzian, 89–90, 92–93, 94–95, 491–492

  horsepower, 16, 17

  hydraulic lifter force, 216

  indicated mean effective pressure, 17

  lash adjuster side load, 82–83

  maximum lift velocity, 71–72, 73

  mean intake gas velocity, 143

  natural frequency, 104–105, 106, 271, 291

  oxidation, 391

  perfect gas equation, 23

Index

Equations *(continued)*
  plastic zone, 347
  potential energy, 330
  pressure angle, 83, 86
  pulse, 70, 73, 179
  quasi-static deflection, 111
  radius of curvature, 197–198
  rocker ratio, 73, 83
  roughness, 440, 442
  shear stress, 335–336, 337
  sliding, 383–384, 389
  slip-slide distance, 81
  spring rate, 278
  spring stress, 286
  spring surge, 106
  strain, 87
  strength, from hardness values, 359, 360*t*
  stress, 91, 421
  surface roughness, 497
  thermal efficiency, 19–20
  torque, 13
  true strain, 353
  true stress, 353
  ultimate tensile strength, 353–354
  valve acceleration, 70, 73
  valve closing velocity, 107
  valve displacement, 69–70
  valve lift, 73, 177, 179
  valve lift curves, 180–190
  valve opening area, 141
  valve seating force, 129
  valve spring force, 72
  valve velocity, 70, 73
  valvetrain effective weight, 101, 103*t*
  valvetrain stiffness, 104
  vibration, 98–99, 100
  volumetric efficiency, 22
  wear rate, 388
  wear volume, 172
  work, 13, 15
Equilibrium distance, 330, 331*f*
Exhaust gas recirculation (EGR), 36, 147
Exhaust gases, 550
Exhaust stroke, 5, 6*f*, 8–9
Exhaust valve, 4, 5*f*, 25, 27–31, 32*t*, 34*f*, 121, 122*f*, 123, 145
  corrosive deposits, 392, 393
  design, 128, 129, 133
  diameter, 138*t*, 139–140
  failures, 546–554
  materials, 148–154, 342
  timing, 35–36, 147–148
  valve stem deposits, 123–124, 128
Exhaust valve guide, and emissions, 299–300
Exhaust-to-intake area ratio, 137
External combustion engine, 1–2

Failure analysis
  cam failures, 554–555
  fatigue cracks, 540
  Fishbone diagrams, 544, 545*f*
  guide failures, 558
  insert failures, 557
  lifter failures, 555–556
  rocker arm failures, 560
  spring failures, 558, 559*f*
  spring retainer failures, 560
  stages of, 538–544
  stem seal failures, 559–560
  valve failures, 546–554
  valve key failures, 561
Fatigue, 371
Fatigue cracks, 540
Fatigue failure, 96, 492–494
Fatigue fracture, 543
Fatigue life prediction, 381–383
Fatigue limit, 372*f*, 373, 374, 376*f*, 380
  relationship with tensile strength, 383, 384*f*
Fatigue resistance, shot peening, 530, 531*f*
Fatigue spalling, 388
Fatigue strength, 161, 164–167, 374, 514
Fatigue testing
  corrosion, 378–380
  methodology, 371–375
  thermal, 375–378, 379*f*
Fatigue wear, 451, 452–453
Federal-Mogul Sintered Products, 241*t*
FeO/Fe_3O_4, 253
Ferrite, 151*t*, 152, 265, 332, 345, 359*t*
Ferritic alloys, 370*t*, 393
Ferrous metals, 373
Fiat
  cam and follower materials, 205*t*
  Super Fire, 59*t*
File hardness test, 358

Film coating, surface hardening, 517–526
Finger follower, 62, 64–65, 79, 80*f*
Finite element analysis (FEA) model, 395, 404
Finite element method (FEM), 123
First law of thermodynamics, 1
Fishbone diagrams, 544, 545*f*
Five-valve cylinders, 138, 139*f*
Flame hardening, 508–509
Flank, 176
Flash temperature, 448
Flat barrel lifters, 224–225
Flat follower, 67*t*, 73, 75, 76, 77*f*, 78, 85, 93, 222
    and cam design, 191*t*, 193, 194*f*, 196–197
Flat follower rocker arm, 268*f*
Flow coefficient, 144–145
Flow stress, 338, 339, 344
Flow testing, 399, 400*f*, 401*f*
Fluoroelastomer, 310
Follower arm, 25, 26
Follower contact, 96, 228*t*
Follower materials, 228*t*
Follower rotation, 75–78
Force, interatomic, 330, 331*f*
Forced vibration, 97
Ford, 3*f*
    cam and follower materials, 205*t*
    Type II valvetrain engines, 65
    Type IV valvetrain engine, 65
    Type V valvetrain engines, 66
    Zetec SE, 59*t*
Ford/Mazda, Type I valvetrain engines, 64
Forging, 157–158, 238, 249, 250*f*
Four-cylinder engine, 34, 35*f*
Fourier equation, 116
Four-stroke combustion engine, structure, 2, 3–10
Four-valve cylinders, 138, 139*f*
Fracture, 363
Fracture mechanics approach to fatigue testing, 371–373
Fracture toughness testing
    hydrogen embrittlement testing, 368–369
    impact testing, 366–367, 368*t*
    linear-elastic methods, 364–365
    nonlinear fracture methods, 366
    stress-corrosion cracking testing, 369–371

Free vibration, 97
Fretting, 386–387, 448, 452, 543
Friction, 94, 272, 445, 454
    measurement, 430
    temperature, 447
Friction laws, 443–444
Friction loss, 46–47, 51, 52
Friction loss factors
    lubricant type, 487, 489*f*
    oil flow rate, 490
    rolling elements, 486–487, 488*f*
    speed, 481–483
    spring load, 485–486, 487*f*
    temperature, 485, 486*f*
    valvetrain type, 483–485
Friction mean effective pressure, 17
Friction testing, 383–386
Frictional horsepower, 17
Fuel, leaded, 393
Fuel consumption, 36, 38, 46, 50, 57*t*
Fuel delivery system, 3*f*, 4
Fuel efficiency, 9, 36, 42, 44, 46, 50, 430, 481
Fuji Ozecks, 241*t*

Galling, 386–387, 493
Garter spring, 300*f*, 304
Gas pressure, 5, 8, 9, 277
Gas velocity, mean intake, 143
Gas temperature, 8
Gasoline engine, cylinder pressure, 50–51
    see also Diesel engine; Spark ignition engine
General Motors
    Displacement-on-Demand roller lifter, 44–45
    Type V valvetrain engines, 66
Glass, 365, 446*t*
GM
    Type II valvetrain engine, 65
    VVT systems, 59*t*
Gold, 332
Gouged, wear classification, 448
Grain boundary, 338, 343
Graphite, 151*t*, 240, 253, 260, 264, 446*t*
Guide failures, 558

Hall-Petch relationship, 344
Hardening. *See* Strengthening
Hardfacing, 134–135, 494, 525–526
　materials, 149*t*, 150*t*, 154–155, 164, 165*f*, 173, 239, 551, 552*f*
　plasma transferred arc, 532–524
Hardness
　correlation with tensile strength, 252
　effect on wear, 453*f*
　relation to tensile strength, 252
　relation to yield strength, 252
　see also Hot hardness
Hardness conversion tables, 360, 361*t*
Hardness testers, 358
Hardness testing
　converting results, 360, 361*t*, 362*t*
　estimating tensile strength from, 359, 360*t*
　microhardness testing, ultrasonic, 358
　testers, 358
　tests, 354–358
Harmonic amplitude, 115–118
Harmonic motion, 97*f*, 98, 99*f*
Harmonic number, 291, 293*f*
Heat efficiency. *See* Thermal efficiency
Heat treatment, 155–159
Hertz theory, 89–93
Hertzian contact pressure, 443
Hertzian stress, 491–492, 498, 501*f*
High-performance engine, 32*t*, 200*f*
High-stress sliding wear, 448
High-temperature metallic wear, 448
High-velocity oxy-fuel (HVOF) spray, 521
Hitachi P/M Co., Ltd., 242*t*
Honda, 3*n*
　cam and follower materials, 205*t*
　Hyper-VTEC system, 44, 45*f*
　i-VTEC, 41–43, 59*t*, 274
　Type II valvetrain engine, 65
　Type III valvetrain engine, 65
　VTEC, 59*t*, 274, 275*f*
Hooke's Law, 89, 338, 404
Horsepower, 13, 14*f*, 16–17, 18
Hot corrosion, 123–124, 392–395
Hot fatigue strength, 161, 164–168
Hot hardness, 154, 161, 163–164, 165*f*, 249, 252, 253*f*, 363
HR_{15N}89/92, 322, 324
Hydraulic actuator, 36, 44–45

Hydraulic lash compensator. *See* Lash compensator, hydraulic
Hydraulic lifter, 44–46, 194, 195, 218*t*
　design considerations, 213–229
　testing, 399–402
　types, 223–226
　see also Lash compensator, hydraulic
Hydraulic lifter failures, 556
Hydraulic pendulum, 55
Hydrocarbon, 35
Hydrocarbon emissions, 304–305, 307*f*
Hydrodynamic lubrication, 457
Hydrofinished (HVI) lubricant base stocks, 459, 462*t*
Hydrogen embrittlement, 542
　chromium-plated valves, 528
　prevention, 369
　testing, 368–369
Hysteresis, 285
Hysteresis loop, 374, 375*f*
Hyundai, Type III valvetrain engine, 65

Idle stability, 46
Ignition system, 3*f*, 4
Ignition timing, 32*t*
Impact testing, 366–367, 368*t*
Inconel 751, 136, 343, 393, 512
　for valves, 149*t*, 150*t*, 153, 154, 159, 160*f*, 161, 162*f*, 163, 164*f*, 165*f*, 167*f*, 169*f*, 170*f*
Indicated horsepower, 16–17, 18
Indicated mean effective pressure (IMEP), 16–17
Indicated power, 16
Induction hardening, 509–509, 510*f*
Inertia, 110, 423
Inflection point, 176
Injection timing, 32*t*
In-line four-cylinder engine, 34, 35*f*
Insert failures, 557
Intake air velocity, 46
Intake stroke, 5, 6*f*, 7, 9
Intake valve, 4, 5*f*, 25, 27–31, 32*t*, 34*f*, 121, 122*f*, 123
　design, 127, 128, 129
　diameter, 138*t*, 139–140
　failure, 546–554
　flow characteristics, 144–145

Intake valve *(continued)*
    materials, 148–152, 156, 157, 174*f*
    timing, 35, 148
Intercooler, 23
Interference angle, valve seat and valve seat
  insert, 131, 236–237, 253–254
Interference fit, valve guide and cylinder head
  bore, 258
Intermetallics, fracture toughness values, 365
Internal combustion engine
    efficiency, 18–23
    four-stroke, 4–10
    performance, 13–17
    size, 11–13
    structure, 3–4
    two-stroke, 10–11
    types, 2
Iron, 151*t*, 240, 332, 542
Iron carbides, 332
Iron nitriding structures, 511*f*
Iron-based alloys
    nitriding, 512–514
    for seal inserts, 240, 251*f*
    strengthening, 343
    for valves, 148, 149*t*, 150*t*, 153, 154–155,
      160, 162*f*, 163*f*, 164, 165*f*, 167*f*, 171, 173
    for valvetrain lash compensators, 229
Isuzu, VVT system, 59*t*
i-VTEC, 41–43, 59*t*, 274
Izod impact test, 366

Jaguar
    cam and follower materials, 205*t*
    VVT system, 59*t*
Jerk. *See* Pulse
Jump, valvetrain dynamic testing, 423–426

Keeper. *See* Valve keys
Keeper groove, 133
    design, 131–132
    failures, 554
    location, 80, 122*f*, 313
    tolerances, 126*t*, 127*f*
Keystone Carbon, 242*t*

Kinematics, 69
    center-pivot rocker arm valvetrain system,
      83–87
    contact mechanics, 87
    direct-acting valvetrain system, 69–72,
      74–79
    end-pivot rocker arm valvetrain system,
      79–83
    rocker arm followers, 73
Kinetic energy, 55, 56, 57
Kinetics, 69, 96
    see also Vibration
Knoop hardness test, 354, 356–357, 359*t*, 361*t*

Lamination wear, 452
Lancia, cam and follower materials, 205*t*
Laser beam surface hardening, 509–510
Laser cladding, 246–249, 524–526
Lash, 78–79, 114, 206–207, 208*f*
    dry lash, 214–215, 222, 226
    effect on impulse force, 426
    and valve tip wear, 495
Lash adjuster, 26, 26*f*, 49*f*, 62, 65, 79, 227*f*
    materials, 226
    measurements, 226–228
    side load, 82–83
    see also Lash compensator
Lash compensator, 69
    dimensions and surface finishes, 222–228
    hydraulic, 195, 207–229
      cam design considerations, 217–219
      compared to mechanical, 191*t*, 195,
        207, 208, 214, 217, 218, 219, 220, 228
      design considerations, 213–229
      dimensions and surface finishes, 224–
        228
      mechanism, 211–213
      oil supply, 212, 213, 215–217
      types, 223–226
      valve dynamics considerations, 219–222
    materials, 229
    mechanical, 206–207, 208*f*
      lifter dimensions and surface finishes,
        222–223
    stresses, 228, 229*f*

Index

Lash loss, 550
Lash measurement, 427, 428*f*
Lattice strain, 340
Lattice system, 331–332, 334–335*t*
Laves phase alloys, 155, 239, 240, 252
LE Jones Company, 242*t*
Lead, 239, 253, 264
Lead corrosion, 393–394
Leak-down, 215, 221–222, 400, 401*f*
Leeb hardness tester, 358
Leonardo da Vinci, 443
Lexus, Type I valvetrain engine, 64
L-head engine, 64
Lift. *See* Valve lift
Lift loss, 221–222
Lift velocity, 177, 178*f*
Lifter failures, 555–556
Lifter return spring, 229
Linear-elastic loading, 364–365
Liquefied petroleum gas (LPG) engine, 240
Liquid nitriding, 511–512
Liquid penetrant inspection, 404, 405*t*, 406*t*,
  411, 412*f*, 412*t*
Load, and wear behavior, 453
Lost motion, 41, 43–46, 57*t*, 274, 276*f*
Lotus, cam and follower materials, 205*t*
Lubricant additives, 457, 459, 463
  effects on wear, 501–503
  synthetic oils, 476–477
  types
    anti-wear agents, 465, 468*t*, 469*t*, 477,
      501–503
    detergents, 466, 468*t*, 469*t*, 477, 503
    dispersants, 466, 468*t*, 469*t*
    foam inhibitors, 466, 468*t*
    friction modifiers, 464, 469*t*
    oxidation inhibitors, 466, 468*t*, 469*t*, 477
    pour depressants, 465, 468*t*
    rust and corrosion inhibitors, 464, 468*t*,
      469*t*, 477, 478*f*
    viscosity modifiers, 463, 468*t*, 469*t*
    wax modifiers, 465
  ZDP, 477, 500, 502–506
Lubricant base stocks
  aromatic, 458, 459
  hydrofinished (HVI), 459, 462*t*

mineral oil, 458, 459, 463, 495, 499, 500,
  501*f*, 502
  naphthenic, 458
  paraffinic, 458
  synthetic, 458, 459, 461*t*, 462*t*
    additives, 476–477
    performance, 474–475
    properties, 475*t*, 477
Lubricant film thickness, 495–497, 498
Lubricant formation, 458–459
Lubricant-related engine malfunctions
  contamination, 478, 479, 504
  corrosion, 480
  deposit formation, 478–480, 505, 506*f*
  oil consumption, 480
  oil thickening, 480
  ring and valve sticking, 478, 480
  wear, 477, 480, 498–507
Lubricants
  and coefficient of friction, 445
  effectiveness testing, 396
  liquid classes, 459, 461*f*, 462*f*
  properties, 460*t*, 472–473, 474*f*
  solid, 239, 252, 253, 264
  temperature, 462*f*, 463, 464*f*, 477
  types, and friction loss, 487, 489*f*
  viscosity, 456, 457, 463, 487, 497, 500
  volatility, 480
  see also Oil
Lubricated wear, 448
Lubrication
  additive effects on wear, 501–503
  and cam lifter failures, 555–556
  film thickness, 495–497, 498*f*
  and guide failures, 558
  and rocker arm failures, 560
  stress effects, 498–499
  temperature effects, 503–504
  types, 457–458
  viscosity effects, 500
Lubrication system, 3*f*, 4
  oil pump, 471–472
  full-pressure system, 467, 470–471
  pressure-splash combination system, 467,
   470
  splash system, 467

Machining tools, 265, 266*f*
Magnesium, 332, 542
Magnetic methodology, nondestructive testing, 404
Magnetic particle inspection, 404, 405*t*, 406*t*, 411–414, 415*f*
Manganese, 148, 152, 156, 158
Manganese-based alloys, 170*f*
Manganese sulfide (MnS), 239, 253, 264
Martensite, 151*t*, 240, 265, 344–345, 359*t*
Martensitic alloys, 370*t*, 393, 429–430, 513*t*
  see also Steels, martensitic
Materials
  area reduction, 161, 163, 164*f*
  atomic forces, 330, 331*f*
  chemistry, 148–155
  crystal structure, 151*t*, 152, 153, 163, 330, 331–332, 334–335*t*
  deformation. *See* Deformation
  strengthening. *See* Strengthening
Maximum cam acceleration, 72
Maximum lift velocity, 71–72, 74–75
Mazda, VVT system, 59*t*
16McCr5, 368*t*, 530*f*
Mean effective pressure (MEP), 16–17
Mean stress, 374
Mechadyne, cam phaser, 37*f*, 38
Mechanical efficiency, 18
Mercedes, cylinder cutout system, 43–44
Mercedes-Benz, Type III valvetrain engine, 65
Metallic bonds, 330
Metallographic examination, for failure analysis, 540
Metal-to-metal sliding wear, 448
Methane engine, 246*t*
Methanol engine, 246*t*
Metrology measurement, for failure analysis, 539
Mica, 253, 264
Microhardness testing, ultrasonic, 358
Mineral oil lubricant base stocks, 458, 459, 463, 495, 499, 500, 501*f*, 502
Mitsubishi, MIVEC, 59*t*
Mitsubishi Material, 242–243*t*
Mixed-film lubrication, 458
16MnCr5, 273
Modulus of elasticity, 351, 352*f*
Mohs scale, 358

Molybdenum, 155, 156, 158, 169, 240, 252, 332, 345, 446*t*, 542
MoS$_2$, 239, 253, 264
Multiple-mass valvetrain model, 101, 102*f*
MWP Pleuco GmbH, 243*t*

21-2N, 149*t*, 150*t*, 152, 158, 160*f*, 162*f*, 163*f*, 164*f*, 165*f*, 167*f*, 169*f*, 170*f*, 342–343, 393
21-4N, 149*t*, 150*t*, 153, 158, 160*f*, 162*f*, 163*f*, 164*f*, 165*f*, 167*f*, 169*f*, 170*f*, 173*f*, 342–343, 393
23-8N, 149*t*, 150*t*, 153, 158, 160*f*, 162*f*, 163*f*, 164*f*, 165*f*, 167*f*, 170*f*, 173*f*, 342–343, 393
Naphthenic lubricant base stocks, 458
Natural frequency, 64*f*, 68
  and critical engine speed, 107, 108, 109*f*, 112, 114*f*
  determination, 422, 427
  equations, 104–105, 106, 271, 291
  multiple-mass valvetrain model, 101, 102*f*
  single-mass valvetrain model, 101, 102*f*
  spring surge, 106, 194, 291
  springs, 290–293
  valvetrain effective weight, 101, 103*t*
  and valvetrain stiffness, 101–102, 104, 271–272
Natural gas engine, 150*t*, 240, 246*t*
Naturally aspirated engine, 23, 28*f*, 29–30
  valves, 124, 127, 147
Ni30, 149*t*, 150*t*, 153, 162*f*, 165*f*, 167*f*
Nickel, 152, 153, 156, 158, 171, 247, 332, 345, 446*t*
Nickel-based alloys, 149*t*, 150*t*, 151, 153–155, 160, 161, 162*f*, 164, 165*f*, 167*f*, 169, 170*f*, 171, 173, 240, 242, 251, 343 370*t*, 543, 512
Nimonic 80A, 149*t*, 150*t*, 154, 159, 160*f*, 162*f*, 163, 164*f*, 165*f*, 167*f*, 343, 380*f*, 393
Nimonic 90, 149*t*, 150*t*, 154, 159, 343
Niobium, 153, 158, 252, 332
Nippon Piston Ring, 243*t*
Nippon Powder Metal, 244*t*
Nissan, 14*f*, 40
  cam and follower materials, 205*t*
  Neo VEL, 59*t*
  VEL system, 50*f*, 51–52
Nitrided, 136, 173, 205*t*, 222, 223, 229, 264
Nitriding, 494, 511–514, 517*t*

Nitrile, 310, 311*f*, 560

Nitrocarburizing, 516, 517*t*

Nitrogen, 152, 153, 158

Nitrogen diffusion, 510–511

Nitrogen oxide (NO$_X$), 36, 57–58

21-4N-Nb-W, 149*t*, 150*t*, 153, 162*f*, 163*f*, 164*f*, 165*f*, 167*f*, 170*f*, 393

Noise, 70, 114, 120, 179, 183, 185, 189, 418

    and excessive wear, 491, 495, 554

    and lash, 207, 427

    and oil aeration, 216, 217

    and oil improvements, 476*t*

    pump-down, 220

    ramp, 194

    surfaces, 202, 259

    valve, 55, 57, 71, 106, 177, 207, 213, 216, 217

    valve stem seals, 299

    and valvetrain dynamic testing, 420, 421–422

Noise, vibration, and harshness (NVH), 26, 67–69, 190, 554, 560

Nondestructive testing

    eddy current testing, 404, 405*t*, 406*t*, 414–416, 417*f*

    electrical methodology, 404

    liquid penetrant inspection, 404, 405*t*, 406*t*, 411, 412*f*, 412*t*

    magnetic methodology, 404

    magnetic particle inspection, 404, 405*t*, 406*t*, 411–414, 415*f*

    radiography, 404, 406*t*

    thermography, 404

    ultrasonic testing, 404, 405*t*, 406*t*, 408–410, 411*t*

    visual-optical testing, 404, 406*t*

    x-ray inspection, 404–408

Nose radius, 176, 190, 191*t*, 198

Notches, effect on test results, 347, 367

NO$_x$ emissions, 306, 307*f*

Nylon, 310, 446*t*

Oil

    aeration, 216–217, 218*f*

    degradation, 504

    filtration, 216

    flow rate, 490

    leakage rates, valve stem seals, 300–301, 302*f*, 303, 306*f*

    metering, testing, 398, 402, 403*f*

    metering rate, 264, 299, 301, 303–304, 305*f*

    performance, and temperature 503–504

    pressure, 215–216

    thickening, 480

    viscosity, 458, 463, 464*f*, 472–473, 474*f*, 487, 497, 500

        relation with temperature, 458, 463, 474*f*

        requirements, 473*t*

Oil grooves, 224, 256, 257*f*

Oil lubrication

    camshaft tube, 201

    and sulfidation corrosion, 394–395

    valve guides, 256, 264

    valve stem surface, 135–136

    valvetrain lash compensators, 212, 213, 215–217, 224

    see also Lubricant; Lubrication

Oil pump

    external gear pumps, 471

    internal gear pumps, 471

    relief valve, 472*f*

    rotary pumps, 471

    vane pumps, 471

Oil reservoir, 212, 213, 215

Opel, cam and follower materials, 205*t*

Open duration, 34*f*

Operating environment, consideration in component bench testing, 395

Otto engine. *See* Gasoline engine; Spark ignition engine

Overhead cam (OHC), 34, 61–66, 72, 76*t*

Overhead cam engines, 204, 205*t*, 209*f*, 210*f*, 219, 220

Overhead valve (OHV), 34, 61–63, 66

Overhead valve engines, 211*f*, 485

Overlap, 27, 29–31, 32*f*, 33, 34–35, 147

Oversquare engine, 12–13

Oxidation, 168–170, 393, 448, 479, 480

Oxidation testing, 391

Oxidative wear, 451–452, 453

Parabolic constant acceleration curves, 183–185

Paraffinic lubricant base stocks, 458

Pearlite, 151*t*, 264, 265, 345, 359*t*

Perfect gas equation, 23
Petrol engine. *See* Gasoline engine; Spark ignition engine
Peugeot, cam and follower materials, 205*t*
Phosphate, 493–494, 528, 529*t*
Physical vapor deposition (PVD), 517–518, 519*t*, 519*f*
Piston, 4, 5*f*, 16, 33*f*
Pitting, 448, 493, 554
PL 7, 173, 174*f*, 243*t*
PL 33, 174*f*, 243*t*
Plasma transferred arc (PTA) hardfacing, 532–524
Plastic deformation, 89, 163, 172, 251, 252, 333, 335, 338, 350, 452
Plastic deformation-type wear, 494
Plastic flow, 172, 337, 338, 339, 345
    representation of characteristics, 351–354
    valve seat inserts, 251–252
    see also Plastic deformation
Plastic strain, 336, 339, 348, 371, 375*f*, 381
Plastic zone, 347–348
Plowing, 358, 444*f*
Plunger spring, 212, 213
PMF 16, 172*f*, 241*t*
Poisson's ratio, 89, 440
Polished, wear classification, 448
Polyacrylate, 310, 311*f*
Polymers, fracture toughness values, 365
Polynomial curves, 180–188
Polytetrafluoroethylene, 310
Poppet valve, 25, 121, 125*f*, 125*t*
Porsche
    Plus system, 59*t*
    Type I valvetrain engine, 64
    VacioCam, 59*t*
    VarioCam, 59*t*
    VarioCam Plus, 46–47
Port diameter, 124, 125*f*, 138
Potassium, 332
Potential energy equation, 330
Powder metal (PM) alloys, 322
    for cams, 201, 203, 204, 206
    for valve guides, 258, 260, 261, 264, 266
    for valve seat inserts, 237–239, 240, 249, 250*f*, 251*f*, 253
Power stroke, 5, 6*f*, 8, 9, 16, 34*f*
Precipitation-hardened stainless steel, 152, 153

Precipitation-hardening stainless steels, stress-corrosion cracking, 370*t*
Press-fit
    valve guides, 258–259, 261*f*
    valve seat inserts, 231, 246, 247*f*, 248
Pressure, effective pressure (MEP), 16–17
Pressure angle, 82, 83, 85–87, 194, 270, 271*f*
Pressure-volume diagram, 7*f*, 15*f*, 16*f*, 19*f*, 20*f*
Primary atomic bonds, 3c30
Proof stress, 338
Proportional limit, 350
Pulse, 70, 73, 105, 112, 114, 179
Pump-down, 220, 221*f*
Pumping mean effective pressure, 17
Pushrod, 224, 268*f*, 269, 320–322
Pushrod-operated system, 205*t*
    cam/follower failures, 554–555
    engine speed, compared with OHC design, 220
    hydraulic lash compensator, 211–212
    lifter, 224
    Type V, 198, 214, 219, 220, 271, 310
    valve lift, 144
Pyromet 31
    for valves, 149*t*, 150*t*, 154, 159, 160*f*, 161, 162*f*, 164*f*, 165*f*, 167*f*, 170*f*, 173*f*, 343, 393

Quartz, 359*t*
Quasi-static deflections, 110–111
Quenched steels, 359
Quill, 175

Racing engine, 138*t*, 200*f*
Radial crush test, 252
Radial flow, 494
Radiography, 404, 406*t*
Radius of curvature, 195–198, 199*f*
Ramps. *See* Camshaft ramps
Rate of loading, 346
Real area of contact, 439–440
Renault, cam and follower materials, 205*t*
Repeatability, component bench testing, 395
Repeated-cyle deformation mechanism, 452
Retainer. *See* Valve spring retainer
Retarding performance, 58
Rig testing, 396, 397*f*, 398*f*, 399*f*, 401*f*, 403*f*

Riken, 244–245*t*

Rocker arm, 25, 26*f*, 41, 42*f*, 43, 44, 69
    center-pivot, 61, 62, 63*f*, 65–66, 67*t*, 73, 74*f*, 76*t*, 266, 268*f*, 269, 270
    end-pivot, 62, 63*f*, 64–65, 67*t*, 76*t*, 266–270, 272*f*, 274, 276*f*
    failures, 560
    fulcrum-mounted, 266, 268*f*
    materials, 273–274
    with roller, compared to sliding, 273
    and valvetrain natural frequency, 427
    and valvetrain stiffness, 270–272

Rocker arm followers, 27, 47–48, 73, 192
    see also Flat followers; Roller followers

Rocker arm stress waves, 422

Rocker contact ratio, 81

Rocker pad, contact travel, 83–84

Rocker ratio, 27, 48–52, 53*f*, 73, 74*f*, 75*f*, 76*t*, 81, 83, 84, 269–270

Rockwell hardness, 132, 133, 157, 158, 354, 355–356, 360, 361*t*

Rod bearing, 5*f*

Rolled-edge fit, 231

Roller barrel lifters, 225–226

Roller followers, 62, 65, 67*t*, 73, 76*f*, 85, 93, 487, 488*f*
    and cam design, 191*t*, 193, 194*f*, 197–198
    and friction, 271–272
    wear testing, 388–390

Roller hydraulic lifters, 486, 488*f*

Roller lifters, 45*f*, 225–226, 486, 488*f*

Roller rocker arms, 49*f*, 272*f*, 273, 486, 488*f*

Rolling contact, 37*f*, 38

Rolling friction, 447

Rolling wear, 388–390, 448

Rolling-element bearings, 486

Rolls-Royce, cam and follower materials, 205*t*

Rotary engine, 2

Roughness, 440–442, 453, 454*f*

Rover, cam and follower materials, 205*t*

Rubber, 446*t*

Rust, 463–464

Rust inhibitors, 464, 468*t*, 469*t*, 477, 478*f*

SAE 20, 262

SAE 1008, 273

SAE 1010, 204, 273, 313, 317

SAE 1018, 317, 322

SAE 1020, 204, 516*f*

SAE 1039, 204

SAE 1045, 515

SAE 1050, 510*f*

SAE 1215, 229, 317

SAE 1457, 133

SAE 1522, 229

SAE 1541H, 149*t*, 150*t*, 160–161, 162*f*

SAE 1547, 132, 148, 149*t*, 150*t*, 156, 157*f*, 162*f*, 165*f*, 166*f*, 167*f*, 368*t*, 393

SAE 2515, 204

SAE 3140, 133, 149*t*, 150*t*, 169, 393

SAE 4100, 229

SAE 4118, 223

SAE 4140, 133, 273, 513*f*, 514*f*

SAE 4140H, 149*t*, 150*t*, 151

SAE 4340, 204, 381*f*, 382*f*, 383*f*, 531*f*

SAE 4615, 204

SAE 4T12, 223

SAE 5100, 229

SAE 5150H, 149*t*, 150*t*

SAE 8520, 204

SAE 8620, 229, 273, 322, 324, 513*f*, 514*f*

SAE 8645, 133, 149*t*, 150*t*, 151, 156, 157*f*, 165*f*, 173

SAE 9310, 229

SAE 52100, 173, 204, 229, 274, 322, 368*t*, 382*t*

SAE J1682, 262*t*, 263*t*

SAE J1692, 239, 240, 241*t*, 252, 239, 240, 241–245*t*, 252

Scanning electronic microscopy (SEM), for failure analysis, 539

Scavenging, 10

Scoring, 493

Scratch hardness tests, 358

Scratched, wear classification, 448

Scrub velocity, 81

Scuffed, wear classification, 448

Scuffing, 386–387, 493, 554

Seat facing alloys, hardfacing, 525–526

Second law of thermodynamics, 18

Secondary atomic bonds, 330

Sensitization, 152

Shear strain, 452, 494

Shear stress, 335–336, 337, 440

Shimless bucket lifter, 222, 223*f*

Shimmed bucket lifter, 223
Shore hardness tests, 358, 361*t*
Shot peening, 285, 290, 297–298, 529–530, 531*f*
Shrunk-fit, 231
Sil 1, 149*t*, 150*t*, 151, 156, 157*f*, 160*f*, 161, 162*f*, 163, 164*f*, 165*f*, 166*f*, 167*f*, 173, 174*f*, 368*t*, 393, 429–430, 513*f*
Sil XB, 149*t*, 150*t*, 152, 164*f*, 165*f*, 174*f*
Silchrome 10, 169*f*
Silicates, 253, 264
Silicon, 155, 156, 158, 169, 248, 252, 393
Silicon carbide, 359*t*
Silicon nitride, 359*t*
Silicone, 310, 311*f*
Silver, 332
Simca, 205*t*
Simple harmonic curve, 188–189
Simulated test models, 395
Simulation testing, for failure analysis, 541
Sine acceleration curves, 189–190
Single-mass valvetrain model, 101, 102*f*
Sintered powder metal alloys, 201, 204, 206, 261
Sintering, 136, 236, 237–238, 249, 262, 250*f*, 260
Six-cylinder engine, 34*f*, 34, 35*f*
Sliding velocity, 81
Sliding wear, 386–390
Slip, 336–338, 389
Slip-slide, 80–81, 85
Sludge, 479
S-N curves, 166, 167*f*, 372*f*, 373–376, 515*f*
Sodium, 124, 134, 332
Sodium potassium, 134
Solubility limit, 341
Sommerfeld number, 457
Soot, 478, 479, 480, 505
Spalled, wear classification, 448
Spalling, 493
Spark ignition engine, 5–9, 10, 31, 123*f*
  compared to compression ignition engine, 9–10, 122–123
  efficiency, 48–50
  see also Gasoline engine
Spark retard, 213
Speed
  and wear behavior, 453

effect on friction, 481–483
Spring, 25, 26*f*, 72, 97–98, 210*f*
  buckling, 288
  failures, 558, 5559*f*
  forms
    barrel, 278
    beehive, 278–279, 315
    conical, 278–279, 315
    cylindrical, 278
    helical, 276, 279, 290*f*, 292*f*
    variable pitch spring, 291–293, 294*f*
  load, and friction loss, 485–486, 487*f*
  load margin, 220
  manufacturing, 289–290
  mass, 106
  materials, 229
  natural frequency, 290–293
  parallelism, 282
  performance, 285–288
  squareness, 282
  stress, 285–286, 289, 294
  types
    Belleville spring, 320, 321*f*
    check ball spring, 209–211*f*, 212
    garter spring, 300*f*, 304
    lifter return spring, 229
    lost motion spring, 276*f*
    plunger spring, 212, 213
  see also Valve spring
Spring acceleration rate, 114
Spring coils, 277*f*
  activity, 284, 293
  coiling direction, 284, 285*f*
  diameter tolerances, 282–283
  end coils, 283, 284*f*
Spring ends, 283, 284*f*, 287–288
Spring force, 211–212
Spring index, 277, 292*f*
Spring rate, 106, 114, 278, 286
Spring relaxation, 288–290
Spring reserve rate, 191*t*
Spring retainer failures, 560
Spring seat seal, 308
Spring surge, 106, 194, 291
Spring wire, tensile strength, 296, 297*f*
Sprocket drive unit, VVT camshaft, 36–38
Square engine, 12–13
422SS, 149*t*, 150*t*, 152, 165*f*, 368*t*

Index

Stainless steels, 148, 149t, 150t, 152, 240, 295, 370t
Starting system, 3f, 4
Steels,
  carbon, 322, 345, 360t, 370t
    for valve keys, 313
    for valve spring retainers, 317
    for valve springs, 295
    for valves, 148, 149t, 150t, 151, 156, 161, 162f, 169
    for valvetrain lash compensators, 229
  chromium-silicon, for valve springs, 295–296
  chromium-vanadium, for valve springs, 295–296
  coefficient of friction data, 446t
  converting hardness results, 362t
  crystal structure, 332
  fracture toughness values, 365
  hardenable, 204, 206, 207f, 229, 313, 317
  low-carbon, 542, 543
  martensitic
    hardening, 156, 157–159, 344–345
    for valve seat inserts, 240
    for valves, 148, 149t, 150t, 152, 161–167, 170f, 173
  nitrided, 205t, 229
  nonhardened steels, 360t
  precipitation-hardened stainless, 152, 153
  quenched, 359
  tempered steels, 359
Stellite™ alloys, 134–135, 154
  Stellite 1, 149t, 150t, 155, 164, 165f, 173
  Stellite 3, 174f, 246t
  Stellite 6, 149t, 150t, 154, 155, 160f, 165f, 170, 173
  Stellite 12, 149t, 150t
  Stellite F, 149t, 150t, 155, 165f, 173
Stem seal failures, 559–560
Stem seal oil metering testing, 402, 403f
Stem welding, 133
Stiffness, 101–102, 104, 218, 270–272, 351, 352f, 423
Straight-line curves, 180, 181
Strain, 87–89, 420–421, 422f
Strain gage testing, 404
Strain hardening, 157
Strain rate, 346

Strain-hardening exponent, 157
Strain-life (-N) approach to fatigue testing, 371–373
Strain-stress curve. See Stress-strain curve
Streibeck curve, 456f
Strengthening, 338–345
  aging heat treatment, 157–159
  aging precipitation strengthening, 152–153, 157, 159, 341–343
  cold forming, 158
  forging, 157–158, 238, 249, 250f
  grain boundary hardening, 340, 343–344
  interstitial strengthening, 340
  particle hardening, 340
  phase transformation hardening, 340, 344–345
  quench hardening, 156, 158, 159t
  solid solution strengthening, 153, 156, 340–342
  strain hardening, 157
  substitutional strengthening, 340
  tempering, 156–157, 159t, 344–345
  work hardening, 340, 343
Stress, 88–89, 91, 381, 555–556
  analysis, valvetrain dynamic testing, 420–421, 422f
  amplitude, 374
  distribution, strain gage testing, 404
  equations, 421
  intensity, 363
Stress profiles, valve head, 123, 124f
Stress ratio, 374
Stress-corrosion cracking (SCC), 542–543, 546, 369–371
Stress-life (S-N) approach to fatigue testing, 371–373, 383f
Stress-strain curve, 333, 338–340, 350
  compared with true stress-true strain curve, 351–354
  grain boundary strengthening, 343
  mechanical testing, 348–354
Stroke-to-bore ratio, 12–13
SUH 3, 149t, 150t, 151
SUH 11, 149t, 150t, 151, 166f, 167f
SUH 35, 149t,150t, 153, 166
Sulfidation corrosion, 123–124, 168, 170–171
Sulfur, 124
Superalloys, 153–154, 159–167, 169, 170f, 365

Supercharged engine, 29–31, 124, 127, 146, 147

Supercharger, 23

Surface

    contaminants, and wear behavior, 453

    fatigue, wear mechanisms, 448–453

    finish, effect on wear, 454

    interactions, 438–442

    profile, 440–442

    roughness, 497

Surface engineering, 507–531

Surface hardening, 507

    carbonitriding, 516, 517*t*

    carburizing, 514–515, 517*t*

    chrome plating, 526–528

    diffusion, 510–511

    electronic beam, 510

    induction, 508–509

    laser beam, 509–510

    nitriding, 511–514, 517*t*

    nitrocarburizing, 516, 517*t*

    phosphate coating, 528, 529*t*

    shot peening, 529–530, 531*f*

    thick film coating, 521–523, 524–526, 532–524

    thin film coating, 517–521

Suzuki, VVT system, 59*t*

Swirl, 236

Synthetic lubricant base stocks, 458, 459, 461*t*, 462*t*, 474–477

T 400. *See* Tribaloy 400

T 800. *See* Tribaloy 800

Talbot, cam and follower materials, 205*t*

Talc, 264

Tantalum, 153, 158, 169, 252, 332

Tappet, 25, 26, 62, 64, 66, 195

    cam curvature, 197

    curvature, and film thickness, 498, 499*f*

    diameter, 71

    faces, and contact stress, 228, 229*f*

    force between cam and, 198, 199*f*

    hydraulic compared with mechanical, 191*t*, 195

    offset, 191*t*

    radius, 72

    tolerances and surface finishes, 223, 224*f*

    see also Bucket tappet

Taylor-von Mises criterion, 542

Teflon, 310, 311*f*

Temperature

    effect on deformation, 345–347

    effect on diffusion rates, 512, 513*f*

    effect on fatigue strength, 375–378, 379*f*

    and friction loss, 485

    liquid lubricants, 462*f*, 463, 464*f*, 477

    and oil performance, 503–504

    shot peening, 530, 531*f*

    sliding surface, 447–448

    valves, 122, 123*f*, 123, 429–430

Temperature-check values, 245

Tempered steels, estimating strength from hardness values, 359

Tempering, 156–157, 159*t*, 344–345, 509, 510*f*

Tensile strength, 333

    correlation with hardness, 252

    materials, 161, 162*f*, 295*t*

    relationship with fatigue limit, 383, 384*f*

    spring wire, 296, 297*f*

    tempering and, 156

    testing, 348–354

    see also Ultimate tensile strength

Tensile stress, 333–335

    chrome plating, 136

    valves, 122–123, 192

Tensile yield strength, 333, 347

Testing equipment, 395

Thermal conductivity, 160–161, 251, 376

Thermal deformation, 156

Thermal efficiency, 50, 51

    compression ignition engine, 20–22

    spark ignition engine, 18–20, 21

Thermal expansion, 129, 160, 249–250

Thermal fatigue, 375–378, 379*f*, 402, 403*f*, 551

Thermal shock, 376–378, 379*f*, 402, 403*f*

Thermal spray coating, 521–523

Thermodynamics, laws of, 1, 18

Thermography, 404

Three-valve cylinders, 138

Thrust face, 175

Tin, 446*t*

Titanium, 252, 332

Titanium alloys, 365, 370*t*

Titanium carbide, 359*t*

Titanium nitride, 359*t*

Top dead center (TDC), 4, 5*f*, 27–28
Torque, 13, 14*f*, 17, 22, 34, 35, 36, 136, 429*f*
Torque curve, 36, 37*f*, 39–40, 43, 47*f*, 52*f*, 248*f*, 273*f*
Torque reduction, 179, 208, 248, 272, 273
Toughness, 363
    see also Fracture toughness testing
    Toyota, 3*n*
    cam and follower materials, 205*t*
    Type I valvetrain engine, 64
    VVT-i system, 47–48, 59*t*, 274
    VVTL-i system, 59*t*
Transient response, 111, 112*f*
Tribaloy, 154, 155, 240
Tribaloy 400, 149*t*, 150*t*, 155, 164, 165*f*, 173, 174*f*
Tribaloy 800, 149*t*, 150*t*, 155
Tribology, 437–531
Trigonometric curves, 188–190
Triumph, cam and follower materials, 205*t*
True stress-true strain curve, 351–354
Tungsten, 153, 154, 155, 156, 158, 169, 252, 332, 542
Tungsten carbide, 359*t*, 446*t*
Turbocharged engine, 23, 31
    valve timing, 147
    valves, 127, 129, 146
Two-stroke combustion engine, structure, 2, 3–4, 10–11
Two-valve cylinders, 138
Type I valvetrain, 62–63, 64, 76*t*, 121
    base circle runout, 219
    cam torque, 484*f*
    and friction, 483, 485
    lash compensation, 207, 209*f*
    lifters, 214, 222, 223
    maximum cam acceleration, 198
    stiffness, 104*t*
    valve guides, 258, 259*t*
    valve rotation, 320
    weight calculation, 103*t*
    see also Direct-acting valvetrain
Type II valvetrain, 26*f*, 62–63, 64*t*, 64–65, 76*t*, 81
    cam torque, 484*f*
    and friction, 485, 488*f*
    lash compensation, 209*f*
    lifters, 214

    rocker arms, 266, 267*f*, 269–270, 272*f*, 488*f*
    stiffness, 104*t*
    valve guides, 259*t*
    valve tips, 81
    weight calculation, 103*t*
    see also End-pivot rocker arm valvetrain
Type III valvetrain, 62–63, 64*t*, 65, 76*t*
    cam torque, 484*f*
    and friction, 485
    lash compensation, 210*f*
    rocker arms, 266, 268*f*, 270
    stiffness, 104*t*
    valve guides, 259*t*
    weight calculation, 103*t*
    see also Center-pivot rocker arm valvetrain
Type IV valvetrain, 62–63, 64*t*, 65–66, 76*t*
    cam torque, 484*f*
    and friction, 485
    lash compensation, 210*f*
    lifters, 222, 224
    rocker arms, 266, 268*f*, 270
    stiffness, 104*t*
    valve guides, 259*t*
    weight calculation, 103*t*
    see also Center-pivot rocker arm valvetrain
Type V valvetrain
    and friction, 485, 488*f*
    base circle runout, 219
    cam torque, 484*f*
    cam/follower failures, 554
Type V valvetrain, 62–63, 64*t*, 66, 76*t*, 83–87
    full-pressure lubrication system, 470
    lash compensation, 211*f*, 212*f*
    lifters, 214, 222, 224, 488*f*
    maximum cam acceleration, 198
    rocker arms, 266, 268*f*, 270, 270–271
    stiffness, 104
    valve guides, 259*t*
    valve stem seal materials, 310
    valvetrain deflection testing, 420
    weight calculation, 101, 103*t*
    see also Center-pivot rocker arm valvetrain

Ultimate tensile strength (UTS), 161, 252, 298
    equations for, 353–354
    as function of temperature, 376, 377*f*
    mechanical testing, 348–350

Ultralight™ valve, 134
Ultrasonic testing, 358, 404, 405*t*, 406*t*, 408–410, 411*t*
Undamped vibration, 97–99
Undersquare engine, 12–13
Unlubricated wear, wear classification, 448

V4 engine, 45–46
V6 engine, 34, 35*f*
V8 engine, 34, 35*f*, 46
Valve
    acceleration, 69, 70, 71*f*
    actuation, 26–34, 57–58
        cam driven, 34–51, 53
        camless system, 52–57
        individual, 54–55
        variable, 46–57
    capacity, 137–139
    closing noise testing, 421–422
    design, 124–148
    dimensions, 124–126
    displacement, 69–70
    duration, 192–193
    failures, 546–554
        Fishbone diagram of root cause analysis, 544*f*
        stress-corrosion cracking, 543
    hollow, 121, 122*f*, 133–134
    internally cooled, 121, 122*f*, 133–134
    length, 124, 126*t*, 127*f*
    materials, 148–174
    motion, 192
    one-piece, 122*f*, 132, 133*f*
    operating characteristics, 122–124
    opening noise testing, 421–422
    overshoot, 111
    rotation, 318, 320
    seat-welded, 122*f*
      stem-welded, 122*f*, 133
      stress distribution, 122–123, 124*f*
    temperature, 122, 123*f*, 123, 429–430
    timing, 27–32, 147–148
        see also Variable valve timing
    tip-welded, 122*f*, 133
    tolerances, 126*t*, 127*f*
    two-piece, 122*f*, 133
    types, 121, 122*f*, 132–134

    undershoot, 112
    velocity, 69, 70, 71
    wafer, 122*f*, 133
    weight, 127, 133
Valve bridge, 322, 323*f*
Valve face failures, 547
Valve fillet, 122*f*, 124*f*, 126*t*, 127, 404, 420, 553
Valve float, 220
Valve flow capacity
    flow area, 137–143
    flow density, 146–147
    flow rate, 139*f*
    flow velocity, 143–144
    pressure differential, 146
    and valve timing, 147–148
Valve flow coefficient, 144–145
Valve gap, 310
Valve guide, 135, 136, 299–300
    dimensions, 257–259, 260*f*
    installation, 256, 259
    materials, 260–266
    tolerances, 256–257, 258*f*
    types, 256, 257*f*
    wear testing, 398, 399*f*
    wear/seizure, 544*f*
Valve head, 64, 71, 122*f*
    cupping, 547
    design, 126–128
    diameter, 124, 125*f*, 125*t*, 126, 127, 128, 131, 137–140
    shapes, 128*f*
    stress profiles, 123, 124*f*
    thickness, 126
    tolerances, 126*t*, 127*f*
    tulipping, 547
Valve keeper. *See* Valve keys
Valve key failures, 561
Valve keys, 316, 321*f*
    diameter, 125*t*, 126*t*, 127*f*
    dimensions and tolerances, 313
    four-bead design, 132, 312–313, 315, 316*f*, 318
    function, 310
    materials, 313
    single-bead design, 311–312, 313*f*, 314–315, 316
    and spring retainer failures, 560

three-bead design, 132, 312–313, 315, 316*f*, 318
  see also Keeper groove
Valve lift, 27, 28*f*, 30*f*, 39*f*, 70–71, 73, 74*f*, 83–84, 108*f*, 116, 176–179
  acceleration, 178*f*, 179
  curves, 178*f*, 192
    bounce, 423–426
    jump, 423–426
    polynomial curves, 180–188
    trigonometric curves, 188–190
  effect on flow area, 142–143
  equations, 177, 179
  and flow coefficient, 144–145
  maximum velocity, 71–72, 74–75
  noise, 422
  and valve seating angle, 130
  velocity, 177, 178*f*
  see also Valve timing diagram; Variable valve lif*t*
Valve lifter. *See* Lash compensator; *specific types*
Valve margin, 122*f*
Valve rotators, 318–320, 321*f*
Valve seat, 121, 122*f*, 123, 127, 140
  burning, 550–551, 552*f*
  eccentricity, 130–131
  guttering, 550–551, 552*f*
  hardfacing, 134–135, 149*t*, 150*t*, 154–155
  laser cladding, 525*f*
  radial cracks, 551
  wear, 172, 173*f*, 174*f*, 550, 551*f*
  wear testing, 396–397, 398*f*
  width, 124, 125*t*, 126, 127*f*
Valve seat angle, 129–131, 142, 233–235, 236–237, 253
Valve seat insert angle, 236–237
Valve seat inserts, 122, 127, 131, 230–255
  cast-in, 231, 240, 247*f*, 249
  clad, 231, 247*f*, 248
  corrosion resistance, 254–255
  and flow coefficient, 144–145
  inside diameter, 233, 236
  installation, 231–232
  lead chamber or radius, 235
  machinability, 254, 255*f*
  materials
    cast alloys, 240
    ceramics, 245–246
    chemistries, 241–245*t*
    coefficient of thermal expansion, 249–250
    combinations with valve materials, 246*t*
    powder metal alloys, 237–240
    properties, 249–255
    wear resistance, 252–255
    wrought alloys, 245
  outside diameter, 233, 235
  press-fit, 231, 246, 247*f*, 248
  and seat angle, 131, 233–235, 236–237, 253
  seat face, 233–235
  swirl-type, 236
  tolerances, 234*f*
  wall thickness, 232–233, 249
  wear, 172, 173*f*, 174*f*, 230, 544*f*
  welded, 231, 247*f*, 249
Valve seating force, 129–130
  valvetrain dynamic testing, 423–426
  valvetrain jump and bounce, 423–427
Valve seating velocity, 106–107, 108*f*, 109*f*, 111, 192
Valve spring
  and acceleration change, 70–71, 176, 192
  coil diameter tolerances, 282–283
  coils, 277*f*, 316
  damping, 293–294
  design considerations, 281–285
  dimensions, 277*f*, 282*f*
  force, 72, 192, 198, 211, 215–216
  forms and configurations, 278–279
  frequency, 190
  friction dampers, 280*f*
  helical compression spring, 276, 282*f*, 284*f*, 287*f*, 292*f*, 316
  jumping speed, 285
  load, 81, 83, 90, 101, 114, 115*f*, 116, 204, 213, 215–216, 220, 277
  materials, 295–298
  minimum load, 285
  quasi-static deflection, 110–111
  static stress, 296
  strain gage testing, 420–421
  strength, 51, 254
  stress, 285–288, 296
  tolerances, 282*f*

Valve spring *(continued)*
    and valve actuation, 53–54
    weight, 103*t*
    see also Spring; Vibration
Valve spring rate, 278*f*
Valve spring retainer, 25, 26*f*
    dimensions and tolerances, 317
    function, 277, 313
    materials, 317
    one-piece, 311, 312, 314–315
    two-piece, 312, 316–317, 319*f*, 320
Valve spring wire, 279–280, 289*f*
Valve stem, 121, 122*f*
    coatings, 135–136, 173
    coefficient of thermal expansion, 129
    design, 128–129
    diameter, 124, 125*f*, 125*t*, 126*t*, 127*f*,
      128–129
    failures, 553–554
    force, 105
    taper, 129
    tolerances, 126*t*, 127*f*
    wear testing, 398, 399*f*
Valve stem seals, 25, 26*f*
    design considerations, 299–310
    dimensions, 308–310
    lubrication, 299
    materials, 310, 311*f*
    oil leakage rates, 300–301, 302*f*, 303, 306*f*
    oil metering rate, 264, 299, 301, 303–304,
      305*f*, 307*f*
    shapes, 300–301
    tolerances, 308–310
Valve timing diagram, 27–29, 30*f*, 32*f*, 34*f*, 39*f*,
    42*f*, 45*f*, 49*f*, 54*f*, 62*f*, 71*f*, 178*f*
Valve tip, 62, 64, 65, 102, 121
    coefficient of friction, 82–83
    contact travel, 83–85
    end, 62, 122*f*
    failures, 554
    location, 80, 122*f*
    materials, 173
    pitting, 174
    slip-slide, 80–81, 85
    type selection, 80
    wear, 174–173
Valve tip weld, 122*f*

Valvetrain
    contact mechanics, 87–96
    dynamics, 96–118
    mechanisms, 3*f*, 4, 25–26
    system timing, 33–34
    types, 61–66, 67*t*, 191*t*
    weight, 101, 103*t*, 200–201
Valvetrain dynamics testing
    acceleration, 418–420, 424, 425*f*, 426
    bounce, 423–426
    deflection, 419–420, 424, 426
    displacement, 418–419, 420
    equipment, 418–419
    jump, 423–426
    noise, 421–422
    software, 418
    strain, 420–421, 422*f*
    stress analysis, 420–421, 422*f*
    velocity, 418–419, 420, 423, 426, 427
    vibration, 418, 420, 421–422
Valvetron, 3
Valvetronic, 48, 49*f*, 51, 59*t*
Vanadium, 124, 156, 158, 252, 332, 359*t*
VANOS, 39*f*, 48, 49*f*, 59*t*
Variable pitch spring, 291–293, 294*f*
Variable valve actuation (VVA), 57*t*, 59*t*
    cam driven, 46–52
    camless systems, 52–57
    rocker arms, 41–43, 47–48, 274, 275*f*
    rocker ratio, 48–52
Variable valve lift, 46–48, 51–52
    by cam profile switching, 40–43
    by lost motion, 43–46
Variable valve timing (VVT), 34–40
Vauxhall, cam and follower materials, 205*t*
Velocity
    valvetrain dynamic testing, 418–419, 420,
      423, 426, 427
    mean intake gas, 143
Vertical valves, 139
Vibration
    critical engine speed, 107–108, 110–118
    damped, 97, 100
    equations, 98–99, 100, 101
    forced, 97
    free, 97
    harmonic motion, 97*f*, 98, 99*f*
    harmonics, 115–118

Index

Vibration *(continued)*
    natural frequency, 101–106, 271
    spring surge, 106, 194, 291
    undamped, 97–99
    valve seating velocity, 106–107, 108*f*, 109*f*,
      111
    valvetrain dynamic testing, 418, 420,
      421–422
Vickers hardness test, 354, 356, 357*f*, 359,
  360, 361*t*
Viscosity (oil), 458, 463, 464*f*, 472–473, 474*f*,
  487, 497, 500
    relation with temperature, 458, 463, 474*f*
Visual-optical testing, 404, 406*t*
Viton, 310, 311*f*
VMS 585, 149*t*, 150*t*, 155, 165*f*
Volatility, lubricant, 480
Volkswagen, cam and follower materials, 205*t*
Volumetric efficiency, 22–23, 146–147
Volvo
    cam and follower materials, 205*t*
    engine brake, 60–61, 62*f*
    VVT system, 59*t*
VTEC, 3, 59*t*, 274, 275*f*
VVT-i system, 3, 47–48, 59*t*, 274
VVTL-i system, 59*t*

Wasploy, 149*t*, 150*t*, 154
Wear
    characteristics, 453–454
    classifications, 448
    component bench testing, 395–402
    effect of hardness on, 453*f*
    measurement, 454–456
    mechanisms, 448–453
    testing, 254, 386–390
Wear rate, 453
Wear transition, 454, 455*f*
Winsert, 245*t*
Work (mechanical), 13, 15, 16–17, 18
Wrought alloys, 245–246, 260, 261, 322

X-782, 149*t*, 150*t*, 165*f*
X-ray inspection, 404–408

Yield strength, 333, 350
    correlation with hardness, 252
    materials, 152, 161, 162*f*, 240
    plastic deformation, 335
    static design based on, 350
    tempering and, 156
    valve seat inserts, 251–252
Yield stress, 338, 342*f*
Yielding, 338
Young's modulus, 89, 351, 404

ZDP, lubricant additive, 477, 500, 502–506
Zinc, 332, 542
Zirconium, 332

About the Author

Since 2004, Dr. Yushu Wang is a Chief Engineer at Huaiji Auto Accessories Inc. and is Director of Guangdong Engine Valve Research and Development in China. He received his B.Sc. from Zhejiang University and his M.Sc. from Tsinghua University in China, both in materials science and engineering.

Dr. Wang was a faculty member in the Mechanical Engineering Department at Tsinghua University before he joined the National Institute of Standards and Technology of the U.S. Department of Commerce from 1987 to 1991 as a guest scientist. After completing his Ph.D. in materials science at the University of Maryland, he joined Oakland University as a Postdoctoral Fellow.

Dr. Wang was employed with Eaton Corporation as a Senior Product Engineer from 1995 to 2004. His expertise lies in the areas of materials, mechanical testing, failure analysis, coatings, manufacturing, and tribology of engine valvetrains. He has authored more than twenty publications and has been credited with seven U.S. patents.